Mechanical Properties of Polymers Based on Nanostructure and Morphology

Mechanical Properties of Polymers Based on Nanostructure and Morphology

edited by

G. H. Michler
F. J. Baltá-Calleja

CRC Press
Taylor & Francis Group
Boca Raton London New York

CRC Press is an imprint of the
Taylor & Francis Group, an **informa** business
A TAYLOR & FRANCIS BOOK

CRC Press
Taylor & Francis Group
6000 Broken Sound Parkway NW, Suite 300
Boca Raton, FL 33487-2742

First issued in paperback 2019

ISBN-13: 978-0-367-39272-7

Library of Congress Card Number 2004063533

Library of Congress Cataloging-in-Publication Data

Mechanical properties of polymers based on nanostructure and morphology / edited by G.H.
    Michler and F.J. Baltá-Calleja.
        p. cm.
    Includes bibliographical references and index.

    1. Composite materials. 2. Nanostructure materials. 3. Polymers. I. Michler, Goerg H. (Goerg
Hannes) II. Baltá-Calleja, F.J. III. Title.

TA418.9C6M397 2005
620.1'1—dc22   2004063533

Preface

The mechanical behavior of polymers has been the subject of considerable research in the past. Mechanical properties are, indeed, of relevance for all applications of polymers in industry, medicine, household, and others. The improvement of properties in general and the better fitting of specific properties to defined applications is a continuous goal of polymer research. Of particular interest is not only the improvement of the special properties themselves, such as stiffness, strength or toughness, but also the combined improvement of usually contradictory mechanical properties (like strength and toughness) in combination with other physical properties (e.g., transparency, flame resistance, conductivity, etc.). The outstanding role of the mechanical properties applies, as well, to many of the applications of polymers in which other properties are those playing the primary role, such as in medicine, optics, electronics, micro-system techniques and others. The defined improvement of the mechanical properties demands a better understanding of the multiple dependence between molecular structure, morphology, polymerization and processing methods on the one hand, and ultimate mechanical properties, on the other; i.e., structure-property correlations. The bridge between the structure, the morphology and the mechanical properties is the micromechanical processes or mechanisms occurring at microscopic level: the so-called field of micromechanics.

Polymeric systems become increasingly complicated and multifunctional if they entail a larger level of structural complexity. In the last two decades the level of interest has gradually shifted from the μm-scale to the nm-scale region. Systems with at least one structural size below 100 nm are considered nowadays as new classes of materials: the so-called *nanostructured polymers, nanopolymers or nanocomposites*. However, nanomaterials in the form of rubber carbon black composites have existed already for nearly one century, and biomedical materials such as bone, teeth, and skin also have been known for millions of years. Thus, although, the class of nanomaterials is not totally new, rapid development of research activity aiming for a better understanding of the basic mechanisms contributing to the properties of this class of remarkable systems has been recently observed. Natural materials, like human bone or seashell (abalone), reveal more and more very complex hierarchical structures with highly specific functions that have been optimized during bio-evolution over very long periods of time. Far-off these biomaterials, in most synthetic polymer blends and composites the hierarchical structure is most often created accidentally during synthesis or processing. Therefore, the mechanical properties of these man-made polymers must be better understood by examining the length scale, architecture and interactions occurring in these synthetic materials.

This volume focuses on selected results concerning the mechanical properties of polymers as derived from the improved knowledge of their structures at the μm- and nm-scale as well as from the interactions (micro- and nanomechanisms) between the complex hierarchical structures and functional requirements. The interest in the topic for this volume arose at the 1998 Europhysics Conference on Macromolecular Physics "Morphology and Micromechanics of Polymers" that was held in Merseburg (Germany) (see special volume of the *Journal of Macromolecular Science-Physics,* Vol. B38, 1999). Several authors of this book contributed as main lecturers to the success of the conference.

The structure of the book is organized as follows:

In the first part, "Structural and Morphological Characterization," the main aspects of the morphology of semicrystalline polymers, as revealed by electron microscopy (*Bassett*) and x-ray scattering techniques (*Hsiao*) are highlighted. Emphasis on the nanostructure of amorphous block copolymers and blends (*Adhikari, Michler*) is also given.

The second part, "Deformation Mechanisms at Nanoscopic Levels," is devoted to describing the main micro- and nanomicroscopic effects and mechanisms occurring in different classes of polymers. First, the influence of molecular variables on crazing and fracture behavior is discussed in the case of amorphous polymers (*Kausch, Halary*). Then, the physical elementary mechanisms including strength, crystal plasticity, orientation processes, and different modes of deformation are illustrated for selected semicrystalline polymers (*Galeski*) and complemented with results from electron microscopic microdeformation tests (*Plummer; Henning, Michler*). Recent results on micromechanical properties, as derived from microindentation hardness studies in different polymers and correlated to nanostructural parameters, are presented (*Baltá-Calleja, Flores, Ania*). Basic aspects of toughness enhancement for particle-modified semicrystalline polymers using model analysis are considered (*van Dommelen, Meijer*). This part ends with an overview about nano- and micromechanical effects in heterogeneous polymers, partly known in industry, partly new or up until now only theoretical possibilities (*Michler*).

The third part, "Mechanical Properties Improvement and Fracture Behavior," offers selected examples of heterogeneous polymers with improved mechanical properties and fracture behavior. Structure-property relationships and mechanisms of toughness enhancement are discussed for rubber-modified amorphous polymers (*Heckmann, McKee, Ramsteiner*) and semicrystalline polymers (*Harrats, Groeninckx*). New aspects of manufacturing, structure development and properties of practical relevance in nanoparticle-filled thermoplastic polymers are given (*Karger-Kocsis, Zhang*) and the state of the art of carbon nanotube and nanofiber-reinforced polymer systems is emphasized (*Schulte, Nolte*). Additionally, novel unusual methods of polymer modifications are based on micro- and nanolayered polymers (*Bernal-Lara, Ranade, Hiltner, Baer*) and hot-compaction of oriented fibers and tapes are also presented (*Ward, Hine*).

In addition to the wide spectrum of properties present in the above polymers, toughness enhancement is a particular aim of many of the discussed modifications. In the different chapters the usual routes of rubber-toughening of amorphous and semicrystalline polymers are completed by effects of toughness enhancement due to nanoparticle and nanofiber modification, micro- and nanolayer production and hot compaction of oriented polymers.

The book is directed particularly at polymer scientists in research institutes and in industry, and should serve as a link between more practical aspects of polymers and the knowledge about the influence of the different levels of structure and morphology on properties. It additionally aims to provide a better understanding on new effects and new possibilities to improve mechanical properties of polymer systems. Therefore, the book will be also helpful for students of polymer physics, chemistry and engineering, as well as those researchers interested in materials science.

F.J. Baltá-Calleja
G.H. Michler

Acknowledgment

The editors gratefully acknowledge the generous support of the Alexander von Humboldt Foundation, Bonn, during the preparation of the present volume.

The Editors

Francisco J. Baltá-Calleja, Ph.D., received his B.Sc. degree in physics at the University of Madrid in 1958. He then started his research work at the University of the Sorbonne in Paris on pioneering NMR studies of organic liquids relating to intermolecular effects. In 1959 he joined the H.H. Wills Physics Laboratory in Bristol, with a Ramsay Memorial Fellowship, to work on crystallization and morphology of synthetic polymers. In 1962 he received a Ph.D. in physics at the University of Bristol. In 1963 he was appointed adjoint professor of electricity and magnetism at the University of Madrid. He spent many years as a research associate and visiting professor in various international institutions, including the Fritz Haber Institute of the Max Planck Society in Berlin; Camille Dreyfus Laboratory, Research Triangle Institute, North Carolina, USA; J.J. Thomson Physical Laboratory, University of Reading; Abt. Experimentelle Physik, University of Ulm; University of Hamburg; University of Leeds; University of Shizuoka, etc. In 1970 he established the Macromolecular Physics Laboratory at the Spanish Research Council, a department for basic polymer research, collaborating with university, industry and international research laboratories. Presently, he is professor of physics at the Institute for

Structure of Matter, CSIC, Madrid, and has been director of this institute (1986–2003) and founder and director of the Centre of Physics (1996–2003), CSIC in Madrid. His research interests focus on interrelating structure, processing and dynamic changes in polymers, by means of x-ray diffraction methods using synchrotron radiation, and properties (micromechanical, electrical, electro-optical, dielectric and magnetic properties) of advanced polymers, liquid crystalline systems and composites. He is the author of about 350 papers and several books (*X-ray Scattering of Synthetic Polymers*, Elsevier, 1989; *Microhardness of Polymers,* Cambridge University Press, 2000; *Block Copolymers,* Marcel Dekker, 2000; etc). He has also been active in organizing several major international meetings in Europe. Recognition of his activities has been considerable, among the most notable being his election in 1988 as chairman of the Solid State Physics group of the Spanish Royal Society of Physics, his election in 1994 as chairman of the Macromolecular Board of the European Physical Society and his election in 1999 as a member of the Royal Academy of Sciences in Barcelona. He has also received the DuPont Research Award in 1994 and the Humboldt Research Award (Germany) in 1995. He is a member of the editorial boards of several international journals including *Journal of Macromolecular Science-Physics, Journal of Polymer Engineering, International Journal of Polymeric Materials,* and *Journal of Applied Polymer Science.* He is a member of the American Physical Society, the European Physical Society, and the Materials Research Society, US. He has also been a consultant to DuPont de Nemours (Luxembourg) and Exxon Mobil (Texas).

Goerg Hannes Michler, Ph.D., received his M.Sc. degree in physics at the University of Halle-Wittenberg in 1968. He then started his work on the morphology of polymers as a research scientist at the Institute of Solid State Physics and Electron Microscopy of the Academy of Sciences in Halle/Saale in cooperation with the chemical industry, first at the Department of Polymer Research, Leuna Werke, Germany (1969–1981), and then as head of the Polymer Physics Group, Chemische Werke Buna, Schkopau, Germany (1981–1990). In 1978 he received a Ph.D. in physics at the University of Halle-Wittenberg and in 1987 the Habilitation and venia legendi. In 1990 he was appointed professor of experimental physics at the Technical University of Merseburg. Presently, he is professor of general materials science at the University of Halle-Wittenberg, founder and director of the Institute of Polymeric Materials and vice director of the Polymer Service GmbH, Merseburg. His research interests focus on the structure-property correlations of materials, especially amorphous and semicrystalline polymers, blends, copolymers, composites, and biomedical materials by means of electron and atomic force microscopy. His special field of interest is the study of toughness enhancement and of the mechanisms of deformation and fracture in polymers based on nanostructure and morphology (micro- and nanomechanics). He has headed several interdisciplinary research projects at the university and between the university and chemical industry on new polymeric materials, polymers with improved mechanical properties, sustainable development of materials and biomedical polymers. He is a consultant to Dow Chemical Germany and Sasol Polymers, South Africa. He is author of more than 190 papers and several books (*Kunststoff-Mikromechanik: Morphologie, Deformations- und Bruchmechanismen*, Hanser, 1992; *Ultramikrotomie in der Materialforschung*, Hanser, 2004). He has also been active in organizing annual symposia and workshops on electron microscopy in materials science and ultramicrotomy in materials science and several major international meetings. His work has been awarded with the Alexander von Humboldt – J.C. Mutis Prize (2002, Ministry of Science and Technology, Spain) and the Paul J. Flory Polymer Research Prize (2003, University of North Texas, USA). He is a member of several international committees including the Scientific Committee of Polymer Physics of the German Physical Society (since 1991), IUPAC Macromolecular Division IV.2 (since 1997), and Macromolecular Board of the European Physical Society (since 1998). He is also a member of the American Chemical Society, the European Physical Society, German Physical Society, and Society of German Engineers.

Contributors

Rameshwar Adhikari
Institute of Materials Science
Martin-Luther-University
  Halle-Wittenberg
Merseburg, Germany

F. Ania
Department of Macromolecular
  Physics
Instituto de Estructura de la
  Materia
CSIC
Madrid, Spain

E. Baer
Department of Macromolecular
  Science and Engineering, and
Center for Applied Polymer
  Research
Case Western Reserve
  University
Cleveland Ohio

David C. Bassett
JJ Thomson Physical
  Laboratory
University of Reading
Reading, United Kingdom

Francisco J. Baltá-Calleja
Department of Macromolecular
  Physics
Instituto de Estructura de la
  Materia
CSIC
Madrid, Spain

T.E. Bernal-Lara
Department of Macromolecular
  Science and Engineering, and
Center for Applied Polymer
  Research
Case Western Reserve
  University
Cleveland Ohio

A. Flores
Department of Macromolecular
  Physics
Instituto de Estructura de la
  Materia
CSIC
Madrid, Spain

Andrzej Galeski
Center for Molecular and
  Macromolecular Studies
Polish Academy of Sciences
Lodz, Poland

G. Groeninckx
Division of Molecular and
  Nanomaterials
Katholieke Universiteit Leuven
Heverlee, Belgium

J.L. Halary
Ecole Supérieure de Physique et
  Chimie Industrielles del la
  Ville de Paris
Paris, France

Charef Harrats
Division of Molecular and
  Nanomaterials
Katholieke Universiteit Leuven
Heverlee, Belgium

W. Heckmann
Polymer Research Laboratory
BASF Aktiengesellschaft
Ludwigshafen, Germany

Sven Henning
Institute of Materials Science
Martin-Luther-University
  Halle-Wittenberg
Merseburg, Germany

A. Hiltner
Department of Macromolecular
  Science and Engineering, and
Center for Applied Polymer
  Research
Case Western Reserve
  University
Cleveland Ohio

P.J. Hine
IRC in Polymer Science and
  Technology
School of Physics and
  Astronomy
University of Leeds
Leeds, United Kingdom

Benjamin S. Hsiao
Chemistry Department
State University of New York at
  Stony Brook
Stony Brook, New York

József Karger-Kocsis
Institute for Composite Materials
Kaiserslautern University of
  Technology
Kaiserslautern, Germany

Hans-Henning Kausch
c/o Science de Base
École Polytechnique Fédérale
  de Lausanne
Lausanne, Switzerland

G.E. McKee
Polymer Research Laboratory
BASF Aktiengesellschaft
Ludwigshafen, Germany

Goerg H. Michler
Institute of Materials Science
Martin-Luther-University
  Halle-Wittenberg
Merseburg, Germany

H.E.H. Meijer
Department of Mechanical
  Engineering
Eindhoven University of
  Technology
Eindhoven, Netherlands

M.C.M. Nolte
Polymer Composites Section
Technische Universität
  Hamburg-Harburg
Hamburg, Germany

Christopher J.G. Plummer
Laboratoire de Technologie des
  Composites et Polymères
École Polytechnique Fédérale
  de Lausanne
Lausanne, Switzerland

F. Ramsteiner
Polymer Research Laboratory
BASF Aktiengesellschaft
Ludwigshafen, Germany

Aditya Ranade
Department of Macromolecular
  Science and Engineering, and
Center for Applied Polymer
  Research
Case Western Reserve
  University
Cleveland Ohio

Karl Schulte
Polymer Composites Section
Technische Universität
  Hamburg-Harburg
Hamburg, Germany

J.A.W. van Dommelen
Department of Mechanical
  Engineering
Eindhoven University of
  Technology
Eindhoven, Netherlands

I.M. Ward
IRC in Polymer Science and
  Technology
School of Physics and
  Astronomy
University of Leeds
Leeds, United Kingdom

Z. Zhang
Institute for Composite
  Materials
Kaiserslautern University of
  Technology
Kaiserslautern, Germany

Contents

PART II   Deformation Mechanisms at Nanoscopic Level

PART III *Mechanical Properties Improvement and Fracture Behavior*

Part I

Structural and Morphological Characterization

1

The Morphology of Crystalline Polymers

D.C. BASSETT

University of Reading, UK

CONTENTS

I. INTRODUCTION

The wide variation in properties of a given polymer according to its processing conditions reflects differences in internal organization. Polymer morphology is the study of this internal organization, primarily by microscopy but complemented by other techniques [1]. It has been and continues to be responsible for establishing the principal elements in our understanding of macromolecular self-organization and thence to establishing structure–property relationships. Its central position in polymer science arises essentially from three causes. First microscopy identifies specific locations of interest and is not restricted, as are nonmicroscopic techniques, to average values. Second, microscopic information is much more detailed and so potentially more informative than that from other sources. Third, the morphological record is particularly rich in crystalline polymers, more so than in other materials, because, in large measure, the long molecules remain where they were placed during crystallization and subsequent treatments such as deformation, allowing the sample's history to be read whereas in atomic and small-molecular solids this information is usually lost.

II. CRYSTALLINE POLYMERS

Polymer morphology is mostly, but not entirely, concerned with crystalline polymers, partly because of their rich record as mentioned but also because the two most economically important synthetic polymers, polyethylene and polypropylene, are built from the two monomers of most fundamental interest. Polyethylene is the closest approach to the ideal linear chain and can be modified to introduce branches through copolymerization while polypropylene is the first α-polyolefine and introduces stereospecificity into the polymer chain.

Crystalline polymers are not uniform solids. In the polarizing optical microscope they reveal polycrystalline textures (Figure 1) which are infinite in their variety, no two specimens being the same. For polymers crystallized from a quiescent melt, these textures are commonly described as spherulitic

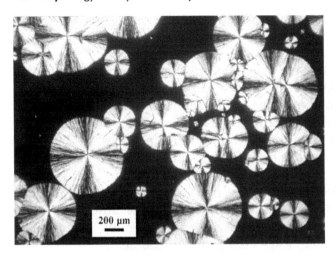

Figure 1 Spherulites of i-polypropylene growing from the melt viewed optically between crossed polars. (From Bassett DC. Phil Trans Roy Soc Lond A 1994; 348:29–43. With permission.)

because, provided there are not too many growth centers (primary nuclei), little spheres (spherulites) are an intermediate stage of development before objects impinge. Prior to attaining a spherical envelope, objects starting at a point or short line commonly pass through embryonic forms which may be sheaflike or polygonal depending on the viewing direction. Especially in circumstances of reduced branching, at low supercoolings, such immature objects are known as axialites and hedrites, respectively. Whether these early forms can mature into spherical entities depends upon there being sufficient space to allow the development, which is controlled by the concentration of primary nuclei: more nuclei mean smaller objects. But whatever stage is reached, the underlying process of spherulitic growth is the same and all objects are constructed on the same principles.

Polymer spherulites are typically constructed on a framework of individual radial lamellae, called dominant, which branch repetitively then diverge at angles ~20°, increasing for thinner lamellae. This repetitive divergence of dominant lamellae is the fundamental reason why an initial single crystal becomes polycrystalline; if adjacent lamellae did not diverge, a multilayer object would keep its initial single-crystal orientation.

Spherulites and lamellae are the two principal morphological entities present in polymers crystallized from a quiescent melt. They and their interrelation will be discussed in detail to provide the initial basis of understanding of structure-property relations in crystalline polymers. (Molecular trajectories have to be inferred from this evidence as they cannot be resolved by any available technique.) Further, complementary discussion concerns the significant modifications occurring when crystallization occurs under flow or stress to give oriented products such as fibers. Finally attention is drawn to contemporary challenges in the field.

III. POLYMER LAMELLAE

Much of our detailed knowledge of individual polymer lamellae comes from early studies of polyethylene crystallized from very dilute solution. These are microns wide, with planar facets, but only 10 to 15 nm thick (Figure 2). The most important finding, with profound implications for polymer science, is that within lamellae long molecules are *chainfolded* [3], i.e., they pass back and forth repetitively between the large (basal) lamellar surfaces, ~100 times for a typical polyethylene molecule, with crystalline stems inside a lamella and folds at its basal surfaces. Chainfolding is typical and characteristic of polymer lamellae whether grown from solution or melt although the proportion of tight folding, in which a molecule returns to an adjacent site, may well vary.

The information contained in lamellar morphology concerns two aspects: their thickness and crystal habit. Polymer lamellae form because they are the fastest means of crystallization. The thickness, ℓ, varies, increasing with crystallization temperature, T_c, according to

$$\ell = 2\sigma_e T_m{}^\circ/\Delta h \Delta T + \delta\ell \tag{1}$$

Here σ_e is the free energy per unit area of the fold surfaces, Δh the enthalpy of crystallization per unit volume of crystal,

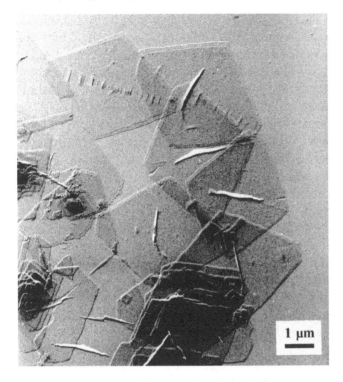

Figure 2 Single crystals of linear polyethylene grown from dilute solution. Notice the well-demarcated {200} sectors in certain of these truncated lozenges. (From Bassett DC. Phil Trans Roy Soc Lond A 1994; 348:29–43. With permission.)

$T_m°$ the melting point of an infinitely thick crystal, $\Delta T = (T_m° - T_c)$ is the supercooling and $\delta\ell$ an increment of thickness. The first term represents the condition when the reduction in free energy (Gibbs function) from crystallization of stems equals the increase due to folding, i.e., when there is no net gain of stability. Stability comes from increasing the stem length, lowering the free energy further, while the energetic cost of the surface is more or less constant. The actual thickness ℓ which results is a compromise between longer lengths which are more stable but crystallize more slowly and shorter lengths which crystallize faster but are less stable and, indeed, unstable below the value of the first term. These processes take place at the growing edge of the crystal, where thickness decreases abruptly, for solution growth, when the crystallization temperature is reduced [4]. SAXS, small angle x-ray scattering, is the preferred means of measuring average lamellar thickness although, unlike microscopy, it is unable to reveal a change in thickness at the growing edge.

The consequences of chainfolding are of major importance for properties because folding interrupts covalent continuity producing van der Waals bonding outside a fold surface, thereby reducing both Young's modulus and strength along the chain direction in a sample. To offset this reduction it is advantageous to retain in lamellae a proportion of molecules which do not fold but pass directly from one lamella to another, so-called tie molecules [5,6]. Various means of achieving this to produce *high-modulus materials* are known and now used in industrial production.

A major property of polymer lamellae is that they are inherently less stable than those of greater thickness. This is reflected in a reduced melting point, T_m, given by

$$T_m = T_m^\circ \, (1 - 2\sigma_e/\Delta h \ell) \qquad (2)$$

The characteristically broad melting range of a crystalline polymer is largely due to the variation in its lamellar thickness. As a consequence, if a sample is heated above its crystallization temperature into its melting range — a process known as annealing — then the thinnest crystals, whose melting point lies below the annealing temperature, will melt and, given sufficient time, will recrystallize (on unmolten lamellae as nuclei) at a higher temperature and thickness than before [7].

Instability may also occur at the crystallization temperature, known as isothermal lamellar thickening with thickness generally increasing logarithmically with the logarithm of elapsed time [8]. This process, which does not normally occur for solution-grown lamellae which retain their as-formed thickness, is generally present at the higher temperatures of melt crystallization. The mechanism cannot be partly melting then recrystallizing a sample as above because thermodynamics does not allow a sample to melt at its crystallization temperature. Instead it involves thickening in the solid state at and behind the growing edge [9], most probably via the well-known strong longitudinal vibrations and associated rotations to which long molecules are subject, especially near their melting point.

Lamellar habits provide unique information on crystallization processes. In some instances this has come from optical and scanning electron microscopy, SEM, but transmission electron microscopy, TEM, alone has the ability to combine imaging with diffraction which was key to the discovery of chainfolding. Chainfolding itself leaves clear evidence of its presence in polymer lamellae notably in sectorization. This phenomenon is the division of a single lamella into distinct regions adjacent to each growth face. Thus a polyethylene lozenge-shaped lamella with four {110} growth faces comprises four {110} sectors with well-marked internal boundaries. When the lozenge is truncated by two {200} faces, as occurs at higher crystallization temperatures, there are an additional two sectors now of the {200} kind. These latter are slightly thinner and, accordingly, melt at a lower temperature than the {110} sectors. Sectors are due to distortion of the crystal lattice by molecules folding parallel to the growth plane and not along nominally equivalent directions. Although the polyethylene structure has two equivalent {110} planes, folding parallel to only one of these in a given sector produces a small difference in spacing, ~0.001 nm between the two {110} planes. In consequence, the lattice loses its orthorhombic symmetry, **a** and **b** axes no longer meet at 90° but at a slightly smaller angle, and the lamella is slightly dished by a ~2° semi-angle rather than planar.

Sectorization as described applies when chains are normal to lamellae. In polyethylene this happens only for low mass materials, otherwise chains are inclined to lamellar normals. The inclination is higher at higher crystallization temperatures, when fold surfaces are {312} from solution and {201} from the melt giving inclinations ~35°. The inclination itself arises from the difficulty of packing folds, of asymmetric shape, in a tight lattice with small cross-sectional area per chain; inclination increases the surface area per fold. But the inclination is not generally uniform across a lamella but differs in different sectors giving solution-grown polyethylene lamellae hollow pyramidal or related three-dimensional habits.

Such habits have been observed directly for crystals in suspension or sedimented on glycerine. They have also been inferred from the molecular orientations in different sectors revealed by dark-field TEM [10]. It has been shown that there are both hollow pyramidal and chair-shaped habits, the latter geometrically related to the former by bisection along the short diagonal, parallel to **b**, then a 180° rotation around the **a** axis and, finally, reconnecting the two halves. The important points are that the two related habits exist and that their shapes remain constant during growth. To explain these, it was proposed that, initially, lamellae began to grow flat then adopted one of the two alternative three-dimensional habits [11]. The situation at lower crystallization temperatures is more complicated with a lesser inclination, of {314} planes, and repeated reversals of slope within a given sector [10] which suggest that inclination was adopted after the lamella formed.

These early suggestions fit well with recent work on melt-crystallized polyethylene lamellae[1] [12], which has shown explicitly that at higher temperatures, $\geq$ 127°C for a linear polymer at 1 bar, lamellae form and stay inclined with {201} fold surfaces. Below this range, fold surfaces form as {001} then transform, with lamellar twisting, to an inclined condition to reduce surface stress (Figure 3). The difference is because fold packing has time to organize before the next molecular layer is added to a lamella for slower growth at higher temperatures but not for faster growth in the lower range. For the latter condition, fold surfaces must form with disorganized fold packing and be rough, more so the faster

---

[1]The technique of permanganic etching allows melt-crystallized lamellae in polyolefins and certain other polymers to be studied systematically in both scanning and transmission electron microscopes. Figures 3 to 9 are all of permanganically etched samples. Details of precise conditions may be found in the original papers but all are variants of those in reference [41].

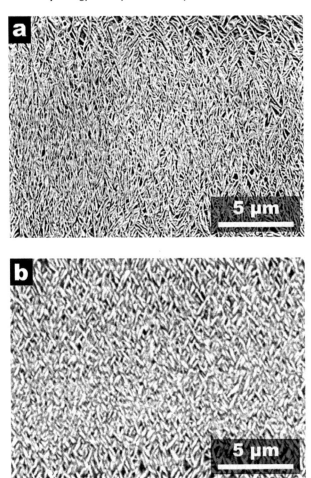

Figure 3 Fold surfaces of polyethylene lamellae form with different initial orientations according to whether or not there is sufficient time for preferred fold packing to be adopted. The effect is shown here for rows (horizontal on the page) of linear polyethylene grown on a fiber as a linear nucleus: (a) after 25 s growth at 123°C the dense central growth near the nucleating fiber has lamellae perpendicular to it but giving way to more open textures at top and bottom; (b) the same polymer grown as a row at 129°C for 3 h has inclined lamellae in a chevron pattern throughout with denser packing at the center. (Detail of figures is given in Abo el Maaty MI and Bassett DC. Polymer 2001; 42:4957–4963. Copyright 2001, with permission from Elsevier.)

the growth rate. This is a further reflection of the kinetic basis of crystallization being driven by the addition of stems which, by increasing length, is always able to overcome the energetic cost of the fold surface, organized or disorganized.

The reductions in free energy available from organization of the fold surface are always liable to be small in relation to those involved in crystallization but, given sufficient time, they may be realized. For example, the alkane lattice will distort to achieve better surface packing. This has recently been demonstrated for the monodisperse centrally branched alkanes which form scrolls at high crystallization temperatures, thereby accommodating all branches on the outer surface of the scroll even though the lattice must be strained in consequence [13].

In general, lamellar habits must conform to the crystal structure, but the precise shape depends upon the growth mechanism. For solution-grown polyethylene, the lozenge habit indicates that {110} are the slowest-growing planes as befits their close-packed condition. The appearance of {200} truncating faces has been interpreted in terms of the different fold surface free energies corresponding to the altered pattern of folding in the two types of sector. The {200} sectors become wider with increasing crystallization temperature and their growth faces also become curved. Ultimately in individual melt-crystallized lamellae grown at 130°C, {110} facets presumed to be present are hard to detect and a lamella consists almost entirely of two {200} sectors with strongly curved outlines [14]. Such curvature may be explained in terms of the lateral spreading rate of steps nucleated on the growth surface [15].

In certain systems, notably α-polypropylene, lath-like lamellar growth is a feature [16]. While in low mass polyethylene laths can extend parallel to a {110} twin plane, because of the niche giving favorable conditions for molecular attachment at the twin boundary, the situation in polypropylene is not so clear and has never been fully explained.

Sectorization is most evident in solution-crystallized lamellae but has been shown in melt-crystallized lamellae, at least to an extent. For example, individual polyethylene

Figure 4   A single crystal of linear polyethylene, within a quenched matrix, grown from the melt at 130°C. The central hole is due to preferential etching around a giant screw dislocation during preparation for TEM. (Reprinted from Patel D and Bassett DC. Polymer 2002; 43:3795–3802. Copyright 2001, with permission from Elsevier.)

lamellae grown at high temperatures tend to form two oppositely inclined sectors across their long diagonal, the more so when crystallized slowly and for longer times (Figure 4). Melt-crystallized i-polystyrene lamellae are dished in a similar way to sectored solution-crystallized lamellae, presumably for the same reason as molecules lying preferentially along the growth plane [17,18]. But this is a relatively unexplored area; it is, for example, by no means clear how sectorization is affected by isothermal lamellar thickening.

   A further important deduction from lamellar habits concerns recent suggestions that metastable states may be the first stage in crystallization from the melt. For example, it has been proposed that the two-dimensional hexagonal phase of polyethylene always precedes crystallization of the orthorhombic phase from the melt [19]. That this does not happen follows from the existence of single crystals of the orthorhombic phase. When the hexagonal phase, which crystallizes as circular lamellae from the melt [20], transforms to the lower symmetry orthorhombic phase, it has three choices as to which transverse axis should become the **b** axis of the orthorhombic cell. As a consequence, lamellae which have converted from the hexagonal to the orthorhombic phases are invariably twinned and in a complex way. Single crystals of the orthorhombic phase must have formed directly from the melt [21]. A similar argument applies to all cases where a metastable phase of higher symmetry is supposed to precede the crystallization of a crystalline phase of lower symmetry.

IV. SPHERULITES

When, growing from the melt, a single polymer lamella branches, for example, around a giant screw dislocation, adjacent layers do not lie parallel as in small-molecular crystals, but diverge (Figure 5). This is the key phenomenon which, by iteration, leads to the formation of spherulites [17]. For long molecules two causes of divergence have been identified, ciliation and rough surfaces [22]. The observation that, in multilayer crystals, adjacent lamellae have linear traces which diverge at roughly constant angles indicates that they are

Figure 5 A similar crystal to that of Figure 4 but in a perspective that shows that adjacent layers on the right-hand side, generated by the central (etched-out) screw dislocation, diverge. (From Bassett DC. Phil Trans Roy Soc Lond A 1994; 348:29–43. With permission.)

growing in stress-free conditions with divergence developing close to the branch point at separations no greater than molecular lengths. This strongly suggests the presence of a mesoscopic repulsion for which ciliation, both transient and permanent, has now been confirmed to be a responsible factor and in considerable detail. The morphology of inorganic spherulites is difficult to observe and is not nearly so well characterized as for polymers but also involves long entities branching and diverging with the latter feature attributed, in ice-based systems, to osmotic pressure. It seems very likely that the mechanisms shown to operate for polymers are particular expressions of general principles applicable to all spherulites, polymeric or not.

In the polymeric context a cilium is an uncrystallized portion of a molecule partly attached to a lamella. If it subsequently crystallizes and so disappears, it is transient but if it remains indefinitely, it is permanent. In both cases cilia will be present outside basal surfaces and will, over distances no greater than molecular lengths, repel an adjacent lamella which enters this space.

The influence of both transient and permanent cilia has been investigated in detail in the monodisperse n-alkanes which are oligomers of polyethylene. First were qualitative observations on $n\text{-}C_{294}H_{590}$ which can form two separate lamellar populations at the same temperature — one with molecules fully-extended and once-folded in the other. Lamellar divergence and spherulitic growth were absent from the former but present in the latter [23]. The contrast is between extensive parallel lamellar stacks for extended molecules and regular divergence of adjacent chainfolded lamellae as in polyethylene and other spherulites. The significance of the change with molecular conformation is that folding implies the presence of transient cilia while there will be none for extended chains which add as a single unit.

In reality, the addition of molecules in extended-chain crystallization is more subtle with polycrystalline textures and lamellar divergence appearing at higher supercoolings. To explain this, one may anticipate the presence of transient cilia with length equal to the excess of molecular length over that of the secondary nucleus at the growth interface, a dimension proportional to supercooling. A quantitative investigation of several shorter monodisperse n-alkanes showed plots of the angle of divergence of adjacent lamellae against supercooling to be linear with slopes decreasing consistently for longer alkanes and a positive intercept ~8° [24]. Transient ciliation, which vanishes at zero supercooling, cannot account for the non-zero intercept which has been suggested to arise from surfaces being formed rough and unable to lie parallel. This is, therefore, a second factor contributing to lamellar divergence and spherulitic growth. It is likely to contribute also to spherulitic growth in linear polyethylene at temperatures below 127°C when fold surfaces form as {001} and are inherently rough [12].

Permanent ciliation is present in co-crystallizing dilute binary blends of monodisperse n-alkanes when the guest molecule is longer than the host. In such cases the presence of permanent cilia supplements previous effects in a consistent way [25]. This supplementation vanishes, however, when the guest molecule is twice as long as the host so that it will crystallize once-folded with transient but no permanent ciliation [26].

Inhomogeneity is an inevitable product of spherulitic growth with dominant lamellae distinct from those which form later between them (Figure 6). In low crystallinity polymers, spherulites may consist largely of dominant lamellae. With higher crystallinity, subsidiary lamellae form isothermally behind the perimeter occupying, with later infilling lamellae, some or all of the space between the dominants. There are correspondingly different properties, mechanical and thermal, related to differences in orientation and lamellar thickness. Most spherulites (but not those of α-polypropylene) melt from the outside in, with dominant lamellae the last to melt and the highest melting point in spherulite centers [28]. In α-polypropylene, lamellar thickening is suppressed by cross-hatching and spherulite centers retain thin lamellae which melt before those at greater radial distance where radial lamellae are more pronounced [29].

A reduction in inhomogeneity, which is usually desirable commercially, may be achieved by reducing spherulite size

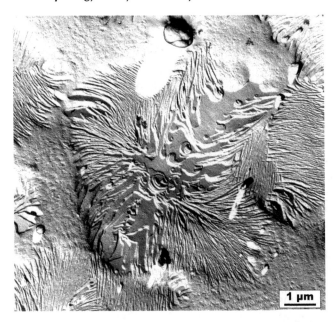

Figure 6 An embryonic spherulite of linear-low-density polyethylene grown at 124°C showing individual dominant lamellae advancing into the melt with subsidiary lamellae filling intervening spaces to the rear. (From Bassett DC. Crystallization of polymers, Dosiere M, ed., Dordrecht: Kluwer, 1993; 107–117. With permission.)

with increased primary nucleation. In the particular case of α-polypropylene, where improved clarity is sought, a sufficient reduction is not easy to achieve and this remains an active area of research. For other polymers the problem is rarely serious.

Spherulitic textures also bring differential response to mechanical stress, in large measure because the different lamellar orientations present respond differently to the applied stress (Figure 7). The link between morphology and mechanical behavior has recently been reviewed [31].

V. BANDED SPHERULITES

Certain spherulites, notably those of polyethylene at lower crystallization temperatures (< ~127°C for linear polymers

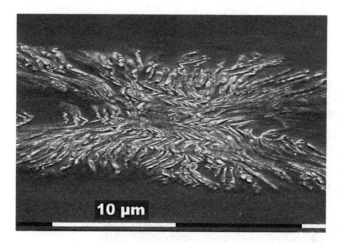

Figure 7 Differential deformation in lamellae within a spherulite of linear polyethylene according to the inclination of lamellae to the horizontal draw direction. (Reprinted from Lee SY, Bassett DC, Olley RH. Polymer 2003; 44:5961–5967. Copyright 2001, with permission from Elsevier.)

[32]) and certain aliphatic polyesters show banding, i.e., an approximately concentric set of extinction rings in the polarizing microscope. These are well understood optically, being due to a twisting of the average orientation around the growth direction but a knowledge of, and explanation for, the associated lamellar texture has been much harder to obtain.

Once again the best-characterized system is polyethylene. In the linear polymer, banding is only found for faster growth, giving lamellae which are S-profiled viewed down **b**, the radial growth direction [33]. The lamellae themselves are slightly twisted but the bulk of the twist is inserted at, and maintained by, successive isochiral giant screw dislocations when lamellar cross-sections are much reduced. The underlying cause of twisting in polyethylene is (partial) relief of surface stress associated with faster growth and the initial {001} fold surfaces leading to the adoption of the S-profile, in which central and edge portions of lamellae are inclined [34].

The origin in relief of fold surface stress is reinforced by observations on branched, linear-low-density polyethylene.

Here banding occurs in two circumstances. One is identical to that for the linear polymer: faster growth, initial {001} fold surfaces and S-profiled lamellae. The other concerns slower growth in which lamellae start to grow with {201} fold surfaces but as lamellar thickness increases isothermally, requiring branches to be brought into and stress fold surface regions, twisting begins. This has a greater period, i.e., a lesser torque, than for faster growth as would be expected for a smaller surface stress [35].

VI. CRYSTALLIZATION UNDER STRESS OR FLOW

Polymers are often crystallized under stress or flow to give an overall molecular orientation and enhanced properties. In such circumstances primary nucleation is linear and the products are row structures, frequently described as *shish-kebabs* (Figure 8). Here the shish is a linear backbone, typically tens of nm across, on which chainfolded lamellae (the kebabs) grow transversely. Similar effects are produced with macroscopic fibers as nuclei of quiescent melts (Figure 9). Carbon fiber can be an effective heterogeneous linear nucleus but if a highly oriented fiber of the host is used, it will probably initiate epitaxial crystallization on itself.

The traditional view of row structures is that the nucleus is formed along and in response to the elongational component of applied stress which it helps sustain [38]. It then acts as a linear nucleus for chainfolded lamellae to grow transversely, with the same chain axis direction, in stress-free conditions. While the nucleation element appears to be true, lamellae do not grow in strictly stress-free conditions. This is shown first by being able to crystallize at appreciable growth rates at higher temperatures than from a quiescent melt and by having different lamellar profiles [39]. Polyethylene lamellae grown under simple shear above 127°C, for example, have been found not to be inclined but to have {001} surfaces. The implication is first that the isothermal supercooling has been increased and, second, that the processes of fold surface organization are adversely affected. These are

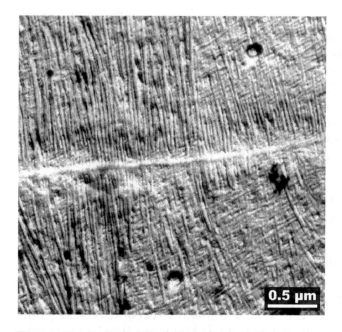

Figure 8   A shish-kebab in polypropylene homopolymer formed, at 140°C, on a pre-existing linear nucleus. (Reprinted from White HM and Bassett DC. Polymer 1997; 38:5515–5520. Copyright 2001, with permission from Elsevier.)

issues which may be clarified in what is now an active area of research.

For a long time, especially when optical microscopy was the principal source of information it was supposed that radial growth in a row structure and a spherulite were essentially the same albeit in different macroscopic geometries. This is not the case. There is a major distinction stemming from differences in the lamellar environment. Whereas the diverging growth in a spherulite creates space in which a lamella may thicken isothermally, space is denied to close-packed lamellae growing parallel in rows. Here thickening is accomplished by some lamellae failing to propagate, so reducing their overall number and allowing thickening into the space so created [40]. This leaves thin lamellae at the origin of the row, which have been unable to thicken in contrast to spher-

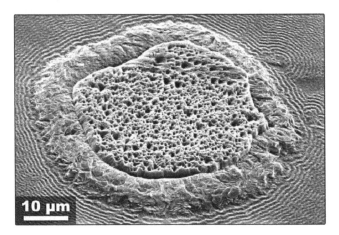

Figure 9 A cross-section through a row structure of linear poly-ethylene grown around a central high-melting polyethylene fiber. (Reprinted from Abo el Maaty MI and Bassett DC. Polymer 2001; 42:4965–4971. Copyright 2001, with permission from Elsevier.)

ulites whose centers contain the thickest lamellae. These are the first to melt [8] in contrast to most spherulites whose centers melt last (except for α-polypropylene whose centers retain thin lamellae and melt first [27]).

VII. FUTURE CHALLENGES

Knowledge and understanding of polymer morphology has been transformed since it became possible to characterize and study lamellae, their habits and associations, systematically by electron microscopy, largely through the advent and application of permanganic etching. Knowledge of the lamellar architecture of spherulites has moved from the presumed to the actual, bringing critical and extensive tests of explanations of their growth processes, a task much facilitated by the availability of the monodisperse long alkanes. Observations of the growth of polyethylene lamellae in row structures have made explicit the reorganization of fold surfaces implicit in kinetically controlled crystallization. This new knowledge now needs to be assimilated, not least into models used to interpret scattering and spectroscopic data, with familiar historic models not being retained uncritically, otherwise the field of structure–property relations will stultify. In particular, it is necessary that explanations of polymeric properties are not merely ascribed to presumed behavior but are examined critically and carefully checked so that the edifice of knowledge can continue to be soundly constructed.

REFERENCES

1. Bassett DC. Polymer morphology. In: Encyclopedia of Polymer Science and Technology, 3rd ed., Hoboken, NJ: John Wiley, 2003; 7:234–261.

2. Bassett DC. Lamellae and their organization in crystalline polymers. Phil Trans Roy Soc Lond A 1994; 348:29–43.

3. Keller A. A note on single crystals in polymers: evidence for a folded chain configuration. Phil Mag 1957; 2:1171–1175.

4. Bassett DC and Keller A. On the habits of polyethylene crystals. Phil Mag 1962; 7:1553–1584.

5. Frank FC. The strength of polymers. Proc Roy Soc Lond A 1964; 282:9–16.

6. Frank FC. The strength and stiffness of polymers. Proc Roy Soc Lond A 1970; 319:127–136.

7. Bassett DC. Principles of Polymer Morphology. Cambridge: Cambridge University Press, 1981.

8. Hoffman JD and Weeks JJ. X-ray study of isothermal thickening of lamellae in bulk polyethylene at the crystallization temperature. J Chem Phys 1965; 42:4301–4302.

9. Abo el Maaty MI and Bassett DC. Evidence for isothermal lamellar thickening at and behind the growth front as polyethylene crystallizes from the melt. To be published, Polymer 2005.

10. Bassett DC, Frank FC, Keller A. Some new habit features in crystals of long chain compounds. Part III. Direct observations of unflattened monolayer crystals. Phil Mag 1963; 8:1739–1751.

11. Bassett DC, Frank FC, Keller A. Some new habit features in crystals of long chain compounds. Part IV. The fold surface geometry of monolayer polyethylene crystals and its relevance to fold packing and crystal growth. Phil Mag 1963; 8:1753–1787.

12. Abo el Maaty MI and Bassett DC. On fold surface ordering and re-ordering during the crystallization of polyethylene from the melt. Polymer 2001; 42:4957–4963.

13. White HM, Hosier IL, Bassett DC. Cylindrical lamellar habits in monodisperse centrally-branched alkanes. Macromolecules 2002; 35:6763–6765.

14. Passaglia E and Khoury F. Crystal growth kinetics and the lateral habits of polyethylene crystals. Polymer 1984; 25:631–644.

15. Mansfield ML. Solution of the growth equations of a sector of a polymer crystal including consideration of the changing size of the crystal. Polymer 1988; 29:1755–1760.

16. Padden FJ and Keith HD. Crystallization in thin films of isotactic polypropylene. J Appl Phys 1966; 37:4013–4020.

17. Bassett DC, Olley RH, Al Raheil IAM. On isolated lamellae of melt-crystallized polyethylene. Polymer 1988; 29:1539–1543.

18. Vaughan AS and Bassett DC. Early stages of spherulite growth in melt-crystallized polystyrene. Polymer 1988; 29:1397–1401.

19. Keller A, Hikosaka M, Rastogi S, Toda A, Barham PJ, Goldbeck-Wood G. The size factor in phase transitions: its role in polymer crystal formation and wider implications. Phil Trans Roy Soc Lond A 1994; 348:3–14.

20. DiCorleto JA and Bassett DC. On circular crystals of polyethylene. Polymer 1990; 31:1971–1977.

21. Bassett DC. On the role of the hexagonal phase in the crystallization of polyethylene. In: Allegra G, ed. Interphases and Mesophases in Polymer Crystallisation. Adv Polym Sci in press.

22. Bassett DC. Polymer spherulites: a modern assessment. J Macromol. Sci. Phys 2003; 42:227–256.

23. Bassett DC, Olley RH, Sutton SJ, Vaughan AS. On chain conformations and spherulitic growth in monodisperse n-$C_{294}H_{590}$. Polymer 1996; 37:4993–4997.

24. Hosier IL, Bassett DC, Vaughan AS. Spherulitic growth and cellulation in dilute blends of monodisperse long n-alkanes. Polymer 2000; 33:8781–8790.

25. Hosier IL and Bassett DC. A study of the morphologies and growth kinetics of three monodisperse n-alkanes: $C_{122}H_{246}$, $C_{162}H_{326}$ and $C_{246}H_{494}$. Polymer 2000; 41:8801–8812.

26. Hosier IL and Bassett DC. Morphology and crystallization kinetics of dilute binary blends of two monodisperse n-alkanes with a length ratio of two. J Polym Sci Part B: Polym Phys 2001; 39:2874–2887.

27. Bassett DC. Lamellae in melt-crystallized polymers. In: Crystallization of Polymers, Dosiere M, ed., Dordrecht: Kluwer 1993; 107–117.

28. Wunderlich B and Melillo L. Morphology and growth of extended chain crystals of polyethylene. Makromol Chem 1968; 118:250–264.

29. Weng J, Olley RH, Bassett DC, Jääskeläinen P. On changes in melting behaviour with radial distance in isotactic polypropylene spherulites. J Polym Sci Part B: Polym Phys 2003; 41:2342–2354.

30. Lee SY, Bassett DC, Olley RH. Lamellar deformation and its variation in drawn isolated polyethylene spherulites. Polymer 2003; 44:5961–5967.

31. Bassett DC. Deformation mechanisms and morphology of crystalline polymers. In: Ward IM, Coates PD, Dumoulin MM, eds. Solid Phase Processing of Polymers. Munich: Hanser, 2000; 11–32.

32. Hoffman JD, Frolen LJ, Ross GS, Lauritzen JI. On the growth rate of spherulites and axialites from the melt in polyethylene fractions: regime I and regime II crystallization. J Res NBS A 1975; 79:671–699.

33. Bassett DC and Hodge AM. On the morphology of melt-crystallized polyethylene. I. Lamellar profiles. Proc Roy Soc A 1981; 377:25–37.

34. Patel D and Bassett DC. On the formation of s-profiled lamellae in polyethylene and the genesis of banded spherulites. Polymer 2002; 43:3795–3802.

35. Abo el Maaty MI and Bassett DC. On celluation and banding during crystallization of a linear-low-density polyethylene from linear nuclei. Polymer 2002; 43:6541–6549.

36. White HM and Bassett DC. On variable nucleation geometry and segregation in isotactic polypropylene. Polymer 1997; 38:5515–5520.

37. Abo el Maaty MI and Bassett DC. On interfering nuclei and their novel kinetics during the crystallization of polyethylene from the melt. Polymer 2001; 42:4965–4971.

38. Peterlin A. Crystallization phenomena. In: Miller RL, ed. Flow-induced Crystallization in Polymer Systems. New York: Gordon and Breach 1979; 1–28.

39. Hosier IL, Bassett DC, Moneva IT. On the morphology of polyethylene crystallized from a sheared melt. Polymer 1995; 36:4197–4202.

40. Abo el Maaty MI. On stable and unstable growth of row structures of linear-low-density polyethylenes. Polymer J 1999; 31:778–783.

41. Olley RH and Bassett DC. An improved permanganic etchant for polyolefines. Polymer 1982; 23:1797–1710.

2

Nanostructure Development in Semicrystalline Polymers during Deformation by Synchrotron X-Ray Scattering and Diffraction Techniques

BENJAMIN S. HSIAO

State University of New York at Stony Brook

CONTENTS

I.   INTRODUCTION

The hierarchical structures in semicrystalline polymers from atomic, nanoscopic, microscopic to mesoscopic scales can be thoroughly characterized by microscopic means such as scanning electron microscopy (SEM), transmission electron microscopy (TEM), atomic force microscopy (AFM) and optical microscopy. However, the *in situ* structure development during polymer processing, information that is critical for optimization of properties, is rather challenging using the microscopy techniques. The combined x-ray scattering and diffraction techniques are, perhaps, the most complementary techniques to microscopy, which provide the structure information in reciprocal space. Small-angle x-ray scattering (SAXS) probes relatively large-scale structures from 2 to 100 nm, while wide-angle x-ray diffraction (WAXD) deals with the atomic structure of crystals (0.1 to 2 nm). With synchrotron radiations, the high power density and small beam divergence of the incident x-ray beam further permit the design of time-resolved SAXS/WAXD experiments [1,2] that can also be coupled with varying polymer processing techniques, such as polymer deformation [3–6], fiber spinning [7–11] and drawing [12–13], shearing and complex flow [14–18].

In this chapter, we will demonstrate two case studies by using synchrotron SAXS/WAXD techniques to characterize the structure and morphology of semicrystalline polymers under deformation. These studies include (1) flow-induced crystallization in polymer melts (isotactic polypropylene, iPP and polyethylene, PE), (2) strain-induced phase transitions and structural development in polymer solids (iPP and poly(ethylene terephthalate), PET). Before we describe the scientific findings of these studies, the combined synchrotron SAXS/WAXD techniques and the unique sample chambers used in these studies are briefly introduced as follows.

II. EXPERIMENTAL TECHNIQUES

A.   Combined Synchrotron SAXS and WAXD Techniques

Both case studies were conducted at the Advanced Polymers Beamline (X27C) of the National Synchrotron Light Source in Brookhaven National Laboratory [1]. This facility is the first synchrotron beamline in the United States dedicated to chemistry/materials research (with emphasis on polymers) using state-of-the-art simultaneous small-angle x-ray scattering (SAXS) and wide-angle x-ray diffraction (WAXD) techniques. The x-ray optics at X27C are simple but unique. After the radiation is emitted from the bending magnet, a double-multilayer (silicon/tungsten) monochromator is used to mono-chromatize the incident beam (Figure 1). This monochromator has been demonstrated to increase the x-ray flux by approximately 10 times when compared with the conventional double-crystal monochromator. Higher orders of harmonics are eliminated by slightly detuning the two multilayers. Although the wavelength is adjustable from 0.07 – 0.2 nm (energy 6–20 KeV) with this monochromator, the wavelength is usually fixed at 0.13 nm to optimize the flux and the SAXS angular resolution. The corresponding energy resolution ($\Delta E/E$) at this wavelength is 1.1%, which is an order of magnitude broader than that from the crystal monochromator. However, this energy resolution is sufficient for SAXS and WAXD measurements of polymer materials. The typical level of x-ray flux at the testing condition is about 9×10^{11} photon/s.

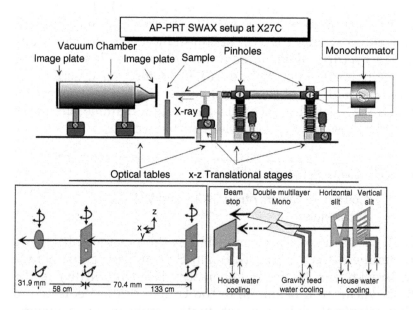

Figure 1 Simultaneous SAXS/WAXD setup at the Advanced Polymers Beamline (X27C), National Synchrotron Light Source, Brookhaven National Laboratory. (Reprinted with permission from Chu B, Hsiao BS. Chem. Rev. 2001; 101(6):1727–1761. Copyright 2001 American Chemical Society.)

The most critical component of this SAXS facility is the collimation system, which directly determines the maximum spatial resolution power. At X27C, we have constructed a three-pinhole collimation system for this purpose (the lower diagram in Figure 1). This system is unique in several ways. (1) The construction cost of the pinhole system is significantly less than its slit counterpart. (2) The alignment of the pinhole system was found to be considerably easier than the slit system (since only half of the variables are needed to control the pinhole system). (3) Several series of pinholes with different resolution power were prepared (their diameters were derived from theoretical calculation), and the changing of different pinholes is also straightforward. (4) All pinholes were manufactured on tantalum substrates (2.5 mm thickness) by a special drilling technique to minimize the parasitic scattering from the pinhole edge. Using the first pinhole of 0.1 mm diameter, we have routinely obtained the maximum spatial resolution of 110 nm from the duck tendon standard.

B. *In Situ* X-ray Shear Apparatus

The rheo-SAXS experiments were carried out by a Linkham CSS-450 high-temperature shearing stage, modified for x-ray diffraction and scattering experiments [15,16]. The synchrotron setup and the shear apparatus are shown in Figure 2. In this apparatus, two parallel plates were used to sandwich the ring-shaped specimen. A Kapton film and a diamond window were used in place of the quartz optical windows on the top and bottom steel plates to support the specimen. The top plate had a narrow aperture hole (3 mm in diameter), which allowed the x-ray beam (0.3 mm diameter) to enter the sample. The bottom rotating plate had open slots (wider than the hole in the top window) for the exit x-ray beam. The sample was sheared by rotating the bottom plate through a precision stepping motor while the top plate remained stationary. The bottom opened slots allowed the scattered x-ray beam to be detected during shear. SAXS/WAXD signals were separately measured during shear under identical experimental conditions. The attainable shear rate of this apparatus is from 0.1

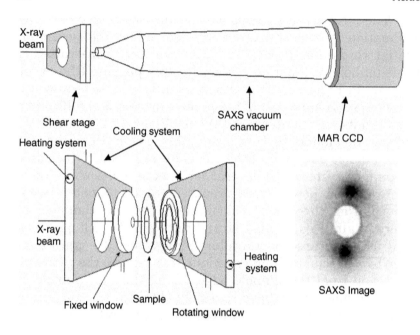

Figure 2 Modified shear apparatus for *in situ* SAXS/WAXD experiments. (Reprinted with pemission from Somani RH, Hsiao BS, Nogales A, Srinivas S, Tsou AH, Sics I, Balta-Calleja FJ, Ezquerra TA. Macromolecules 2000;33(25):9385–9394. Copyright 2000, American Chemical Society.)

– 100 s^{-1} and the typical operating temperature range is from 25–300°C.

C.   *In Situ* X-ray Tensile-Stretching Apparatus

Dynamic studies of polymer films and fibers during tensile deformation have become very common in synchrotron experiments. These studies yield important information on the changes of structure and morphology during deformation, which can be related to the properties of polymers. The major requirement of the apparatus is that it should provide symmetrical stretching, which guarantees that the focused x-ray can illuminate the same position on the sample during deformation. Otherwise, the sample detection position will be changed continuously, which can lead to uncertainty in conclusion. In our laboratory, we have modified a tabletop non-symmetrical stretching device to provide symmetrical deformation [4–6]. The modification can be briefly described as follows (Figure 3). The tensile stretching apparatus was a modified version of model 4410 from Instron Inc. and had a load capacity of 500 Newtons. The maximal distance between the two grips was about 460 mm. The sample could be heated to a temperature up to 300°C with a custom-designed sample chamber. The stretching speed could be adjusted from 0.2 mm/min up to 1000 mm/min. The modified stretching unit adopted a custom-built vertical translational stage, which provided translational motion opposite to the programmed stretching with the same speed.

III.   FLOW-INDUCED CRYSTALLIZATION IN POLYMER MELTS

The subject of flow-induced crystallization remains to be one of the most important problems in polymer processing today, and it is still not yet fully understood. The major hurdles in the past were mainly due to the lack of suitable *in situ* characterization tools to determine the structures at the earliest stages of flow-induced crystallization. Using advanced

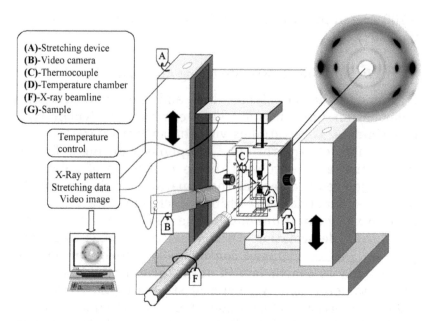

Figure 3 Schematic diagram of stretching apparatus (with temperature control chamber) setup for *in situ* x-ray study. (Reprinted with pemission from Yeh F, Hsiao BS, Sauer BB, Michel S, Siesler HW. Macromolecules 2003;36(6):1940–1954. Copyright 2003, American Chemical Society.)

synchrotron x-ray techniques, these hurdles have been effectively overcome.

The selected sample studies involved shear-induced crystallization in iPP and PE melts. These studies aimed to investigate several fundamental issues of the flow-induced crystallization subject, including (1) the pathway of forming nanostructured scaffolds (shish-kebab crystallization precursor structures) in undercooled polymer melt by flow prior to full scale crystallization; (2) the relationships among the internal material parameters (chain length, length distribution and chain branching, or the relaxation time spectrum), the external flow parameters (temperature, strain and strain rate, or stress), and the formation of the initial precursor structures, and (3) flow-induced polymorphism.

A. Flow-Induced Shish-Kebab Precursor Structures of Isotactic Polypropylene Near Nominal Melting Point

When crystallized from the melt or solution, iPP always adopts a 3_1 or 3_2 helical confirmation in the crystalline structure. The helical chains can pack in different cell structures with specific helical hand registrations giving rise to different polymorphs of iPP, i.e., the α, β and γ crystal forms, depending on the crystallization conditions [19–22]. In 1959, Natta et al. [23] pointed out the existence of a mesomorphic form of iPP when the sample was quenched from melt in ice water. The mesomorphic form has a degree of order intermediate between the amorphous phase and the crystalline phase. Recently, de Jeu et al. observed that the smectic phase persists at high temperatures in the molten state (about 20°C above its nominal melting temperature) [24].

Deformation studies of supercooled iPP melts near or above its nominal melting point have yielded some interesting findings. For example, Kornfield and coworkers [25] used a rheo-optical technique to study the influence of short term shearing in iPP melt at 175°C. They observed a distinct rise in melt birefringence after cessation of shear. As the optical techniques cannot elucidate the molecular information, we have carried out *in situ* rheo-SAXS and rheo-WAXD experiments to study a similar iPP at the same temperature (175°C) [26,27]. In our studies, the iPP melt was subjected to a step-shear (shear rate = 60 s^{-1}) at two different shear durations (0.25 and 5 s). Under both conditions, the imposed deformation was found to generate oriented layered structures that were stable at the experimental temperature, as evidenced by *in situ* SAXS patterns obtained after shear, where no detectable WAXD crystalline reflections were seen. In the following sample study, results from a similar rheo-x-ray study at 165°C will be discussed [28]. The slightly lower experimental temperature (still near the nominal melting point of iPP – 168°C) enabled us to study some details of the shear-induced crystallization precursor structures

A Ziegler-Natta iPP homopolymer was used in this study. Its molecular weights were: M_n = 92,000 g/mol, M_w = 368,000 g/mol and M_z = 965,000 g/mol. *In situ* rheo-SAXS and -WAXD measurements were carried out in the beamlines X27C at the NSLS, BNL, USA. A 2D MAR CCD x-ray detector (MARUSA) was employed to collect time-resolved SAXS and WAXD patterns in the shear experiments. The polymer melt was subjected to shear (a shear rate of 60 s^{-1} and a shear duration, ts, of 5 s) immediately after the temperature dropped to 165°C. Two-dimensional patterns (2D) (SAXS or WAXD) were collected continuously: before, during, and after cessation of shear. All x-ray images were corrected by sample absorption and beam fluctuation.

1.   SAXS Observation

The SAXS patterns obtained under a shear condition of t_s = 5 s are illustrated in Figure 4, which clearly show the emergence of an equatorial streak immediately after shear (pattern

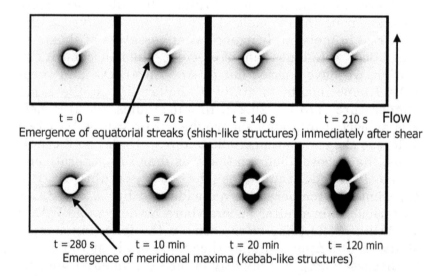

t = 0   t = 70 s   t = 140 s   t = 210 s   Flow

Emergence of equatorial streaks (shish-like structures) immediately after shear

t = 280 s   t = 10 min   t = 20 min   t = 120 min

Emergence of meridional maxima (kebab-like structures)

Figure 4 2D SAXS patterns of iPP melt before and at selected times after shear (shear rate = 60 s^{-1}, shear duration = 5 s, T = 165°C).

at t = 70 s). The streak becomes stronger with increasing time. The equatorial streak can be attributed to the formation of microfibrils (or shish), bundles of parallel chains consisting of either mesophase or crystalline entities parallel to the flow direction. Also, it can arise from isolated, narrow shish-kebab entities that can form in a short time period after shear. The meridional maxima emerge after the development of equatorial streak, as can be seen in the SAXS patterns at later times: t = 600, 1200 and 7200 s in Figure 4. The meridional scattering, which exhibits three equal-distant scattering maxima at t = 7200 s (120 min), can be attributed to the development of regular spacing crystalline lamellae (or kebabs).

We used the following procedure to deconvolute the contributions of oriented SAXS intensity of the shish (I_{shish}) and kebab fractions (I_{kebab}) by sectional integration of the oriented SAXS pattern [28]. Time evolution of intensities I_{shish}, and I_{kebab} calculated by this method at selected times after shear are illustrated in Figure 5. It is seen that I_{shish} rises immediately after shear (Figure 5, left), while I_{kebab} increases only after a short induction time, about 210 s (see inset of Figure 5, right). Although I_{shish} rises rapidly immediately at initial times, its rate of increase diminishes at the later times. In contrast, I_{kebab} increases at a much greater rate than I_{shish}. It reaches a significantly higher value, about 24 times higher than I_{shish} at t = 7200 s after shear.

The rise in I_{shish} can be attributed to the continuous ordering of chains in the shish structures (that results in the increase of electron density contrast) as well as the shish growth along the flow direction after cessation of flow (that results in the increase of the shish volume). These observations are consistent with the results of a recent study by Petermann, et al. [29], who argued that the shish crystal growth is an autocatalytic process. That is, the shish growth is due to self-orientation of the molecules in the growth front of the crystal tip and does not necessarily require an external flow field. The increase in I_{shish} also implies that the connectivity between the linear nuclei along the flow direction can significantly increase with time. After the formation of shish, oriented crystals are initiated from the linear nuclei and grow

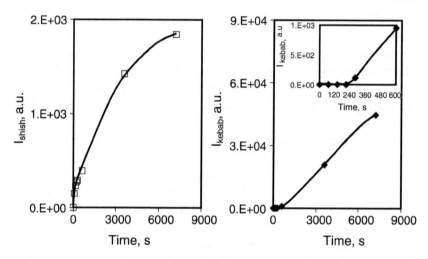

Figure 5 Time evolution of SAXS integrated intensities, I_{shish} (left) and I_{kebab} (right) after shear (shear rate = 60 s^{-1}, t_s = 5 s, T = 165°C). (Reprinted with pemission from Somani RH, Yang L, Hsiao BS, Agarwal P, Fruitwala H., Tsou AH. Macromolecules 2002;35(24):9096–9104. Copyright 2002, American Chemical Society.)

perpendicularly to the flow direction, forming layered lamellar structures. This is seen by the emergence of meridional maxima in the SAXS patterns after a short time delay. The increase in I_{kebab} (Figure 5, right) is found to be much greater than that in I_{shish} at the later stages.

2.   WAXD Observation

Figure 6 shows the linear WAXD intensity profiles taken from the equatorial slice of the 2D WAXD patterns at selected times after shear (left), and a selected WAXD pattern at the end of the experiment (t = 7200 s) (right), which is dominated by the appearance of the α-crystals for iPP. The (110) reflection of the α-crystals is observed at about 600 s after shear. The mass fraction of crystallites in the melt can be estimated by the standard peak-fitting procedure of the integrated WAXD intensity profile. As the crystal reflection signals are very weak, particularly at the beginning of the process, only a few selected

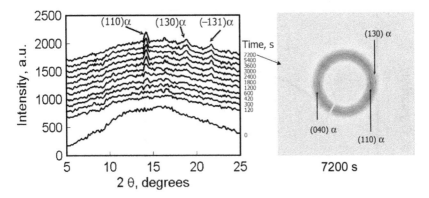

Figure 6 1D WAXD intensity profiles along the equator at selected times after shear (shear rate = 60 s^{-1}, t_s = 5 s, T = 165°C) (left); 2D WAXD pattern of iPP melt after shear at t = 7200 s (the equatorial peaks were indexed by the α-crystal form) (right).

profiles at later times can be analyzed by the curve-fitting method with a fair degree of accuracy. The percent crystallinity was estimated by subtracting the area under the fitted profile of the amorphous halo from the total area, and the result is shown in Figure 7. It is seen that even at t = 7200 s after shear, the measured crystallinity is only about 2.3%. Such a low crystallinity can be attributed to two reasons: (1) the chosen temperature (165°C) is near the nominal melting point of typical iPP crystals, and (2) only the shear-induced oriented crystals that are stable at such a high melt temperature can contribute to the crystallinity. At t = 7200 s after shear, while a very small fraction of isotropic crystals may be stable above the nominal melting point, both SAXS and WAXD results suggest that the melt consists mostly of oriented crystals.

3. Crystalline Fractions in Shish and Kebab Structures

The mass fractions of crystals in shish and kebab structures: X_{shish} and X_{kebab}, can be estimated from (1) the measured crystallinity X_c by WAXD, and (2) the ratio of SAXS intensity between the shish (I_{shish}) and kebab structures (I_{kebab}). If we

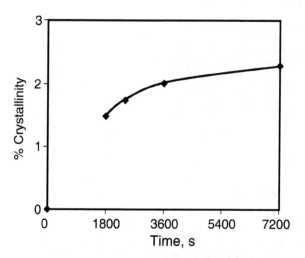

Figure 7 Development of crystallinity in iPP melt at selected times after shear (shear rate = 60 s^{-1}, t$_s$ = 5 s, T = 165°C). (Reprinted with pemission from Somani RH, Yang L, Hsiao BS, Agarwal P, Fruitwala H., Tsou AH. Macromolecules 2002;35(24):9096–9104. Copyright 2000, American Chemical Society.)

assume that the SAXS intensity is directly proportional to the amount of crystallinity, which is reasonable at the initial stages of crystallization [30], the following relationships for X_{shish} and X_{kebab} can be written in terms of the results obtained from the SAXS and WAXD data.

$$X_{shish} + X_{kebab} = X_c \text{ (WAXD)} \qquad (1)$$

$$X_{kebab}/X_{shish} = I_{kebab}/I_{shish} \text{ (SAXS)} \qquad (2)$$

The mass fraction of the total oriented crystals, X_c, at t = 7200 s after shear as estimated from WAXD is 2.3%; and the corresponding value of I_{kebab}/I_{shish} ratio from SAXS is 24. Thus, at t = 7200 s after shear, the values of X_{shish} and X_{kebab} in iPP melt at 165°C are 0.1% and 2.2%, respectively. Although the calculated value of X_{shish} is too low to be statistically meaningful, the order of magnitude must be correct. It is clear that only a very small fraction of crystalline structure exists in the shish. The kebabs later developed have a much higher crystallinity.

4.   Possible Pathways for Development
         of Shish-Kebab Crystallization
         Precursor Structures

Combined SAXS and WAXD results suggest that a scaffold or network of precursor structures form at the early stages of crystallization by flow. This scaffold contains the shish structure with a linear assembly of primary nuclei having excellent connectivity along the flow direction, and the kebab structure with folded chain lamellae having poor lateral connectivity. This semi-connected network is consistent with the recent rheo-optics results obtained by Winter et al. [31–32] and Kornfield et al. [25], where precrystallizing melt exhibits a gelation behavior with strong strain-induced birefringence under shear. A model for the structure of polymer nuclei at the early stages of crystallization in polymer melt under flow conditions is schematically presented in Figure 8. Diagram A represents the polymer melt before shear; where molecules are in the "random coil" state. Diagram B illustrates the melt structure immediately after shear. We adopt the proposed mechanisms of polymer orientation and primary nuclei growth by Petermann, et al. [29]. The applied flow immediately produces bundles of parallel chain fragments, which become thermodynamically stable, forming primary nuclei (mesomorphic or crystalline). The primary nuclei growth can be maintained or enforced by a self-induced orientation of the molecules in front of the growing tip. However, if a chain is incorporated in two adjacent growth fronts, the restraints such as entanglements or branching can stop the growth process. We envision that as the adjacent nuclei grow, they are bound to be connected by some long chains, producing a large local stress distribution in the amorphous chains surrounding the nuclei. These nuclei may be rearranged into a linear array (we term the shish structure) in order to minimize the stress concentration as seen in diagram B. This is consistent with the observations of equatorial streaks (a result of uncorrelated arrays of the shish structures) in SAXS patterns immediately after shear.

         Primary nuclei in the shish structure provide nucleation sites for the lateral growth of folded chain α-form crystals in iPP.

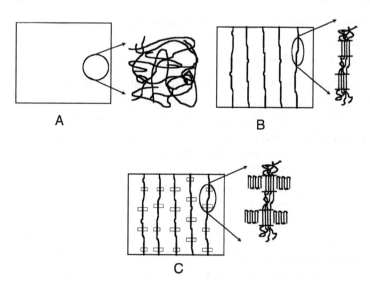

Figure 8 Schematic representation of flow-induced precursor structures at different stages: (A) before shear, (B) formation of precursor structures containing linear nuclei (shish), (C) formation of shish-kebab morphology through secondary nucleation from the primary nuclei. (Reprinted with pemission from Somani RH, Yang L, Hsiao BS, Agarwal P, Fruitwala H., Tsou AH. Macromolecules 2002;35(24):9096–9104. Copyright 2002, American Chemical Society.)

The mechanism of this process has been clearly demonstrated by the recent modeling works of Muthukumar et al. [33] They observed that the development of the shish-kebab morphology is related to the stretch-coil transition of isolated chains under extensional flow. The stretched chains crystallize into shish, while the coiled chains form lamellae and then adsorb to the shish forming the kebabs. The local inhomogeneity in polymer concentration and the change of flow rate can significantly alter the population of stretched and coiled conformations, thus affecting the onset of shish-kebab morphology.

In a different simulation effort, Hu et al. [34] also demonstrated that a single chain can induce transcrystallization, which implies that the lowest limit of the shish diameter is the stem of a single chain. We argue that the deposited chains can be either of lower molecular weight species with greater mobility or high molecular weight species with prevailing chain connection, both at the state of 'random coiled'. The folded-chain crystals grow perpendicularly to the flow direction and form lamellae of critical dimensions (kebabs) (Diagram C). This is consistent with the SAXS observation that a scattering maximum emerges along the flow direction after a short time (t ~ 280 s) and grows with time. The crystal reflection signals could only be detected in the corresponding WAXD pattern at 600 s after shear. The fast rise in the SAXS intensity (I_{kebab}) may be due to the large density contrast between the lamellar crystals and the surrounding melt. As the crystallinity is low even at later times after shear, the connectivity between the kebabs in the initial stages is probably low. The topology of the initial crystalline lamellae such as density and orientation will dictate the morphology of the final product.

B.   Flow-Induced Crystallization and Polymorphism in Isotactic Polypropylene at Large Supercooling

In situ synchrotron SAXS [16] and WAXD [35] experiments were carried out to study the crystal structural and morphological development of the same iPP sample at a large degree of supercooling (crystallization temperature at 140°C) after step-shear. The melt was subjected to a relatively short shear strain of 1428% (shear duration time was about 1 order of magnitude less than the previous case study) at three different shear rates (10, 57, and 102 s^{-1}). In this study, the precursor structures were overwhelmed by the crystallization process and cannot be easily detected after shear.

1.   SAXS Observation

Figure 9 illustrates a representative series of 2D SAXS patterns of iPP at 140°C before and after application of a step shear (102 s^{-1} shear rate and 1428% strain). The pattern of the initial amorphous melt (t = 0) consists of a very weak diffuse scattering profile from the isotropic melt, indicating the absence of any detectable structures and/or preferred ori-

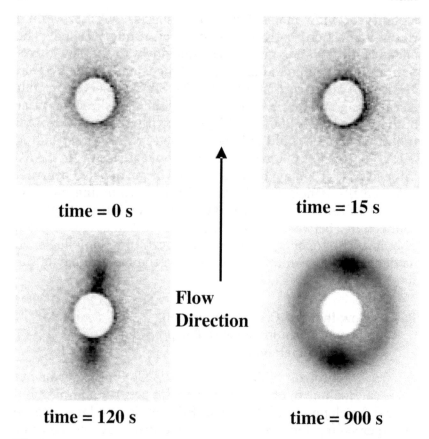

time = 0 s   time = 15 s

Flow
Direction

time = 120 s   time = 900 s

Figure 9 Selected SAXS images collected during deformation of iPP at T = 140°C (step shear with rate = 102 s^{-1}, strain = 1428% and t_s = 0.14 s).

entation. The pattern collected at 120 s after shear shows the appearance of meridional maxima (only a single scattering peak is seen) but without any sign of equatorial streak. The discrete meridional reflections were observed almost immediately after shear. The meridional maxima imply that the formation of crystal lamellar stacks in this weakly sheared melt. The intensity of discrete reflections in the consecutive images increases gradually, due to the growth of lamellar structures. The SAXS pattern at 900 s after shear shows that the oriented SAXS maxima are better defined but they are superimposed on a strong isotropic scattering ring. This is due to the further crystallization from the unoriented chains.

The half-times of crystallization for isotactic polypropylene at 140°C under different shear rates (10, 57, and 102 s^{-1}) were determined from the total scattered intensity profiles. At t = 0, the value of the total scattered intensity, I(0), is due to the amorphous, noncrystalline melt. The increase in total scattered intensity, I(t) − I(0), is then directly proportional to the growth of the crystallites in the polymer. The total scattered intensity reaches a plateau, I(s) (steady-state value), at the end of crystallization. The fraction of crystallized material, X_c, can be approximated as,

$$X_c = [I(t) − I(0)]/[I(s) − I(0)]. \qquad (3)$$

where the half-time of crystallization was taken as the time corresponding to X_c = 0.5. Figure 10 illustrates the effect of shear rate on both oriented fraction (to be discussed later) and half-time of crystallization for iPP at 140°C. At low shear rates (10 s^{-1}), the half-time for crystallization is slightly longer than those at high shear rates (57, 102 s^{-1}). However, it is interesting to note that the half-time at 10 s^{-1} is about two orders of magnitude lower than that of quiescent state, even under such a weak shear.

2.   WAXD Observation

A representative series of 2D WAXD patterns of iPP at 140°C, before and after application of step shear (102 s^{-1} shear rate and 1430% strain) are shown in Figure 11. The initial pattern (t = 0 s) shows diffuse scattering without any crystal reflections arising from a completely amorphous melt. In the WAXD pattern at 30 s after shear, sharp oriented α-crystal reflections are clearly observed. The azimuthal breadths in the intensity of crystal reflections are relatively narrow, indicating that the orientation of the crystals is high. In the WAXD pattern at t = 60 s, the (300) reflection of β-crystals can be clearly identified. The intensity of both α- and β-crystal reflections becomes stronger with time. The observation of the initial (300) reflection suggests that the first formed β-crystals are oriented;

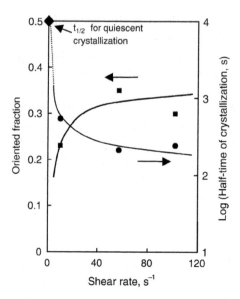

Figure 10 Fraction of oriented material and half-time ($t_{1/2}$) of crystallization as a function of shear rate (at constant strain = 1428%) in iPP at 140°C. (Reprinted with pemission from Somani RH, Hsiao BS, Nogales A, Srinivas S, Tsou AH, Sics I, Balta-Calleja FJ, Ezquerra TA. Macromolecules 2000;33(25):9385–9394. Copyright 2000, American Chemical Society.)

however, the diffraction patterns obtained at t = 120, 345, and 945 s after shear show that the (300) reflection becomes isotropic. In addition, it is seen that the average intensity of the (300) reflection becomes much stronger with time compared to the intensities of the reflections from the α-crystals.

The integrated 1D WAXD intensity profiles at t = 1185 s after shear (corresponding to the end of crystallization) at different shear rates (10, 57, and 102 s⁻¹) are shown in Figure 12. The corresponding time evolution of the percent contribution of the β-crystal growth at various shear rates is shown in Figure 13. The variation in the intensities of the (110) and (040) reflections from the α-crystals and the (300) reflection from the β-crystals, as a function of the shear rate can be clearly seen in Figure 12. At a shear rate of 10 s⁻¹, both the (110) and (040) reflections (of α-crystals) are stronger than

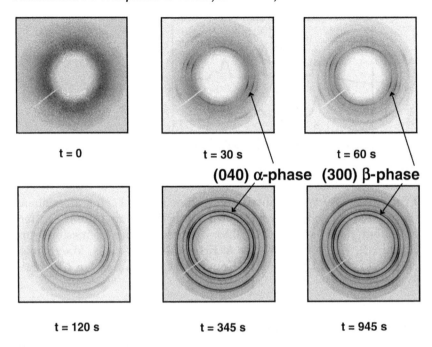

t = 0   t = 30 s   t = 60 s

(040) α-phase   (300) β-phase

t = 120 s   t = 345 s   t = 945 s

Figure 11   2D WAXD patterns of iPP at 140°C before and after step-shear (shear rate = 102 s⁻¹, strain = 1430%) (t = 0 – image of amorphous melt before shear; t = 30 s – image showing oriented α-crystal reflections; t = 60 s – image first showing reflection of β-crystals). (Reprinted with pemission from Somani R, Hsiao BS, Nogales A, Fruitwala H, Tsou A. Macromolecules 2001;34(17): 5902–5909. Copyright 2001, American Chemical Society.)

the relatively weak (300) reflection (of β-crystals). Thus, the contribution of α-crystals to the total crystalline phase is much higher than that of β-crystals at low shear rates. At high shear rates (57 and 102 s⁻¹) the strength of (300) reflection from the β-crystals is stronger than that of the (110) and (040) reflections from the α-crystals. Figure 13 shows that the time of inception and the growth of the β-crystals are slightly earlier for the shear rates of 57 and 102 s⁻¹ than for the shear rate of 10 s⁻¹. In addition, the percent contribution of β-crystals is only 20% at the low shear rate (10 s⁻¹), compared to 65 – 70% at high shear rates (57 and 102 s⁻¹). Figures 12 and

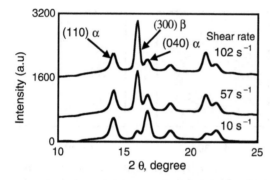

Figure 12 1D WAXD intensity profiles at t = 1185 s after shear (long after the end of crystallization) of iPP melt at 140°C at different shear rates (strain = 1430%). (Reprinted with pemission from Somani R, Hsiao BS, Nogales A, Fruitwala H, Tsou A. Macromolecules 2001;34(17):5902–5909. Copyright 2001, American Chemical Society.)

13 show that the difference in the contribution to the total crystalline phase from the β-crystals at the different shear rates is substantial; however, the difference is not significant between the shear rates of 57 and 102 s^{-1}. These results clearly show that both inception time and amount of β-crystals are shear-rate dependent.

In crystallization of iPP under quiescent conditions, the spherulites consist of only unoriented α-crystals. In the sheared iPP melt (without nucleating agents) both oriented α-crystals and β-crystals are seen. However, it is found that the oriented α-crystals form first, immediately after shear, and the β-crystals appear soon afterwards. These results clearly indicate that the oriented α-crystals are responsible for the growth of β-crystals, which have also been observed by Varga, et al. [36,37]. They reported that the "surface" of growing, oriented α-crystal assemblies such as the cylindrites could generate nucleation sites for the growth of β-crystals (in the form of spherulites). Thus, the amount of the β-crystals depends on the amount of oriented α-crystals in the melt. The amount of β-phase at low shear rates is lower than that at high shear rates (Figure 13).

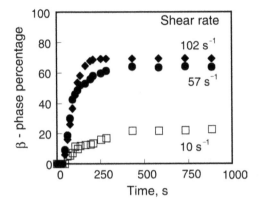

Figure 13 The ratio of the β-crystal to the total crystalline phase and its growth in the iPP melt at 140°C after shear at different shear rates (strain = 1430%). (Reprinted with pemission from Somani R, Hsiao BS, Nogales A, Fruitwala H, Tsou A. Macromolecules 2001;34(17):5902–5909. Copyright 2001, American Chemical Society.)

C. Concept of Critical Orientation Molecular Weight in Flow-Induced Crystallization of iPP

The behavior of flow-induced crystallization can be explained by the concept of the stretch-coil transition, first proposed by de Gennes [38]. Based on this concept, Keller et al. [39–41] have proposed the relationship of $\dot{\varepsilon} \propto (M^*)^{-\beta}$ for the case of elongational flow-induced crystallization in polymer solutions, where $\dot{\varepsilon}$ is the critical elongational rate, M^* is the critical orientation molecular weight defining the onset of the stretch-coil transition and β is a factor that was found to be equal to 1.5 in polyethylene solution. We have proposed that similar behavior also occurs in shear-induced crystallization experiments, which is illustrated in Figure 14. When the polymer melt, containing a broad chain length distribution, is subjected to a shear flow field of a particular rate $\dot{\gamma}$ and temperature, only the longer chains will remain oriented after shear (although with different degrees of extension) with a low end cut-off at M^*; while the rest of the molecules will

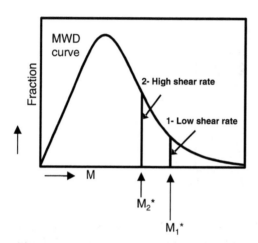

Figure 14   Schematics showing the effect of imposed shear conditions (high and low shear rate) on the shift in the location of the critical orientation molecular weight M*. (Reprinted with pemission from Somani RH, Hsiao BS, Nogales A, Srinivas S, Tsou AH, Sics I, Balta-Calleja FJ, Ezquerra TA. Macromolecules 2000;33(25): 9385–9394. Copyright 2000, American Chemical Society.)

relax rapidly into the initial random coil state and remain unstretched. The result is a bimodal distribution of chain molecules represented by a dual population of oriented and unoriented (stretched and coiled) chains. The value of M* shifts to a higher value at low shear rates, following the relationship of $M^* = K(\dot{\gamma})^{-\alpha}$ (K is a constant), which is shown by lines 1 (low $\dot{\gamma}$) and 2 (high $\dot{\gamma}$) in Figure 14.

Using results in Section III.B, the values of M* and α (exponent) have been examined for shear-induced crystallization of iPP at 140°C using the following procedures. The total scattered intensity (I_{total} [s,ϕ] where $s = 2\sin\theta/\lambda$ is the scattering vector, 2θ is the scattering angle, and ϕ is the azimuthal angle) from the SAXS pattern under shear (Figure 9) was first deconvoluted into two components: (1) $I_{unoriented}$[s], scattering from the randomly distributed crystalline lamellae, and (2) $I_{oriented}$[s,ϕ], scattering from the oriented lamellae, with a 2D image analysis method outlined elsewhere [43]. The orientation fraction was calculated as the ratio between $I_{oriented}$[s,ϕ]

and $I_{total}[s,\phi]$) (its value at different shear rates is shown in Figure 10). We assumed that the oriented fraction value is proportional to the fraction of the molecular weight distributions (MWD from GPC chromatogram) above M*. As a result, the following values have been calculated: K = 6.6 × 10^5 g/mol, α = 0.15 (under a step-shear flow at 140°C).

We have further verified the concept of M* for the case of shear-induced crystallization in a series of iPP blends with different MWD [42]. The experiments were carried out in the undercooled melts of iPP and blends of different molecular weights (sample A: M_n = 3.19 × 10^4 g/mol, M_w = 1.48 × 10^5 g/mol; sample I: M_n = 4.37 × 10^4 g/mol, M_w = 3.09 × 10^5 g/mol) at 150°C. The samples were subjected to a low shear strain (1428%) at a fixed shear rate (57 s^{-1}) and results are shown in Figure 15. At the given shear conditions, M* (≈ 2.5 × 10^5 g/mol) was found to be independent of the molecular weight distribution. This finding confirmed that only the polymer molecules having a molecular weight above the critical value M* can remain stretched after shear, where polymer chains with molecular weight lower than M* would relax back into the random coiled state. The degree of chain extension can be represented as the ratio of the radius of gyration in the direction parallel to the deformation field (R_g∥) to the radius of gyration in the perpendicular direction to the deformation field (R_g⊥); in the case that R_g∥ >> R_g⊥, a large fraction of chains is stretched.

D.   Verification of Flow-Induced Precursor Structures in Bimodal Polyethylene Blends

Development of shear-induced crystallization precursor structure has been further confirmed by *in situ* rheo-SAXS and -WAXD techniques using binary polymer blends of high and low molecular weight polyethylenes near their nominal melting temperatures (120°C) [43]. Two low molecular weight polyethylene copolymers, containing 2 mol% of hexene, with weight average molecular weights (M_w) of 50,000 (MB-50k) and 100,000 (MB-100k), and polydispersity of 2, were used as the non-crystallizing matrices. A high molecular weight poly-

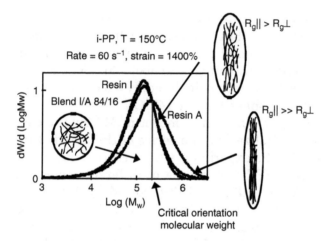

Figure 15 The constant value of M* determined by SAXS in different iPP samples and their blends under shear at 150°C. (Reprinted from Nogales A, Hsiao BS, Somani RH, Srinivas S, Tsou AH, Balta-Calleja FJ, Ezquerra TA. Polymer 2001;42(12): 5247–5256. Copyright 2001, with permission from Elsevier.)

ethylene homopolymer with M_w of 250,000 (MB-250k) and polydispersity of 2 was used as the crystallizing minor component. Two series of model blends, MB-50k/MB-250k and MB-100k/MB-250k, each containing weight ratios of 100/0, 97/3, 95/5 and 90/10, were prepared by solution blending to ensure thorough mixing at the molecular level.

In this study, we have obtained several new findings of the development of crystallization precursor structures by shear (rate = 60 s^{-1}, duration = 5 s, T = 120°C) prior to the occurrence of full scale crystallization. The unique feature of these blends is that the low molecular weight matrix (MB-50k or MB-100k) does not crystallize under the experimental conditions, where only the high molecular weight additive can form the scaffold of precursor structure. These new insights can be summarized as follows. (1) Both chemical composition and applied shear flow can significantly affect the crystallization kinetics of the crystallizing high molecular weight component (MB-250k), but in a very different fashion since the viscosity (thus the relaxation time spectrum) of the high

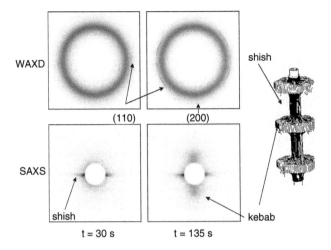

Figure 16 Selected 2D WAXD and SAXS patterns to illustrate development of "shish-kebab" morphology in PE blends. (From Yang L, Somani RH, Sics I, Kolb R, Fruitwala H, Ong C. Macromolecules 2004; 37:4845–4859. Copyright 2004, American Chemical Society.)

molecular weight component is changed cooperatively with the viscosity of the matrix. When the viscosity of the matrix is low, the viscosity of the crystallizing high molecular weight component also becomes lower, leading to faster crystal growth under both quiescent and flow condition. In contrast, when the viscosity of the matrix is high, the viscosity of the crystallizing high molecular weight component becomes higher, which favors the development of higher nucleation density induced under flow (e.g., the composition of 10 wt% of MB-250k exhibited the fastest crystallization rate in the MB-100k/MB-250k blends). (2) In the MB-100k/MB-250k (90/10) blend, both rheo-SAXS and –WAXD patterns showed the formation of distinct shish-kebab morphology (Figure 16 — the initial development of the equatorial "streak" in SAXS and the simultaneous appearance of two equatorial (110) reflections in WAXD clearly suggest the formation of a shish structure with a length on the order of 1000 Å and a diameter on the order of 100 Å based on the crystallite size analysis of the equatorial (110) peak using the Scherrer equation); while the rest of the blends only exhibited oriented lamellar (kebab) morphology without the presence of shish. The observed shish structure can be attributed to the extended chain crystallization of stretched PE chain segments. In addition, all formed kebabs exhibited a twisted lamellar structure, typically seen in PE crystallites formed from chains of no orientation. We speculate that the observed shish is related to the high values of relaxation time spectrum in MB-250k, where the corresponding high values of local stress promote the rapid alignment of chains. In other blends, as the stretched chains do not aggregate rapidly, no detectable shish was observed. However, the pathway of the shish-kebab or the net kebab formation clearly follows the concept of stretch-coil transition first proposed by Keller et al. for elongation of polymer solutions [39–41] and later demonstrated by Muthukumar et al. using simulation tools [33].

IV. NANOSTRUCTURAL DEVELOPMENT IN POLYMER SOLID DURING DEFORMATION

Structural and morphological changes during deformation of two polymer solids: iPP and poly(ethylene terephthalate), PET, are demonstrated here. As both polymers exhibit mesomorphic structures during deformation, our emphasis will be placed on the behavior of strain-induced phase transitions and hierarchical structural changes during uniaxial deformation.

A.   Deformation of Semicrystalline iPP

There have been many studies on the structure, morphology and mechanical property and their relationships for iPP. As discussed earlier (Section III.A.), the mesomorphic structure in iPP is a unique feature. Miller [44] suggested that the order existing in the iPP mesophase could be of the type described by Hosemann [45] as "paracrystalline." This idea was reproposed by Zannetti et al. [46–47] later. Wyckoff [48] found a certain degree of correlation between the adjacent helices in the mesophase, suggesting short-range three-dimensional (3-D) ordering in the structure. Gailey and Ralston [49] pointed out that the partially ordered phase of iPP was composed of small hexagonal crystals that were 50 to 100 Å in size. Gomez et al. [50] studied the polymorphism of iPP using high-resolution solid-state 13C-NMR. They found that the packing of the 3_1 helices in the mesomorphic form was similar (at least at a very local scale) to that of the iPP chains in the β-crystal form. On the other hand, Bodor et al. [51] assumed the mesomorphic form to be composed of microcrystals of the α form and suggested that the small crystal size broadening to the x-ray detection limit was responsible for the typical x-ray diffraction pattern. This possibility was also considered by Farrow [52]. Wunderlich and Grebowicz [53,54] proposed that the conception of conformationally disordered crystal was perhaps more appropriate to attribute to the mesomorphic form. This explanation was based on the hypothesis that the mesomorphic form of iPP might have a frozen liquid-like structure, in which the threefold helices, as obtained from the melt, consisted of short-chain segments with opposite signs of helical structure. Corradini et al. [55,56] considered the various models corresponding to pseudohexagonal (as in the β form) and monoclinic (as in the α form) crystals, as well as several other disordered models possessing characters of both forms from deformation studies. They compared the x-ray intensities derived from these models with the experimental x-ray data and concluded that the mesomorphic form of iPP was not composed of small pseudohexagonal crystals, but of much more disordered bundles of chains.

Although many studies on the subject of mesophase in iPP have been carried out, there are still some unresolved issues in relation to the structure and morphology of the mesophase, particularly during deformation. The objective of this sample study was to reexamine this subject with advanced synchrotron SAXS and WAXD techniques and to demonstrate that quantitative fractions of crystal, mesomorphic and amorphous phases can be extracted from 2D WAXD patterns using a unique image analysis method. Figure 17 illustrates the 2D WAXD patterns of a semicrystalline iPP fiber with different draw ratios at room temperature. Figure 17A represents the pattern of the original fiber without draw-

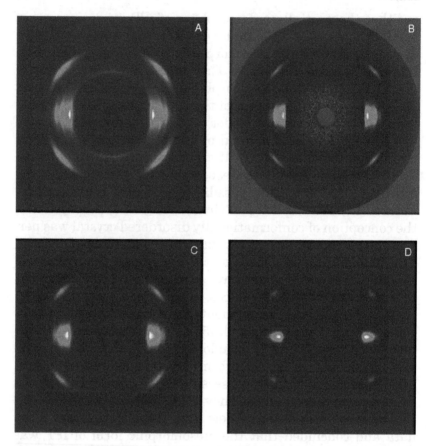

Figure 17 WAXD patterns of isotactic polypropylene fiber deformed at room temperature and different draw ratios (A: 1.0, B: 1.5, C: 2.0, D: 2.75).

ing, which shows that the initial fiber is partially oriented. The three strong peaks located on the equator are the characteristic of the α-form crystal, which can be indexed as the (110), (040) and (130) reflections, respectively. Consistent with the literature, these three peaks are located at $2\theta = 14.2°$, $17.0°$ and $18.8°$ [57]. However, it is noted that these three peaks are not clearly separated. There could be two reasons. One is that the formed crystals might be defective. The other is that there is the presence of the mesomorphic form mixed with the α-crystal form in the original fiber. With the increasing draw ratio, the azimuthal spread of the reflection becomes narrower, indicating that the crystal orientation increases with the draw ratio. It is interesting that the superposition of the three α-form equatorial peaks also becomes more severe such that it is impossible to calculate the unit cell parameters for the fiber with larger draw ratios. At a draw ratio equal to 2.5, the three equatorial peaks completely disappear, resulting in one broad peak. This broad equatorial peak and the weak but distinct four off-axis peaks (at the first layer line) (Figure 17D) are often regarded as WAXD fingerprints of the mesomorphic form in iPP [58]. From the 2D WAXD results, we conclude that the defective α-form crystals are converted into the mesomorphic form with increasing draw ratio at room temperature.

As discussed earlier in Section III.C, the same 2D image analysis method [16] was used to deconvolve the mass fractions of the constituting phases from the 2D WAXD data. Figure 18 shows the fractions of the crystal, mesomorphic and amorphous phases as a function of draw ratio at room temperature based on this analysis. At small draw ratios (less

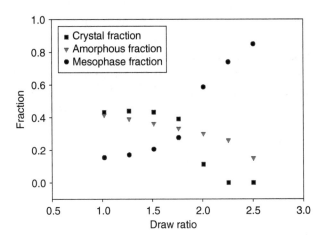

Figure 18 Fractions of crystal, mesomorphic and amorphous phases as a function of draw ratio at room temperature.

than 1.5), the crystal fraction didn't change much, but the mesophase increased and the amorphous phase decreased gradually. Above draw ratio 1.5, the crystal fraction decreased dramatically and the mesophase increased quickly, too. The amorphous phase still decreased as before. Our explanation is that the molecular mobility of polymer chains is small at room temperature. So the crystallization does not easily occur at room temperature. When the draw ratio is small, the energy provided by the strain is not enough to deform the crystal structure even for the defective crystals. However the chains in the amorphous phase would pack along the draw direction and orient with some kind of ordering, which could form the mesomorphic modification. When the draw ratio is large enough that the energy from the drawing could deform the defective crystals, the lateral registration of some crystals could be destroyed. At room temperature, three-dimensional ordering of the crystals could not be destroyed completely by drawing. As a result, the defective crystals disappeared and formed the mesomorphic modification.

Figure 19 shows the 2D SAXS patterns of iPP fiber with different draw ratios at room temperature. Figure 19A is the pattern of iPP fiber before drawing, which exhibits a two-bar pattern at the first layer line position of the meridian. This indicated that α-form crystals, probably in the form of layered lamellae, are present in the fiber. The intensity difference between the two-bar scattering and the background becomes weaker with increasing draw ratio, which suggests that the lamellar ordering is gradually destroyed. When the draw ratio reaches 2.0, the meridional two-bar pattern completely disappears, replaced by a strong scattering streak at the equator (Figure 19C). The equatorial streak can be attributed to the scattering from the fibril superstructure, which is related to the strain-induced mesophase in iPP (results from WAXD).

B.   Deformation of the Isotropic Amorphous PET

The recent simulation work by Frenkel et al. [59] indicated that the isotropic-nematic transition occurs when the aspect

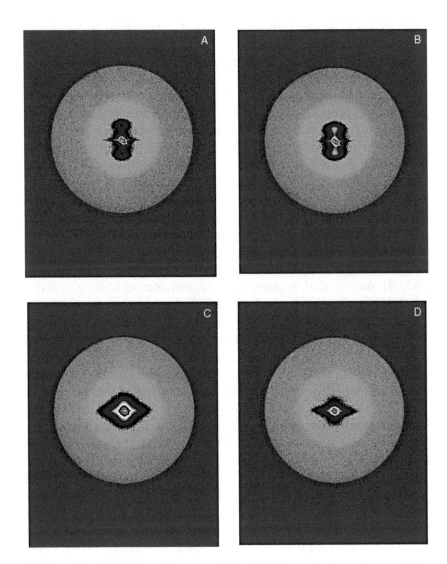

Figure 19 SAXS patterns of isotactic polypropylene fiber deformed at room temperature and different draw ratios (A: 1.0, B: 1.5, C: 2.0, D: 2.75).

ratio (L/D, where L represents the length and D represents the diameter) of the molecule is in the range of 3 to 4. In the case of PET, its persistence length is about 1.33 nm and the average molecular diameter is about 0.66 nm based on the experimental work by Imai et al. [60]. This suggests that the aspect ratio of the PET molecule is around 2, which falls short of being a liquid crystalline polymer. However, the existence of mesomorphic phases has been well documented in oriented PET samples. Bonart was the first scientist who reported the formation of nematic and smectic phases during the tensile stretching of PET using conventional x-ray diffraction method [61,62]. Recently, with the use of high intensity synchrotron x-rays, more detailed features of mesomorphic phases in oriented PET have been revealed by different research groups. For example, Windle et al. reported the formation of a transient smectic phase in oriented fibers made of random PET and PEN (polyethylene naphthalene-2,6-dicarboxylate) copolymers [63,64]. Asano et al. reported the appearance of a smectic order at 60°C having a spacing of 10.7 Å during the annealing of cold-drawn amorphous PET films [65]. Blundell and coworkers observed the smectic A structure during the fast extension of PET. They proposed that the smectic structure is a precursor of crystalline based on the simultaneous appearance of the triclinic crystalline peak and disappearance of the smectic peak [66–70]. Our group reported the smectic C phase during the deformation of amorphous PET film below the glass transition temperature (T_g) at 50°C [71]. We observed that the mesophase developed immediately upon the neck formation. As the mesophase contained a sharp meridional peak (001') (d = 10.32 Å), which was smaller than the monomer length in the typical triclinic unit cell (c = 10.75 Å), we concluded that the chains in the mesophase formed an inclined smectic C structure.

Thus, PET can be considered as a "marginal" liquid crystalline polymer, because its mesomorphic structures are not obvious in the unoriented state, but are very distinct in the oriented state. We thus conclude that the dynamic pathway is the key to dictate the phase transition in PET. In the following case studies, the phase transition and structure development in an initially isotropic PET sample below and above its glass transition temperature (T_g) at 70 and 90°C, respectively, are discussed. The chosen sample had M_w of 35,000 g/mol and polydispersity around 2.0. Minimum amounts of antimony (the catalysis for polymerization) and phosphate (the additive to enhance the heat durability) were used to prepare this sample. Thus the sample could be viewed as pure polymer that would not decompose under high molding temperatures. The sample was first molded into a dumbbell-shape at 270°C followed by rapid quenching with ice water (0°C) and showed no detectable crystallinity (by x-ray and DSC).

1.   Strain-Induced Phase Transition and
        Structural Development below T_g

The relationships among the structure, morphology and load-strain curve measured at 70°C can be divided into four zones (I-IV) for discussion. Figure 20 shows the load-strain curve in conjunction with selected WAXD images (corrected by the Fraser method [72] — these images are illustrated in undistorted reciprocal space). Zone-I (strain 0–55%) represents the stage where isotropic amorphous chains are oriented by deformation where the isotropic-nematic transition begins to take place. The final WAXD image in Zone I exhibits a distinct nematic phase, thereby the main event in Zone I can be attributed to generation of the nematic phase. In Zone II (strain 55–200%), sharp meridional (001') peaks appear, which are characteristics of the smectic phase. The increase of the (001') intensity indicates that the population of the smectic phase increases with strain in Zone II. In Zone III (strain 200–360%), the equatorial (010) reflection appears, which is indicative of the occurrence of the triclinic crystalline phase. Thus the main feature in Zone III is the transition of crystalline phase from the smectic phase. The final stage is Zone IV (strain 360–500%), where the development of the crystalline phase becomes dominant but at a relatively reduced rate.

Figure 21 illustrates selected corresponding SAXS images collected during *in situ* deformation at 70°C. Although only one image was shown in Zone I, it is interesting to note that this image exhibits a strong equatorial streak which is

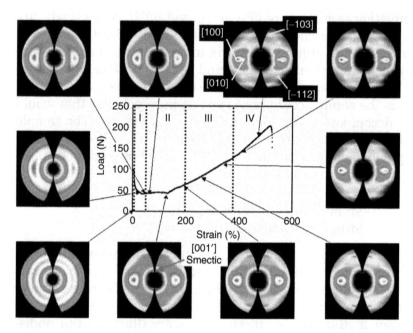

Figure 20 Selected WAXD images of PET (normalized with the Fraser correction [72]) during the collection of load-strain curve at 70°C. (Reprinted with permission from Kawakami D, Hsiao BS, Burger C, Ran S, Avila-Orta C, Sics I, Kikutani T, Jacob KI, Chu B. Macromolecules 2005; 38:91–103. Copyright 2005, American Chemical Society.)

drastically different from the SAXS image of the undeformed sample, where only diffuse scattering was seen. The equatorial streak may be attributed to two possibilities: (1) the formation of microvoids (crazes) [73,74] and (2) the formation of fibrillar superstructure [4,75–77]. Judging by the relatively weak scattered intensity (the void scattering is usually several orders stronger) and the transparent appearance of the deformed sample, we conclude that the equatorial streak in SAXS is mainly due to the formation of fibrillar superstructure, which persists through Zone II. The SAXS images in Zone III exhibit extra scattering features on the meridian. The meridional scattering, showing a maximum value that corresponds to a long spacing around 100 nm, reflects the

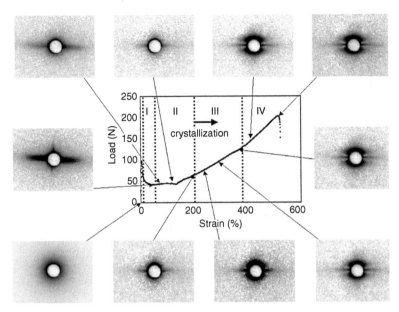

Figure 21 Selected SAXS images of PET during the collection of load-strain curve at 70°C. (Reprinted with permission from Kawakami D, Hsiao BS, Burger C, Ran S, Avila-Orta C, Sics I, Kikutani T, Jacob KI, Chu B. Macromolecules 2005; 38:91–103. Copyright 2005, American Chemical Society.)

emergence of a very loosely arranged layered lamellar structure. The observed scattering maximum is quite different from the typical SAXS peak observed in fully crystallized PET samples, which usually exhibits a long period only around 10 nm. The morphology in Zone IV is similar to that in Zone III. Based on combined results from WAXD and SAXS, we argue that the layered structure is formed within the fibril superstructure. As the length of the fibril is around several hundred nanometers, only several layers of crystalline lamellae are formed in the fibril, where the average width of the crystal layer is about several nanometers.

The strain-induced phase transition diagram for an amorphous PET sample during deformation at 70°C is illustrated in Figure 22. We note that this diagram is illustrated not from a thermodynamic perspective but from a "dynamic"

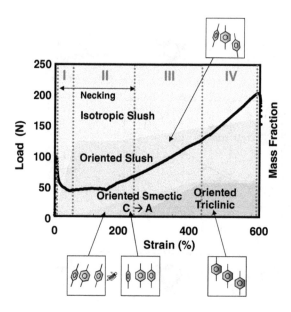

Figure 22   Strain-induced phase diagram of PET during deformation at 70°C. (Reprinted with permission from Kawakami D, Hsiao BS, Burger C, Ran S, Avila-Orta C, Sics I, Kikutani T, Jacob KI, Chu B. Macromolecules 2005; 38:91–103. Copyright 2005, American Chemical Society.)

perspective. The mass fraction of each phase was derived from the WAXD data, which would change at different temperature and deformation rates. In Figure 22, contrary to the classical concept of strain-induced isotropic-crystal transition in PET, we argue that several phase transitions (isotropic-nematic, nematic-smectic and smectic-crystal) occur sequentially. It is interesting to see that the fraction of the nematic phase is the largest throughout the deformation, and all four phases coexist at high strains. The schematic diagrams for the pathways of hierarchical structural development and their relationships with the phase transitions in different zones (I, II and III) are illustrated in Figure 23A-C, respectively. The representative SAXS/WAXD patterns with major scattering/diffraction features are also included in these diagrams.

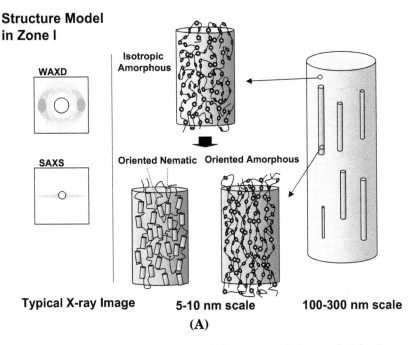

Figure 23 Schematic diagrams of phase transition and structure development pathways as well as corresponding SAXS/WAXD patterns during PET deformation at 70°C in (A) Zone I (B) Zone II and (C) Zone III (drawing not to scale). (Reprinted with permission from Kawakami D, Hsiao BS, Burger C, Ran S, Avila-Orta C, Sics I, Kikutani T, Jacob KI, Chu B. Macromolecules 2005; 38:91–103. Copyright 2005, American Chemical Society.) Continued.

Zone-I (Figure 23A). We assume that the initial sample consists of only isotropic amorphous chains. Upon deformation, some amorphous chains will become oriented and can transform into the nematic phase. The aggregation of the nematic phase seems to be a corporative behavior with strain, which will grow more preferably along the machine direction and result in a fibrillar superstructure (the length of the fibril is several hundred nanometers). The yield point of the load-strain curve coincides with the first detection of the isotropic-nematic transition. At the end of Zone I, the nematic phase

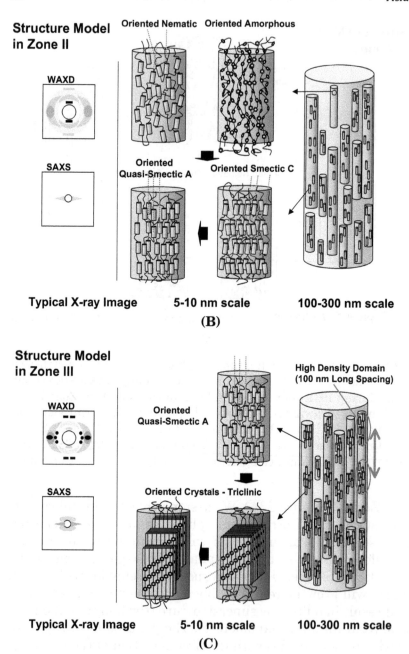

Figure 23 Continued.

becomes dominant in the sample (mass fraction about 70%), where the applied load also reaches a low plateau level.

Zone II (Figure 23B). Within the fibrillar superstructure, the nematic-smectic C phase transition is observed. The formation of the smectic domain probably forms the precursor structure for the later development of crystalline lamellae. However, as the density contrast between the nematic and smectic phases is very low, no meridional scattering feature is seen in SAXS. An apparent "plastic deformation" stage is observed during the nematic-smectic transformation process, which indicates that the plastic flow behavior is coupled with the transitions between the liquid-crystalline phases. With the increase in strain, the tilt angle of the smectic C phase decreases, forming a structure similar to the quasi-smectic A phase. The transformation between the smectic C to quasi-smectic A phase seems to increase the applied load, marking the initial stage of strain-hardening.

Zone III (Figure 23C). Triclinic crystalline phase is formed from the strained quasi-smectic A phase by chain sliding. The formation of the triclinic crystalline phase in the fibrillar superstructure leads to a crystalline lamellar structure with a long period of about 100 nm. SAXS can detect the lamellar structure because of the improved density contrast between the crystalline and nematic phases. Upon the formation of crystalline phase, the remaining smectic phase converts back into the stable smectic C structure, probably due to the relief of the local stress. The initial structure of the crystalline phase is relatively defective, which is dominated by the preferred arrangement of benzene sheet formation. The network of the defective crystallites enhances the mechanical properties and generates a near linear load-strain relationship.

2.   Structural Development above T_g

Deformation studies of amorphous PET above its T_g are relatively rare because of the experimental difficulties at high temperatures. The typical study dealing with this subject was usually carried out in two steps: (1) deformation at high temperatures, and (2) subsequent quenching to preserve the structure in the deformed sample for characterization (we termed this the step-quenching process) [78–80].

The structural development of an amorphous PET sample during uniaxial deformation above T_g (at 90°C) was studied by *in situ* synchrotron WAXD. Results indicated that the structural development can also be categorized into three zones, designated as I, II and III. In Zone I, the oriented mesophase is induced by strain, where the applied load remains about constant. In Zone II, crystallization is initiated from the mesophase through nucleation and growth, where the load starts to increase marking the beginning of the strain-hardening region. In Zone III, the stable crystal growth process is facilitated by strain-induced orientation until the breaking of the sample, where the ratio between load and strain remains about constant.

Overall, the structure and property relationship during deformation above T_g is generally similar to that below T_g (Section IV.B.1). However, two major differences were also observed: (1) the absence of the yielding behavior, (2) the lack of clear liquid-crystalline transitions. The dominant feature in the structural development above T_g is the strain-induced crystallization, responsible for the strain-hardening behavior, which is described as follows.

Figure 24 shows that the displacements of two reflections (–103) and (003) change continuously with the applied strain. The displacement was defined as the azimuthal angle between the peak position and the meridional axis after correcting the effects of the curvature of the Ewald sphere using Fraser's method [72]. It was found that the displacement of the (–103) peak decreased, while that of the (003) peak increased with strain. The changes in the azimuthal displacements of these reflections can be attributed to the deformation and/or the rotation of the PET unit cell. The effect of the unit cell rotation can be evaluated as follows. If the molecule axis (i.e., the crystallographic c axis) is tilted with respect to the fiber axis (which is actually known to occur in some samples of PET), while both crystallographic a* and b* axes remain perpendicular to the c axis, then as a consequence of the tilt, either the (h00) or the (0k0) and most likely the mixed

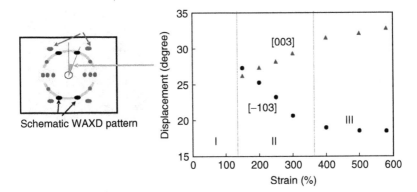

Figure 24 The schematic WAXD pattern illustrating the peaks of (003) and (–103) (left) and the displacement angles of diffraction peaks (003) and (–103) during deformation at 90°C (right). (Reprinted with pemission from Kawakami D, Ran S, Burger C, Fu B, Sics I, Hsiao BS. Macromolecules 2003:36(25):9275–9280. Copyright 2003, American Chemical Society.)

(hk0) reflections on the equator should split up and show a bimodal azimuthal distribution. This feature was not seen in the WAXD patterns, thus the effect of unit cell rotation must be very small. As a result, we have neglected the unit cell rotation as a possible source for the changes of the (–103) and (003) azimuthal displacements.

Concerning the unit cell deformation, the triclinic PET unit cell offers six adjustable lattice parameters. Our strategy to analyze the unit cell deformation was as follows. The majority of the parameters were kept constant at their ideal values in the stable crystalline PET structure. In addition, the changes of unit cell angles by shearing deformation were assumed to be more prompt to occur than the changes of unit cell dimensions by deforming crystalline chains. We found that a good description of the azimuthal changes in (–103) and (003) could be obtained by keeping a, b, c, and γ constant (a = 4.56 Å, b = 5.94 Å, c = 10.75 Å and γ = 112 °) [83] and using α and β as variables. This approach gave us two variables to solve with two observable quantities. This approach is acceptable in this study because we are not too concerned with the

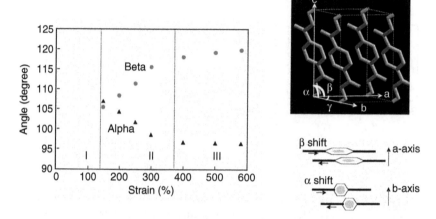

Figure 25 The changes of α and β angles in the unit cell during deformation (left) and the schematic diagram of PET triclinic unit cell structure calculated by Cerius 2. (Lattice Parameters: a = 4.56 Å, b = 5.94 Å, c = 10.75 Å, α = 98.5°, β = 118°, g = 112° [83]) as well as the molecular diagrams representing the α and β shifts. (Reprinted with pemission from Kawakami D, Ran S, Burger C, Fu B, Sics I, Hsiao BS. Macromolecules 2003:36(25):9275–9280. Copyright 2003, American Chemical Society.)

precise determination of the deformed lattice constants; rather, we are looking for clues to the molecular mechanisms taking place during the process of strain-induced crystallization. In other words, while the actual values of a, b, c, and γ may vary slightly during stretching, the changes in α and β are the most dominant parameters in Zone II. The changes of α and β are shown in Figure 25. It was seen that the β angle increased while the α angle decreased in Zone II. The β shift represented the sliding of the (100) plane or the benzene ring in the unit cell. The initial value of β (105°) was closer to 90° rather than the equilibrium value (118°) in the triclinic structure, suggesting that the benzene molecules stacked more perpendicularly to each other at the initial stage of crystal formation. With the increase in strain, the benzene molecules slipped past each other probably due to the shearing motion and eventually settled into a stable triclinic structure.

Upon the initiation of crystallization, the applied load increased immediately. The load increase can be attributed to the formation of a three-dimensional (3D) network of imperfect crystallites, immersed in a continuous matrix containing a random amorphous phase and oriented mesophase. The greater the concentration of the crystallites became, the larger the load developed. The crystal registration along the benzene sheet, which was indicative of the (010) peak, appeared to form first, whereas the growth along the benzene stacking direction appeared to develop later. During this zone, several processes seemed to proceed simultaneously with increasing strain: the crystal growth along all three directions, the crystal perfection and the crystal orientation. The crystal perfection process can be primarily followed by the changes of two unit cell angles (α and β) in the triclinic structure as discussed earlier. Finally, the end of Zone II can be marked by the stabilization of the crystal structure and concentration, where the load is found to be linearly proportional to strain afterwards.

ACKNOWLEDGMENTS

The author gratefully thanks the National Science Foundation (DMR0098104) for the support of this work. The author also acknowledges the assistance of C. Avia-Orta, D.F. Fang, L. Liu, A, Nogales, S.F. Ran, I. Sics, S. Toki, R.H. Somani, L. Yang, and X.H. Zong for experimental assistance. In particular, the author wishes to thank Profs. B. Chu and F. Baltá Calleja for their insightful comments on several projects.

REFERENCES

1. Chu B, Hsiao BS. Small angle x-ray scattering of polymers. Chemical Rev. 2001;101(6):1727–1761.

2. Bras W, Ryan AJ. Sample environments and techniques combined with small angle x-ray scattering. Adv. in Colloid and Interface Sci. 1998;75:1–43.

3. Hughes DJ, Mahendrasingam A, Martin C, Oatway WB, Heeley EL, Bingham SJ, Fuller W. An instrument for the collection of simultaneous small and wide angle x-ray scattering and stress–strain data during deformation of polymers at high strain rates using synchrotron radiation sources. Rev. Sci. Instrum. 1999;70(10):4051–4054.

4. Ran S, Fang D, Zong S, Hsiao BS, Chu B, Cunniff PM. Structural changes during deformation of Kevlar fibers via on-line synchrotron SAXS/WAXD techniques. Polymer 2000;42(4):1601–1612.

5. Toki S, Sics I, Ran S, Liu L, Hsiao BS. New insights into structural developments in natural rubber during uniaxial deformation by *in situ* synchrotron x-ray diffraction. Macromolecules 2002;35(17):6578–6584.

6. Yeh F, Hsiao BS, Sauer BB, Michel S, Siesler HW. Structure studies of a polyurethaneurea elastomer under deformation. Macromolecules 2003;36(6):1940–1954.

7. Cakmak M, Teitge A, Zachmann HG, White JL. On-line small-angle and wide-angle x-ray scattering studies on melt-spinning poly(vinylidene fluoride) tape using synchrotron radiation. J. Polym. Phys. Polym. Phys. 1993;31:371–381.

8. Terrill NJ, Fairclough JPA, Towns-Andrews E, Komanschek BU, Young RJ, Ryan AJ. Density fluctuations: the nucleation event in isotactic polypropylene crystallization. Polymer 1998;39(11):2381.

9. Samon JM, Schultz JM, Hsiao BS, Seifert S, Stribeck N, Gurke I, Saw C, Collins G. Structure development during the melt spinning of polyethylene and poly(vinylidene) fibers by *in situ* synchrotron small- and wide-angle x-ray scattering techniques. Macromolecules 1999;32(24):8121–8132.

10. Nolan, SJ, Broomall CF, Bubeck RA, Radler MJ, Landes BG. Monofilament drawing device for *in situ* x-ray scattering studies of orientation development in polymeric fibers. Rev. Sci. Instr. 1995;66:2652–2657.

11. Ran S, Burger C, Fang DF, Zong XH, Cruz S, Hsiao BS, Chu B, Bubeck RA, Yabuki K, Teramoto Y, Martin DC, Johnson MA, Cunniff PM. *In situ* structural development during PBO solution spinning by synchrotron WAXD/SAXS studies. Macromolecules 2002;35(2):433–439.

12. Hsiao BS, Kennedy AD, Leach RA, Chu B, Harney P. Studies of structure and morphology development during the heat-draw process of Nylon 66 fibers by synchrotron x-ray diffraction and scattering techniques. J. Appl. Cryst. 1997;30:1084–1095.

13. Ran S, Zong X, Fang D, Hsiao BS, Chu B, Phillips RA. Structural and morphological studies of isotactic polypropylene fibers during heat/draw deformation by *in situ* synchrotron SAXS/WAXD. Macromolecules 2001;34(8):2569–2578.

14. Hamley IW, Pople JA, Gleeson AJ, Komanschek BU, Towns-Andrews E. Simultaneous rheology and small-angle scattering experiments on block copolymer gels and melts in cubic phases. J. Appl. Crystallogr. 1998;31(6):881–889.

15. Hongladarom K, Ugaz V, Cinader D, Burghardt WR, Quintana JP, Hsiao BS, Dadmum MD, Hamilton W, Butler PD. Birefringence, x-ray scattering and neutron scattering measurements of molecular orientation in shear liquid crystal polymer solutions. Macromolecules 1996;29(16):5346–5355.

16. Somani RH, Hsiao BS, Nogales A, Srinivas S, Tsou AH, Sics I, Balta-Calleja FJ, Ezquerra TA. Structure development during shear flow induced crystallization of iPP: *in situ* small angle x-ray scattering study. Macromolecules 2000;33(25):9385–9394.

17. Li L and de Jeu WH. Shear-induced smectic ordering as a precursor of crystallization in isotactic polypropylene. Macromolecules 2003;36(13):4862–4867.

18. Caputo FE, Burghardt WR, Krishnan K, Bates FS, Lodge TP. Time-resolved small-angle X-ray scattering measurements of a polymer bicontinuous microemulsion structure factor under shear. Phys. Rev. E: Statistical, Nonlinear, and Soft Matter Physics 2002;66(4–1):041401/1–041401/18.

19. Caldas V, Brown GR, Nohr RS, MacDonald JG, Raboin LE. The structure of the mesomorphic phase of quenched isotactic polypropylene. Polymer 1994;35(5):899–907.

20. Lotz B, Wittmann JC. Structural relationships in blends of isotactic polypropylene and polymers with aliphatic sequences. J. Polym. Sci., Polym. Phys. 1986;24:1559–1575.

21. Lotz B, Wittmann JC. The molecular origin of lamellar branching in the α (monoclinic) form of isotactic polypropylene. J. Polym, Sci., Polym. Phys. 1986;24:1541–1558.

22. Lotz B, Graff S, Wittmann JC. Crystal morphology of the γ (triclinic) phase of isotactic polypropylene and its relation to the α phase. J. Polym, Sci., Polym. Phys. 1986;24:2017–2032.

23. Natta G, Peraldo M, Corradini P. Smectic mesomorphic form of isotactic polypropylene. Rend. Accad. Naz. Lincei 1959; 26:14–17.

24. Li L and de Jeu W. Shear-induced smectic ordering in the melt of isotactic polypropylene. Phys. Rev. Lett. 2004; 92(7):075506/1–075506/3.

25. Kumaraswamy G, Issaian AM, Kornfield JA. Shear-enhanced crystallization in isotactic polypropylene. 1. Correspondence between *in situ* rheo-optics and *ex situ* structure determination. Macromolecules 1999;32:7537–7547.

26. Somani RH, Yang L, Hsiao BS. Precursors of primary nucleation induced by shear in isotactic polypropylene. Phys. A, Statistical Mechanics and Its Applications 2002;304(1-2):145–157.

27. Somani RH, Nogales A, Srinivas S, Fruitwala H, Tsou AH, Hsiao BS. Orientation-induced crystallization in polymers — a case study of isotactic polypropylene in shear. Proceed. Int. Conf. on Flow Induced Crystallization of Polymers, Salerno, Italy 2001;14-17:21–26.

28. Somani RH, Yang L, Hsiao BS, Agarwal P, Fruitwala H., Tsou AH. Shear-induced precursor structures in isotactic polypropylene melt by *in situ* rheo-SAXS and -WAXD studies. Macromolecules 2002;35(24):9096–9104.

29. Lieberwirth I, Loos J, Petermann J, Keller A. Observation of shish crystal growth into nondeformed melts. J. Polym. Sci., Part B, Polym. Phys. 2000;38:1183–1187.

30. Wang ZG, Hsiao BS, Sirota EB, Agarwal P, Srinivas S. Probing the early stages of polymer crystallization by simultaneous small- and wide-angle x-ray scattering. Macromolecules 2000;33(3):978–989.

31. Pogodina NV, Soddiquee SKS, Van Egmond JW, Winter HH. Correlation of rheology and light scattering in isotactic polypropylene during early stages of crystallization. Macromolecules 1999;32:1167–1174.

32. Pogodina NV, Winter HH. Polypropylene crystallization as a physical gelation process. Macromolecules 1998;31(23):8164–8172.

33. Dukovski I and Muthukumar M. Langevin dynamics simulations of early stage shish-kebab crystallization of polymers in extensional flow. J. Chemical Phys. 2003;118(14):6648–6655.

34. Hu W, Frenkel D, Mathot VBF. Simulation of shish-kebab crystallite induced by a single prealigned macromolecule. Macromolecules 2002;35:7172.

35. Somani R, Hsiao BS, Nogales A, Fruitwala H, Tsou A. Structure development during shear flow-induced crystallization of ipp: *in situ* wide-angle x-ray diffraction study. Macromolecules 2001;34(17):5902–5909.

36. Varga J and Karger-Kocsis J. Rules of supermolecular structure formation in sheared isotactic polypropylene melts. J. Polym. Sci., Part B, Polym. Phys. 1996;34 (4):657–670.

37. Varga J and Karger-Kocsis J. Interfacial morphologies in carbon fiber-reinforced polypropylene microcomposites. Polymer 1995;36 (25):4877–4881.

38. De Gennes PG. Coil stretch transition of dilute flexible polymers under ultrahigh velocity gradients. J. Chemical Phys. 1974;0(12):5030–5042.

39. Keller A and Kolnaar HWH. Flow-induced orientation and structure formation, Chapter 4 in Processing of Polymers (Vol. 18) of Materials Science and Technology, A Comprehensive Treatment (ed. R. W. Cahn, P. Hassen and E. J. Kramer), VCH Publisher, Weinheim, p.189 (1997).

40. Pope DP, Keller A. A study of the chain extending effect of elongational flow in polymer solutions. Colloid. Polym. Sci. 1978;256:751–756.

41. Miles MJ, Keller A. Conformational relaxation time in polymer solutions by elongational flow experiments. 2. Preliminaries of further developments. Chain retraction; identification of molecular weight fractions in a mixture. Polymer 1980;21:1295–1298.

42. Nogales A, Hsiao BS, Somani RH, Srinivas S, Tsou AH, Balta-Calleja FJ, Ezquerra TA. Shear-induced crystallization in blends of isotactic polypropylene with different molecular weight: *in situ* synchrotron small- and wide-angle x-ray scattering studies. Polymer 2001;42(12):5247–5256.

43. Yang L, Somani RH, Sics I, Kolb R, Fruitwala H, Ong C. Shear-induced crystallization precursor studies in model polyethylene blends by in-situ rheo-SAXS and rheo-WAXD. Macromolecules 2004;37:4845–4859.

44. Miller RL. Existence of near-range order in isotactic polypropylenes. Polymer 1960;1:135–143.

45. Hosemann R. Paracrystalline fine structure of natural and synthetic proteins. Visual method for the determination of the oscillation tensors of the cell edges. Acta Crystall. 1951;4:520-530.

46. Zannetti R, Celotti G, Fichera A, Francesconi R. Structural effects of annealing time and temperature on the paracrystal-crystal transition in isotactic polypropylene. Makromolek. Chem. 1969;128:137–142.

47. Zannetti R, Celotti G, Armigliato A. Relations between radial atomic distribution curves and the mechanism for the paracrystal-crystal transition of isotactic polypropylene. Eur. Polym. J. 1970;6:879–889.

48. Wyckoff HW. X-ray and related studies of quenched, drawn, and annealed polypropylene. J. Polym. Sci. 1962;62:83–114.

49. Gailey JA, Ralston PH. The quenched state of polypropylene. Plast. Engineers Trans. 1964;4:29–33

50. Gomez MA, Tanaka H, Tonelli E. High-resolution solid-state carbon-13 nuclear magnetic resonance study of isotactic polypropylene polymorphs. Polymer 1987;28:2227–2232.

51. Bodor G, Grell M, Kallo A. Determination of the crystallinity of polypropylene. Faserforsch. Textil-Tech. 1964;15:527–532.

52. Farrow G. Measurement of the smectic content in undrawn polypropylene filaments. J. Appl. Polym. Sci. 1965;9:1227–1232.

53. Wunderlich B and Grebowicz J. Thermotropic mesophases and mesophase transitions of linear, flexible macromolecules. Adv. Polym. Sci. 1984;60/61:1–59.

54. Grebowicz J, Lau JF, Wunderlich B. The thermal properties of polypropylene. J. Polym. Sci., Polym. Symp. 1984;71:19–37.

55. Corradini P, Petraccone V, De Rosa C, Guerra G. On the structure of the quenched mesomorphic phase of isotactic polypropylene. Macromolecules 1986;19:2699–2703.

56. Corradini P, De Rosa C, Guerra G, Petraccone V. Comments on the possibility that the mesomorphic form of isotactic polypropylene is composed of small crystals of the b crystalline form. Polym. Commun. 1989;30:281–285.

57. Huang MR, Li XG, Fang BR. β nucleators and β crystalline form of isotactic polypropylene. J. Appl. Polym. Sci. 1995;56:1323–1337.

58. de Candia F, Iannelli P, Staulo G, Vittoria V. Crystallization of oriented smectic polypropylene. I. Thermally induced crystallization. Colloid & Polym. Sci. 1988;266(7):608–613.

59. Bates MA and Frenkel D. Phase behavior of two-dimensional hard rod fluids. J. Chem. Phys. 2000;112:10034–10041.

60. Imai M, Kaji K, Kanaya T, Sakai Y. Ordering process in the induction period of crystallization of poly(ethylene terephthalate). Phys. Rev. B. 1995;52:12696–12704.

61. Bonart von R. Paracrystalline structures in poly(ethylene terephthalate). Kolloid Zeitschrift & Zeitschrift fuer Polymere 1966;213:1–11.

62. Bonart R. Crystalline and colloidal structures during elongation and plastic deformations. Kolloid Zeitschrift & Zeitschrift fuer Polymere 1969;231(1–2):438–458.

63. Welsh GE, Blundell DJ, Windle AH. A transient liquid crystalline phase as a precursor for crystallization in random co-polyester fibers. Macromolecules 1998;31:7562–7565.

64. Welsh GE, Blundell DJ, Windle AH. A transient mesophase on drawing polymers based on polyethylene terephthalate (PET) and polyethylene naphthoate (PEN). J. Mater. Sci. 2000;35:5225–5240.

65. Asano T, Balta-Calleja FJ, Flores A, Tanigaki M, Mina MF, Sawatari C, Itagaki H, Takahashi H, Hatta I. Crystallization of oriented amorphous poly(ethylene terephthalate) as revealed by X-ray diffraction and microhardness. Polymer 1999;40:6475–6484.

66. Blundell DJ, Mahendrasingam A, Martin C, Fuller W. Formation and decay of a smectic mesophase during orientation of a PET/PEN copolymer. J. Mater. Sci. 2000;35:5057–5063.

67. Mahendrasingam A, Blundell DJ, Martin C, Fuller W, MacKerron DH, Harvie JL, Oldman RJ, Riekel C. Influence of temperature and chain orientation on the crystallization of poly(ethylene terephthalate) during fast drawing. Polymer 2000;41:7803–7814.

68. Blundell DJ, Mahendrasingam A, Martin C, Fuller W, MacKerron DH, Harvie JL, Oldman RJ, Riekel C. Orientation prior to crystallisation during drawing of poly(ethylene terephthalate). Polymer 2000;41:7793–7802.

69. Mahendrasingam A, Martin C, Fuller W, Blundell DJ, Oldman RJ, MacKerron DH, Harvie JL, Riekel C. Observation of a transient structure prior to strain-induced crystallization in poly(ethylene terephthalate). Polymer 2000;41:1217–1221.

70. Blundell DJ, MacKerron DH, Fuller W, Mahendrasingam A, Martin C, Oldman RJ, Rule RJ, Riekel C. Characterization of strain-induced crystallization of poly(ethylene terephthalate) at fast draw rates using synchrotron radiation. Polymer 1996;37:3303–3311.

71. Ran S, Wang Z, Burger C, Chu B, Hsiao BS. Mesophase as precursor for strain-induced crystallization in amorphous poly(ethylene terephthalate) film. Macromolecules 2002;35(27):10102–10107.

72. Fraser RDB, MacRae TP, Suzuki E. An improved method for calculating the contribution of solvent to the x-ray diffraction pattern of biological molecules. J. Applied Cryst. 1978;11(6):693–694.

73. Brown HR, Mills PJ, Kramer EJ. A SAXS study of a single crack and craze in plasticized polystyrene. J. Polym. Sci. Polym. Phys. 1985;23(9):1857–1967.

74. Brown HR, Kramer EJ. Craze microstructure from small-angle x-ray scattering (SAXS). J. Macromole. Sci., Phys. 1981;B19(3):487–522.

75. Grubb DT, Prasad K, Adams W. Small-angle x-ray diffraction of Kevlar using synchrotron radiation. Polymer 1991;32:1167–1172.

76. Grubb DT and Prasad K. High-modulus polyethylene fiber structure as shown by x-ray diffraction. Macromolecules 1992;25:4575–4582.

77. Ran SF, Zong XH, Fang DF, Hsiao BS, Chu B, Cunniff PM, Phillips RA. Study of the mesophase in polymeric fibers during deformation by synchrotron SAXS/WAXD. J. Mater. Sci. 2001;36:3071–3077.

78. Salem DR. Development of crystalline order during hot-drawing of poly(ethylene terephthalate) film: influence of strain rate. Polymer 1992;33:3182–3188.

79. Salem DR. Crystallization kinetics during hot-drawing of poly(ethylene terephthalate) film: strain-rate/draw-time superposition. Polymer 1992;33:3189.

80. Gorlier E, Haudin JM, Billion N. Strain-induced crystallisation in bulk amorphous PET under uni-axial loading. Polymer 2001; 42:9541–9549.

81. Kawakami D, Ran S, Burger C, Fu B, Sics I, Hsiao BS. Mechanism of structural formation by uniaxial deformation in amorphous poly(ethylene terephthalate) above glass temperature. Macromolecules 2003:36(25):9275–9280.

82. Kawakami D, Ran S, Burger C, Fu B, Sics I, Hsiao BS, Kikutani T. Structural formation of amorphous poly (ethylene terephthalate) during uniaxial deformation above the glass temperature. Polymer 2004;45(3):905–918.

83. Daubery RD, Bunn CW, Brown CJ. The crystal structure of polyethylene terephthalate. Proc. R. Soc. Lond. A 1954;226:531–542.

84. Kawakami D, Hsiao BS, Burger C, Ran S, Avila-Orta C, Sics I, Kikutani T, Jacob KI, Chu B. Deformation-induced phase transition and superstructure formation in poly(ethylene terephthalate). Macromolecules 2005;38:91–103.

3

Nanostructures of Two-Component Amorphous Block Copolymers: Effect of Chain Architecture

RAMESHWAR ADHIKARI and GOERG H. MICHLER

Institute of Materials Science,
Martin-Luther-University Halle-Wittenberg

CONTENTS

I.   INTRODUCTION

Block copolymers are important examples of nanostructured
heterogeneous polymers and lie at the focus of intensive
research activities in contemporary macromolecular science
and technology. This is attributable to a wide range of fasci-
nating fundamental issues associated with the understanding
of self-assembly processes and their potential application possibilities in nanotechnology [1].

The self-assembly processes in these materials leading
to the formation of well-ordered nanostructures are a consequence of the intramolecular phase separation between the
dissimilar chains linked together by means of a covalent bond
[2–4]. In general, the nature of microphase-separated morphology of two-component diblock copolymers, at sufficiently
high molecular weight and low polydispersity, is determined
by relative composition of the constituents. However, it is of
practical importance to control the total composition of the
copolymer and the morphology to be formed independently.

This possibility may be provided by the modification of molecular architecture of block copolymers [5–10], which will be the main focus of this chapter.

In order to properly characterize the influence of modified chain architecture on the morphology of block copolymers, a systematic experimental study on model systems by combination of different techniques is essential. In particular, direct imaging techniques such as electron microscopy and atomic force microscopy are useful to study structure and morphology of these heterogeneous polymers. For this purpose, we have chosen the polystyrene-*block*-polybutadiene-*block*-polystyrene (SBS) triblock copolymers and their derivatives by keeping the overall composition constant (polystyrene content ~70%) and changing the chain architecture. These copolymers, owing to the widely separated glass transition temperature (T_g) of the constituent phases, provide a broad range of service temperatures [3]. The ordered microphase-separated structures endow them with outstanding mechanical (stiffness, strength, toughness, etc.) and optical (transparency) properties. At room temperature, the flexible rubbery polybutadiene blocks ($T_g \sim -100°C$) are anchored on both sides by the glassy polystyrene blocks ($T_g \sim +100°C$). Therefore, these materials behave as a cross-linked rubber at ambient conditions and allow a thermoplastic processing at higher temperature [3].

This chapter is organized as follows. First, an overview of structure-property correlation of styrene/butadiene block copolymers will be given, with special emphasis on various ways of morphology control (Sections II.A and II.B). After briefly introducing the main features of the materials, the sample preparation and the characterization techniques in the experimental section (Section III), the experimental results highlighting the impact of molecular architecture on phase behavior of styrene/butadiene block copolymer systems, will be discussed (Section IV). The results will be supplemented by the impact of processing on nanostructure evolution. Finally, self-assembled nanostructures of block copolymer produced by controlled radical polymerization techniques will be introduced (Section V), which is emerging as one of the potential trends in contemporary nanotechnology.

II. BLOCK COPOLYMER MORPHOLOGY AND STRUCTURE-PROPERTY CORRELATIONS

A. Mechanical Behavior and Morphology Control

The block copolymer nanostructures that are formed in the solid state are of practical interest. So it is of fundamental importance to control these structures to achieve the goal of developing materials with application-relevant mechanical properties [3,11,12]. The details on the block copolymer morphology and thermodynamics of microphase separation have been collected in recent reviews [4,13,14].

In AB diblock copolymers (e.g., polystyrene-*block*-polyisoprene [SI] diblocks), the nature and the dimension of the microphase-separated structures are usually adjusted by changing the composition and molecular weight of the constituents at constant interaction parameter (χ_{AB}). In the strong segregation limit, with increasing polystyrene (PS) content, body centered cubic spheres, hexagonally arranged cylinders and three-dimensional (3-D) "gyroid" network of polystyrene (PS) domains dispersed in the matrix of polyisoprene (PI) were observed in polystyrene-*block*-polyisoprene (SI) diblock copolymers [4]. With further increase in PS content, the alternating layers of PS and PI lamellae, and then the structures mentioned above were found in the reversed order. Figure 1 (top) shows schematically the basic morphologies of the classical two-component block copolymers, which are responsible for different kinds of mechanical behavior. Figure 1 (bottom) shows the stress-strain curves of SBS triblock copolymers having different compositions. One can notice that their tensile properties are directly coupled with the nature of nanostructures. The mechanical behavior of SBS triblock copolymers can be broadly classified into three groups:

1. Rubber-elastic behavior (curve 1): At lower PS content (i.e., if the PS forms the dispersed phase), the block copolymers, like most of the commercial thermoplastic elastomers, deform homogeneously (rubber-like behavior) under tension [3].

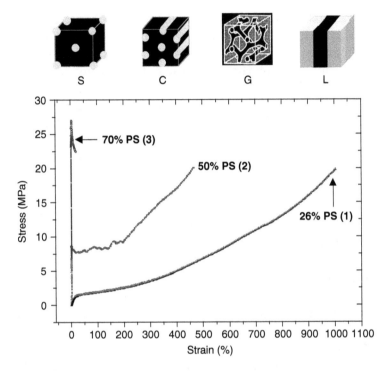

Figure 1 Schematics of stable nanostructures observed in two-component styrene/diene block copolymers (top) and tensile stress-strain behavior of solution cast SBS triblock copolymers as a function of composition (bottom).

2. Ductile behavior (curve 2): As the block copolymer approaches a compositional symmetry forming alternating layers of PS and PB phases; the macroscopic neck-formation and drawing prevails during tensile deformation [10].
3. Brittle behavior (curve 3): With increasing styrene content, as the morphology reverses (i.e., PB domains in PS matrix), the yield stress increases and elongation at break drastically decreases. The block copolymer breaks in a quite brittle manner due to localization of deformation in the form of crazes [15].

The general picture of deformation discussed above is valid only if the component blocks are well phase-separated and the flexible rubbery blocks possess sufficient physical "cross-links" on either side of the molecules. The strong physical cross-linking of the rubber phase is provided by the glassy nature of outer blocks. In diblock copolymers consisting of glassy/rubbery block chains, the rubber phase is not sufficiently networked because the rubbery molecules are 'cross-linked' only at one end at room temperature. As a result, diblock copolymers show relatively brittle behavior even at high molecular weights [16,17].

An approach of fine-tuning of block copolymer nanostructures is also provided by blending a two-component block copolymer with constituent homopolymers or other block copolymers [4,13,18]. It was shown for both AB diblock and ABA triblock copolymers that different microphase-separated morphologies may be produced by simply mixing two block copolymers having highly asymmetric but complementary compositions [18–20].

TEM images showing two SBS triblocks having PS volume content equal to 0.28 and 0.74 are given in Figure 2a and Figure 2c, which show, as expected PS cylinders in PB matrix and PB cylinders in PS matrix, respectively. A 50:50 (weight/weight) mixture of these block copolymers would yield a nearly symmetric composition. A lamellar morphology corresponding to this composition is observed (Figure 2b). Thus this route makes it possible to design all the possible block copolymer morphologies by simply mixing two block copolymers of varying compositions.

It should, however, be mentioned that the morphology to be formed in the binary block copolymer blends does not exactly correspond to that corresponding to the composition range for that particular morphology in a neat block copolymer. Neither, the block copolymer molecules are always miscible. The influential factors are the symmetry and the molecular weight ratio of the copolymer molecules (to be discussed later).

The relative composition of the constituents of a two-component block copolymer can be altered by incorporating

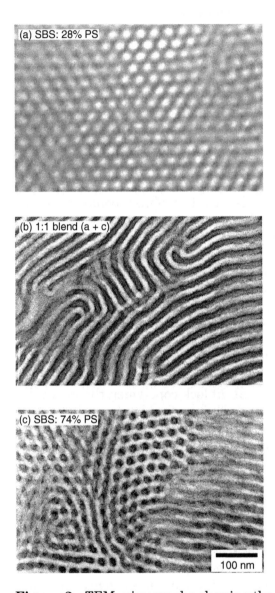

Figure 2 TEM micrographs showing the morphology of two SBS block copolymers having nearly complementary composition (a, c) and a 50:50 blend of them (b); OsO_4 staining makes the PB phase appear dark.

the corresponding homopolymers into the block domains. This leads to a change in curvature of the interface and eventually to a morphology transition [4,13]. In this case, the molecular weight of the added homopolymer relative to that of the corresponding block plays a key role in the phase behavior. Depending upon the molecular weight ratio ($\alpha = M_{homopolymer}/M_{block}$), an interplay between microphase- and macrophase-separation occurs [4]. In order that the homopolymer be accommodated into the corresponding domain of the block copolymer, the molecular weight of the former should be sufficiently smaller than that of the latter.

Comprehensive studies on the phase behavior of the blends of AB diblock copolymer and A homopolymer have been performed by Hashimoto and co-workers [13]. For the case $\alpha < 1$, transition from lamellar morphology to cylindrical and then to spherical was found on addition of PS homopolymer to polystyrene-*block* polyisoprene (SI) diblock copolymer. The observed morphological transitions were discussed on the basis of changes in interfacial curvature and packing density resulting from the variation in chain configuration (discussed by Hasegawa and Hashimoto [13]; see schematic illustration in Figure 3).

In the pure lamellar SI diblock copolymer, the molecular volume is symmetrical, and the interface between the PS and polyisoprene (PI) nanodomains is flat (Figure 3a). When a low molecular weight PS homopolymer is added to SI diblock, the PS blocks swell due to uniform solubilization of PS chains leaving the PI blocks unaffected (Figure 3b). In this case, the PS block chains should stretch or the PI chains should be compressed in order to attain the constant segmental volume of each phase. Since the latter processes are entropically unfavorable, a curved interface evolves to maintain a uniform packing density (Figure 3c). Consequently, as the volume fraction of added PS increases, the interface has convex curvature toward the majority phase (i.e., PS phase).

The possibility of designing different block copolymer morphologies by addition of low molecular weight homopolymers is useful for the applications where mainly the nature of the nanostructures is of importance. However, from the view

PS-PI/PS mixture

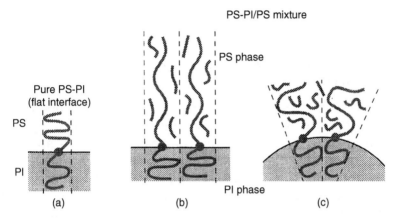

Figure 3   Schematics of the morphology transition in block copolymer/homopolymer blends if $M_{\text{PS-block}} > M_{\text{hPS}}$. (Taken from Hasegawa H, Hashimoto T. Self assembly and morphology of block copolymer systems. In: Aggarwal SL, Russo S, eds. Comprehensive Polymer Science, Suppl 2. London: Pergamon, 1996. pp. 497–539. Copyright 1996, with permission from Elsevier.)

of mechanical properties, this may weaken the entanglements and worsen the strength and ductility of the products.

Three-component ABC block copolymer systems, due to the presence of more interaction parameters (χ_{AB}, χ_{BC}, χ_{AC}), form a rather new variety of microphases, which may allow novel routes to developing materials having new property profiles [22–26]. More complex morphologies have been recently predicted in three-component multiblock copolymers which are yet to be experimentally confirmed [27,28].

The morphologies entirely different from the equilibrium ones can be obtained by altering the processing conditions (e.g., using different solvents, application of external fields, etc.). For example, application of shear field in the melt may alter the nature of the morphology to be formed [29] and align the microphase separated structures leading to a "single crystal-like" texture [30–32] and anisotropic deformation behavior. Additionally, via incorporation of specific inorganic substances into a particular block, the nanostructures can be transformed into organic/inorganic hybrid nanocomposites [21,33], which may open new horizons for controlling mechanical properties of heterophase polymers for specific high-tech applications.

B.   Architectural Modification of Block Copolymers

The modified molecular architecture of block copolymers may significantly alter their phase behavior [5–10]. Via this process, the restriction of changing composition (e.g., in an AB diblock) to achieve different morphologies can be overcome. A brief overview of this route of morphology control follows.

Architectural modification consists of designing block copolymer chains into a variety of topologies, controlling the symmetry of the constituent blocks and incorporating a tapered interface between the incompatible chains. Based on the study of morphology and physical properties of a large variety of block and graft copolymers, Hadjichristidis and co-workers showed a pronounced shift in phase behavior of those systems with respect to corresponding diblock [7,34,35]. For example, lamellar morphology was observed in a styrenic miktoarm star copolymer in a composition range in which a cylindrical morphology would be expected for a diblock analogue [35].

With the development of new synthetic routes, experimental studies on more complex architectures began to emerge, which, in turn, inspired new theoretical studies on the impact of molecular architecture on morphological behavior. Milner calculated the phase diagram of asymmetric miktoarm star block copolymers and demonstrated that the stability window for a particular morphology is dramatically shifted as a function of copolymer architecture [5]. Milner's theory predicts successfully the phase behavior of miktoarm star and graft block copolymers.

Asymmetric A_1BA_2 type copolymers (where A and B are glassy and rubbery blocks, respectively; and $M_{A1}/M_{A2} \neq 1$, where M represents the molecular weight) are of special technical importance. In such block copolymers, the shorter glassy blocks may serve to enhance the deformability of the products while the longer ones endow the copolymers with higher levels

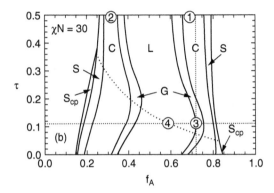

Figure 4 Phase diagram of an asymmetric ABA triblock copolymer calculated by Matsen. τ and f_A stand for asymmetry parameter and volume fraction of component, respectively; the locations denoted by ①, ②, ③ and ④ will be discussed with respect to the AFM micrographs presented in Figure 6. (From Matsen MW. J Chem Phys 2000; 113:5539–5544. With permission.)

of strength [8,36]. Recently, Matsen has examined the phase behavior of A_1BA_2 triblock copolymer melt using self consistent field theory (SCFT) and demonstrated that a drastic shift in order-order phase boundaries relative to the symmetric ABA triblock copolymer occurs in asymmetric ones [6]. The phase diagram of an asymmetric ABA triblock copolymer at intermediate segregation regime ($\chi N = 30$) is presented in Figure 4 [6].

Matsen introduced an asymmetry factor (τ) whose magnitude lies between two extreme values: 0 (for AB diblock) and 0.5 (for symmetric ABA triblock). As the length of the outer A chains become dissimilar, the order-order transition (OOT) lines are shifted toward higher overall A volume fraction. For an intermediately segregated system (for example, with $\chi N = 30$, see Figure 4 [6]), this theory predicts lamellar morphology for an A_1BA_2 triblock copolymer up to $f_A > 0.70$ (where f_A stands for the volume fraction of component A) with the value of τ equal to 0.10 which corresponds to the A_1/A_2 ratio of approximately 5. At higher asymmetry, a fraction of the short A blocks is allowed even to be pulled out of the A domains, as the continuous extraction of *A* blocks will reduce the stretching energy of the *B* domain chains [6]. When the asymmetry becomes sufficiently large in an A_1BA_2 triblock, shorter outer A blocks begin to pull out of their domains. Although unfavorable interactions occur when an *A* block leaves its domain, this is more than compensated for by the fact that its *B* block can relax. The phase behavior of the asymmetric triblocks was found to be dominated by the standard lamellar (L), gyroid (G), cylindrical (C), and spherical (S) morphologies (see Figure 4), but there is also an unusually large region where, as opposed to the normal body centered cubic (bcc) packing, closed-packed spheres (S_{cp}) are predicted at very high asymmetry.

Depending on the extent of asymmetry and A/B volume fractions, different locations (1, 2, 3 and 4) have been marked in Figure 4, which will be referenced later while discussing the morphology of block copolymers having different molecular architectures (see schemes of architectures in Figure 6 and AFM images in Figure 7).

Introduction of a tapered or a statistical chain between the incompatible blocks may further modify the block copolymer phase behavior [37–42]. It has been, in general, shown that presence of a tapered or statistical chain between the incompatible blocks results in a broadened interface due to enhanced mixing at this region. Recently, the phase behavior of *normal* and *inverse* tapered block copolymers was studied, and it was demonstrated that an *inverse* tapered sequence leads to higher compatibility (wider interfacial width) than the *normal* ones [42].

A direct consequence of the architectural modification is its impact on the deformation behavior of respective block copolymers. The latter is reflected not only by the formation of a wide range of morphologies at constant composition [9] but also by enabling a variation in effective physical cross-link sites and in the ratio of bridge-to-loop conformations at constant morphologies [11,12]. Thermoplastic elastomers based on graft copolymers having tetra-functional branch points were found to possess improved mechanical properties than their triblock analogues [11]. Furthermore, the ABABA pentablock copolymers (higher bridge/loop ratio) showed noticeably pronounced higher ductility than their ABA tri-block (lower bridge/loop ratio) counterparts [12].

The review of the literature works outlined here demonstrates that there exists an inherent relationship between the molecular parameters (chain architecture, nature of interface, chain topology etc.) and the phase behavior. Thus, mechanical and micromechanical behavior of these materials are dictated not only by the kind of microphase-separated structures, their size and orientation, grain size etc., but also are coupled with the chain architecture and processing conditions. The schematic illustrations presented in Figure 5 provide an overview of important parameters, which control the solid-state morphology of block copolymer systems and thereby affect directly or indirectly their end-use properties.

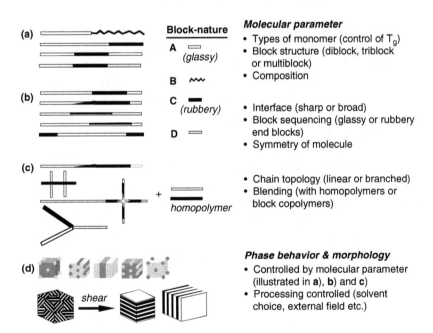

Figure 5   Schematic summary of various molecular and morphological parameters affecting the morphology formation and the deformation micro-mechanisms of block copolymers.

The symmetry of end blocks and the location of glassy blocks in the molecular backbone might play a decisive role in the mechanical behavior of the copolymers. For example, SBS triblock copolymers become tougher when styrene outer blocks are made more and more asymmetric [8]. Indeed, the kinds of monomers chosen as the block chains determine the extent of chemical incompatibility and hence the degree of segregation. The latter has an important influence on the block copolymer phase behavior. Last but not least, molecular weight and molecular weight distribution, etc., should be optimized in order to achieve a good balance of mechanical and rheological properties.

III.   EXPERIMENTAL

A.   Materials and Sample Preparation

The characteristic data of the investigated block copolymers are listed in Table 1. The block copolymers possess nearly identical chemical composition (styrene volume fraction ~0.70) but differ in the chain architecture. Molecular architecture of the block copolymers studied is schematically outlined in Figure 6. Polystyrene (PS) and polybutadiene (PB) are represented by white and dark areas, which also reflect the contrast in the AFM (as well as TEM) images, respectively, discussed in this work. The linear copolymer LN1, a neat SBS triblock, has symmetric styrene end blocks (i.e., the outer PS blocks are of equal lengths), which are separated from the butadiene center block by a sharp interface. Sample LN2, in contrast, has a tapered transition (shown by an oblique line between PB and PS blocks in Figure 6), and comprises asymmetric PS end blocks, the larger block being about five times longer than the shorter one. The copolymer LN3 has analogous structure as LN2, but contains a random copolymer of PS and PB (PS-co-PB) as center block instead of a pure PB block. The molecular weight of longer polystyrene blocks in these asymmetric block copolymers is in the range of 60,000 – 90,000 g/mol. The sample LN4 consists of short symmetric PS end blocks (M_n ~ 18,000 g/mol) connected by a rubbery

TABLE 1   Characteristics of Investigated Block Copolymer Samples

Samples[a]	M_n (g/mol)[b]	M_w/M_n [b]	Φ_{PS} [c]	Remarks
LN1	82,000	1.07	0.74	Symmetric SBS triblock, neat transition with the pure PB mid-block, linear architecture [9]
LN2	93,000	1.13	0.74	Asymmetric S_1BS_2 triblock copolymer, ($S_1{\neq}S_2$), tapered transition [9]
LN3	127,300	1.10	0.74	Structure similar to LN2, $S_1(S/B)S_2$ structure ($S_1{\neq}S_2$), S/B is a random copolymer of PS and PB [45]
LN4	116,000	1.20	0.65	Linear symmetric S(S/B)S triblock copolymer, mid block structure similar to that of LN3 [8,9]
ST3	85,700	2.10	0.74	Highly asymmetric star architecture, each arm with $S_1(S/B)S_2$ structure ($S_1{\neq}S_2$), S/B is a random copolymer of PS and PB [45,47]

[a] ST, star block; LN, linear block copolymer.
[b] Weight average (M_w) and number average (M_n) molecular weights were determined by gel permeation chromatography.
[c] Total styrene volume fraction determined by Wijs double bond titration.

block made up of PS-co-PB. The total volume fraction of polystyrene as outer block is about 0.32.

The star-shaped ST3 molecules have approximately four asymmetric arms each, one of them being much longer than the others. The molecular weight of the longest PS blocks lies in the same range as longer PS blocks of LN2 and LN3. Like LN3, a random PS-co-PB exists as a soft block instead of a pure polybutadiene chain in ST3.

Each sample was dissolved in toluene to prepare about 3% solution. The solution was poured into a flat glass dish and the solvent was allowed to evaporate in about two weeks. Each film was dried in air for several days and finally annealed for 48 h at 130°C in a vacuum oven. The thickness of the films was about 0.5 mm. Details on the synthesis, structures and properties of these polymers may be found elsewhere [8,9,43–48].

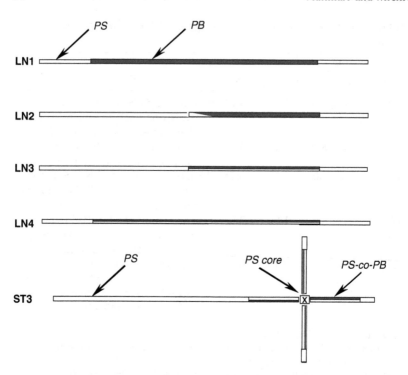

Figure 6 Schemes showing the molecular structure of the block copolymers studied; dark and white colors stand for PB and PS phases respectively; the oblique lines between the blocks represent the tapered transition.

B.   Techniques

Transmission Electron Microscopy (200 kV TEM, Jeol) was used to image the nanostructures of the samples using ultrathin sections (about 50–70 nm thick) ultramicrotomed from the cast film. The polybutadiene phase was selectively stained by osmium tetroxide (OsO_4). Morphological studies using dip-coated films were carried out by means of *Atomic Force Microscopy (AFM;* Multimode atomic force microscope, Digital Instruments Inc.) using tapping mode.

    Tensile Testing was performed at room temperature (23°C) using a universal tensile machine at a cross-head speed of 50 mm/min. Total length of the dog-bone shaped tensile bars punched out of the solution cast films was 50 mm each. At least six samples were tested in each case.

Differential scanning calorimetry (DSC) measurements were performed with a Mettler DSC 820 in the temperature range from $-120°C$ to $+150°C$ with a rate of $10°C/min$ using the cycle heating – cooling – heating. The heat flow and the second derivative of the heating scans were used for the determination of glass transition temperatures. The weights of the studied samples were approximately 10 mg each.

IV. INFLUENCE OF CHAIN ARCHITECTURE ON NANOSTRUCTURE EVOLUTION

A. Block Copolymer Morphologies as Revealed by AFM and TEM

The representative AFM phase images of the block copolymers studied are given in Figure 7a–e. As the films were cast slowly from neutral solvent (toluene) and annealed above the glass transition temperature of both the component phases, these micrographs illustrate their near-equilibrium morphologies. In the pictures given in Figure 7, brighter and darker areas correspond to glassy (PS) and rubbery (PB or PS-co-PB) phases, respectively. For the purpose of comparison, the morphology of a linear symmetric SBS triblock copolymer (LNX, $\Phi_{PS} \sim 0.28$; a commercially available SBS thermoplastic elastomer, Kraton D-1102), is presented in Figure 7f. The micrographs of different samples are arranged in such a way that the variation of morphologies follows the order: PS matrix (LN1) → Lamellae (LN2, LN3) → bicontinuous-like (ST3) → PS domains (LN4, LNX).

Owing to their symmetric architecture, the ratio of length of outer PS blocks is unity, and the value of asymmetry parameter (τ) is 0.5 for both LN1 and LNX. With respect to their composition, LN1 and LNX can be allocated the positions ① and ② of the phase diagram in Figure 4, respectively. Note, that these locations are valid for the strongly segregated PS and PB chains of LN1 and LNX as well since the shift in

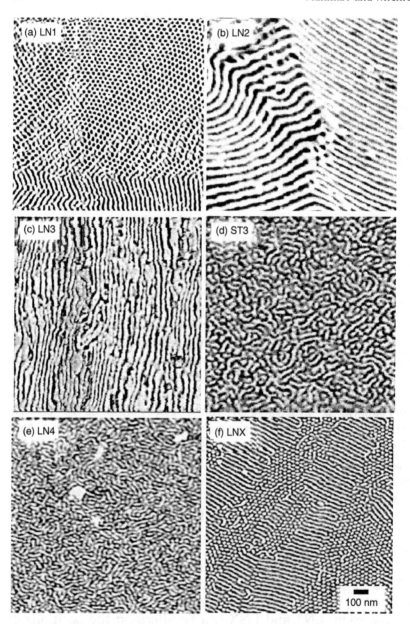

Figure 7 AFM phase images showing microphase-separated morphologies of the block copolymers studied; hard and soft phases appear bright and dark, respectively.

phase behavior is significant only at lower values of τ. Thus, the block copolymers LN1 (Φ_{PS} ~ 0.74) and LNX (Φ_{PS} ~ 0.28) show expected cylindrical morphologies with PS and PB matrices, respectively. However, a change in molecular architecture may appreciably modify the classical picture of morphology formation by overcoming the precondition of altering composition to change the microphase-separated structure. Therefore, the asymmetric block copolymers LN2, LN3 and ST3 show entirely different morphologies in spite of identical composition as LN1. LN2 and LN3 possess clearly a lamellar morphology (Figure 7b,c) while ST3 possesses a co-continuous structure (Figure 7d).

Due to presence of a tapered chain at interfacial region (e.g., LN2 in Figure 7b) or statistical copolymer (e.g., LN3 in Figure 7c and ST3 in Figure 7d) as middle-block, the interfacial tension will be decreased and the copolymers are finally driven towards weaker segregation. Assuming the intermediate segregation of these copolymers, the asymmetric block copolymer LN2 can be conveniently placed at the point of intersection of two-dotted lines in Figure 4 (denoted by ③), hence explaining the formation of lamellar morphology due to asymmetric architecture. In LN3, the volume fraction of polystyrene as outer blocks is approximately 0.66 (total PS volume content 0.74), and the length ratio of longer to the shorter PS blocks is much higher than one. This copolymer falls at a location denoted by ④ in Figure 4, thus confirming the architecture-induced lamellar morphology.

Additionally, in ST3, LN3 and LN4, due to presence of a PS-co-PB as rubbery block, the actual volume fraction of PS as hard phase is considerably decreased. This favors the formation of morphologies corresponding to lower overall polystyrene content. The volume fraction of outer PS blocks is about 0.32 in LN4 (i.e., the soft/hard volume ratio is 68/32). Due to the presence of symmetric architecture, this copolymer can also be placed at a location somewhere near ② in the phase diagram given in Figure 4. Hence, hexagonal ordered PS cylinders dispersed in a matrix of PS-co-PB can be expected. Actually, the PS domains dispersed in a rubbery matrix were observed. This morphology resembles the structure of classical

SBS thermoplastic elastomers, which contain about 28% polystyrene (Figure 7f), i.e., the composition is nearly reverse of LN4. However, a lattice of ordered structures is missing. Additionally, besides the presence of domains having a diffuse boundary, the order-disorder transition temperature (T_{ODT}) was highly depressed [8]. With these characteristics, the linear block copolymer LN4 represents a weakly segregated system.

Interestingly, the largest deviation was observed in ST3 whose overall composition and mid-block constitution is identical to that of LN3. Owing to star-shaped architecture and large polydispersity, the morphologies predicted by the theory of asymmetric block copolymer cannot be assigned to ST3. However, one may anticipate that a large part of short PS chains (especially those from PS core and short outer blocks) can be mixed to the rubbery PS-co-PB blocks. The notion of intermixing of a part of PS chains with the rubbery phase in asymmetric block copolymers is also supported by the dynamic mechanical analyses of several block copolymers [48]. An impression of the co-continuous morphology of ST3 can be obtained more conveniently from a TEM micrograph presented in Figure 8, which reveals that the PS struts (gray

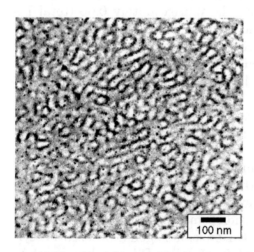

100 nm

Figure 8 High magnification of a TEM image showing co-continuous morphology of ST3; OsO_4 staining makes the PB phase appear dark.

domains) form an interpenetrating network embedded in the PS-co-PB matrix. This morphology closely resembles the 'gyroid' phase observed in block copolymer systems.

B.   SAXS Results

Supplementary information on the nanostructure of the block copolymers can be readily achieved by using small angle x-ray scattering (SAXS) methods. Representative one-dimensional SAXS curves of some of the linear block copolymers studied [49], illustrating the dependence of the scattering intensity function ($I.q^2$) with the wave vector $q = 2\pi s$ ($s = 2\sin\theta/\lambda$ and $\lambda = 0.15$ nm) are illustrated in Figure 9. In accordance with our microscopic investigations, the block copolymers yield, in spite of having nearly identical chemical composition, totally different SAXS patterns. The values for the nanostructure spacings, as determined by SAXS (LN1: 26; LN2: 32 nm and LN4: 27 nm) are in good agreement with those determined by microscopy (compare with Figure 7).

In the SAXS curve for the sample LN1 (Figure 9) higher order reflections, in the ratio $1{:}\sqrt{3}{:}\sqrt{7}$ of a Bragg's periodicity at $q_0 = 0.241$ nm^{-1} are present. Even a shoulder at $\sqrt{4}\ q_0$ can be clearly observed. In good agreement with the AFM results, the scattering maxima (Figure 9a) confirm the expected cylindrical morphology of the block copolymer LN1. On the other hand, in LN2 (Figure 9b), the scattering maxima appear in the ratio 1:2:3:4 suggesting a lamellar morphology. The weak higher order reflections result from polygranular structure of the solution cast films, in which the nanostructures have no preferential alignment.

In case of sample LN4 (Figure 9c), the scattering curve lacks well-defined higher-order maxima, which makes an unambiguous assignment of a particular block copolymer nanostructure difficult. Nevertheless, one can make a tentative assessment of the scattering maxima which are in the ratios $1{:}\sqrt{2}{:}\sqrt{3}{:}\sqrt{4}$ of a first order periodicity $q_0 = 0.229$ nm^{-1}, suggesting that the morphology LN4 is close to that of a cubic arrangement (spherical morphology) of microphase separated structures.

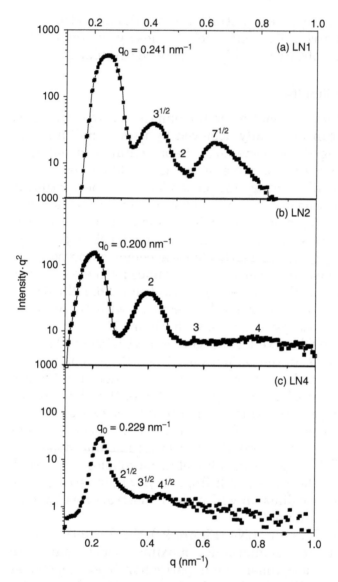

Figure 9 SAXS intensity curve of linear block copolymers LN1, LN2 and LN4. (Courtesy of Dr. Marc Langela, Max Planck Institute for Polymer Research, Mainz. Struktur und Rheologische Eigenschaften von PS-PI und PS-PB Blockcopolymeren. Thesis, University of Mainz, Germany, 2001.)

The morphology and nanostructures presented in Figure 7, Figure 8, and Figure 9 were obtained by changing the block copolymer architecture for PS content of approximately 70%. Particularly, all the basic morphologies observed in block copolymer systems were produced at a narrow composition range. Thus our experimental results demonstrate that the architectural modification of the block copolymers allows an independent control of their microphase morphology and composition.

C.   Molecular Mobility and Mechanical Properties

Owing to different architecture of the block copolymers including different interfacial structure and symmetry of the outer blocks, the mobility of the constituent chains is very different. As a result, the unusual microphase separation behavior was found to be reflected in a shift of their phase behavior.

The molecular mobility can be well characterized by the location of the glass transition temperature (T_g) of the components. Figure 10 illustrates the differential scanning calorimetric (DSC) plots for the linear block copolymers. Each sample shows two distinct glass transition temperatures: one at the lower temperature regime corresponding to polybutadiene homopolymer ($T_{g\text{-}PB}$) and the other one at the higher temperature side of the DSC plots corresponding to the polystyrene homopolymer ($T_{g\text{-}PS}$). It is obvious that the glass transition temperature T_g of PS phase ($T_{g\text{-}PS}$) in LN1, LN2 and LN3 lies at about +100°C, which matches the T_g of pure PS homopolymer. It suggests, in agreement with the sharply separated PS microphases observed by microscopic techniques (see Figure 7 and Figure 8), that polystyrene exists as pure material in these polymers. However, the magnitude of $T_{g\text{-}PS}$ appears very broad in LN4. The broad range of $T_{g\text{-}PS}$ values, which may result from the intermixing of the relatively shorter outer PS blocks ($M_n \sim 18,000$ g/mol) with the middle PS-co-PB rubbery block, suggests that a pure polystyrene phase is practically absent in LN4.

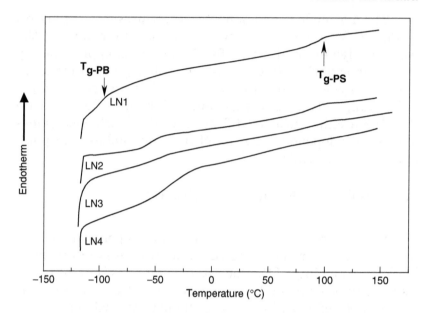

Figure 10 DSC scans for the linear block copolymers studied; second heating scans recorded at a rate of 10 K/min.

As the incompatible PS and PB chains in LN1 molecules are separated by a sharp interfacial region, the measured glass transition temperatures are those of the constituent homopolymers. The $T_{g\text{-PB}}$ measured at −98°C in this sample corresponds to that of trans-1,2-polybutadiene.

In other block copolymers, a significant shift of T_g of the soft PB phase ($T_{g\text{-PB}}$) toward higher temperature occurs in sequence LN1 (−98°C) → LN2 (−53°C), LN3 (−50°C) → LN4 (−34°C). Generally, the main factors affecting the T_g values are thermal histories of the sample, system pressure, diluent types and concentration, molecular weight and distribution, and polymer structure [50]. Because the block copolymers were prepared at identical conditions, only the latter two factors appear to be important. Furthermore, the total molecular weight of the polybutadiene chains in the samples studied does not differ much from each other, and is far above 10,000 g/mol, i.e., the range where the molecular weight has only a negligible influence on glass transition temperature.

We mentioned that the $T_{g\text{-}PB}$ values in the block copolymers having modified architecture are much higher than that
of normal polybutadiene homopolymer. It should be noted that
the microstructures of the PB phase in the investigated block
copolymers is not very different. Thus the shift of $T_{g\text{-}PB}$ towards
higher temperatures results essentially from the incorporation of PS chain segments into the PB phase. The higher $T_{g\text{-}PB}$
and the corresponding broad glass transition region in the
samples LN3 and LN4 are obviously connected with the presence of PS-co-PB as rubbery center block. In LN2, the tapered
transition at the interfacial region allows an intermixing
between the PS and the PB chains. This alone, however, does
not explain the drastic increase of $T_{g\text{-}PB}$ to about $-55°C$ from
about $-98°C$. This is an indication of the probable interference
of a fraction of shorter outer PS block chains into the PB
chains in accordance with the prediction of Matsen [6]. This
argument should also hold in case of asymmetric block copolymer LN3. The shift of the $T_{g\text{-}PB}$ toward a significantly higher
temperature, indicative of the incorporation of PS chain segments into the PB phase, was also observed in other asymmetric styrene/butadiene block copolymers even having neat
interfacial structure [44,48].

An important consequence of the architectural modification of block copolymers is that a wide range of mechanical
properties can be tailored at constant composition. For example, the tensile stress-strain curves of some of the block copolymers containing PS-co-PB as rubbery block (LN3, LN4 and
ST3) are compared in Figure 11 [51]. Note that these block
copolymers have total PS volume content of 65–74% (Table
1). LN3 and ST3 exhibit yielding at a strain of about 4% that
represents the onset of plastic deformation of the polymers.
The yield stress is about 10 MPa. After the yield point, the
deformation is accompanied by a so-called *cold drawing* process until about 100% and then, the stress level rises monotonically in both block copolymers. Both copolymers (with
lamellar [LN3] or co-continuous [ST3] morphology) show a
ductile behavior characterized by a large degree of plastic
deformation. The curves of LN3 and ST3 (74 vol% PS) are
similar to that of an SBS triblock copolymer having 50 vol%

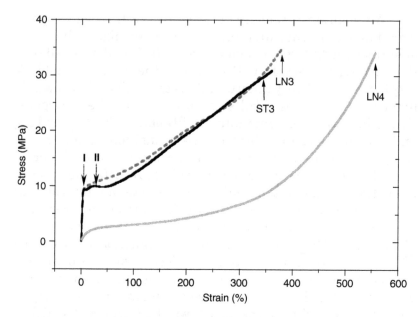

Figure 11 Tensile stress-strain diagrams of some of the solution cast block copolymer samples. Tensile testing was carried out at a rate of 50 mm/min at 23°C. (Taken from Adhikari R, Huy TA, Buschnakowski M, Michler GH, Knoll K. New J Phys 2004; 6:28(1–20). With permission.).

PS (see curve 2 in Figure 1). With a symmetric architecture, they would have shown the brittle deformation behavior represented by curve 3 of Figure 1.

The sample LN4 shows, as expected from its phase morphology, a tensile behavior similar to that of an SBS thermoplastic elastomer [3] at room temperature (compare with curve 1 of Figure 1). This represents a predominantly entropy-elastic behavior. A shoulder (corresponding to a yield point where the plastic deformation of glassy domains begins) at a strain of about 10% was observed. The sample showed an elongation at break of several hundred percent and a large degree of strain recovery.

A closer inspection of the initial part of stress-strain diagram of sample ST3 reveals that there are two different yield points located at a strain of about 4% and 20% (as indicated by I and II in Figure 11). Independent of sample preparation methods and thickness of tensile bars, the phenomenon of "double yielding" was always observed in ST3. The presence of these well separated yield points makes its mechanical deformation behavior resemble that of different polyethylene (PE) samples discussed in the literature (details on yielding processes in ST3 in [45]).

To summarize, the tensile properties of the block copolymers studied show a wide variety of mechanical properties, which would be expected for classical SBS triblock over a composition range of 25–75% PS. These properties are mainly dictated by the nanostructures formed in the solid state. Hence, it can be concluded that the mechanical behavior of the block copolymers studied are affected *directly* by the nature of microphase-separated morphology, whereas molecular architecture of the block copolymers plays an *indirect* role.

Before concluding this section, the influence of molecular architecture on micromechanical behavior of the block copolymers deserves mentioning. Unlike the craze-like deformation zones formed in symmetric block copolymer having 74 vol% PS [9], yielding and drawing of the microphase separated glassy phase (lamellae or co-continuous glassy phase) was found in the asymmetric block copolymers during tensile deformation [10,51]. In the lamellar block copolymers having PS lamella thickness in the range of 20 nm, a large homogenous plastic deformation of PS lamellae (called *thin layer yielding*) leading to the high ductility was observed [10][1]. Similar effects were reported earlier by Fujimora et el. [52] in symmetric styrenic block copolymers.

The orientation of the block copolymer nanostructures using an external force leads to a pronounced anisotropy in the mechanical properties. The extent to which the applied external forces affect the mechanical properties depends

---

[1]See Chapter 10.

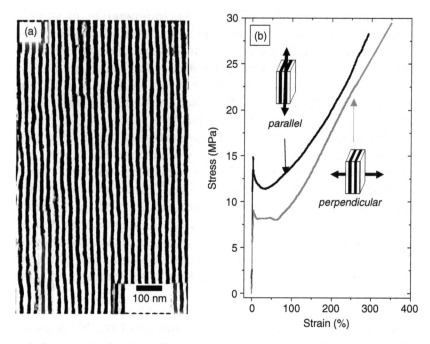

Figure 12 (a) Representative TEM micrograph of a lamellar tri-block copolymer (LN3) prepared by extrusion showing the layers aligned in the direction of applied shear; (b) Tensile stress-strain diagrams of some extruded films loaded parallel and perpendicular to the lamella orientation direction.

largely on whether and how far the nonequilibrium nano-structures are deviated from the equilibrium ones. Among the block copolymers studied, the linear asymmetric ones were found to form highly oriented morphologies under the influence of steady shear during the extrusion process. Figure 12 shows a TEM micrograph (Figure 12a) of extruded LN3 and its stress-strain curves (Figure 12b) on subjecting the extruded films undergo tensile deformation parallel and per-pendicular to the orientation direction.

Attributable to the steady shear, the lamellar nanostruc-tures are aligned along the direction of applied shear. It is found that the continuity of the lamellae extends over several micrometers. It should be, however, noted that the alignment of the lamellae is not uniform through the thickness of extruded films. The reason is the variable stress field across the thickness of the die used in the film processing (details in [53]).

As for the solution cast samples, independent of the loading direction, the deformation of each sample is characterized by a well-defined yield point. The yield points are rather sharp in the oriented samples. Deformation of the film parallel to the orientation direction gives rise to different values of yield stress and Young's modulus than that for the perpendicular deformation, which is in agreement with the recent results on other oriented linear block copolymers prepared by roll casting [30–32]. In earlier studies, the differences in mechanical properties have been discussed in light of composite models. During the parallel deformation, the load is mainly received by the glassy lamellae, which leads to the higher value of yield stress and Young's modulus. The necking and drawing of glassy lamellae (*thin layer yielding* mechanism) accompanied by a shearing in the rubbery lamellae predominate at high deformation during parallel deformation. On loading the sample perpendicular to the lamellar orientation direction, the applied load is first concentrated in the rubbery phase that leads to a decrease in yield stress. At high deformations, the principal deformation mechanism is the formation of typical *chevron morphology* (for details, see Chapter 10)

D.   Nanostructure of Binary Blends of Block Copolymers

As mentioned before, a method for controlling the solid-state morphology of the block copolymers is to mix two block copolymers having different compositions. It was pointed out that the molecular weight ratio (r) of the constituent two-component block copolymers dictates the interplay between microphase-separation and macrophase-separation in their binary blends. It was shown that macrophase separation occurs, if the value of r is higher than 10, forming macrophase separated lamellae with different periods. In contrast, if $r \leq 5$, the blend components were found to mix in molecular level forming a uniform nanostructure [13,18–20,54].

The route of morphology control by mixing two different block copolymers having comparable molecular weights has been shown to work effectively in the blends of two AB diblock copolymers [19] and also that of two symmetric ABA triblock copolymers [18]. However, new mechanical properties, other than that expected for the known classical morphologies, cannot be achieved. It would be interesting either to induce demixing of the block copolymers or to induce the partial miscibility between the block copolymer molecules even when $r \sim 1$. Via the latter way, one may be able to produce materials with a new property profile. In this section, we will show that binary blends containing an asymmetric block copolymer provide new variables to control the block copolymer super-lattices and thereby their mechanical properties.

The TEM micrographs in Figure 13 show the morphology of a blend of an asymmetric (LN3) block and a symmetric SBS (LNX) triblock copolymer, which have the total styrene volume fraction of 0.74 and about 0.28, respectively. Both the copolymers have linear molecules and possess comparable molecular weight (~100,000 g/mol). As described earlier, LN3 and LNX have lamellar and cylindrical (PS cylinders) morphologies, respectively (see Figure 7). The weight ratio of LN3/LNX in the blend was 65/35. There are two different areas: a PS rich region having lamellar arrangement of nanostructures (matrix); and a PB rich region having cylindrical arrangement (dispersed macrophase). Hence, these regions represent the morphology of the individual blend components. The obvious message is that the linear block copolymers LN3 and LNX, in spite of the comparable molecular weights, are immiscible.

Earlier electron microscopic studies by Spontak et al. indicated a new possibility of fine-tuning block copolymer nanostructures by controlling the architecture of the blend constituents [55]. It was shown that an equimolar blend of a diblock copolymer and an octablock copolymer, both of identical composition and molecular weights, was immiscible leading to the formation of a macrophase separated grain of one

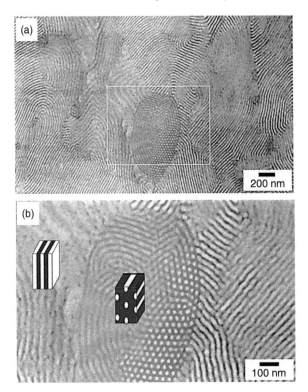

Figure 13 TEM micrographs showing micrphase-separated areas in the blends of two linear block copolymers LN3 and LNX, the weight ratio of LN3 to LNX is 65/35; OsO_4 staining makes the PB phase appear dark; a part of (a) is magnified in (b).

block copolymer embedded in the matrix of the other. In contrast to the behavior of a blend consisting of symmetric block copolymers (e.g., Figure 2), where it was possible to adjust a particular morphology by mixing a definite amount of a block copolymer to its compositionally complementary counterpart, the blend components are macrophase separated in the present case. This is a clear signal of the fact that the macrophase separation can be induced by the asymmetric architecture of one of the blend partners. Indeed, the ratio of long PS block in LN3 (M_w ~ 60,000 g/mol) and PS blocks of LNX (M_w ~ 12,000 g/mol) is approximately 5. This value is close to the molecular weight ratio (r) for the macrophase separation in the binary diblock copolymer blends reported in the literature ($r \sim 6$, Yamaguchi et al. [54]). Thus, the architecture-induced macrophase separation observed in the blends investigated may be regarded as being similar to the "*macrophase separation induced by microphase separation*" suggested by Hashimoto et al. [56]. An additional contribution towards the macrophase separation in the present case is offered by widely separated compositions of each SBS triblock copolymer. However, it should be noted that the value of r with respect to total molecular weight of each polymer is approximately unity for the block copolymers studied here.

In the present case, the macrophase separation is also favored in part by the presence of PS-co-PB random copolymer in LN3 that is essentially immiscible with a pure PB center block of the copolymer LNX. Such a macrophase separation was also observed in other binary blends containing at least an asymmetric block copolymer as blend component such as LN2/LNX blends [57]. Therefore, the macrophase separation is predominantly induced by the block copolymer chain architecture rather than the structure of the rubbery center block.

E.   Block Copolymer/Homopolymer Blends

Due to higher production costs, styrene/butadiene block copolymers are seldom used as pure materials. They are rather mixed with other polymers such as polystyrene for the manufacture of packaging films, injection-molded parts, etc. As introduced earlier, the length of the homopolymer chains relative to that of the corresponding block of the block copolymer plays a vital role in the phase behavior of the block copolymer/homopolymer blends [4,13]. The low molar mass homopolymers (e.g., PB or PS), which are easily assimilated by the corresponding block domains, can be used to change the dimension of a given morphology and even to produce new morphologies corresponding to a different phase volume ratio. Hence, the method of blending with homopolymers can also be helpful to locate the position of a "new" morphology in the block copolymer phase diagram.

TEM images in Figure 14 show the morphology of the blends of a star block copolymer (ST3) and two different homopolymers – polybutadiene (PB004) and polystyrene (PS015 and PS190). PB004, PS015 and PS190 stand for corresponding homopolymers having weight average molecular weights (M_w) of 4,000 g/mol, 15,000 g/mol and 190,000 g/mol, respectively.

An addition of 10 wt% low molecular weight polybutadiene (PB004; $M_w \sim 4,000$ g/mol), which is compatible with the rubbery PS-co-PB phase, increases obviously the soft phase volume fraction enabling the formation of morphology corresponding to lower overall PS volume content than ST3. One can clearly notice some regions of hexagonal arrangement of light polystyrene cylinders in Figure 14a. In contrast, an addition of 20 wt% low molecular weight polystyrene (PS015, $M_w \sim 15,000$ g/mol) causes the swelling of PS block domains, and results in the formation of a nanostructure corresponding to higher PS content than ST3. The accommodation of the PS015 chains in the PS block domains of the block copolymer is favored by higher molecular weight of PS block chains ($M_n \sim 60,000 - 90,000$ g/mol). The structure of this blend is clearly a lamellar one (Figure 14b). In fact, the mixing entropy of PS increases with the ratio $M_{PS\text{-}block}/M_{hPS}$ [13]. Since this ratio is quite high in the ST3/PS015 blend, the hPS molecules are able to penetrate deeply into the PS domains and modify the interfacial curvature leading to the lamellar morphology (see discussion of Figure 3).

These observations suggest that the morphology of pure ST3 (see Figure 8) is somewhat intermediate between that of a cylindrical and of a lamellar morphology. Since the stable morphology observed within this window is the "gyroid" phase, the equilibrium microphase-separated structures of ST3 can be regarded as being, at least, very close to the "gyroid" morphology as mentioned above.

If $M_{PS\text{-}block} < M_{hPS}$, the mixing entropy of the homopolymer chains is drastically decreased, and these chains become less successful to wet the copolymer brush effectively. The molecular weight of the PS190 ($M_w \sim 190,000$ g/mol) is much higher than the corresponding block in the block copolymer. Thus the

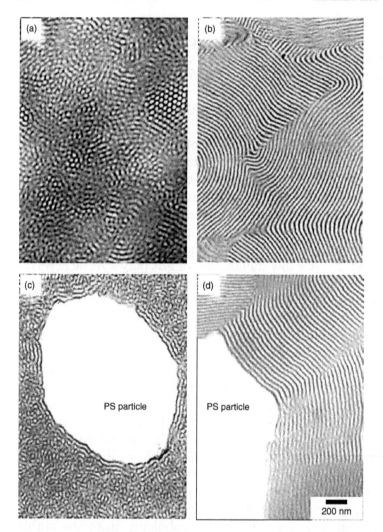

Figure 14   TEM micrographs showing the bulk morphology of blends comprising ST3 and homopolymers having variable molecular weights: a) ST3 + 10% PB004, b) ST3 + 20% PS015, c) ST3 + 20% PS190 and d) ST3 + 20% PS015 + 20% PS190; OsO_4 staining makes the PB phase appear dark. (Partly reproduced from Adhikari R, Lebek W, Godehardt R, Löschner K, Michler GH. Kautschuk-Gummi-Kunststoffe 2004;57:90–94.)

PS homopolymer chains and PS block become less miscible. As a result, the hPS chains are completely expelled from the microdomains. Therefore, in a blend of ST3 with 20 wt% PS190, a morphology comprising the macrophase-separated PS particles (see Figure 14c) surrounded by the ST3 matrix is formed.

The addition of more PS015 to ST3 would lead first to a thickening of PS lamellae and then finally to the formation of PB domains. At higher PS015 content, the resulting product would behave in a brittle manner due to weakening of the entanglements in the PS phase and presence of PS matrix. On the other hand, the addition of more PS190 would lead to stronger macrophase-separation that again may, as a consequence, worsen the mechanical properties. One strategy to maintain the toughness of the blends at a higher PS content would be to keep the ease of macrophase-separation as low as possible and allow the formation of plastically deformable matrix, i.e., combining the morphologies of Figure 14b and Figure 14c. This goal can be easily achieved by using a multimodal polystyrene homopolymer.

Figure 14d is a representative TEM image of a blend of ST3 and 40 wt% bimodal PS (i.e., 20% PS015 and 20% PS190), which is exactly a combination of Figure 14b and Figure 14c, in which the added PS mixture is partly accommodated by the PS block domains forming lamellae and is partly macrophase-separated in the form of PS particles. It was shown that this special morphology allows an intensive plastic deformation of the blend under tensile loading. The lamellar morphology enables the drawing of the matrix and the macrophase separation does not critically reduce elongation at break [58].

Finally, it should be noted that the discussion outlined above in the case of solution cast samples may not be valid for the mixtures produced by extrusion or injection molding. The processing conditions may further dramatically influence the morphology of the binary block copolymer/homopolymer blends (see next section).

F.   Influence of Processing Conditions
         on Nanostructure

As introduced earlier, the ways in which the block copolymers
and their blends with other polymers are processed have a
great implication on the nanostructures formed. One important
consequence of application of an external field is the alignment
of block copolymer nanostructures [30–32,44]. The processing
induced non-equilibrium structures were found to be very different from the equilibrium ones [53,58]. For example, the
injection molded blends of a lamellar SBS block copolymer and
PS having molecular weight higher than that of the PS block
of the copolymer showed lamellae-like arrangement of the
nanostructures at all the compositions [44] (see Chapter 10).

  In the preceding section, we showed that macrophase
separation between ST3 and PS190 occurs under equilibrium
conditions (see Figure 14c) due to incompatibility between
ST3 and PS190 chains. At high PS content, in contrast to the
morphology dominated by PS layers (PS matrix) observed in
blends of lamellar SBS copolymer and PS homopolymer [44],
a typical "droplet-like" structure was found in these blends
when subjected to steady shear [53,59]. Figure 14 shows the
representative TEM micrographs of ST3 and its blends with
PS190 prepared by extrusion (details in [53]).

  The morphology of ST3 (Figure 15a), similar to its equilibrium structure (compare with Figure 8), consists of co-
continuous arrangement of hard (PS) and soft (PS-co-PB)
phases. In the blends, with increasing PS content, some of
the PS domains (see white islands in Figure 15b–d) are
swelled leading to the formation of uniformly distributed
"droplets" of PS phase. While the dimension of rubbery phase
made up of PS-co-PB changes insignificantly, the size of the
PS droplets increases. Even when the added PS is in majority
(e.g., 60 wt% PS190, Figure 15d), the PS phase still forms a
co-continuous network with the block copolymer nanostructures, substantiating the notion of improved compatibility of
added PS with ST3.

  The typical droplet-like morphology observed in
ST3/PS190 blends seems to result from the dispersion of a

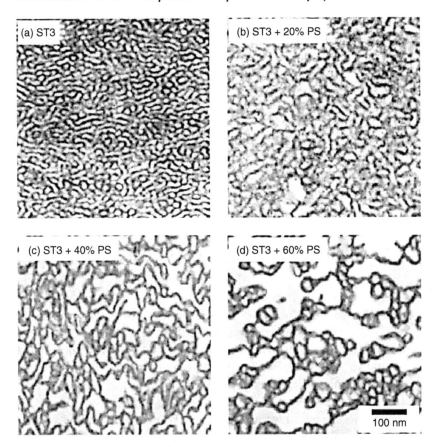

Figure 15 Representative TEM micrographs of extruded ST3/PS190 blends; OsO_4 staining makes the PB phase appear dark.

high molecular weight fraction of the added PS as well as the block copolymer architecture. The dispersion is mainly affected by the processing conditions which hinder the macrophase separation. The macrophase separation between ST3 and added PS190 is prevented by applied shear and limited time available during the rapid cooling of the polymer strands. That the average thickness of the thinner PS domains also increases slightly with increasing added PS content (Figure 15) suggests that a portion of low molar mass fraction of the added PS has been added to the longest PS blocks of the star block copolymer. This is in agreement with the conditions for the solubilization of a homopolymer in the corresponding block of the block copolymer discussed in the literature [13].

The results on nonequilibrium morphologies obtained in ST3/PS190 blends have provided strong evidence to the notion that the nature of processing-induced nanostructures in such systems are highly coupled with the block copolymer chain architecture. One important aspect that deserves mentioning in this regard is that these new droplet-like nanostructures induced by block copolymer architecture and processing contribute significantly to the toughness and transparency of the materials.

V. SUMMARY AND OUTLOOK

At sufficiently high molecular weight and narrow polydispersity, the phase behavior and the mechanical properties of styrene/butadiene based triblock copolymers are generally governed by their chemical composition. However, it has been known for some time that a change in the chain architecture at constant composition can have a dramatic influence on their microphase separation phenomena and deformation behavior. One relevant conclusion of our experimental results is that modification of molecular architecture of the block copolymers allows controlling of their morphology and composition independently.

To explore the impact of molecular architecture on the microphase separation and micromechanical properties, we have studied a series of styrene/butadiene block copolymers. At a constant molecular weight (ca. 100,000 g/mol) and a constant composition (styrene volume fraction ~ 0.70), all the basic morphologies observed in diblock copolymers were found via a change in chain architecture. As a result, without altering the overall composition, mechanical properties could be adjusted over a wide range. The results show that architectural and interfacial modification may provide a novel route toward controlling mechanical properties of these nanostructured materials.

Of particular importance appear the styrene-rich (PS content > 70%) SBS block copolymers with a lamellar or co-continuous morphology because of their high ductility. If the thickness of the PS lamellae (or PS struts in the co-continuous network) falls below a critical value of about 20 nm, a homogeneous plastic yielding of the otherwise brittle PS appears. This effect of a thin layer yielding results in a drastically improved toughness and is discussed in detail in Chapter 10.

We have further extended the study of the impact of asymmetric block copolymer architecture on the phase behavior of binary block copolymer blends and block copolymer/homopolymer blends. An architecturally induced macrophase separation was observed in the binary block copolymer blends. By suitably choosing the processing conditions and homopolystyrene molecular weight, it was possible to control the superstructure of the block copolymer/PS blends.

The practical application of the SBS block copolymers has two major drawbacks. First, the unsaturated polybutadiene block is very sensitive to photodegradation and can be easily oxidized by atmospheric oxygen unless the inherent double bonds are previously hydrogenated. The second disadvantage is the low service temperature (60–70°C) of these polymers that is limited by the glass transition temperature of polystyrene ($T_{g\text{-PS}} \sim 100°C$). Thus one would be interested to produce block copolymers, which offer a broader service temperature range and are not as susceptible to thermal and radiation-induced degradation as butadiene or isoprene-containing polymers, by means of relatively convenient polymerization techniques (e.g., "living" radical polymerization [59]). In this respect, the PMMA-*block*-PnBA-*block*-PMMA triblock copolymers (PMMA: poly(methyl-methacrylate) and PnBA: poly(n-butylacrylate)) can be a good alternative to solve the drawbacks mentioned above. The higher T_g of PMMA (~120°C) than that of PS in an SBS counterpart may clearly enhance the upper service temperature range of the copolymer. On the other hand, the rubbery nBA block ($T_g \sim -50°C$) is insensitive to ultraviolet radiation and more resistant to thermal oxidation.

The controlled radical polymerization has proved a logical and workable alternative to anionic polymerization in the last years [59,60]. This technique involves the chain growth polymerization that proceeds in the absence of irreversible chain transfer and chain termination reactions provided that the initiation is fast and complete. Via these techniques, block copolymers having well-defined block sequences and architecture can be synthesized. For example, PMMA-*block*-PnBA-*block*-PMMA with different hard/soft ratios were produced in semi-technical scale. AFM examination demonstrated the existence of well-defined microphase-separated structures similar to those of ionic block copolymers (see Figure 16, details in [60]).

In Figure 16, the morphology of two different block copolymers is presented which contain 41 and 32 wt% PMMA. In accordance with the phase diagram [4] established for the monodisperse block copolymer systems, alternating lamellae (Figure 16a) and PMMA domains dispersed in PnBA matrix (Figure 16b) are discernible.

The new class of block copolymers can possess a significantly high application potential, provided the mechanical properties are optimized to make them capable to compete with classical styrenic block copolymers in the market. Detailed analyses of structure-property correlations, especially with the aim of optimization of mechanical properties of these block copolymers, should be the focus of future studies.

ACKNOWLEDGMENTS

This work was funded by the Kultusministerium des Landes Sachsen-Anhalt (Project: *Neue Funktionswerkstoffe auf der Grundlage schwachentmischter Blockcopolymere*). Special thanks are due to Dr. R. Godehardt for his support during all sets of AFM experiments. The authors further are indebted to Mr. M. Buschnakowski, Ms. S. Goerlitz, Mr. S. Henning, Dr. T.A. Huy, Dr. M. Langela, and Mr. W. Lebek for their collaboration in the frame of this project. We are thankful to Dr. K Knoll (BASF-Aktiengesellschaft, Ludwigshafen) for collaboration and for the supply of block copolymer samples. We

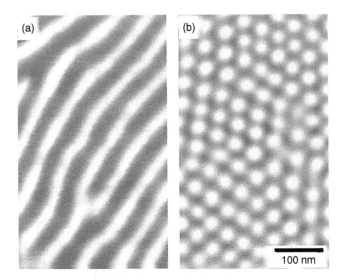

Figure 16 AFM phase images showing the morphologies of PMMA-*block*-PnBA-*block*-PMMA triblock copolymers: (a) 41% PMMA and (b) 32% PMMA; PMMA and PnBA phases appear light and dark, respectively (details to follow in Kraberg et al.[60]).

further cordially acknowledge the collaboration with the groups of Prof. M. Arnold, Prof. W. Grellmann and Prof. H.J. Radusch from our university. Dr. Adhikari is further grateful to the Max-Buchner-Forschungs-Stiftung for providing a research scholarship (MBFSt 6052).

REFERENCES

1. Lodge TP. Block copolymers: Past successes and future challenges. Macromol Chem Phys 2003;204:265–273.

2. Bates FS, Fredrickson GH. Block copolymer thermodynamics: Theory and experiment. In: Holden G, Legge NR, Quirk RP, Schroeder HE, eds. Thermoplastic Elastomers, 2nd ed., Munich: Hanser Publishers, 1998. pp.336–364.

3. Holden G. Understanding Thermoplastic Elastomers. Munich: Hanser Verlag, 2000. pp.15–35.

4. Hamley IW. The Physics of Block Copolymers. Oxford: Oxford Science Publications, 1998.

5. Milner ST. Chain architecture and asymmetry in copolymer microphases. Macromolecules 1994;27:2333–2335.

6. Matsen MW. Equilibrium behavior of asymmetric ABA triblock copolymer melts. J Chem Phys 2000;113:5539–5544.

7. Lee C, Gido SP, Poulos Y, Hadjichristidis N, Tan NB, Trevino S F, Mays JW. H-shaped double graft copolymers: Effect of molecular architecture on morphology. J Chem Phys 1997;107:6460–6469.

8. Knoll K and Nießner N. Styrolux and Styroflex — From transparent high impact polystyrene to new thermoplastic elastomers. Macromol Symp 1998;132:231–243.

9. Adhikari R, Michler GH, Huy TA, Ivankova E, Godehardt R, Lebek W, Knoll K. Correlation between molecular architecture, morphology and deformation behavior of styrene/butadiene block copolymers. Macromol Chem Phys 2003;204:488–499.

10. Michler GH, Adhikari R, Lebek W, Goerlitz S, Weidisch R, Knoll K. Morphology and micromechanical deformation behavior of styrene/butadiene block copolymers: I. Toughening mechanism in asymmetric star block copolymers. J Appl Polym Sci 2002;85:683–700.

11. Weidisch R, Gido SP, Uhrig D, Iatrou H, Mays J, Hadjichristidis N. Tetrafunctional multigraft copolymers as novel thermoplastic elastomers. Macromolecules 2001;34:6333–6337.

12. Mori Y, Lim LS, Bates FS. Consequences of molecular bridging in lamellae-forming triblock/pentablock copolymer blends. Macromolecules 2003;36:9879–9888.

13. Hasegawa H, Hashimoto T. Self assembly and morphology of block copolymer systems. In: Aggarwal SL, Russo S, eds. Comprehensive Polymer Science, Suppl 2. London: Pergamon, 1996. pp. 497–539.

14. Sakurai S. Control of morphology in block copolymers. Trends Polym Sci 1995;3:90–98 and Sakurai S. Block copolymer morphology. Trends Polym Sci 1997;5:210–212.

15. Koltisko B, Hiltner A, Baer E. Crazing in thin films of styrene-butadiene block copolymers. J Polym Sci-B: Polym Phys 1986;24:2167–2183.

16. Weidisch R, Michler GH. Correlation between phase behavior, mechanical properties and deformation mechanisms in weakly segregated block copolymers. In: Balta Calleja FJ, Roslaniec Z, eds., Block Copolymers. New York: Marcel-Dekker Publishers, 2000. p.215–249.

17. Schwier CE, Argon AS, Cohen RE. Crazing in polystyrene-polybutadiene diblock copolymer containing cylindrical poly-butadiene domains. Polymer 1985;26:1985–1993.

18. Sakurai S, Isobe D, Okamoto S, Nomura S. Control of morphologies and mechanical properties in binary blends of elastomeric polystyrene-block-polybutadiene-block-polystyrene triblock copolymers. J Macromol Sci-Phys 2002;4:387–396.

19. Vilesov AD, Floudas G, Pakula T, Melenevskaya EY, Birshtein TM, Lyatskaya YV. Lamellar structure formation in the mixture of two cylinder-forming block copolymers. Macromol Chem Phys 1994;195:2317–2326.

20. Koizumi S, Hasegawa H, Hashimoto T. Ordered structures in blends of block copolymers: 3. Self-assembly in blends of sphere- and cylinder-forming copolymers. Macromolecules 1994;27:4371–4381.

21. Mauritz KA, Storey FS, Mountz DA, Reuschle DA. Poly(styrene-b-isobutylene-b-styrene) block copolymer ionomers (BCPI), and BCPI/silicate nanocomposites. 1. Organic counterion: BCPI sol–gel reaction template. Polymer 2002;43:4315–4323.

22. Abetz V, Goldacker T. Formation of superlattices via blending of block copolymers. Macromol Rapid Comm 2000;2:16–34.

23. Jiang S, Gopfert A, Abetz V. Novel morphologies of block copolymer blends via hydrogen bonding. Macromolecules 2003;36:6171–6177.

24. Mogi Y, Nomura M, Kotsuji H, Matsuhita Y, Noda I. Superlattice structures in morphologies of the ABC triblock copolymers. Macromolecules 1994;27:6755–6760.

25. Shibayama M, Hasegawa H, Hashimoto T, Kawai H. Microdomain structure of an ABC-type triblock copolymer of polystyrene-poly[(4-vinylbenzyl)dimethylamine]-polyisoprene cast from solution. Macromolecules 1982;15:274–280.

26. Stadler R, Auschra C, Beckmann J, Krappe U, Voigt-Martin I, Leibler L. Morphology and thermodynamics of Poly (A-block-B-block-C) triblock copolymers. Macromolecules 1995;31:3080–3097.

27. Drolet F, Fredrickson GH. Combinatorial screening of complex block copolymer assembly with self consistent field theory. Phys Rev Lett 1999;83:4317–4320.

28. Bohbot-Raviv Y, Wang Z-G. Discovering new ordered phases of block copolymers. Phys Rev Lett 2000;85:3428–3431.

29. Bates FS, Koppi KA, Tirrell M, Almdal K, Mortensen K. Influence of shear on the hexagonal-to-disorder transition in a diblock copolymer melt. Macromolecules 1994;27:5934–5936.

30. Cohen Y, Albalak RJ, Dair BJ, Capel MS, Thomas EL. Deformation of oriented lamellar block copolymer films. Macromolecules 2000;33:6502–6516.

31. Honeker CC, Thomas EL. Perpendicular deformation of a near-single-crystal triblock copolymer with a cylindrical morphology, 2. TEM. Macromolecules 2000;39:9407–9417.

32. Dair BJ, Honecker CC, Alward DB, Avgeropoulos A, Hadjichristidis N, Fetters LJ, Capel MS, Thomas EL. Mechanical properties and deformation behaviour of the double gyroid phase in unoriented thermoplastic elastomers. Macromolecules 1999;32:8145–8152.

33. Simon PFW, Ulrich R, Spiess HW, Wiesner U. Block copolymer-ceramic hybrid materials from organically modified ceramic precursors. Chem Mater 2001;13:3464–3486.

34. Hadjichristidis N, Pispas S, Pitsikalis M, Iatrou H, Vlahos C. Asymmetric star polymers: Synthesis and properties. Adv Polym Sci 1999;142:71–127.

35. Hadjichristidis N, Iatrou H, Behal SK, Chludzinski JJ, Disco M M, Garner RT, Liang K, Lohse DJ, Milner ST. Morphology and miscibility of miktoarm styrene-diene copolymers and terpolymers. Macromolecules 1993;26:5812–5815.

36. Ma J-J, Nestegard MK, Majumdar BD, Sheridan MM. Asymmetric star block copolymers: Anionic synthesis, characterization and pressure sensitive adhesive performance. ACS Symp Series 1998;696:159–166.

37. Bühler F, Gronski W. Block copolymers with controlled interphase width: Effects of interphase and composition on domain dimensions. Makromol Chem 1986;187:2019–2037.

38. Sameth J, Spontak RJ, Smith SD, Ashraf A, Mortensen K. Microphase separated tapered triblock copolymers. J de Physique 1993;3:59–62.

39. Laurer J, Smith SD, Sameth J, Mortensen K, Spontak RJ. Interfacial modification as route to novel bilayered morphologies in binary block copolymer/homopolymer blends. Macromolecules 1997;30:549–560.

40. Zielinski JM and Spontak RJ. Thermodynamic consideration of triblock copolymers with a random middle block. Macromolecules 1992;25:5957–5964.

41. Aksimentiev A, Holyst R. Phase behaviour of gradient copolymers. J Chem Phys 1999;11:2329–2339.

42. Hodrokoukes P, Floudas G, Pipas S, Hajdichristidis N. Microphase separation in normal and inverse tapered block copolymers of polystyrene and polyisoprene: 1. Phase state. Macromolecules 2001;34:650–657.

43. Knoll K. Anionische blockcopolymere. In: Becker GW, Braun D, Gausepohl H, Gellert R, eds. Kunststoff-Handbuch: 4. Polystyrol. Munich: Hanser Verlag, 1996. pp.145–166.

44. Adhikari R, Michler GH, Knoll K. Morphology and micromechanical behavior of styrene/butadiene block copolymers and their blends with polystyrene. Macromol Sym 2003; 198:117–134.

45. Adhikari R, Buschnakowski M, Henning S, Huy TA, Goerlitz S, Lebek W, Godehardt R, Michler GH, Lach R, Geiger K, Knoll K. Double yielding in a styrene/butadiene star block copolymer. Macromol Rapid Commun 2004;25:653–658.

46. Adhikari R, Michler GH, Lebek W, Goerlitz S, Weidisch R, Knoll K. Morphology and micromechanical deformation behavior of styrene/butadiene block copolymers: II. Influence of molecular architecture of asymmetric architecture. J Appl Polym Sci 2002;85:701–713.

47. Geiger K, Knoll K, Langela M. Microstructure and rheological properties of triblock copolymers under extrusion conditions. Rheologica Acta 2002;41:345–355.

48. Huy TA, Hai LH, Adhikari R, Weidisch R, Michler GH, Knoll K. Influence of interfacial structure on the phase behavior and deformation mechanism of SBS block copolymers. Polymer 2003;44:1237–1245.

49. Langela M. Struktur und Rheologische Eigenschaften von PS-PI und PS-PB Blockcopolymeren. Thesis, University of Mainz, Germany, 2001.

50. Andrews RJ, Grulke EA. Glass transition temperatures of polymers. In: Brandrup J, Immergut EH, Grulke EA, eds. Polymer Handbook, 4th Ed. John Wiley & Sons, 1999, Chapter VI. pp. 193–277.

51. Adhikari R, Huy TA, Buschnakowski M, Michler GH, Knoll K. Asymmetric PS-block-(PS-co-PB)-block-PS block copolymers: morphology formation and deformation behavior. New J Phys 2004;6:28(1–20). (www.njp.org).

52. Fujimora M., Hashimoto T, Kawai H. Structural change accompanied by plastic-to-rubber transition of SBS block copolymers. Rub Chem Technol 1978;51:215–224.

53. Buschnakowski M, Adhikari R, Michler GH, Danuningrat S, Geiger K, Knoll K. Asymmetric PS-block-(PS-co-PB)-block-PS star block copolymer and its blends with polystyrene homopolymers: Influence of the extrusion process on the morphology and deformation behavior. Polymer 2005, under preparation.

54. Yamaguchi D and Hashimoto T. A phase diagram for the binary blends of nearly symmetric diblock copolymers. 1. Parameter space of molecular weight ratio and blend composition. Macromolecules 2001;34:6495–6505.

55. Spontak RJ, Fung JC, Braunfeld MB, Sedat JW, Agard DA, Ashraf A, Smith SD. Architecture-induced phase immiscibility in a diblock/multiblock copolymer blend. Macromolecules 1996;29:2850–2856.

56. Hashimoto T, Koizumi S, Hasegawa H. Ordered structure in blends of block copolymers. 2. Self-assembly for immiscible lamella-forming copolymers. Macromolecules 1994;27:1562–1570.

57. Adhikari R, Lebek W, Godehardt R, Löschner K, Michler GH. Makrophasen-separierte Blends aus Styrol-Butadien-Blockcopolymeren: Morphologie und Deformationsverhalten. Kautschuk-Gummi-Kunststoffe 2004;57:90–94.

58. Adhikari R, Buschnakowski M, Godehardt R, Lebek W, Michler GH, Baltá-Calleja FJ, Knoll K. Asymmetric PS-b-(PS-co-PB)-b-PS block copolymers: Morphology and deformation behavior of blends with hPS. Polym Adv Technol 2005;DOI:10.1002/pat.577.

59. Davis KA and Matyjaszewski K. Statistical, gradient, block and graft copolymers by controlled/living radical polymerizations. Adv Polym Sci 2002;159:107–152.

60. Kraberg T, Adhikari R, Arnold M, Michler GH. PMMA-b-PnBA-PMMA triblock copolymers via semi-technical ATRP: Synthesis and characterization. Under preparation, 2005.

Part II

Deformation Mechanisms at Nanoscopic Level

4

Crazing and Fracture in Amorphous Polymers: Micromechanisms and Effect of Molecular Variables

H.H. KAUSCH

Institut des Matériaux, Ecole Polytechnique
Fédérale de Lausanne (EPFL)

J.L. HALARY

Ecole Supérieure de Physique et Chimie
Industrielles de la Ville de Paris (ESPCI)

CONTENTS

I. INTRODUCTION

Good ultimate properties are among the most important pre-requisites for the successful use of a polymer material, no matter whether the mechanical, the optical or some specific functional properties are to be exploited. For this reason, the deformation and fracture behavior and the means for their improvement have always been studied intensively [1–4]. The mechanical strength of an isotropic thermoplastic polymer derives primarily from the van der Waals attraction between chain segments. Nevertheless strength and toughness evidently depend on the molecular properties of the chosen material, on molecular packing (density, phase structure, micro-morphology), on the way stresses are transmitted between them (through cohesive forces, cross-links or entanglements) and on the nature and intensity of relaxation (and/or damage) mechanisms.

The general characteristics of macromolecules, weak cohesive interaction, long and highly anisotropic, frequently carefully engineered chain backbones and the high local mobility give rise to an enormous potential for structural organization: from the apparently homogeneous amorphous glasses to the complex morphology of, e.g., spherulitic semicrystalline polymers or hairy-rod molecular composites. In this contribution, macroscopically homogeneous amorphous thermoplastic polymers are studied: methyl methacrylate glutarimide copolymers and two series of amorphous semi-aromatic polyamides. The (molecular) nature of the different deformation mechanisms can be identified by comparing the dynamic and fracture behavior of these systematically chemically modified (glassy) polymers. In the absence of crystalline regions or other forms of superstructure, our systems can be modeled as a well-entangled amorphous *physical network*. The *elementary* deformation (and/or damage) mechanisms to be considered are conformational changes, segmental slip, void formation, disentanglement and chain scission. The strongly time- and temperature-dependent dynamics of these mechanisms determine the modes of deformation and fracture (through, e.g., crazing, shear yielding, creep and/or crack propagation). We will first discuss the competition between (and changes in the relative importance of) elastic straining, segmental slip and void formation as a function of strain amplitude and loading rate.

II. COMPETITION BETWEEN ELASTIC EXTENSION, SHEAR AND CAVITATION

A well-entangled, macroscopically homogeneous, non-oriented (ductile) amorphous polymer subjected to uniaxial stress typically shows the stress-strain curve reproduced in Figure 1. Up to the limit of linear viscoelastic response (at strains $\varepsilon = \Delta L/L$ between $0.1\% < \varepsilon < 1\%$ depending on temperature), the isotropic sample is deformed in a mechanically reversible manner by elastic extension (predominantly by an increase of the intersegmental distances) accompanied by uncorrelated intersegmental shear displacements and conformational changes. The latter two mechanisms give rise to anelastic deformation (time- and strain-dependence of the elastic moduli), to stress relaxation and creep. The local microstructure begins to be modified by void and/or craze initiation and correlated segmental shear displacements at strains larger than 2 to 4%. The ultimate strain and the rupture mechanism strongly depend on material and experimental conditions.

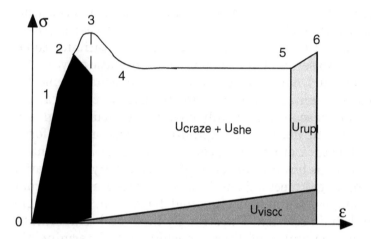

Figure 1 Typical stress-strain curve of a ductile thermoplastic polymer showing the uniform elastic with traces of anelastic deformation (0–1), the anelastic deformation of the isotropic sample (1–2), a region of void and/or craze initiation (2–3), the yield point (3), and the regions of necking (3–4), cold drawing (4–5), and extension of drawn material to final rupture (6). The contributions of the different modes of deformation to the dissipated energy are indicated.

In any sample loaded beyond region 0–1, strain-dependent energy dissipation occurs through the mentioned inter- and intra-segmental mechanisms but also through formation of voids (cavitation). We consider the formation of voids to be of particular importance since it modifies the local state of stress and facilitates subsequent heterogeneous modes of deformation such as crazing. It is for this reason that we wish to quantify the relative importance of the above deformation mechanisms. This can be done in a conventional tensile experiment by measuring the uniaxial strain $\Delta L/L = \varepsilon_{un}$ in stress direction and the lateral strain ε_{lat} in the perpendicular direction. We know that *three* components contribute to the uniaxial elongation of a stressed sample: the one-dimensional *elastic* strain component $\varepsilon_{un,el}$, the *plastic* shear strain ε_{shear} and the *cavitational* strain component ε_{cav}:

$$\varepsilon_{un} = \varepsilon_{un,el} + \varepsilon_{shear} + \varepsilon_{cav} \tag{1}$$

In the limit of small strains (< 5%) we obtain the volume strain $\Delta V/V_0 = \varepsilon_{vol}$ from the axial (ε_{un}) and lateral (ε_{lat}) strain components by:

$$\varepsilon_{vol} = (1 + \varepsilon_{un})(1 + \varepsilon_{lat})^2 - 1 \qquad (2)$$

Since shear is not contributing to a change in volume, the relative increase in sample volume ε_{vol} can only have *two* origins, elastic dilatation ($\varepsilon_{vol,el}$) and void formation before and during crazing (cavitation, $\varepsilon_{vol,cav}$):

$$\varepsilon_{vol} = \varepsilon_{vol,el} + \varepsilon_{vol,cav} \qquad (3)$$

It can be assumed that the effect of cavitation on linear strain, $\varepsilon_{un,cav}$, is identical to that on volume strain, thus $\varepsilon_{un,cav} = \varepsilon_{vol,cav} = \varepsilon_{cav}$.

The uniaxial *elastic* deformation can be obtained from the tangent E_{el} of the stress-strain curve at zero strain; if uniaxial and lateral elastic deformations are related by Poisson's ratio v, we have:

$$\varepsilon_{vol, el} = \varepsilon_{un,el}(1 - 2v) = (1 - 2v)\sigma_{un}/E_{el} \qquad (4)$$

where σ_{un} is the applied uniaxial stress. The shear component is thus derived as:

$$\varepsilon_{shear} = \varepsilon_{un} - \varepsilon_{un,el} - \varepsilon_{cav} = \varepsilon_{un} + \varepsilon_{un,el}(2v) - \Delta V/V \qquad (5)$$

In Figure 2 we demonstrate the relative importance of the three components to uniaxial (linear) strain $\varepsilon_{un} = \Delta L/L$ of pure (non-modified) poly(methylmethacrylate) (PMMA) of high molecular weight (M_w) obtained at RT at a strain rate of $4 \cdot 10^{-4}$ s^{-1}.

It is clearly seen that the elastic component ε_{elast} (light gray surface) dominates the extension of this material. The shear contribution, $\varepsilon_{shear}/\varepsilon_{un}$ (white area), increases steadily from 0 to about 0.45 at a uniaxial strain of 5%. Cavitation begins abruptly at a strain of about 1% and increases gradually only accounting for a small fraction (less than 1/10 of the linear strain) at rupture. The voids formed in pure PMMA must be rather small compared to the wavelength of visible light since no stress whitening is observed and the loss in light transmittance amounts to no more than 2% [5]. It should

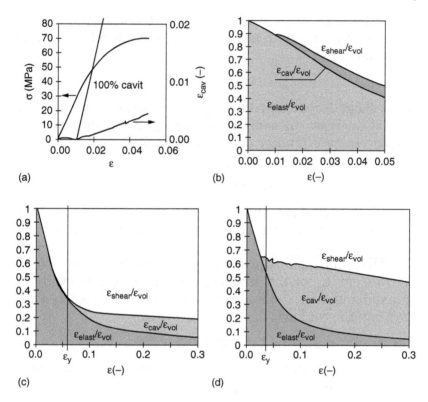

Figure 2 (a) Stress-strain curve and cavitational strain of non-modified poly(methylmethacrylate) (PMMA) of high molecular weight (M_w) at RT and a strain rate of $4 \cdot 10^{-4}$ s^{-1}; (b) the relative contributions of elastic deformation, cavitation and shear to uniaxial strain of high MW-PMMA; (c) and (d) the same contributions for a rubber modified PMMA (30 vol% of three-shell particles) at strain rates of $5 \cdot 10^{-4}$ s^{-1} and $3.5 \cdot 10^{-2}$ s^{-1}, respectively. (After Béguelin P. Thèse No 1572, Ecole Polytechnique Fédérale de Lausanne, Lausanne Suisse 1996.)

be noted that in such a stiff material at the indicated experimental conditions $\varepsilon_{elas} > \varepsilon_{shear} > \varepsilon_{cav}$.

The relative importance of the three components of strain strongly depends on intrinsic and extrinsic variables. At this point we wish to mention the effects of rubber modification and rate of deformation. The presence of rubber particles favors shear at the detriment of elastic dilatation. The inverse is true with increasing rate of deformation: the shear component decreases in favor of cavitation and — to some extent — elastic dilatation [5]. In PMMA modified by three-layer core-shell particles (30% by weight) cavitation starts at about the yield point and its relative intensity increases gradually up to a value of $\varepsilon_{cav}/\varepsilon_{un} = 0.12$ at a uniaxial strain of 30%. A dramatic increase in cavitational strain to a value of $\varepsilon_{cav}/\varepsilon_{un} = 0.40$ (again at a uniaxial strain of 30%) is observed at the higher strain rate of $3.5 \bullet 10^{-2}$ s^{-1} (which corresponds to a cross-head displacement rate of 130 mm/min). It should be noted that strain rates in impact loading are of the order of 40 s^{-1}, that is three orders of magnitude higher, which highlights the strong influence of cavitation on impact resistance [5].

The above experiments show that at small strains the shear contribution to ε_{un} increases more or less linearly with applied stress σ_{un}. This behavior is in agreement with the Eyring equation of the rate of flow of a *particle* over an energy barrier U_0:

$$K = k_1 - k_2 = 2k_0 \sin h(\sigma_{un} \bullet V/RT) \qquad (6)$$

where k_1 and k_2 are the rate constants for a particle jump in the forward and backward directions respectively, k_0 is the rate of decay of the activated state, which is proportional to exp $(-U_0/RT)$ and V is the activation volume.

The formation of voids obeys a different kinetics. For a void of radius r to be formed two conditions must be fulfilled. The hydrostatic component of the external stress tensor must be larger than the negative pressure $2\gamma_s/r$ exercised by the secondary bonds, so that the void does not collapse. Following Fond et al. [6] we assume that a void is stable, if the energy liberated by the nucleation of the void is at least equal to the energy of the newly formed surface, $4\pi r^2 \gamma_s$, where γ_s is the surface tension of the cavitating material. It follows from the above conditions that voids of large radii are stable at low hydrostatic stresses, but that their nucleation would require a prohibitively large activation energy $4\pi r^2 \gamma_s$. Void formation does not occur, therefore, at small stresses.

It is the principal objective of this contribution to investigate the important role of the primary molecular parameters (chain configuration, architecture and molecular weight) on macroscopic strength and toughness. So far our discussion has essentially been concerned with the materials response at small deformation, i.e., with anelastic behavior and the inception of heterogeneous deformation through formation of voids and crazes (region 1–3 in Figure 1). It is evident that the toughness of a craze forming amorphous polymer depends on the stability (and on the number) of the formed crazes. Thus we have to investigate the effect of the above micromechanical mechanisms and of slip, disentanglement and/or scission of chains on the mode of craze breakdown. First we will briefly review, therefore, our present understanding of crazing in amorphous polymers.

III. NUCLEATION OF VOIDS AND CRAZES IN ENTANGLEMENT NETWORKS

Hsiao and Sauer described for the first time half a century ago the scientifically as well as technically highly intriguing phenomenon of crazing. The delicate and complex structure of a craze in an amorphous polymer together with the difficulties to analyze stresses and strains within a craze and along its profile have delayed for decades a complete understanding of craze formation, propagation and breakdown. In the late sixties and early seventies notable progress was achieved by the discovery of the fibrillar microstructure of crazes through Kambour in 1968. The seminal studies in this period of, e.g., Haward, Sternstein, Williams, Marshall, Döll, and Argon have greatly helped to understand the elementary deformation processes and to predict the toughness and the lifetime of crazeable polymers. The resulting criteria were based on *stress bias, critical surface strain or stress intensity factor* respectively (reviewed in [2,3]).

In the last 20 years notable progress has been made concerning the molecular origins of crazing, which is comprehensively discussed in the two special volumes on *Crazing in Polymers* [7,8] and in a number of review articles (e.g., Wu

[9], Plummer [10], Kausch [11]). In the context of this contribution some features are of particular relevance. Thus we
wish to emphasize that a solid amorphous polymer has to be
considered as an entanglement network (with an entanglement density, v_e). The presence of entanglements is an essential prerequisite of ultimate strength and has a strong
influence on craze nucleation.

There is general agreement that the first step in craze
nucleation involves the genesis of voids in regions of considerable tensile stresses (in the neighborhood of a stress-con-
centrator and/or a plastically deformed local region of a glassy
polymer). The formation of voids as the first step of heterogeneous deformation is a common phenomenon in all types of
polymers. It is observed in small elastomeric modifier particles [6], highly stretched rubbers, in amorphous and semicrystalline polymers and — in the form of croids — even in
rubber-modified, cross-linked resins [2–4]. In all cases voiding
tends to relieve the triaxial constraints. What is different from
one polymer to the other is the second step, the response of
the material. Rubbers are torn apart, glassy PMMA and polystyrene (PS) craze, melt-drawn polypropylene (PP) shows a
hard-elastic response, poly(oxymethylene) (POM) develops
craze-like structures, and poly(vinyl chloride) (PVC) stress-
whitens by more or less randomly formed voids. Evidently
these differences are related to molecular structure and morphology. In *semicrystalline* polymers voids form between crystal lamellae and their effects are to a certain extent shielded
by the presence of the stiff and rigid crystal lamellae [11]. In
elastomers it is the high chain flexibility, which allows for
rapid void expansion controlled by void size and a surface
work parameter, which includes surface tension, visco-elastic
losses due to the expansion, and tearing energy [6]. The same
parameters are controlling the cavitation of elastomeric modifier particles. The presence of rubbery particles in *modified
glassy polymers* leads to a (random) multiplication of cavitation sites, which are activated at lower stresses than σ_y;
energy is subsequently dissipated by *generalized* shear of the
matrix between voids. Shear between particles also gives rise
to a special phenomenon; if the stress concentration caused

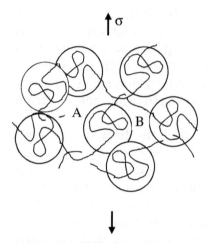

Figure 3 The Kausch model of craze nucleation. An entangled network is schematically represented by statistically coiled sub-chains of mass M_e between entanglements; their radii of gyration are shown as circles (see text for significance of A and B).

by a cavitated particle favors the cavitation of adjacent mod-ifier particles, then the contours of maximum shear stress will be marked by a line of cavitated particles. This has first been observed in *rubber-modified cross-linked resins* (Sue [12], Lu [13]) and in view of its similarity to a voided, craze-like struc-ture, Sue has called this feature a *croid* [12].

Our main interest here is concerned with *amorphous polymers*. We know that depending on the type of polymer and the experimental environment (especially temperature and strain rate), two modes of deformation can be observed: craze formation or homogeneous plastic deformation accom-panied or not by stress-whitening. In order to explain these differences and to understand the influence of molecular and environmental variables, Kausch [14] has used the model represented in Figure 3.

The micro-structural features essential in craze initia-tion are the increase of intersegmental distances with exten-sion, the increase and spatial fluctuation of free volume and its accumulation in sites of lower entanglement density (region A in Figure 3), and the resulting heterogeneous distribution of local stresses. The latter can be relieved by formation of voids as well as by shear displacement. Their competition and relative importance is clearly shown by the above dilatation experiments (Figure 3B). As to be expected, the elastic dilatation dominates in a rigid and fragile polymer well below T_g, but it should be noted, that even in such polymers *shear* is present from the very beginning of extension and its contribution to linear strain is more important than that of cavitation. It seems to be obvious that voids will be formed preferentially in sites of lower entanglement density (region A in Figure 3). Kausch has shown that initially voids can be accommodated without breaking any entanglements [14]. An isolated void is not yet a craze nucleus, however. If the stress is released at this stage, the void is likely to be closed. According to the Kausch model, a craze is nucleated once the stress concentration effect exerted by a void (growing in site A) leads to the preferential creation of adjacent voids in a plane perpendicular to the largest component of stress σ (as in site B). The cooperative creation of voids is made more difficult if the stress transfer between neighboring sites is weakened (by shear deformation of the coiled subchains) or if the resistance to cavitation is high (because an important number of entangled subchains is crossing the future craze plane). This is equivalent to saying that the higher the intrinsic segmental mobility and the higher the density of entanglements, the more difficult the formation of distinct crazes in a polymer. Wu has convincingly demonstrated the fundamental positive correlation between craze initiation stress σ_c and entanglement density v_e (Figure 4) [9].

IV. CRAZE GROWTH AND TOUGHNESS

After initiation the craze grows by the transformation of the glassy matrix material of a primordial region into fibrillar matter and by drawing virgin material into the craze fibrils [10,15]. The ligament extension λ can be quite important (up to the natural draw ratio λ_n) and is accompanied by the cre-

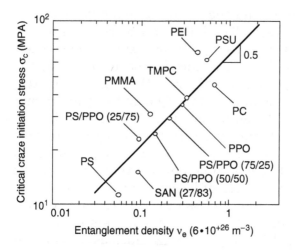

Figure 4   Correlation between craze initiation stress σ_c and entanglement density ν_e (in brackets: the relative copolymer composition). (After Wu S. Polym. Int. 1992; 29: 229–247.)

ation of new surface, which cannot be accommodated by a coherent, entangled network. Craze growth necessarily involves a loss of entanglements, which can occur by any mechanism, which reduces the number of entangled subchains, such as chain scission, forced reptation and slippage against van der Waals forces. The same mechanisms are also involved in the breakdown of the craze fibrils. The total energy dissipated in this process (from craze nucleation to breakdown) is essentially controlled by the craze stress and the maximum craze width. Brown [16] has related the toughness G_c of a polymer failing through the propagation and breakdown of a single craze to molecular variables. Assuming that craze fibril rupture (at a maximum fibril stress σ_f) is the main sample breakdown mechanism he obtains:

$$G_c = \left(\frac{\sigma_f^2 2\pi D}{S\lambda^2} \right) \left(\frac{E_2}{E_1} \right)^{1/2} \left(1 - 1/\lambda \right) \qquad (7)$$

where D is the fibril diameter, S the stress at the craze-bulk interface, and E_2 and E_1 are the elastic tensile moduli of the craze parallel and normal to the fibril direction. The toughness G_c predicted by this "chain scission model" scales with the square of σ_f, which is directly proportional to the number of effectively entangled chains in a fibril and their "strength" f_s. (In this context the *strength*, f_s, is the *force* to break a chain, which depends somewhat on temperature and rate of straining. For linear chains as PE or PA6, chain strengths of the order 3.3 to 4 nN have been indicated [17]). In his model Brown assumes that *all the effectively entangled chains, which intersect a unit cross-section of the virgin material are drawn into a craze fibril* [16]. Thus the strength σ_f of the load bearing strands should be equal to $\Sigma \bullet f_s \lambda$ with Σ being the number of chains, which intersect a unit cross-section of the virgin material. According to this model Σ and σ_f are proportional to the square root of the entanglement density v_e. Thus one derives from Equation 7 that G_c is proportional to v_e. This agrees with the experimental verification of Kausch and Jud [2,18], who have concluded from their crack healing experiments, that up to saturation the strength, G_c, of an interface depends linearly on the number of formed entanglements, which is equivalent to the critical stress intensity factor K_{Ic} scaling with $(v_e)^{1/2}$. Both of the above models, the *chain scission* and the *entanglement* model, have been derived from investigating the strength of *newly formed interfaces* [16,18]. On the other hand Wu [9] had established the craze initiation stress of *solid virgin* polymers and the associated K_{Ic} scale linearly with v_e. In order to assess which of these scaling laws describes toughness more accurately, Gensler [19] has accordingly plotted the data of Wu and his own data for 13 rapidly cooled amorphous and semicrystalline polymers (Figure 5). These plots show that the entanglement density can certainly be considered as the most important variable; it is difficult, however, to decide which scaling law applies [the scaling of K_{Ic} with $(v_e)^{1/2}$ gives the slightly better approximation]. Nevertheless, it is clearly seen that the correlation between K_{Ic} and entanglement density is not unique, other variables such as the relative distance from the glass transition temperature and the length and the intrinsic mobility of the chains will have an influence on toughness as well (as will be demonstrated in the following

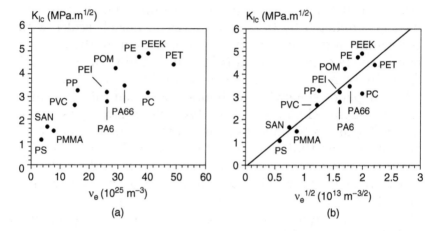

Figure 5 The critical stress intensity factor K_{Ic} of different rapidly cooled polymers as a function of entanglement density v_e. (Adapted from Gensler R. Thèse No 1863, Ecole Polytechnique Fédérale de Lausanne, Lausanne Suisse 1998.)

section). These influences explain the apparent contradiction that the strength of a newly formed interface is sometimes assumed to be proportional to v_e and in other experiments to $(v_e)^{1/2}$.

V.   THE EFFECT OF TEMPERATURE ON CRAZE INITIATION MECHANISMS

In his classical investigations Kambour had established for amorphous polymers a linear correlation between the craze initiation strain ε_c and the product CED•ΔT, where CED is the cohesive energy density and ΔT the difference between the glass transition temperature T_g and the experimental temperature T_{test} [20]. As stated above, craze initiation involves different microdeformation mechanisms (such as void formation, chain scission and slip). Their kinetics being quite different, it has to be expected, therefore, that not only the *level* of craze initiation stress but also the *nature* of the dominant mechanism will change as a function of temperature and molecular composition. In order to investigate this

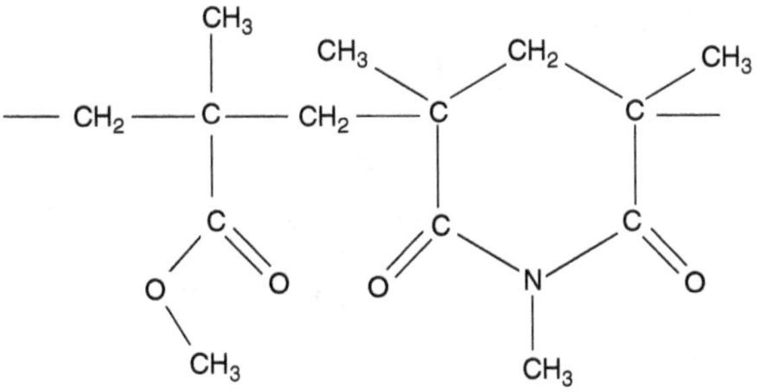

Figure 6 Chemical structure of a methyl methacrylate(MMA)–N-methyl-glutarimide diade.

behavior in more detail, we have selected MMA-glutarimide random copolymers (MGLUT, Figure 6). The principal variable in the MGLUT copolymers is their composition, the glutarimide content, which varied from 36 mol% (MGLUT36) to 76 mol%. Within this range the molecular weight M_w increased slightly from 76 to 110 kg mol^{-1}. All of these copolymers are amorphous, the glass transition temperature T_g increasing from 134 to 151°C [21,22].

When straining thin films (of a copolymer containing 76 mol% of glutarimide, MGLUT76) scission crazing was observed at low temperatures (Figure 7a), crazing and formation of deformation zones in different forms in a large temperature range around and above 50°C (Figure 7b), disentanglement crazing at more elevated temperatures (Figure 7c) and homogeneous deformation at T close to T_g (Figure 7d).

The observations made with two copolymers of different compositions have been compiled in Figure 8 in the form of temperature-deformation mechanism maps. Dark zones indicate notable shear deformation. It is well seen that the molecular composition has an influence on the occurrence of a particular deformation mechanism (disentanglement crazes are absent in the 64/36 material) and especially on the transition temperatures, at which one form changes into another; it should be noted that in the MMA-rich MGLUT36, deformation zones do not appear below temperatures of around 50°C, whereas they are observed at 20°C in MGLUT76 [21–22].

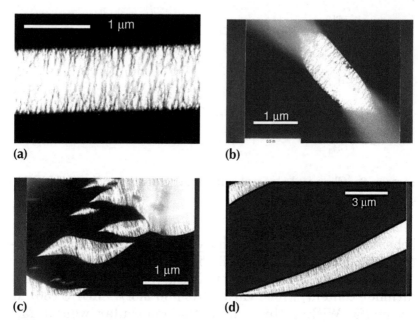

(a)   (b)

(c)   (d)

Figure 7 Microdeformation mechanisms observed in thin films of MGLUT76 strained at a rate of $2 \bullet 10^{-3}$ s^{-1}. Chain scission crazing at 0°C, shear blunting at 50°C, multiple crazing at craze tips (90°C) and disentanglement crazing (140°C). The glass transition occurs at 151°C. (Adapted from Tézé L. Thèse de doctorat, Université Pierre et Marie Curie, Paris, Nov. 10, 1995.)

The observed phenomena can be explained using the current theory on craze initiation and growth [1–3,7–10,16,21]. Crazes grow by transformation of glassy matrix material into fibrillar craze matter, which necessarily involves a loss of entanglements. As mentioned above, there are three mechanisms by which the number of entanglements can be reduced: chain scission, forced reptation and slippage against van der Waals forces. These three mechanisms have quite different kinetics. At low temperatures (in the MGLUT system at $T_g - T > 100$ K) the entangled chains show little mobility, they are held so tightly that they would rather break than slip. Chain scission requires very high stresses (to break a C-C bond a force of 4 nN is required which corresponds to

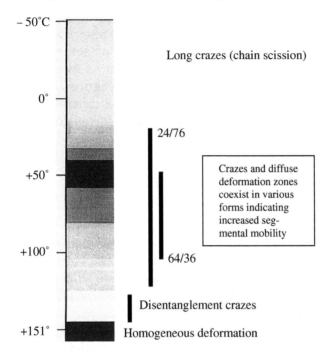

Figure 8 Temperature-deformation mechanism map of MGLUT76. Dark zones indicate notable shear deformation. For comparison, the temperature range, where crazes and diffuse deformation zones (DZ) are observed together, is also shown for MGLUT36 (dashed line).

a stress acting on the chain cross-section of about 6 GPa [17]). Such stresses are only found in regions of high-stress concentration (at craze tips and within the process zones at the fibril-bulk interface). This stress geometry gives rise to (long and thin) crazes of high aspect ratio. With increasing chain mobility overstressed chains are able to transfer part of the load and to escape scission temporarily. Craze fibrils become stronger and longer, the rate of deformation of craze tips and at the craze-bulk interface decreases, which favors further orientation and the formation of (diffuse) shear deformation zones (DZ). In the intermediate temperature region, therefore, mixed forms of plastic instability are found (Figure 7C). Above

80°C chain mobility is sufficiently high so as to favor disentanglement by forced reptation and crazing becomes the dominant deformation mechanism. In MGLUT76 at some 15 to 30 K below T_g it is the only active mechanism as evidenced by the appearance of the sharp-tipped crazes (Figure 7D). On the other hand, very close to T_g (151°C) stress relaxation is too rapid to permit any form of localized deformation; the sample deforms homogeneously [21,22].

VI. MOLECULAR CHARACTERISTICS OF THE INVESTIGATED SEMI-AROMATIC POLYAMIDES

The effect of molecular variables can be studied using well-characterized samples. Thus molecular weight (M_w) effects on mechanical properties become apparent when comparing mono-disperse samples of different M_w. To investigate the effect of specific molecular groups — such as presence or absence of double bonds, aromatic or bulky substituents, or hydrogen bonds — we have used two series of semi-aromatic polyamides: the first, referred to as SAPA-A, is based on lactam-12 sequences, terephthalic and/or isophthalic acid residues and 3,3'-diamino 2,2'-dimethyl dicyclohexylmethane residues (Figure 9). Here the principal molecular variables are the molecular weight M_w, the relative amounts of the lactam-12 sequences (y) and the configuration of the phenyl ring linkages ($x_T = 0$ designates the meta- and $x_T = 1$ the para-position), which influence entanglement molecular weight M_e and entanglement density ν_e.

The second series, referred to as SAPA-R, was based on 2-methyl 1,5-pentanediamine and terephthalic and/or isophthalic acid (Figure 10). The configuration of the phenyl ring linkages and the small mobility of the pentanediamine group will have to be considered.

Like many other (amorphous) polymers, the materials SAPA-A and -R show basically the same sequence of deformation mechanisms. However, the transition temperatures and the relative importance of the different mechanisms depend on molecular composition and influence significantly

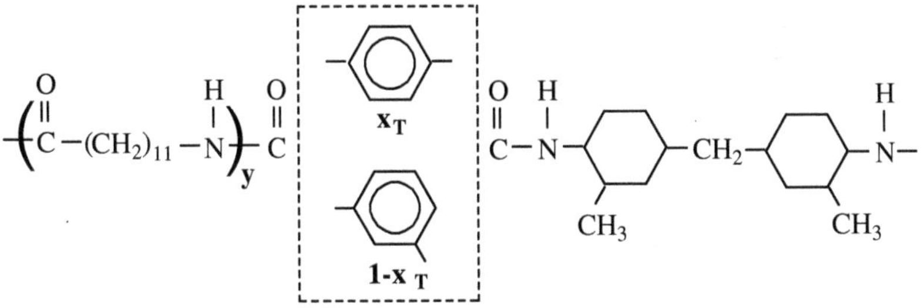

Figure 9 The chemical structure of *Semi-Aromatic PolyAmides* (SAPA-A). (See Table 1 for characterization and designation of the samples.)

TABLE 1 Molecular and Mechanical Characterization of *Semi-Aromatic PolyAmides* (SAPA-A)

Sample (designation)	y	x_T	M_w (g.mol^{-1})	T_α^* (°C)	ρ^* (kg.m^{-3})	M_e (g.mol^{-1})	$10^{-26} v_e$ (m^{-3})	K_{Ic}^* (MPa.m$^{1/2}$)
A-1.8I	1.8	0	22000	130	1042	2700	2.3	2.3
A-1.8T(23)	1.8	1	23000	137	1042	3000	2.1	2.45
A-1I(26)	1	0	26000	161	1055	2800	2.25	2.35
A-1T$_{0.7}$I$_{0.3}$(23)	1	0.7	23000	171	1057	3100	2.0	2.45
A-1T$_{0.7}$I$_{0.3}$(32)	1	0.7	32000	171	1057	3050	2.0	2.5

* T_α = temperature at maximum of dynamic loss modulus at 1 Hz; ρ = density at 25°C; K_{Ic} = critical stress intensity factor at 20°C.

Data from [23–25].

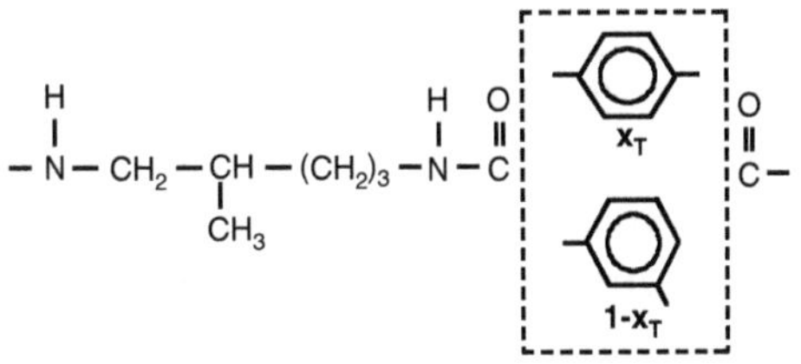

Figure 10 Chemical structure of SAPA-R. (See Table 2 for characterization and designation of used samples).

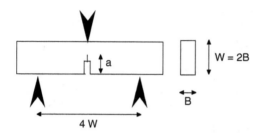

Figure 11 Three-point bending specimen for fracture toughness testing (a, crack length; B, specimen thickness; W, width).

the toughness of these materials, thus making them excellent candidates for a systematic study of the effect of molecular variables on ultimate mechanical properties [23–25].

VII. EFFECT OF MOLECULAR VARIABLES ON TOUGHNESS

A.   Toughness Testing

In their work Brulé et al. [24] have characterized the toughness of the different materials by the critical stress intensity factor K_{Ic}. Plain strain fracture tests were performed in mode I on standard three point bending samples (Figure 11), whose critical dimensions satisfy the criteria of the *ISO draft* [26].

For good reproducibility the machined notch had been sharpened by a pre-crack, introduced with the help of a specially designed falling weight apparatus each time using a fresh razor blade. Both the length of the pre-crack and the sharpness of the crack tip were checked for quality prior to testing by optical microscopy [23–25]. The samples were then loaded in an MTS 810 testing machine at the desired temperatures at a constant crosshead displacement rate of 1 mm/min. The K_{Ic} values were calculated from the expression:

$$K_{Ic} = f(a / W) \frac{P_{\max}}{BW^{1/2}} \qquad (8)$$

where f(a/W) is a standard correction function [26,27] and P_{max} the maximum recorded load. It is useful to remember that the tip of the pre-crack consists of a craze, whose stability and tendency to transform into a plastic zone determine the fracture behavior; in case of brittle fracture, P_{max} designates the load where unstable crack propagation begins; in case of ductile fracture, it is the load where the increase due to continued straining of the cracked sample and the decrease due to stress relaxation and crack growth are in equilibrium. The critical energy release rate, often simply called *fracture energy* G_{Ic}, is obtained from

$$G_{Ic} = \frac{U_i}{BW\phi} \tag{9}$$

where U_i is the area under the load-displacement curve integrated up to P_{max}, and Φ is another, tabulated correction function [26,27]. Representative data of toughness development as a function of temperature obtained for two semi-aromatic polyamides SAPA A-1$T_{0.7}I_{0.3}$ of different molecular weight are shown in Figure 12.

B. Effect of Chain Length and Molecular Composition

Attention is drawn to the fact that the fracture resistance of the low M_w material ((M$_w$/M$_e$ = 7.4) decreases at first slowly, then more rapidly. The high M_w material (with M_w/M_e = 10.5), however, shows three different regions of slightly decreasing (a), then increasing (b) and finally strongly decreasing (c) fracture energy. This behavior will be analyzed in the next section.

At low temperature (T < 0°C) the two SAPA A-1$T_{0.7}I_{0.3}$ materials break in a brittle manner by unstable extension of a craze, with craze growth and breakdown controlled by chain scission. Figure 12 reveals that in this temperature region there is no effect of molecular weight. It must be concluded that the overstressed chain segments of even the lower M_w material are so tightly anchored that they will rather break than slip. With increasing temperature, the stress for craze

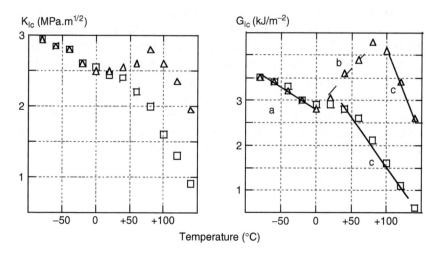

Figure 12 Toughness (K_{Ic}) and fracture energy G_{Ic} of two SAPA A-1T$_{0.7}$I $_{0.3}$ materials having a molecular weight of respectively 23 (□) and 32 kDa (r). Three different regions (a, b, c) are observed (see text for discussion). (Adapted from Brûlé B. Thèse de doctorat, Université Pierre et Marie Curie, Paris, 15 Sept. 1999; and Brulé B, Monnerie L, Halary JL. In: Blackman BRK, Pavan A, Williams JG. eds., Fracture of Polymers, Composites and Adhesives II, ESIS TC4, Amsterdam: Elsevier, 2003; p. 15–25.)

initiation by chain scission decreases slightly (region a) and with it P_{max} and K_{Ic} [24,10]. With further increasing temperature, chains become more mobile, and the transformation of matrix material into fibrillar matter in the process zone occurs at an ever-decreasing stress. As before, it is argued that an increasing number of chains escape chain scission. In the low M_w material the chains are too short, however, to stabilize the process zone and the fibrils will fail eventually, $K_{Ic}(T)$ decreases (region c). The increased mobility has a different effect in high M_w material, since the increasing number of chains that escape chain scission lead to longer and stronger fibrils, P_{max} and thus $K_{Ic}(T)$ increase within this region (b). At still higher temperatures, however, (region c, at $T_g - T < 70$ K) disentanglement becomes the dominant mechanism even for the longer chains and $K_{Ic}(T)$ decreases. The energy release

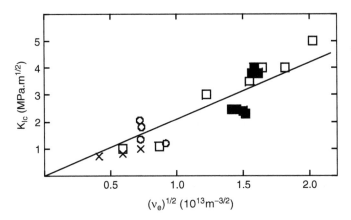

Figure 13 Compilation of room temperature toughness values as a function of entanglement density for different amorphous polymers. □ data from Wu [9], unpublished data from Halary on N-cyclohexylmalimide MMA copolymers (CMAL), ■ data from Brûlé et al. (Adapted from Brûlé B. Thèse de doctorat, Université Pierre et Marie Curie, Paris, Sept. 15 1999; and Brulé B, Monnerie L, Halary JL. In: Blackman BRK, Pavan A, Williams JG. eds., Fracture of Polymers, Composites and Adhesives II, ESIS TC4, Amsterdam: Elsevier 2003; p. 15–25. With permission.) (See text for discussion).

rate $G_{Ic}(T)$ shows a similar behavior although region b of the high M_w material is more pronounced (due to the fact that G_{Ic} ~$1/E(T)$).

Besides chain length, the intrinsic molecular variable of strongest influence on toughness is entanglement density. The available $K_{Ic}(v_e)$ data compiled in Figure 13 show that on the average the data correspond to the good correlation between K_{Ic} and v_e first proposed by Wu [9]. However, the three families studied, MGLUT, and SAPA-A and -R, show systematic deviations, which can be ascribed to particular structural characteristics of the molecules involved.

Within the MGLUT series K_{Ic} increases with increasing glutarimide content although the entanglement density remains constant. Tézé [21] and Tordjeman et al. [28] have related this increase of K_{Ic} to a strengthening of the cooperative character of the β-relaxation motion with increasing glu-

TABLE 2   Molecular and Mechanical Characterization of *Semi-Aromatic PolyAmides* (SAPA-R)

Sample (designation)	x_T	M_w (g.mol^{-1})	T_α (°C)	r (kg.m^{-3})	M_e (g.mol^{-1})	$10^{-26} v_e$ (m^{-3})	K_{Ic} (MPa.m$^{1/2}$)
R-I	0	18000	141	1194	2750	2.6	3.8
R-T$_{0.5}$I$_{0.5}$	0.5	23000	145	1196	2900	2.5	4.0
R-T$_{0.7}$I$_{0.3}$	0.7	22000	147	1196	2950	2.45	3.8

Data from [23–25].

tarimide content. The improved cooperativity leads to an increase of the β-relaxation peak and — as we have seen for thin films (Figure 8) — to a lowering of the transition temperature from scission crazing to formation of diffuse deformation zones. In bulk samples at room temperature, a similar shift from unstable to stable fracture occurs with increasing glutarimide content, which is accompanied by a noticeable increase of toughness [21].

The series SAPA-A and -R are distinguished by their much higher entanglement density as compared to MGLUT; as is to be expected, R is tougher than A. Within these series, however, there is little (or even negative) variation of K_{Ic} with v_e. Looking at the molecular structure (Table 2), one notes that it is systematically the terephthalic chains that have the higher K_{Ic} despite their lower entanglement density. From dynamic mechanical analysis (DMA) one knows that the terephthalic samples have a much stronger β-relaxation peak. As shown by Beaume et al. by ^{13}C-solid state NMR [29] the cooperative character of the β-relaxation motion is strengthened by the capacity of the para-disubstituted phenyl rings to execute π-flip motions. For evident steric and energetic reasons, the isophthalic samples cannot undergo such π-flip motions.

The fact that a cooperative interaction between the molecular motions of neighboring groups within a chain intensifies the β-relaxation peak has been shown for many polymers [1,21,23,24,25,30]. Plummer et al. have studied bisphenol-A polycarbonate derivatives composed of different alternating blocks of varying lengths [31]. From the obtained DMA spectra they conclude that the β-relaxation involves an in-chain cooperative motion extending across 6-9 repeat units and that this motion is also influential in activating the disentanglement crazing at elevated temperatures [31]. It is this capacity of the chain backbone to rapidly relax axial stresses, which also is responsible for the positive correlation between the position of the maximum of the β-peak and toughness, which exists for the majority of amorphous or semicrystalline polymers [32].

VIII. CONCLUSIONS

Structural and dynamic analysis as well as fracture mechanical methods applied to systematically chemically modified (glassy) polymers permit identification and explanation of the effect of the principal intrinsic variables' configuration, chain length and entanglement density. The competition between the elementary deformation mechanisms, chain scission, segmental slip and disentanglement determines the mode of fracture and the toughness of amorphous polymers. The dominant mode changes with temperature and is strongly influenced by the intrinsic variables. For long chains ($M_w > 9\ M_e$) toughness depends most strongly on entanglement density and the intensity of sub-T_g relaxations, with in-chain cooperative motions playing an important role. The present studies on the internal toughening parameters of a polymer have a particular significance in view of the recent investigations of Grein [33]. This author has shown (for rubber-toughened polypropylene) that particle modification facilitates craze and crack *initiation*, but the essential energy dispersion occurs during *propagation*, which requires the relaxation of the triaxial state of stress by local *matrix* deformation mechanisms.

ACKNOWLEDGMENTS

The authors are indebted to their colleagues and collaborators Ph. Béguelin, B. Brulé, R. Gensler, L. Monnerie, C.J.G. Plummer and L. Tézé, for fruitful discussions and to ATOFINA and

RHODIA for having provided samples and for their continued interest.

REFERENCES

1. Haward RN. ed. The Physics of Glassy Polymers. London: Appl. Science Publ. Ltd. 1973.

2. Kausch HH. Polymer Fracture. 2nd ed. Heidelberg-Berlin: Springer 1987.

3. Michler GH. Kunststoff-Mikromechanik. Carl Hanser München-Wien 1992.

4. Kausch HH, Heymans N, Plummer CJ, Decroly P, Matériaux Polymères: Propriétés Mécaniques et Physiques, Principes de Mise en Oeuvre. Lausanne: Presses Polytechniques et Universitaires Romandes 2001.

5. Béguelin P. Approche expérimentale du comportement mécanique des polymères en sollicitation rapide. Thèse No 1572, Ecole Polytechnique Fédérale de Lausanne, Lausanne Suisse 1996.

6. Fond C, Lobbrecht A, Schirrer R. Polymers toughened with rubber microspheres: An analytical solution for stresses and strains in the rubber particles at equilibrium and rupture. Int. J. of Fracture 1996; 77: 141–159.

7. Kausch HH. ed. Crazing in Polymers. Adv. in Polymer Sci. Vol 52/53 Heidelberg-Berlin: Springer 1983.

8. Kausch HH. ed. Crazing in Polymers. Adv. in Polymer Sci. Vol 91/92 Heidelberg-Berlin: Springer 1990.

9. Wu S. Control of intrinsic brittleness and toughness of polymers and blends by chemical structure: A review. Polym. Int. 1992; 29: 229–247.

10. Plummer, CJG. Crazes and deformation zones in thermoplastic polymers. Curr. Trends in Polym. Sci. 1997; 2: 125–155.

11. Kausch HH, Gensler R, Grein C, Plummer CJG, Scaramuzzino P. Crazes in semicrystalline thermoplastics. J. Macromol. Sci. 1999; B38: 803–815.

12. Sue H-J, Yang PC, Puckett PM, Bertram JL, Garcia-Meitin EI. Crazing and dilatation band formation in engineering thermosets. In: Toughened Plastics II, Riew CK and Kinloch AJ, eds., Adv. Chem. Ser. 1996; 252: 161–175.

13. Lu Fan A. Mechanical properties and toughening mechanisms in epoxy systems. Thèse No 1391, Ecole Polytechnique Fédérale de Lausanne, Lausanne Suisse 1995.

14. Kausch HH. Polymer Fracture. 2nd ed. Heidelberg-Berlin: Springer 1987, p. 347.

15. Kramer EJ. Microscopic and molecular fundamentals of crazing. In: Kausch HH. ed Crazing in Polymers. Adv. in Polymer Sci. Vol 52/53 Heidelberg-Berlin: Springer 1983, pp. 1–56.

16. Brown HR. A molecular interpretation of the toughness of glassy polymers. Macromolecules 1991; 24: 2752–2756.

17. Kausch HH. Polymer Fracture. 2nd ed. Heidelberg-Berlin: Springer 1987, p. 133.

18. Jud K, Kausch HH, Williams JG. Fracture mechanics studies of crack healing and welding of polymers. J. Mater. Sci. 1981; 16: 204–210.

19. Gensler R. The effect of thermooxidative degradation on the mechanical performance and the microstructure of polypropylene. Thèse No 1863, Ecole Polytechnique Fédérale de Lausanne, Lausanne Suisse 1998.

20. Kambour RP. Correlations of dry crazing resistance of glassy polymers with other pysical properties. Polym. Comm. 1983; 24: 292–296.

21. Tézé L. Relation entre les mouvements moléculaires, les micromécanismes de déformation et la fracture dans des copoolymères statistiques à base de polymetacrylate de méthyle. Thèse de doctorat, Université Pierre et Marie Curie, Paris, Nov. 10, 1995.

22. Plummer CJG, Kausch HH, Tézé L, Halary JL, Monnerie L. Microdeformation mechanisms in methyl mathacrylate-glutarimide random copolymers. Polymer 1996; 37: 4299–4305.

23. Brûlé B. Relation entre caratéristiques moléculaires et propriétés mécaniques de polyamides amorphes et de polycarbonates. Thèse de doctorat, Université Pierre et Marie Curie, Paris, Sept. 15 1999.

24. Brulé B, Monnerie L, Halary JL. Analysis of the fracture behavior of amorphous semi-aromatic polyamides. In: Blackman BRK, Pavan A, Williams JG. eds., Fracture of Polymers, Composites and Adhesives II, ESIS TC4, Amsterdam: Elsevier 2003, pp. 15–25.

25. Brûlé B, Kausch HH, Monnerie L, Plummer CJG, Halary JL. Polymer 2003; 44: 1181-1192.

26. ISO 13586-2 (1998). Determination of fracture toughness (G_{Ic} and K_{Ic}) for plastics. A LEFM approach.

27. Grellmann W and Seidler S. Deformation and fracture behaviour of polymers. Berlin-Heidelberg: Springer 2001.

28. Tordjeman P, Tézé L, Halary JL, Monnerie L. On the plastic and viscoelastic behavior of methylmethacrylate-based random copolymers. Polym. Eng. Sci. 1997; 37: 1621.

29. Beaume F, Brûlé B, Halary JL, Lauprêtre F, Monnerie L. Secondary transitions of aryl-aliphatic polyamides. IV. Dynamic mechanical analysis. Polymer 2000; 41: 5451–5459.

30. Halary JL and Monnerie L. Int. Conf. on Deformation, Yield and Fracture, Cambridge, April 2003.

31. Plummer CJG, Soles CL, Xiao C, Wu J, Kausch HH, Yee AF. Effect of limiting chain mobility on the yielding and crazing behavior of bisphenol-A-polycarbonate derivatives. Macromolecules 1995; 28: 7157–7164.

32. Ramsteiner F. Zur Schlagzähigkeit von Thermoplasten. Kunststoffe 1983; 73: 148–153.

33. Grein C. Relation entre la structure et les propriétés mécaniques de polypropylène modifié choc. Thèse de doctorat 2341, Ecole Polytechnique Fédérale de Lausanne (EPFL), Lausanne 2001.

5

Strength and Toughness of Crystalline Polymer Systems

ANDRZEJ GALESKI

Center of Molecular and Macromolecular
Studies, Polish Academy of Sciences

CONTENTS

I.   INTRODUCTION

The definitions of strength and toughness are well established in the mechanics of materials. The basic dependencies for the strength and toughness of polymers were discussed by Young [1] in an older review. Crystalline polymers, which are the object of this report, are complicated systems, with an amorphous phase interlaying with crystalline lamellae, and with most of the macromolecular chains engaged in both phases. The strength and toughness of polymer crystalline systems are interdependent, due to several affecting phenomena such as crystal plasticity, cavitation and molecular orientation.

High toughness, i.e., the ability of a polymer to exhibit large plastic deformability and high resistance to an impact without failure, is the most desired property of a material or product. However, the toughness is not a unique material property because it is influenced by the type of load (shear, tension, compression, bending, twisting, tearing), shape of an item, scratches, notches etc., aside from experimental conditions such as temperature, pressure, load rate and the material's properties: molecular weight, polydispersity, packing, chain entanglements, crystallinity, heterogeneity and several other parameters.

The major contribution to toughness derives from plastic deformation of the material which is manifested by the ductile behavior. Plastic deformation itself is a composed phenomenon: it concerns the crystalline as well as amorphous phases. The ductility is expressed by decreasing the stress-strain curve at a *yield stress*. Yield can be caused either by *multiple crazes* or by *shear yielding*. In the first case, the crazes have to be initiated in a relatively large volume of the material in order to contribute significantly to the overall deformation. Shear yielding is the plastic flow without crazing. Crazing, a unique phenomena occurring in polymers usually below the glass transition temperature, are highly localized zones of plastic dilatational deformation [e.g., 2,3,4,5]. Edges of crazes are spanned by highly drawn elongated fibrils called *tufts,* usually having the length of a fraction of 1 μm, depending on the molecular weight of a polymer, the diameter of several nanometers and confined to a small volume of the material. The tufts can carry the load applied to the material and preserve the integrity of a craze. In brittle materials, crazes are initiated at surfaces while the brittle fracture originates from breaking tufts of crazes initiated from the surface. Crazing occurs mostly in amorphous polymers although it has been also observed in crystalline polymers in which crazes are propagated between lamellae through spherulite centers as well as through the material between spherulites [6,7]. Localized crazes initiate, propagate and break down at the stress below that necessary to stimulate shear yielding

Shear yielding can be observed in a wide range of temperatures but only if the critical shear stress for yielding is lower than the stress required to initiate and propagate crazes. Ductile deformation requires an adequate flexibility of polymer chain segments in order to ensure plastic flow on the molecular level. It is long known that the macromolecular chain mobility is a crucial factor deciding on either brittle or ductile behavior of a polymer [8,9,10]. An increase in the yield stress of a polymer with a decreasing temperature is caused by the decrease of macromolecular chain mobility, and vice versa, the yield stress can serve as a qualitative measure of macromolecular chain mobility. The temperature and strain rate dependencies of the yield stress are described in terms of relaxation processes, similarly as in linear viscoelasticity. Also the kinetic elements taking part in yielding and in the viscoelastic response of a polymer are similar: segments of chains, parts of crystallites, fragments of amorphous phase. However, in crystalline polymers above the glass transition temperature the yield stress is determined by the yield stress required for crystal deformation and not by the amorphous phase. The behavior of crystals differs from that of the amorphous phase because the motion of macromolecular chains within the crystals is subjected to severe constraints, making the displacement of neighboring chains much more difficult. Since the mobility of kinetics elements taking part in a plastic deformation is lower at a lower temperature, the energy dissipated increases and produces instabilities; at those places micronecks are formed because locally the temperature increases. The rate of plastic deformation increases drastically in micronecks, and the material may quickly fracture. At a higher temperature the mobility of kinetics elements is higher, so less energy is dissipated and the local temperature increase is lower. The neck is then stable, tends to occupy the whole gauge length of the sample and the material exhibits a tough behavior.

The necessary condition for high plastic deformation is the possibility of motions of kinetic elements in a time scale as it follows from the deformation rate. The relaxation times and the activation energies are the parameters describing the kinetics of the conformation motions of fragments of macromolecules taking part in the deformation. In crystalline polymers there are essentially three processes which are of particular importance: the first process is connected with the presence of the crystalline phase, the second corresponds to relaxations related to defects in the crystalline phase and the third relates to motions of short segments in the amorphous phase related to the glass transition [9].

Both massive crazing and shear yielding dissipate energy; however, shear yielding, dissipating the energy more efficiently [9], is often favored over crazing, especially under uniaxial stress, elevated temperature or slow deformation. Switching between crazing and shear yielding is not obvious as it depends also on additional factors such as shape of an article and the presence of notches or scratches. The material will deform according to the most ductile mechanism which is well explained by the Ludwig-Davidenkov-Orovan hypothesis [e.g., 11,12,13] demonstrated in Figure 1. The deforma-

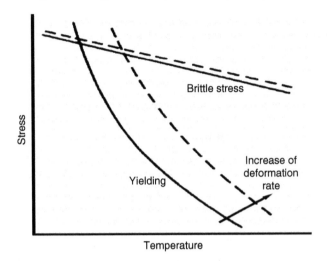

Temperature

Figure 1 Ludwig-Davidenkov-Orovan plot explaining the temperature dependence of ductility and brittleness of a material. Redrawn after Ward and Hadley.

tion of polymeric materials starts usually at scratches, notches or internal defects because these places are the zones of local stress concentration, sometimes much above the applied stress. The toughening concept of polymeric materials is based on the activation of such plastic deformation mechanisms which are triggered at a stress lower than that required for the operation of surface and internal defects. Consequently, one of the important means of toughening appears to be a significant lowering of the yield stress of a material [14]. Many brittle polymers are toughened by incorporating rubbery particles [e.g., 3]. Depending on the character of a polymer, temperature, deformation rate and deformation mode, the toughening occurs either via multiple crazing like in HIPS and ABS or by preferred shear yielding like in polypropylene and polyamide 6 modified by dispersed elastomer particles.

Strength of unoriented crystalline polymers is mainly related to the yield phenomenon and crystal plasticity. Different means of load application activate different material responses, including massive cavitation and plastic deformation of the crystalline and amorphous phases.

A tremendous increase of tensile strength of crystalline polymers was obtained in the past by imposing molecular orientation and taking advantage of carbon-carbon bonds strength. In the description below, the strength, toughness and their interrelation will be addressed with a particular emphasis on cavitation, crystal plasticity and plastic deformation of crystalline and amorphous phases. Plastic deformation and plastic flow lead to a high molecular orientation of both phases and to an increase in a material's strength and toughness.

There are two main modes of behavior of solidified polymer under load: a *brittle,* represented by a short, nearly linear dependence of stress-strain followed by a fracture and a *ductile* characterized by plastic yielding and plastic flow. Brittle behavior is usually the result of highly concentrated crazing [e.g., 2,3,4]. Crazing in amorphous polymers is the phenomenon involving a massive production of voids.

The formation of voids is also a common phenomenon in crystalline polymers, in which quite often a high negative pressure is generated due to local stress concentration. The main reason of cavitation in crystalline polymers is the misfit between mechanical compliances of heterogeneous elements and the surroundings producing an excessive negative pressure. The cavitation appears to be another mechanism of tough response of the material [e.g., 4,15]. The cavitation dissipates a not very large amount of energy, though it enables the surrounding material to undergo further intensive crazing or shear yielding.

Crystalline polymers show relatively high toughness as compared to amorphous glassy polymers. In particular, crystalline polymers exhibit high toughness at a temperature well above T_g.

II. CAVITATION DURING PLASTIC DEFORMATION OF POLYMERS

It is often observed that the plastic tensile deformation of crystalline polymers causes a significant amount of cavitation.

One of the signs of cavitation is a sudden polymer whitening near the yield point. A density change is associated with cavitation: when a possible crystallinity change is accounted for, the reminder is the change due to the formation of pores. The pores usually occupy up to several vol% of the material.

In polyamide 6 the cavities are formed in bulk during tensile plastic deformation in a form of rods with sizes reflecting the thickness (2–3 nm) and the width (20–80 nm) of polyamide interlamellar material as seen in Figures 2a, 2b and 2c [16]. In those experiments the samples were subjected to plastic deformation in tension and still under stress immediately cooled down below their glass transition temperatures, followed by osmium tetroxide fixation. The cavities, which were formed, caused chain scission preferentially in places with mechanical mismatch of adjacent stacks of lamellae. In those places OsO_4 was chemically bonded to radicals formed during deformation, cross-linking the material and producing a strong electron density contrast.

Formation of radicals during plastic deformation was first demonstrated by Peterlin with electron spin resonance in the early 1970s [17]. The stained spots are concentrated in equatorial zones of spherulites, near the spherulite borders and close to contact points of three spherulites. Since the deformation of spherulites is non-uniform, deformation misfit between regions of different effective plastic resistance readily develops inside spherulites during deformation. This gives rise to local stress concentrations over the regions of plastic inhomogeneities, i.e., lamellae, packets of lamellae. Other intriguing features of stained spots are their small sizes and their orientation along radii of spherulites. The dark spots seen in sediments centrifuged from the dissolved deformed OsO_4 fixed samples have sizes of 20–80 nm in length and they split into two to four closely aligned thin rods (see Figure 2c). The overall size of the dark spots and the number of rods in a single spot suggest their close relation to the domains of parallel packed lamellae of undeformed material, in particular to the thickness of domains (40–60 nm) and the number of lamellae (two to four) in a single domain (see [18]). Since the domains are organized into a layered structure with alter-

(a)

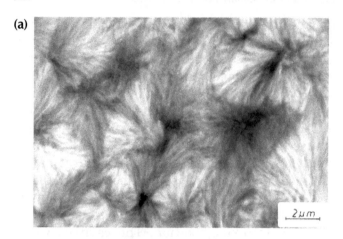

(b)

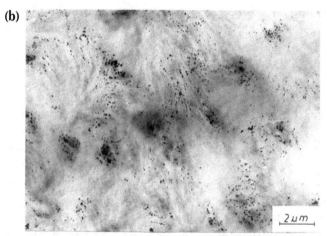

Figure 2 Electron micrographs of ultra thin sections of: a) undeformed compression molded sample of polyamide 6; b) transverse
section of a deformed polyamide compression-molded sample, extension ratio 2 (samples in Figures 2a–b were infiltrated with OsO_4,
sectioned and stained with phosphotungstic acid); and c) electron
micrograph of deposited sediment from dissolved polyamide 6 sample which had been deformed and infiltrated with OsO_4. (Reprinted
with permission from Galeski, A., Argon, A.S., Cohen, R.E. Macromolecules. 1989, 21, 2761–2770. Copyright 1989, American Chemi-
cal Society.)   Continued.

(c)

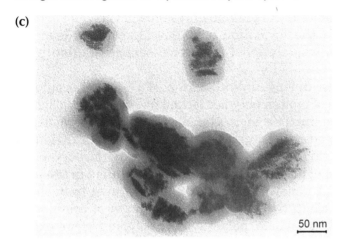

50 nm

Figure 2   Continued.

nating crystalline and amorphous phases, they are mechanically anisotropic in the directions parallel and perpendicular to the lamellar surface. Hence, the strain suffered by each domain depends on its orientation with respect to applied stress and orientation of neighboring packets of lamellae. Local cavitation and/or chemical damage may occur if misfit related stresses exceed the cohesive strength of the material. The dark rods, as seen in Figure 2c, correspond to disrupted interlamellar amorphous layers in a domain.

Cavitation associated with chain scission was found to be a massive phenomenon in several other bulk polymers: polyamide 66, RIM nylon, poly(methylene oxide) [19]. Haudin, G'Sell et al. [20] observed by light microscope the formation of groups of cavities during drawing of polypropylene film. The groups of pores were of sizes ranging up to a few microns and most of them were generated in the yield zone on both sides of the neck region. Others have postulated that pores are produced, on the basis of evidence obtained in light microscopy [21,22] and diffusion experiments [23]. Schaper et al. [22] pointed out that while the density decreases and voiding appears in the early stages of deformation, around the yield point, later the void structure disappears, leaving no direct evidence of its previous presence in the deformation process. Often the formation of nanoscale cavities leads to crazing which was observed in several semicrystalline polymers [24].

Cavitation during crystallization of polymers occurs in micro-pockets of molten polymer locked by surrounding spherulites because further melt transformation to more dense crystals proceeds without fresh melt supply. The negative pressure developed leads to cavitation. Cavitation during crystallization of polymers was extensively studied in the past [25,26,27,28]. A general conclusion can be drawn from these studies that, cavitation in amorphous phases of commodity polymers (polypropylene, polyethylene, polymethylene oxide, etc.) above their glass transition temperatures requires the negative pressure at the level of −5 to −20 MPa, unlike unpurified low molecular weight liquids cavitating without much difficulty. The reason of high absolute value of negative pressure for cavitation in polymers is most probably the macromolecular chain entanglement. It appears then that the increase of the molecular weight of a polymer will most probably impede the cavitation by increasing the cohesive strength of the amorphous phase. However, this hypothesis has never been checked experimentally.

It is reasonable to assume that the physics of cavitation occurring in the amorphous phase above glass transition temperature during deformation of crystalline polymers and cavitation during crystallization requires the negative pressure at a similar level of −5 to −20 MPa.

The immediate conclusion is that the cavitation during deformation can be observed only in those polymers in which the yield stress for crystals deformation is higher than the stress required for cavitation. Otherwise the crystals will deform earlier relaxing the stress and a cavitational pore will not appear. This necessary condition can be written in the form:

$$\sigma_{(hkl)[001]} > p_{cav} (1 - \nu)/3\nu \tag{1}$$

where $\sigma_{(hkl)[001]}$ is the critical resolved shear stress for the easiest slip in crystals (usually chain slip), p_{cav} is the negative pressure required for cavitation and v is the Poisson ratio of the amorphous phase which is around 0.5. From the above condition it follows that cavitation during deformation can be expected in such polymers as nylons (the easiest slip for nylon 6 is 16.24 MPa [29]), polypropylene ((010)[001] slip at around 22–25 MPa [30,31]), poly(methylene oxide) [19]. In polyethylene the cavitation can occur only if thick crystals are present (see the discussion of yield stress of polyethylene as a function of crystal thickness below in Section IV). No cavitation is expected during deformation of low-density polyethylene and quenched high-density polyethylene, both having usually thin lamellar crystals, but cavitation can be found in HDPE which is slowly cooled. Again the visible effect of cavitation of the material during deformation is its whitening.

Cavitational voids are having sizes on the nanoscale level [16], hence, there is a problem with their stability. Surface tension is exerted on each pore with a tendency to close the pore. In order to preserve a pore, a three-dimensional (3D) tensile stress, σ_{3D}, is required (which is a negative pressure, p) at the level reciprocally proportional to the radius of a pore:

$$\sigma_{3D} = p > 2\tau/r \qquad (2)$$

where τ is the surface tension and r is the size of the pore. It follows then that the smallest pores are healed readily while larger pores can be preserved only if an adequate negative stress is maintained. In order to detect pores, one should preserve them by facilitating the stress or by cooling the material quickly below its glass temperature.

The internal cavitation observed in tension experiments has been referred to as "micro necking" by Peterlin (see, e.g., [32]). Such micro necking had been considered for a long time to be essential for large-strain deformation of polymers with chainfolded crystals. It was supposed that micro necking removes kinematical constraints between lamellae and allows them to untangle. In other papers of Peterlin [e.g., 33], in which the micro necking model is defined and developed, cavitation is not consider explicitly. However, from the picture of microfibrillar structure formation in crystalline samples as

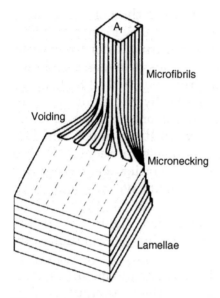

Figure 3 Imaginative picture of a cavitating semicrystalline polymer under tensile deformation. (Redrawn after Peterlin, A. J. Mater. Sci. 1972, B6(4), 583–598.)

imagined by Peterlin (in References [32,33]), it follows that the cavitation is an essential feature of the drawing of crystalline polymers. The imaginative drawing in Figure 3 presents a semicrystalline polymer during tensile deformation.

III. DEFORMATION MECHANISMS IN CRYSTALLINE POLYMERS

It is presently thought that the material consists of lamellar crystals which are separated from each other by a layer of amorphous polymer and are held together by tie molecules through the amorphous phase [e.g., 34]. There are three, currently recognized, principal modes of deformation of the amorphous material in semicrystalline polymers: interlamellar slip, interlamellar separation and lamellae stack rotation [35,36]. Interlamellar slip involves shear of the lamellae parallel to each other with the amorphous phase undergoing shear. It is a relatively easy mechanism of deformation for the material above T_g. The elastic part of the deformation can be almost entirely attributed to the reversible interlamellar slip. Interlamellar separation is induced by a component of tension or compression perpendicular to the lamellar surface. This type of deformation is difficult since a change in the lamellae separation should be accompanied by a transverse contraction and the deformation must involve a change in volume. Hard elastic fibers are found to deform in such a way. When the lamellae are arranged in the form of stacks embedded in the amorphous matrix, the stacks are free to rotate under the stress. Any other deformation of the amorphous phase requires a change in the crystalline lamellae; the amorphous material is then carried along with the deforming crystalline material.

Taking into consideration the yield behavior of semicrystalline polymers, there are two conflicting approaches concerning the crystals. The first presumes that the process of deformation is composed of a simultaneous melting and recrystallization of polymer under adiabatic conditions [37,38]. The second, developed mainly by Young [e.g., 39,40], uses the idea derived from the classical theory of crystal plasticity. The deformation of polymer crystals is considered in terms of dislocation motion within the crystalline lamellae, similarly to slip processes observed in metals, ceramics and low molecular crystals. Crystallographic slips are not processes occurring simultaneously over the whole crystallographic plane. A great role of line and screw dislocations is played in activation and propagation of a slip. The plastic deformation of polymer crystals, like the plastic deformation of crystals of other materials, is generally expected to be crystallographic in nature and to take place without destroying the crystalline order. The only exception to this is a very large deformation, when cavitation and voiding lead to unravelling the folded chains and completely break down the crystals, new crystals may form with no specific crystallographic relationship with the original structure [41]. Polymer crystals can deform plastically by crystallographic slip, by twinning and by martensitic transformation. The slip mechanism is the most important one since it can produce

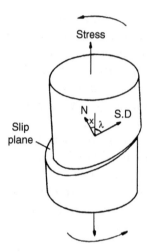

Figure 4 Definition of a slip system: slip plane and slip direction.

larger plastic strains than the other two mechanisms. A slip system in a crystal is the combination of a slip direction and a slip plane containing that direction as shown in Figure 4. The notation for the slip system is (hkl)[$h_1k_1l_1$] where (hkl) is the slip plane while [$h_1k_1l_1$] is the slip direction. A single slip system is only capable of producing a simple shear deformation of a crystal. A general change of a shape of a crystal requires the existence of five independent slip systems [42]. Polymer crystals rarely possess this number of independent slip systems. However, under the right conditions the deformation of bulk material can occur without voiding or cracking perhaps because amorphous regions between lamellae allow for a certain amount of adjustment. In polymer crystals, the slip plane is restricted to planes which contain chain direction. That is because covalent bonds remain unbroken during deformation.

In polymers two types of slip can occur: chain slip, i.e., slip along the chains and transverse slip, i.e., the slip perpendicular to the chains, both slips occurring in planes containing the chains. The general rule that applies to slip deformation is that the plane of the easiest slip tends to be a close-packed plane in the structure and the slip direction is a close-packed direction. Hence, in crystalline material, it is possible to predict certain mechanical properties associated with crystallographic slip directly from the crystallographic unit cell [43]. In folded-chain polymer crystals, the folds at the surface of crystals may in addition impose some restraint on the choice of a slip plane; usually a slip will be able to occur only parallel to the fold plane. An implication of the geometry of the slip process is that a crystal undergoing a single slip will rotate relative to the stress axis as it is seen in Figure 5. For a single slip the slip direction in the crystal rotates always toward the direction of maximum extension: in uniaxial tension it rotates toward the tensile axis while in uniaxial compression away from the compression axis. The angle through which the crystal rotates is a simple function of the applied strain [42]. It must be mentioned that the slip takes place when the resolved shear stress on the slip plane reaches a critical value known as a critical resolved shear stress. The critical resolved shear stresses for slips are now well known only for a few polymers. They were measured using samples of rather well defined texture. First measurements performed on polyethylene sample with fiber texture subjected to annealing at high pressure for increasing the crystal thickness yielded the critical shear stress of 11.2 MPa for undisclosed crystal thickness [35]. However, with fiber symmetry a combination of easiest slip systems could act simultaneously disturbing the true value of the critical shear stress. The most exact data concerning critical resolved shear stresses for possible slip systems in polyethylene were obtained by Bartczak et al. [44]. They used single crystal textured polyethylene obtained by plane strain compression. The measurements in uniaxial tension, uniaxial compression and simple shear for samples cut out at various orientations delivered the following: for polyethylene orthorhombic crystals [44] the most active slip system is (100)[001] chain slip at 7.2 MPa, the second is (100)[010] transverse slip at 12.2 MPa, and the third is (010)[001] chain slip at 15.6 MPa. The forth slip system for polyethylene orthorhombic crystals was predicted as (110)[001] with the estimated critical resolved shear stress greater than 13.0 MPa [44], however, it was never observed separately.

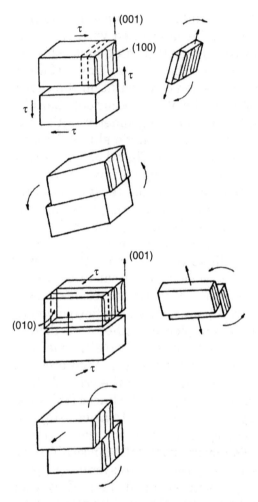

Figure 5 The rotation of crystal fragments due to slip: marked by arrows. The resolving of the shear on a plane due to simple tension or compression is also illustrated.

For polyamide 6 α crystals the slip systems are: (001)[010] chain slip at 16.24 MPa, (100)[010] chain slip at 23.23 MPa and (001)[100] transverse slip [29].

Relatively little attention was paid to the plastic deformation of other semicrystalline polymers [45,46]. In particular, there are only a few papers [47,48] describing the investigations of the yield behavior and plastic resistance of oriented iPP.

For the determination of critical shear stress of one of the most important deformation mechanisms of iPP crystals, namely the crystallographic slip in the (100) planes along chain direction, i.e., (100)[001] chain slip [49,50], the biaxially oriented film was used [51].

According to the theoretical predictions [49] and experimental studies [50], the easiest slip system in iPP crystals is (010)[001] slip, while (100)[001] and (110)[001] systems have higher critical resolved shear stress. The studies of the above-mentioned slip system can be made by the investigation of the yield behavior of the specimens deformed in tension with the tensile axis oriented in those specimens at various angles to the orientation direction. One can expect that for a certain range of this angle only the (100)[001] slip system will be activated due to proper orientation of crystallites providing high shear stress on the (100) plane in [001] direction, while other deformation mechanisms will remain inactive due to much smaller resolved shear stresses in appropriate directions [44,52]. The analysis of the yield stress of such samples would give the value of the shear stress necessary to activate the (100)[001] slip [45,53,44,36]. The critical resolved shear stress for (100)[001] slip for α crystals of polypropylene was determined at the level of 22.6 MPa. Similar studies for oriented iPP in tension and compression were performed by Shinozaki and Groves [47], but they used the samples of uniaxially oriented iPP with a fiber symmetry, so that the critical shear stress they determined, 25MPa, was an average over slips in plane oriented around the fiber axis.

Crystal orientation by channel die compression of PET was studied by Bellaire et al. [54]. They have found that the macromolecular chains' orientation in PET along the flow direction and the texture development are the results of possible crystallographic slips having the following glide planes and directions: (100)[001] chain slip and (100)[010] transverse slip and (010)[001] chain slip. The probable sequence of activities of these slips is the following: the (100)[001] chain slip being the easiest, (100)[010] slip and a sluggish (010)[001] chain slip [54]. Values for critical resolved shear stress for those slips were not determined.

The stress-induced martensitic transformation is a transformation from one crystallographic form to another form and associated by a displacement of chains to new positions in the new crystallographic cell in order to accommodate the deformation. An example of martensitic transformation from orthorhombic to monoclinic form was found in oriented polyethylene with well-defined texture subjected to uniaxial compression. The martensitic transformation was also found in other polymers: in poly(L-lactic acid) [55] and in nylon 6 with the α form transforming to the γ form [56].

Twinning may occur in crystals of sufficiently low symmetry: cubic symmetry excludes twinning while orthogonal symmetry allows for twinning. Hexagonal crystal structure allows for twinning of low molecular weight materials while in polymer crystals of hexagonal symmetry the basic twinning plane would be perpendicular to chains and therefore forbidden. Twinning along other planes in hexagonal crystals is not possible because of their high symmetry. In polyethylene of orthorhombic crystal symmetry, the twinning is expected along (110) and (310) planes [57]. Only (110) plane twinning was found in bulk polyethylene. Twinning along (310) plane is blocked in bulk polyethylene because the fold plane is the (110) plane. In contrast, in rolling in a channel at a high rate and to a high compression ratio, the texture of HDPE sample consist of two components [58]. One of them is the (100)[001] component, while the two others are rotated by ±53° around the rolling direction coinciding with the position of (310) poles clearly indicating the {310} twinning of the basic (100)[001] component. The twinning occurs on unloading, when the sample leaves the deformation zone between the rolls. The partial recovery of the strain produces a tensile stress along direction of loading. Twinning is activated at high strain rates because the sample does not have sufficient time for stress relaxation, while at high compression ratio the material is highly oriented and contains no more folds in (110) planes. It was estimated earlier that the critical resolved shear stress in the twin plane to activate twinning is around 14 MPa [59]. Therefore the tensile stress generated along LD on unloading must be at least 28 MPa. Such stress is apparently generated on unloading only when high deformation rate is applied during deformation.

Besides polyethylene the twinning was found only in a few polymers, including isotactic polypropylene: twinning along (110) plane [50].

Stress-induced martensitic transformation and twinning alone are not responsible for large strain deformation.

From the presented review of mechanisms of plastic deformation of amorphous and crystalline phases, it follows that the easiest is the deformation of the amorphous phase since it requires very little stress; the crystallographic mechanisms of plastic deformation need larger stresses. Therefore it is expected that first the amorphous phase is deformed and then crystallographic mechanisms are activated. An illustrative experimental evidence of this prediction is presented in Figure 6 where the Hermans orientation parameters for the amorphous and crystalline phases of a series of tensile deformed isotactic polypropylene samples are mapped out [60]. The data points are based on the measurements of birefringence, infrared dichroism and wide angle x-ray diffraction. It is seen from this figure that at first the amorphous phase becomes oriented while the crystalline phase remains very slightly oriented or even oriented in the transverse direction (the reason for such behavior of polypropylene is the presence of cross-hatched lamellae). When the draw ratio is further increased, the amorphous phase becomes almost entirely stretched out and then the crystalline phase begins to orient. Finally, at high draw ratio the macromolecular chain fragments embedded in both phases become highly oriented.

It may be concluded that most of plastic deformation of both crystalline and amorphous phases occurs due to shear stresses, and the shear contributes greatly to plastic deformation. It is then obvious that the great amount of plastic deformation is usually found at an acute angle with respect to applied tensile or compressive forces.

There are several ways of achieving plastic deformation of macroscopic samples. Those deformation methods in which

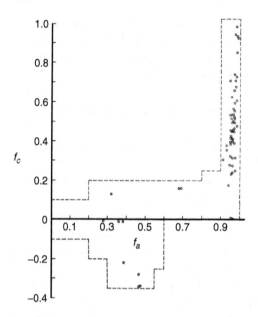

Figure 6   Orientation parameters of the amorphous, f_a, and crystalline, f_c, phases for a series of isotactic polypropylene samples uniaxially deformed to a different degree of orientation. Data on the orientation parameters are obtained by the measurements of birefringence, infrared dichroism and wide angle x-ray diffraction. The region surrounded by the dotted lines represents briefly the course of deformation starting from f_a and f_c equal 0. (From Kryszewski, M., Pakula, T., Galeski, A., Milczarek, Pluta, M. Faserforschung und Textiltech 1978, 29, 76–85. With permission.)

some amount of hydrostatic pressure is generated in the material prevent cavitation of the material. Because of the absence of cavitation, the material undergoes plastic deformation via shear yielding; cavitation is damped. Also crazing is not preferred under hydrostatic pressure as it involves volume increase and the production of empty spaces between tufts.

In a macroscopic sample of a semicrystalline material subjected to stress few or all of the presented mechanisms of plastic deformation are orchestrated. Some of them are preferred, because of low shear stress required, and show up in early stages of deformation; others are activated in later stages under larger stress. The intensity of a particular mechanism may also change, if for example the possibility of a certain slip is already exhausted or that the other slip mechanism rotated the crystals in such a way that the process mentioned is not now possible. Together with the complicated aggregated supermolecular structure, the process of the deformation in semicrystalline polymers is complicated and not easy to track.

IV. PLASTICITY OF POLYMER CRYSTALS

The detailed knowledge concerns the system of crystallographic slips operating in polymer crystals, critical resolved shear stresses, twinning, martensitic transformation and succession of activation of individual slip mechanisms [45,61,62].

The magnitude of the shear stress needed to move a dislocation along the plane was first determined by Peierls [63] and Nabarro [64]. For the orthorhombic unit cell of a crystal, it varies exponentially with the ratio of both unit cell axes perpendicular to the macromolecular chain direction. For a (100) plane being the closed-packed plane (**a** axis larger than **b** axis), the shear stress reaches the minimum. That is because for closed-packed planes, the interplanar bonds are weaker which results in lower activation energy and shear stress. The result is that the dislocations tend to move in the closest packed planes and in the closest packed direction because the Peierls-Nabarro force is smaller for dislocations with a short Burgers vector.

As this mechanism of dislocation propagation is widely accepted for slip mechanisms in semicrystalline polymers, there are doubts about how the dislocations are generated. The density of dislocation in crystals is estimated at the level from 10^5 to $10^8/cm^2$ [65]. This number is not enough to give rise to a fine slip most often observed during polymer plastic deformation. There must be a way in which new dislocations are generated within crystals. In metals and other large crystals the identified source of dislocations is a multiplication mechanism known as Frank-Read source [66]. The Frank-

Read mechanism involves a dislocation line locked on both ends. When the shear stress is applied above a certain critical value, the dislocation line will move to form first a semicircle and then the line spiral around the two locked ends, forming finally the dislocation ring which will continue to grow outward under the applied stress. At the same time the original dislocation line has been regenerated and is now free to repeat the whole process. In this way a series of dislocation rings is generated indefinitely. In polymer crystals this elegant process is not active because the crystals are usually too thin for the dislocation line to spiral up and to form a dislocation loop.

Due to a peculiar structure of polymer crystals (long molecules which lie across the crystal thickness, fold and re-enter or take part in forming adjacent crystals), there are some restrictions which reduce the number of possible slip systems. Theoretical calculations demonstrated [67] and experiments confirmed [44] that polyethylene lamellae deform easily by slip in the direction of the c-axis. There is a problem of the origin of the dislocations for activation of the slip mechanism. Shadrake and Guiu [67] pointed out that in the case of PE the energy necessary for creation of a screw dislocation with the Burgers vector parallel to chain direction can be supplied by thermal fluctuations. It was shown that the change in the Gibbs free energy, ΔG, (i.e., the energy which must be supplied by thermal fluctuations) associated with the creation of such dislocation under applied shear stress, τ, is equal:

$$\Delta G = \frac{Kb^2 l}{2\pi} \ln\left(\frac{r}{r_0}\right) - \tau b l r, \tag{3}$$

where l is the stem length; b is the value of the Burgers vector; K is the shear modulus of a crystal; r is the radius of dislocations (the distance from dislocation line to the edge of lamellae); and r_0 is the core radius of dislocations.

There are still doubts if the model can be applied over the whole range of temperature, i.e., from the temperature of glass transition to the onset of melting process. The dispute concerns the upper temperature of validity of this approach. Crist [68] suggested that the temperature of γ and α relaxation processes of PE are the limits of applicability of the model. Young [39,40] and Darras and Seguela [69] have used that approach to model the yield behavior of bulk crystallized annealed polyethylene at a much higher temperature. Other authors [70] reported the existence of a transition in the range from −60°C to 20°C, depending on the material and strain rate, above which Young's model cannot be applied. They pointed out that there is a relationship between the transition temperature and β-relaxation and suggested that below the transition temperature the yield process is nucleation controlled, while above it is propagation controlled. However, Galeski et al. have shown that in plane strain compression for HDPE at 80°C the beginning of yielding is mainly associated with (100) chain slip within crystalline lamellae. Moreover, it was shown that for linear polyethylene only fine slip occurs below the deformation ratio of 3 associated with chain tilt and thinning of lamellae. Only at higher deformation the widespread fragmentation of those thinned to one-third lamellae takes place. That is because further thinning becomes unstable — much like layered heterogeneous liquids responds by capillary waves and breakup of stacks of layers.

Many authors identified two distinct yield points in PE deformed in the tensile mode which are not seen in other deformation modes not producing cavitation. Gaucher-Miri and Seguela [71] tried to clear up the mechanism of these two processes on the micro-structural level, as a function of the temperature and strain rate. Seguela [72] proposed that the driving force for the nucleation and propagation of screw dislocations across the crystal width relies on chain twist defects that migrate along the chains' stems and allow a step-by-step translation of the stems through the crystal thickness. The motion of such thermally activated defects is responsible for α crystalline relaxation.

When the above-mentioned problems were investigated by others, the crystal thickness of PE was controlled by changing the cooling rate and crystallization temperature, using copolymers of PE characterized by different degrees of branching or by crystallizing PE either from solution or from melt. The applied procedures allowed obtaining PE orthorhombic crystals with thickness over the range from 3 to 35 nm. Crystallization of PE under elevated pressure evolved further by us [73] makes it possible to obtain much thicker crystals. The method exploits the pseudo-hexagonal mobile phase of PE at a certain range of pressure and temperature. The details of how to obtain the samples with crystals of various thicknesses due to crystallization under high pressure are described in [73,74]. The great advantage of the approach is that the series of samples with various crystal thicknesses is obtained from the same polymer. Most of the previously reported studies of the structural changes caused by deformation were performed in a tensile mode, guided by obvious technological stimuli to explain processes associated with orientation by drawing. However, from the fundamental point of view the deformation by compression is more important. Uniaxial compression has the great advantage because the deformation is nearly a homogenous process and occurs without any significant deformation instabilities such as necking and cavitation. In the reported study [74,75] the samples prepared by high pressure crystallization were characterized by crystal thickness, covering the range from 20 up to 150 nm. Figure 7 presents a typical true stress-true strain curve obtained in uniaxial compression for HDPE with crystal thickness of 150 nm uniaxially compressed at a room temperature. After the usual initial elastic response below a compression ratio of 1.05–1.07 there is only a single yield and a region of intense plastic flow which sets in at a compression ratio of 1.12, followed by strain hardening.

The changes in the yield stress with increasing crystal thickness are reported in Figure 8. The yield stress increases with crystal thickness up to 40 nm. This part of the data agrees well with the published results of Brooks and Mukhtar [76], covering the range of lamellae thickness from 9.1 to 28.3. Beyond the region explored in the past, the yield stress still increases in the range up to 40 nm and then, above 40 nm, the dependence on crystal thickness abruptly saturates at the level of 29.5, 35 and 37 MPa for initial compression rates of 0.000055, 0.0011 and 0.0055 s^{-1}, respectively. The increase of the rate of compression increases the yield stress in the region of crystal thickness below 40 nm and also above 40nm. How-

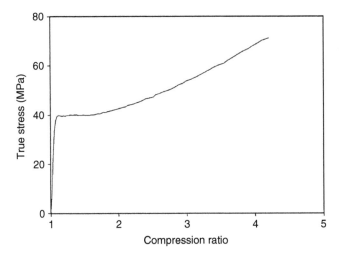

Figure 7 Typical curve true stress–compression ratio of high density polyethylene crystallized at elevated pressure (M_w = 120 000, M_w/M_N = 3.4, density 0.952 g/cm^3 in pellets, MFI of 2.3 @ 190°C/2.16 kg). Uniaxial compression at room temperature with the initial compression rate 0.0055 s^{-1}, mean crystal thickness 119 nm. (From Kazmierczak, T., Galeski, A. Plastic deformation of polyethylene crystals as a function of crystal thickness. Proceedings of VIth ESAFORM Conference on Material Forming, University of Salerno, Italy, 2003.)

ever, the saturation of the yield stress occurs at 40 nm independently of the compression rate.

These observations imply that above 40 nm the crystals' thickness is no longer the decisive factor for the yield and that some other mechanism overtakes the control of the yielding process. Any "coarse slip" and other inhomogeneities in the course of compression which could decrease the yield stress did not occur as evidenced in SEM examination by chain tilt and homogeneous lamellae thinning up to the compression ratio above 2. It can be concluded then that the observed yielding process is a "fine slip" process.

According to Young's model of the yield process, the yield stress should increase with crystal thickness, which in fact is observed up to the crystal thickness of 40 nm. However, above 40 nm the dependence of the yield stress quickly levels off.

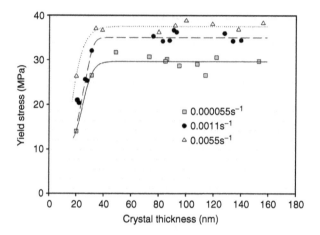

Figure 8 Yield stress vs. crystal thickness of high density polyethylene samples crystallized at elevated pressure and temperature (M_w = 120 000, M_w/M_N = 3.4, density 0.952 g/cm^3 in pellets, MFI of 2.3 @ 190°C/2.16 kg). Uniaxial compression at room temperature with three different initial compression rates as marked on the plot. (From Kazmierczak, T., Galeski, A. Plastic deformation of polyethylene crystals as a function of crystal thickness and compression rate. Proceedings of 12th International Conference on Deformation, Yield and Fracture of Polymers, Cambridge: UK, April 7–10, 2003. With permission.)

Recently, the rate-controlling mechanisms of crystal plasticity in semicrystalline polymers were fundamentally reconsidered [77,78]. The widely accepted mechanism of Young of monolithic nucleation of screw dislocations from edges of crystalline lamellae predicting an increase in plastic resistance with increasing lamella thickness was reexamined and modifications were made. Two new models of nucleation of both edge and screw dislocation half loops from lamella faces that are independent of lamella thickness were proposed. These two new modes of dislocation nucleation explain well the observed transition from a plastic resistance increasing with lamella thickness to one of constant resistance above a lamella thickness of ca. 35 nm in polyethylene. They also present a background to explain the temperature- and strain-rate dependence of the plastic resistance of polyethylene and predict the observed levels of activation volumes.

V. STRENGTH OF CRYSTALLINE POLYMERS

The strength of a polymer chain depends on the strength of –C-C- bonds. Evaluation of the strength of a single macromolecule requires quantum mechanics. Theoretical strength of a –C-C- bond can be estimated on the basis of a two-atom molecule from the dissociation energy of unstressed C-C bonds which is 335 kJ/mol [79]. For such dissociation energy of a C-C bond, the strength should be at the very high level of 360 GPa [79,80]. However, the thermal fluctuations lowers the stress at bond breakdown by lowering of the activation energy of C-C bond dissociation and the bond strength to one-third of its uncorrected values, it is 104 kJ/mol and 126 GPa, respectively [79,81]. In reality the achievable strength is estimated at the level of 60–100 GPa [79]. For comparison, unoriented polystyrene shows the strength at 40–60 MPa only, ordinary steel 0.3–0.6 GPa, while the best steel 2 GPa. These values are far below the expected value for the strength of a single polymer chain. The theoretical elastic modulus for C-C bonds is 405 GPa [79]. Ultimate strain of –C–C–C–C– chain can be as high as 40%, also due to conformational changes of –C–C–C– triads. Theoretical strength of a polyethylene fiber composed of parallely aligned extended chains is also at the level of 126 GPa. However, the load is usually transferred to the core of a material through its surface, which means that the surface is subjected to shear stresses. The chains should be bonded to each other in order to withstand the shear. Crystalline bonds, besides the knots of entanglements, can bond the macromolecular chains to each other. Then, the resistance to shear of polymer crystals along the chain direction is equal to the critical resolved shear stress for crystallographic slip and it is usually of the order of ten or a few tens of MPa and is much less than the tensile strength of single polymer chains. It appears then that in order to take advantage of the full strength of a well-oriented polymeric fiber, the surface through which the load is transferred to the fiber core should be substantial. For example, for a 10 μm thick fiber having the tensile strength of 100 GPa and the lateral shear strength of 100 MPa, the load should be uni-

formly transmitted through the lateral surface along 10 mm.
For equally strong thicker fiber or rod, the gripping length
for making full use of its strength increases proportionally to
its thickness: for 1 cm thick well-oriented rods it approaches
unrealistic 10 m!. The conditions for stress transfer to the
core of highly oriented fibers or rods are not so severe if the
string forms a closed loop. Then the compressive and tensile
stresses are acting at the loop's support.

Although the oriented polymers can withstand much
higher tensile stress than the best steel, the problem with
polymers is that they have much higher anisotropy than steel.

A premature fracture occurs in real systems, which is
connected with defects of regular packing and chain ends.
Unfortunately local stresses at these places are not equal to
the macroscopic stress and are 2 to 3 times higher. Local
fracture starts at those locations decreasing the maximal
strength in real systems to 20–50 GPa at an ultimate strain
of 17 to 24%.

Another problem is in achieving a perfect parallel orientation of polymer chains. Usually molecular orientation is
obtained by a strong deformation of a polymer either by spinning from the melt or deforming a solidified polymer. The
factor obstructing the elongation is the presence of chain
entanglements. For example, the average distance between
entanglement knots along the chain for linear polyethylene
is 1200–1300 daltons [82,83] which indicates that the entanglement knot is every 80–90 mers along the chain obstructing
the ideal orientation.

For the last thirty years the main effort in achieving high
strength polymeric materials was to extend and align macromolecular chains into a flawless parallel register. Due to a
low lateral strength of oriented polymers, most of the
attempts were limited to thin fibers because the surface to
cross-section ratio is advantageous for the stress transfer. The
simplest way seemed to be melt spinning; however, the problem with melt spinning is that macromolecules undergo intensive relaxation during extension and recoil back. Lowering
the temperature of spinning (or extrusion) slows the retraction but polymers undergo flow-induced crystallization

obstructing the process. It appeared that in order to obtain high chain orientation, the drawing should be performed in a separate step. In the 1970s, Ward and coworkers [84–86] analyzed the drawability of polyethylene in a solid state, and they made a major breakthrough in achieving high strength materials by melt spinning of fibers followed by solid-state drawing. By optimizing the composition and parameters of spinning and drawing, polyethylene fibers with elastic modulus around 75 GPa and the strength of 1.5 GPa were obtained. Melt spinning is used now routinely for increasing the tensile strength of many polymers [87,88].

However, the melt spinning and subsequent drawing are subjected to severe limitations. With increasing molecular weight, which should be beneficial for achieving high strength, the spinning process becomes difficult due to a strong increase of the viscosity, while the drawability in the solid state decreases — it becomes difficult to extend significantly solidified fibers. Several other means of achieving high strength polymeric materials were explored, among them solution spinning and flash spinning — the technique with pressurized polyethylene solution [89,90].

Chain extension of dilute solutions in elongational flow field was addressed first theoretically by deGennes [89] who indicated that there should be a certain critical value of the velocity gradient upon which a solute polymer coil will abruptly unwind. The issue was then intensively explored experimentally [e.g., 92]. In the careful experimental studies of monodisperse samples by Keller and Odell [93], it was concluded that an isolated chain can be fully and abruptly stretched if the strain rate exceeds a certain critical value which is related to the molecular weight: $d\varepsilon/dt \propto M^{-1.5}$ where M is the molecular weight. For polydisperse polymers only high molecular weight fraction can easily be extended in the elongational field while the remaining part stays coiled in the solution. Chain extension in a solution can be made permanent if the extension is followed by crystallization. Upon cooling, the extended chains crystallize in the form of fibrous structures called "shish" while the coiled macromolecules crystallize around extended chains in the form of chain folded crystals called "kebab." The shish-kebab structure due to lamellar overgrowth is not optimal for stiffness and strength — only the elastic modulus up to 25GPa can be achieved [94].

Fibrous structure without an excessive amount of kebabs was obtained by Zwijnenburg and Pennings [94] by a so-called surface growth technique. A seed fiber is immersed in a dilute solution of polyethylene subjected to strong shear between the rotating inner cylinder and the wall of a vessel. A tape-like fibrous material can be drawn at a low speed from the solution. Later a modification of the Pennings method by Mackley and Sapsford [95] allows to obtain PE fibrous tapes having the modulus of 60 GPa with the rate of several meters per minute.

A gel spinning process was developed by Smith and Lemstra at DSM [96,97,98]. A semi-dilute solution is spun into water from the extruder. A gelly filament still consists of a large amount of a solvent. Ultra drawing is possible after removal of the solvent. A simple model derived by Smith and Lemstra [99] from rubber elasticity assuming a network of chain entanglements explains the capability of gelly filaments of ultradrawing: the entanglement density is reduced upon dissolution and the maximum draw ratio is greatly enhanced in comparison to melt crystallized polyethylene. The entanglement model explains that a relatively low concentration solution is needed to remove entanglements before drawing. The topic of deformability and entanglements was further addressed in several papers, e.g., [100]. Due to further developments in efficient mixing of double screw extruders and in the temperature gradient drawing, the solutions for gel spinning are now more concentrated. Several high strength and high modulus commercial fibers are based on the gel spinning process of ultra high molecular weight polyethylene: Dyneema by DSM and Toyobo, and Spectra by Allied Signal.

A solvent-free path for disentangling polyethylene chains prior to ultra drawing was explored intensively in the past. One of the approaches is to collect precipitated single crystals from dilute solution. The other route is to prepare nascent single crystals of polyethylene directly in the polymerization reactor. By low temperature polymerization single chain-folded crystals are formed on the surface of the catalyst. The macro-molecular chains are nearly fully disentangled [101,102].

Bieser and others at Dow Chemical [103] discovered a method of disentangling macromolecular chains by applying high shear just prior to spinning by passing the polymer melt through sintered metal filter media, multiple stacks of fine mesh screens or similar shear-inducing media. As the result, an improved melt fiber spinning is achieved with a minimal fiber breakage, especially for thin fibers. The method is very effective for relatively high molecular weight polymers.

Chainfolded crystals with disentangled macromolecules show the same high melt viscosity upon melting as the entangled melt; no memory effect from previous history of either polymerization or solution crystallization was noticed. Furthermore, after melting, the advantageous ultra drawing feature is entirely lost. This confusing phenomenon was considered by Barham and Sadler [104] in the studies of melting of chainfolded crystals. Using neutron scattering and deuterated polyethylenes, they measured the changes in radius of gyration. For chain folded crystals, the radius of gyration is rather low and upon melting it suddenly inflates to the equilibrium value of a random coil. The expansion of chains is very rapid (at a molecular weight of 189,000 the equilibrium radius of gyration is reached within less than 4 s) and pays no attention to the neighboring chains, which is in contrast to the reptation theory based on the reptation tube formed by fragments of neighboring chains. The phenomenon associated with the melting of chainfolded crystals is termed "coil explosion," as it is graphically explained in Figure 9, and it happens independently of the molecular weight. In the view of this concept it follows that the chain re-entangling occurs suddenly. However, the problem remains open as to whether the entanglement also assumes its equilibrium density instantaneously.

Lemstra, van Aerle and Bastiaansen [105] proposed an alternative model for the loss of drawability upon melting which involves local disorder only. Upon melting, the chains immediately adopt random coil conformation as mentioned above, and during recrystallization the adjacent reentries are replaced by more random order within crystals. The shearing

Figure 9 A computer-generated sketch illustrating typical trajectories of a molecule of 390,000 molecular weight before and after melting. (Reprinted from Barham, P.J. and Sadler, D.M. Polymer, 32, 393–395. Copyright 1991, with permission from Elsevier.)

and unfolding of the crystals is more difficult since it requires now a great degree of cooperativeness between many chain segments from different locations within a crystal and in different crystals. Lemstra's approach emphasizes the importance of adjacent reentry for easy crystal shearing, fragmentation and unfolding.

Galeski and others [106] grew chain-extended polyethylene crystals under high pressure in order to disentangle fully the macromolecules. No "coil explosion" occurs upon melting of chain-extended crystals but rather recoiling, i.e., extended chains assume coiled conformation by fast contraction instead of expansion, as it occurs during melting of chainfolded crystals with adjacent reentry. The contraction to coiled conformation does not cause a deep entangling. Only loose entanglements are formed during melting, and the melt is disentangled for a period of time required for the tube renewal by reptation. Subsequent crystallization is faster in regimes I and II of crystallization as the crystallizing chains do not need to disentangle. Nonetheless, a similar mechanism of randomization of folding is in operation as in recrystallization of a melt from chainfolded disentangled crystals obtained from solution or from polymerization reactor.

From the above consideration it appears that the disentangled material should not be heated above the melting point before drawing. Being below melting point, the disentangled material or nascent powders can be calendered or hot-compacted and then drawn into fibers or tapes since they are extraordinarily ductile. The drawability of nascent polyethylene is similar to solution-crystallized polymer when the material is sintered or hot compacted between rolls.

Kanamoto et al. [107] developed a two-stage process of drawing for the reactor-nascent polyethylene with the elastic modulus above 100 GPa and the strength of 1.0–1.4 GPa. The process relies on solid-state coextrusion of compacted powder film up to the draw ratio of about 6 followed by tensile drawing at elevated temperature.

Nippon Oil Co. developed a three-stage process for the production of super strong tapes. The process consisted of compaction, roll drawing and tensile drawing [108]. The products are characterized by high elastic modulus around 120 GPa and the tensile strength below 2.0 GPa.

Hot compaction/sintering is usually conducted slightly below the melting temperature. This process appears to be necessary for chain disentangled powders from polymerization reactor or for single crystals from dilute solution because it generates their high drawability. There must be some mechanism acting during sintering which enables the stress transfer between neighboring crystals. In fact Lemstra and others [109,110] have discovered that due to chain stem diffusion across crystal interfaces, the doubling of lamellar thickness occurs. Displaced stems produce an adhesion between crystals required for efficient stress transfer. High drawability of individual crystals is not affected because the stem diffusion does not occur across crystal planes containing polymer chains.

High drawability of polyethylene is achieved also thanks to the absence of specific interactions between chains such as hydrogen bonding. The lack of strong interactions between chains results, however, in a very high mechanical anisotropy of all high modulus and high strength polyethylene fibers.

An attempt to increase the transverse strength by irradiation and cross-linking was made by Pennings et al. [111,112], Ward et al. [113] and Lemstra and Keller et al. [114]. However, a significant loss of longitudinal strength was noticed upon irradiation, e.g., for a dose of 100 Mrads of electron irradiation the tensile strength of polyethylene fibers decreased from 3 GPa to approximately 0.8 GPa.[1]

VI. CAVITY-FREE DEFORMATION

While the picture of a cavitating polymer is reasonable in tensile deformation, it is not correct for modes of deformation in which a positive normal stress component prevents the formation of cavities. It has been shown that micro necking is not essential for the development of nearly perfect single crystal textures for several semicrystalline polymers that result from plane strain compression in a channel die [62]. Although plane strain compression is kinematically very similar to drawing of wide strips, the pressure component, which arises due to compression, prevents cavitation. The WAXS, SAXS, TEM and light microscopy observations indicated that although some degree of inhomogeneous deformation in the form of localized shear bands occurs, the crystalline and associated amorphous regions of the material undergo a continuous series of shear-induced morphological transformations without any cavitational process.

A gel-spinning process of polyethylene is a cavity-free operation due to an easy drawability of folded-chain crystals

---

[1]Concerning the increase of transverse strength due to hot compaction of fibers see Chapter 16.

with adjacent reentry. An example of the products other than fibers is the development of ultra strength films by gel spinning by Lemstra [115]. The film exhibits the elastic modulus above 50 GPa. The indication of the cavity-free deformation is the gloss of the surface of the film and its opacity below 10%.

A range of solid-state processes for obtaining high orientation in thicker than fibers species was applied to polymers. Some of them were adapted from metal processing while others were elaborated solely for polymer deformation and orientation.

Tensile drawing has been known for a long time. It is characterized by the natural draw ratio which is intimately connected with the necking. The natural draw ratio decreases with the increase of the molecular weight; for linear polyethylene having M_w = 200 kDa it decreases to 8. A careful thermal treatment, including slow crystallization at a high temperature, causes the draw ratio for linear polyethylene to jump over 30 and the elastic modulus to 70 GPa [116]. It is now rather evident that the reason for these changes is a partial disentanglement of chains due to slow crystallization.

Hydrostatic extrusion is known for metal forming. A billet of a material is made to flow through a converging die by application of pressurized liquid. This has the advantage over ram-extrusion that the friction between the billet and the walls is eliminated. Large and small diameter polymer products can be obtained by this method [116]. Again a careful thermal treatment of the billet is necessary to achieve a high extrusion ratio, a modulus of 60 GPa for extrusion ratio of 25 can be achieved for polyethylene. Hydrostatic extrusion was not so successful with other polymers.

Die-drawing was one of the earliest processes applied to polymers. The Leeds die-drawing process was elaborated in the 1980s. It relies on extrusion of a billet of a polymer through a die with simultaneous drawing of the extrudate from the die. Recently die-drawing was applied to the continuous production of a polymer core for wire ropes [117]. The process was modified by introducing the second stage of drawing in a cooled fluted conical die. The first application of these polyethylene cores was for carrying ropes in elevators.

Die-drawing was employed with success for deformation of other polymers: polypropylene sheets [118] and monofilaments of polyoxymethylene [119].

Rolltrusion is a process of mechanical properties enhancement by combining the rolling and subsequent drawing [120,121]. For example, in the case of polypropylene, the strength enhancement in the drawing direction results in tensile strength of 524 MPa while in the direction perpendicular to rolls of 104 MPa as compared to 27.6 MPa of unoriented polypropylene [122]. Several other polymers were subjected to rolltrusion, e.g., poly(aryl ether ether ketone) (PEEK) [123].

In all the above experiments, the positive pressure component of acting stresses prevents cavitation.

VII. ROLLING WITH SIDE CONSTRAINTS

Among other known methods of plastic deformation, rolling is one of the simplest ways of producing high preferred orientation. Due to a high pressure component, cavitation is not observed except for the edges of a rolled sheet. Rolling is an attractive process of plastic deformation since it could be designed as a continuous process. However, for wider strips, the force required to significantly roll the material often increases to an unacceptably high level, while for narrow strips of polymeric materials there is an additional component of a transverse deformation, which causes a less sharp texture of the oriented material. The side effect of a transverse deformation is the formation of fissures, cracks and cavitation at edges of a rolled material because there is no compressive stress component and relative freedom for distortion.

A novel method of obtaining highly oriented polymeric materials, being a combination of channel die and rolling, is by rolling with side constraints [124]. The process relies on rolling of a material inside a channel on the circumference of a roll with another roll having the thickness matching closely the width of the channel (see Figures 10a and 10b). The side constraints are the side walls of the channel. The other roll is serving as a plunger. The system of rolls with a channel

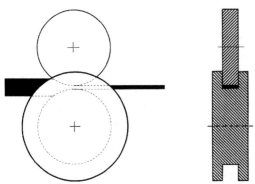

Figure 10a A scheme illustrating the principles of rolling with side constraints.

Figure 10b View of the laboratory 4-roll rolling apparatus used in this study.

produces conditions for plain strain compression of the rolled material. Plain strain compression is known to produce single crystal (or twinned) texture of compressed materials. The advantage of rolls with side constraints is the possibility of compressing relatively wide, thick and long shapes in a continuous manner in a neck-free fashion. The resulting shapes or rods may have a considerably high cross-section area.

Constraints enforce the deformation with no or little volume change and produce a high positive pressure component; the cavitation process is strongly inhibited and no cavities are formed during plastic deformation of polymeric materials.

The presence of compressive component and constraints completely changes the process of plastic deformation as compared to tensile drawing [125]. This effect is illustrated in a comparative experiment: isotactic polypropylene bulk sample was deformed in tension, and in the second test a similar sample was deformed in a channel die. The difference is clearly seen in Figure 11 where the respective true stress-true strain curves are plotted. Without constraints, an intense cavitation restricts the strength to only 120 MPa, and its fracture is initiated in microfibrils one by one, while the same material deformed in channel die in a cavity-free manner responds with the stress of nearly 300 MPa at a similar deformation ratio. Also due to the presence of microfibrils, the cavitated material has little transverse strength due to loose connection between microfibrils. The strength in the transverse direction of the material deformed in a channel die is at the level of strength of undeformed material, i.e., around 35 MPa.

The two examples of rolling with side constraints, polypropylene and high density polyethylene, described below, illustrate the growth of strength and toughness due to the evolution of texture by crystallographic slips and rotation.

A.   Case of Isotactic Polypropylene (iPP)

The rolling with side constraints of iPP bars in the constructed apparatus was performed with the rates from the range from

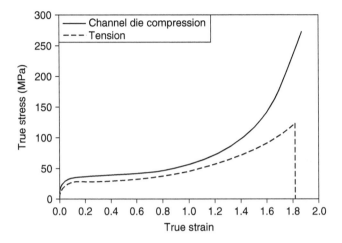

Figure 11 True stress-strain curves for isotactic polypropylene Malen P B200 (melt flow index 0.6 g/10 min, density 0.91 g/cm³, Orlen SA, Plock, Poland). a) strained at the rate of 5 mm/min; b) compressed in a channel die. (From Morawiec, J., Bartczak, Z., Kazmierczak, T., Galeski, A. Mater. Sci. Eng. 2001, A317, 21–27. With permission.)

0.5 to 4 m/min at room temperature, and at a various temperatures from the range from 90 to 140°C [126].

The mechanical properties of a series of rolled polypropylene bars having various compression ratio are characterized by stress-strain curves presented in Figure 12. These representative stress-strain curves of the specimens cut out from the bars deformed at 120°C were obtained in tension along the direction of the molecular orientation which coincides with the direction of rolling. These data show continuous increase of both modulus and ultimate tensile strength with increasing deformation ratio, which is a typical effect of orientation. The maximum strength at the level of 340 MPa is achieved for the sample with the compression ratio of 10.4. In Figure 12 the stress-strain curves for low-carbon steel and concrete-reinforcing steel 18G2 are also plotted for comparison. It is seen that the strength of rolled iPP (DR = 10.4) is

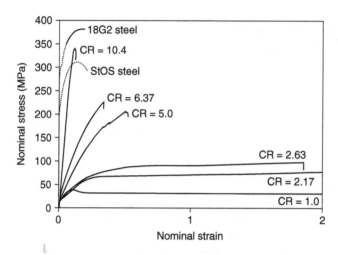

Figure 12 Stress-strain curves of rolled polypropylene at the rate of 4m/min at room temperature and at 120°C. All tensile tests were performed at room temperature with constant crosshead speed of 2 mm/min corresponding to the initial deformation rate of 5%/min. For comparison stress-strain curves of unoriented iPP and two types of steel used for concrete reinforcement: low carbon and hardened are shown. (From Bartczak, Z., Morawiec, J., Galeski, A. J. Appl. Polym. Sci. 2002, 86, 1413–1425. With permission.)

comparable with that for steel, although its elastic modulus is one order of magnitude lower.

Impact properties of the rolled bars probed by means of an instrumented notched Izod test performed at room temperature show that the undeformed specimen as well as the specimen of lowest compression ratio break completely and show the Izod impact strength at the level of several kJ/m². The other samples having higher compression ratios do not break, and only a limited fracture and delamination starting at the notch is observed. The force and the energy at peak during impact can be connected with initiation and propagation of fracture, while the remaining energy is dissipated mostly for bending of the unbroken portion of the sample. The energy data demonstrate that more energy is consumed by the latter process than by cracking. The force at peak increases steadily with the increase of the deformation ratio. In contrast, the peak and total energy dissipated on the deformation ratio reach the maximum near the compression ratio of 5 for which the smallest area of the fracture is observed. The energy dissipated during impact of the sample having the compression ratio of 5 corresponds to impact strength above 170 kJ/m². It is an almost 40-fold increase as compared to an unoriented sample of iPP.

B. Case of High-Density Polyethylene (HDPE)

Rolling of HDPE bars was performed in the rolling apparatus with the linear speed of 200 mm/min at room temperature, 90 and 110°C [58,127]. Neither necking nor cavitation phenomena were observed during rolling at these conditions. Deformation was homogenous in the entire strain range studied, similar to that observed in compression in a channel-die. Samples with the deformation ratio approaching 4 became translucent, in contrast to opaque bars of virgin HDPE. Figure 13 shows representative stress-strain curves of the specimens cut out from the bars of HDPE oriented to various compression ratios by rolling with side constraints. These data show that the elastic modulus increases with increasing deformation ratio, while the ultimate stress of oriented samples is approximately proportional to the second power of this ratio. The strain at break of oriented material decreases substantially with increasing deformation ratio.

The toughness of oriented bars of polyethylene was probed by means of notched Izod impact test. The impact tests revealed that the samples do not fracture. The specimens bend and some delamination occurred along the rolling plane perpendicular, starting from the tip of the notch. The plane of fracture and delamination coincide with the plane of preferred orientation of (100) plane of the crystallites. The (100) plane of orthorhombic crystals of polyethylene is the most densely packed plane, and therefore the force to cleavage the crystals along it is the lowest similar to the lowest plastic resistance of a (100)[001] slip system. When the specimens were struck along constraint direction no fracture occurred and the impact

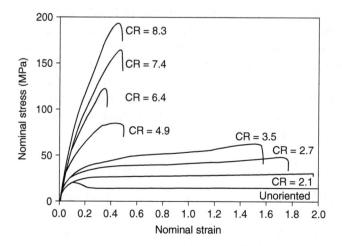

Figure 13 Stress-strain curves of HDPE bars oriented to various compression ratios by rolling with side constraints. Tension along rolling direction at room temperature with a constant crosshead speed of 2 mm/min corresponding to the initial deformation rate of 5%/min. (From Bartczak, Z., Morawiec, J., Galeski, A. J. Appl. Polym. Sci. 2002, 86, 1405–1412. With permission.)

resulted in specimen bending. In the vicinity of the notch tip some very localized delamination in several (100) planes was observed. The energy delivered by the impact hammer was dissipated mainly for the bending of the specimen.

Impact tests revealed extremely high toughness of the rolled bars, which is the result of their strong quasi-single crystal texture [58]. The dissipated energy is approaching the value of 200 kJ/m², which is nearly 15 times higher than the strength of 14 kJ/m² determined for unoriented material. The dependence of the total dissipated energy on the deformation ratio passes through a broad maximum near the compression ratio of 5.

VIII. CONCLUSIONS

The toughening in most of the polymer systems relies on main mechanisms of promotion of energy dissipative processes that delay or entirely suppress fracture processes originating from imperfection of internal structure or scratches and notches. The advantage of having cavitation is that the most energy dissipative processes, crazing and shear yielding, occur at a reduced stress level. In crystalline polymer systems the tough response, besides cavitation and crazing, is of crystallographic nature. Crystallographic slips are the main plastic deformation mechanisms.

In almost all cases cavitation either makes possible further toughening by activating other mechanisms or contributes itself to the plastic response of the polymer.

However, the cavitation during plastic deformation greatly reduces the strength of polymeric materials. The cavity-free deformation, such as in the plane-strain compression mode, leads to oriented polymeric materials with the strength and toughness much higher than the material oriented by deformation with no constraints. The side constraints imposed on the material during its compression help to prevent unwanted cavitation as well as to produce the material with well-defined and sharp texture.

The new method of rolling inside a channel placed on the circumference of another roll, reported here, resembles a plane-strain compression in a channel die. Rolling in a channel is the mode of cavity-free deformation due to the constraints formed by sidewalls of a channel. Shapes of isotactic polypropylene and polyethylene rolled in a channel show a significant increase in tensile strength and the modulus, which to a great extent, depends on the temperature of rolling. It was evidenced that the deformation process of polyethylene by both the compression in a channel die and the rolling with side constraints proceeds in a very similar fashion. The method of rolling with side constraints has, however, a big advantage over the other methods of processing of solidified polymers since it allows for a continuous production of oriented material of unlimited length with high output rate. Moreover, the rolling with side constraints seems to be better than the conventional rolling, in which the relatively weak side constraints result merely from friction forces between the material and rolls only. It limits the conventional rolling to the production of comparatively thin sheets or films. In contrast, the constraint rolling allows for the production of bars or profiles with relatively large cross-sections — in the laboratory set-up we were able to produce long oriented bars of the 12×12 mm^2 cross-section with the speed of 4 m/min. Such materials may become a very attractive engineering material of superior mechanical properties.

One of the explored possibilities for the increase of yield stress of a polymer, and hence its strength, is to increase its crystal's thickness. The yield stress increases initially with the increase of crystal thickness. In the case of polyethylene, the crystal thickness can be controlled by high-pressure crystallization. The yield sets in at the stress sufficient to activate crystallographic slips. However, the increase of the yield stress with the increase of crystal thickness in polyethylene samples shows a remarkable saturation above 40 nm. No further increase of the yield stress is observed. In addition, the yield stress displays the dependence on the deformation rate. In spite of the yield stress dependence of deformation rate, the "magic" thickness of 40 nm is independent of the deformation rate. Since the crystallographic slips are controlled by generation and propagation of mobile crystallographic dislocations, it is recognized that in polyethylene with thin crystals the monolithic mobile dislocations generated by thermal fluctuations are responsible for its plastic resistance. For crystals thicker than 40 nm two new modes of nucleation of both edge and screw dislocation half loops from lamella faces that are independent of lamella thickness were proposed. These two new modes of dislocation nucleation account for the observed transition from a plastic resistance increasing with lamella thickness to a steady resistance above a lamella thickness of 40 nm in polyethylene. The temperature- and strain-rate dependences of the plastic resistance of polyethylene and observed activation volumes are explained based on these two nucleation processes.

ACKNOWLEDGMENTS

Grant KBN 7 T08E 055 22 from the State Committee for Scientific Research (Poland) and Centre of Molecular and

Macromolecular Studies, Polish Academy of Sciences are acknowledged for the financial support of the work. The author expresses his appreciation to co-workers for numerous discussions during preparation of this paper.

REFERENCES

1. Young R.J., Strength and toughness, In: Comprehensive Polymer Science: The Synthesis, Characterization, Reaction and Applications of Polymer, Allen G., Ed., Vol.2 Polymer Properties, Booth C., Price C., Eds., Pergamon Press: Oxford UK, 1989, 511–532.

2. Argon, A.S., Cohen, R.E. Crazing and toughness of block copolymers and blends. Adv. Polym. Sci. 1990, 91, 92, 301–351.

3. Kinloch, A.J., Young, R.J. Fracture behaviour of polymers. Applied Sci. Pub. London, New York, 1983, 147–178.

4. Bucknall, C.B. Fracture and failure of multiphase polymers and polymer composites. Adv. Polym. Sci. 1978, 27, 121–148.

5. Kramer, E.J., Berger, L.L. Fundamental processes of craze growth and fracture. Adv. Polym. Sci. 1990, 91, 92, 1–68.

6. Olf, H.G., Peterlin, A. Cryogenic crazing of crystalline, isotactic polypropylene. J. Colloid Interface Sci. 1974, 47, 621–35.

7. Horst, J.J., Spoormaker, J.L. Mechanism of fatigue in short glass fiber reinforced polyamide6. Polym. Eng. Sci. 1996, 36, 2718–2726.

8. Boyer, R.F. Mechanical relaxation spectra of crystalline and amorphous polymers. Materials distributed by Boyer, R.F. during the Symposium. Midland Macromolecular Institute Symposium on Molecular Basis of Transitions and Relaxations, Midland, Michigan, Feb.1975.

9. Ferry, J.D. Viscoelastic Properties of Polymers., 2nd Ed., J.Wiley: New York, 1970.

10. Galeski, A. Dynamic mechanical properties of crystalline polymer blends. The influence of interfaces and orientation. e-Polymers 2002, (026), 1–29.

11. Orovan, E. Fracture and strength of solids. Rept. Prog. Phys. 1948–49, 12, 85–232.

12. Vincent, P.I. The tough-brittle transition, In: Thermoplastic Polymers, 1960, 1, 425–444.

13. Ward, I.M, Hadley, D.W. An Introduction to the mechanical properties of solid polymers. Wiley: Chichester, New York, 1993, 271–276.

14. Piorkowska, E., Argon, A.S., Cohen, R.E. Size effect of compliant rubbery particles on craze plasticity in polystyrene. Macromolecules 1990, 23, 3838–48.

15. Kramer, E.J. Microscopic and molecular fundamentals of crazing. Adv. Polym. Sci. 1983 52, 53, 275–334.

16. Galeski, A., Argon, A.S., Cohen, R.E. Changes in the Morphology of Bulk Spherulitic Nylon 6 Due to Plastic Deformation. Macromolecules 1989, (21), 2761–2770.

17. Peterlin, A. Molecular model of drawing of polyethylene and polypropylene. J.Mater.Sci. 1972, B6 (4), 583–598.

18. Galeski, A, Argon, A.S, Cohen, RE. Morphology of nylon 6 spherulites in bulk. Makromol. Chem. 1987, (188),1195–1204.

19. Galeski, A, Krasnikova, N.P. unpublished results.

20. Duffo, P., Monasse, B., Haudin, J.M., G'Sell, C., Dahoun, A. Rheology of polypropylene in the solid state. J. Mat. Sci. 1995, (30), 701–711.

21. Weynant, E., Haudin, J.M., G'Sell, C.J. *In situ* observation of the spherulite deformation in polybutene-1(Modification I). J. Mater. Sci. 1980, (15), 2677–2692.

22. Schaper, A., Hirte, R., Ruscher, C. The electron microscope characterization of the fine structure of nylon 6. II. On the rearrangement of the structure during cold-drawing. Colloid. Polym. Sci. 1986, (264), 668–675.

23. Peterlin, A. Dependence of diffusive transport on morphology of crystalline polymers J. Macromol. Sci. Phys. 1975, B(11), 57–87.

24. Kausch, H.H., Gensler, R., Grein, C., Plummer, C.J.G., Scaramuzzino, P. Crazing in Semicrystalline Polymers. J. Macromol. Sci. Phys. 1999, B(38), 803–815.

25. Galeski, A., Koenczoel, L., Piorkowska, E., Baer, E. Acoustic emission during crystallization of polymers. Nature 1987, (325), 40–41.

26. Pawlak, A., Piorkowska, E. Effect of Negative Pressure on Melting Behaviour of Spherulites in Thin Films of Several Crystalline Polymers. J. Appl. Polym. Sci. 1999, (74), 1380–1385.

27. Piorkowska, E., Nowacki, R. Cavitation during isothermal crystallization of iPP and POM, In: NATO Science Series, vol. 84. Liquids Under Negative Pressure, In: Imre, A.R., Maris, H.J., Williams, P.R. Eds., Kluwer Academic Publisher: Dordrecht, Boston, London, 2002, 137–144.

28. Nowacki, R., Kolasinska, J., Piorkowska, E. Cavitation During Isothermal Crystallization of Isotactic Polypropylene. J.Appl.Polym.Sci. 2001, (79), 2439–2448.

29. Lin, L., Argon, A.S. Deformation Resistance in Oriented Nylon 6. Macromolecules 1994, (25), 4011–4024.

30. Shinozaki, D., Groves, G.W. The plastic deformation of oriented polypropylene: tensile and compressive yield criteria. J. Mater. Sci. 1973, (8), 71–78.

31. Bartczak, Z., Galeski, A. Yield and Plastic Resistance of α-Crystals of Isotactic Polypropylene. Polymer 1999, (40), 3677–3684.

32. Peterlin, A. Plastic deformation of polymers with fibrous structure. Colloid. Polym. Sci. 1975, (253), 809–823.

33. Peterlin, A. Plastic deformation of crystalline polymers, In: Polymeric Materials Baer, E. Ed., Metals Park Ohio: American Society for Metals, 1975, 175–195.

34. Keller, A. Polymer crystals. Rep. Prog. Phys. 1968, (31), 623–704.

35. Bowden, P.B., Young, R.J. Critical Resolved Shear Stress for [001] Slip in Polyethylene. Nature 1971, (229), 23–25.

36. Haudin, J.M. Plastic deformation of semicrystalline polymers, In: Plastic Deformation of Amorphous and Semi-crystalline Materials. Escaig, B., G'Sell, C., Eds., Les Editions de Physique: Paris, 1982, 291.

37. Flory, P.J., Yoon, D.Y. Molecular morphology in semicrystalline polymers. Nature 1978, (272), 226–229.

38. Gent, A.N., Madan, S. Plastic Yielding of Partially Crystalline Polymers. J. Polym. Sci. Polym. Phys. Ed. 1989, (27), 1529–1542.

39. Young, R.J. A dislocation model for yield in polyethylene. Phil. Mag. 1976, (30), 86–94.

40. Young, R.J. Screw Dislocation Model for Yield in Polyethylene. Mater. Forum 1988, (11), 210–216.

41. Peterlin, A. Radical formation and fracture of highly drawn crystalline polymers. J. Macromol. Sci.-Phys. 1971, (6), 490–508.

42. Kelly, A., Groves, G.W. Crystallography of Crystal Defects, Longman: London, UK, 1970, 1–275.

43. Cottrell, A.H., Dislocation and Plastic Flow in Crystals, Oxford University Press: London, 1953, Chapter 3.

44. Bartczak, Z., Argon, A.S., Cohen, R.E. Deformation Mechanisms and Plastic Resistance in Single-Crystal-Textured High-Density Polyethylene. Macromolecules 1992, (25), 5036–5053.

45. Bowden, P.B., Young, R.J. Deformation mechanisms in crystalline polymers. J. Mater. Sci. 1974, (9), 2034–2051.

46. Ward, I.M., Hadley, D.W., An Introduction to the Mechanical Properties of Solid Polymers, Wiley: New York, 1993, 232–245.

47. Shinozaki, D., Groves, G.W. The plastic deformation of oriented polypropylene: tensile and compressive yield criteria. J. Mater. Sci. 1973, (8), 71–78.

48. Caddell, R.M., Raghava, R.S., Atkins, A.G. A yield criterion of anisotropic and pressure dependent solids such as oriented polymers. J. Mater. Sci. 1973, (8), 1641–1646.

49. Aboulfaraj, M., G'Sell, C., Ulrich, B., Dahoun, A. *In situ* observation of the plastic deformation of polypropylene spherulites under uniaxial tension and simple shear in the scanning electron microscope. Polymer 1995, (36), 731–742.

50. Bartczak, Z., Martuscelli, E. Orientation and properties of sequentially drawn films of an isotactic polypropylene, hydrogenated oligocyclopentadiene blends. Polymer 1997, (38), 4139–4149.

51. Bartczak, Z., Galeski, A. Yield and Plastic Resistance of α-Crystals of Isotactic Polypropylene. Polymer 1999, (40), 3677–3684.

52. Lee, B.J., Argon, A.S., Parks, D.M., Ahzi, S., Bartczak, Z. Simulation of Large Strain Plastic Deformation and Texture Evolution in High-Density Polyethylene. Polymer 1993, (34), 3555–3575.

53. Lin, L., Argon, A.S. Review: Structure and plastic deformation of polyethylene. J. Mater. Sci. 1994, (29), 294–323.

54. Bellaire, A., Argon, A.S., Cohen, R.E. Development of texture in poly(ethylene terephthalate) by plane-strain compression. Polymer 1993, (34), 1393–1403.

55. Eling, B., Gogolewski, S., Pennings, A.J. Biodegradable materials of poly(L-Lactic acid): 1 Melt-spun and solution n-spun fibres. Polymer 1982, (23), 1587–1593.

56. Galeski, A., Argon, A.S., Cohen, R.E. Morphology of bulk nylon 6 subjected to plane strain compression. Macromolecules 1991, (24), 3953–3961.

57. Lewis, D., Wheeler, E.J., Maddams, W.F., Preedy, J.E. Comparison of twinning produced by rolling and annealing in high- and low-density polyethylene. J. Polym. Sci. 1972, A-2 Notes, (10), 369–373.

58. Bartczak, Z. Deformation of high density polyethylene produced by rolling with side constraints I Orientation behavior. J. Appl. Polym. Sci. 2002, (86), 1396–1404.

59. Young, R.J., Bowden, P.B. Twinning and martensitic transformations in oriented high-density polyethylene. Phil. Mag. 1974, (29), 1061–1073.

60. Kryszewski, M., Pakula, T., Galeski, A., Milczarek, P., Pluta, M. Über einige Ergebnisse der Untersuchung der Korrelation zwischen der Morphologie und den mechanischen Eigenschaften von kristallinen Polymeren. Faserforschung und Textiltech 1978, (29), 76–85.

61. Bartczak, Z., Cohen, R.E., Argon, A.S. Evolution of Crystalline Texture of High-Density Polyethylene During Uniaxial Compression. Macromolecules 1992, (25), 4692–4704.

62. Galeski, A., Bartczak, Z., Argon, A.S., Cohen, R.E. Morphological alterations during texture producing plastic plane strain compression of high density polyethylene. Macromolecules 1992, (25), 5705–5718.

63. Peierls, R. The size of dislocations. Proc. Phys. Soc. 1940, 289, (52) 34–37.

64. Nabarro, F.R.N. Dislocations in a simple cubic lattice. Proc. Phys. Soc. 1947, (59), 256–272.

65. Honeycomb, R.W.K. The Plastic Deformation of Metals, Edward Arnold Ltd: London, UK, 1968, 41–47.

66. Frank, F.C., Read, W.T. Multiplication processes for slow moving dislocations. Phys. Rev. 1950, (79), 722–723.

67. Shadrake, L.G., Guiu, F. Dislocation in polyethylene crystals: Line energies and deformation modes. Phil. Mag. 1974, (34), 565–581.

68. Crist, B. Yielding of semicrystalline polyethylene: a quantitative dislocation model. Polym. Comm. 1989, (30), 69–71.

69. Darras, O., Seguela, R. Tensile Yield of Polyethylene in Relation to Crystal Thickness. J.Polym.Sci.Part B. Polym.Phys. 1993, (31), 759–766.

70. Brooks, N.W.J., Ducket, R.A., Ward, I.M. Temperature and Strain-Rate Dependence of Yield Stress of Polyethylene.J. Polym. Sci. Part B. Polym. Phys. 1998, (36), 2177–2189.

71. Gaucher-Miri, V., Seguela, R. Tensile yield of polyethylene and related copolymers: Mechanical and structural evidences of two thermally activated processes. Macromolecules 1997, (30), 1158–1167.

72. Seguela, R. Dislocation Approach to the Plastic Deformation of Semicrystalline Polymers: Kinetic Aspects for Polyethylene and Polypropylene. J. Polym. Sci. Part B. Polym. Phys. 2002, (40), 593–601.

73. Kazmierczak, T., Galeski, A. Transformation of Polyethylene Crystals by High Pressure. J. Appl. Polym. Sci. 2002, (86), 1337–1350.

74. Kazmierczak, T., Galeski, A. 2003 Plastic deformation of polyethylene crystals as a function of crystal thickness. Proceedings of VIth ESAFORM Conference on Material Forming, University Salerno, Italy.

75. Kazmierczak, T., Galeski, A. Plastic deformation of polyethylene crystals as a function of crystal thickness and compression rate. Proceedings of 12th International Conference on Deformation, Yield and Fracture of Polymers, Cambridge: UK, Apr. 7–10, 2003.

76. Brooks, N.W.J., Mukhtar, M. Temperature and stems length dependence of the yield stress of polyethylene. Polymer 2000, (41), 1475–1480.

77. Kazmierczak, T., Galeski, A., Argon, A.S. Plastic deformation of polyethylene crystals as a function of crystal thickness and compression rate. Polymer, in press.

78. Argon, A.S., Galeski, A., Kazmierczak, T. Rate mechanisms of plasticity of semi-crystalline polyethylene, Polymer, in press.

79. Prevorsek, D.C. Ultimate properties, uniaxial systems, In: Encyclopedia of Polymer Science and Engineering, Suppl. Vol. Wiley: New York, 1989, 803–821.

80. van der Werff, H., Pennings, A.Z., Tensile deformation of high strength and high modulus polyethylene fibers. Colloid. Polym. Sci. 1991, (269), 747–763.

81. Zhurkov, S.N., Vettegren, V.J., Korsukov, V.E., Noval, J.J. Frequency of fracture as a function of temperature and tensile stress. Proceedings of the 2nd International Conference on Fracture, Brighton London, UK: Chapman and Hall, Ltd., 1969, 545.

82. Raju, V.R., Smith, G.G., Martin, G., Knox, J.R., Graessley, W.W. Properties of amorphous and crystallizable hydrocarbon polymers. J. Polym. Sci. Polym. Phys. Ed. 1979, (17), 1183–1195.

83. Cassagnau, P., Montfort, J.P., Martin, G., Monge, P. Rheology of polydisperse polymers: relationship between intermolecular interactions and molecular weight distribution. Rheol. Acta 1993, (32) 156–167.

84. Capaccio, G., Ward, I.M. Ultra-High-Modulus Linear Polyethylene through Controlled Molecular Weight and Drawing. Polym. Eng. Sci. 1975, (15), 219–224.

85. Capaccio, G., Crompton, T.A., Ward, I.M. The Drawing Behavior of Linear Polyethylene. I. Rate of Drawing as a Function of Polymer Molecular Weight and Initial Thermal Treatment. J. Polym. Sci., Polym. Phys. Ed. 1976, (14), 1641–1658.

86. Capaccio, G., Crompton, T.A., Ward, I.M. The Drawing Behavior of Linear Polyethylene. II. Effect of Draw Temperature and Molecular Weight on Draw Ratio and Modulus. J. Polym. Sci., Polym. Phys. Ed.1980, (18), 301–309.

87. Mezgani, K., Spruiell, J.E. High Speed Melt Spinning of Poly(L-Lactic Acid) Filaments. J. Polym. Sci., Polym. Phys. Ed. 1998, (36), 1005–1012.

88. Gupta, P., Schulte, J.T., Flood, J.E., et al. Development of High-Strength Fibers from Aliphatic Polyketones by Melt Spinning and Drawing. J. Appl. Polym. Sci. 2001, (82), 1794–1815.

89. Bond, E.B., Spruiell, J.E. Melt Spinning of Metallocene Catalyzed Polypropylene. II. As Spun Filament Structure and Properties. J. Appl. Polym. Sci. 2001, (82), 3237–3247.

90. Suh, J., Spruiell, J.E., Schwartz, S.A. Melt Spinning and Drawing of 2-Methyl-1,3-Propanediol-Substituted Poly(Ethylene Terephtalate). J. Appl. Polym. Sci. 2003, (88), 2598–2606.

91. deGennes, P.G. Coil-Stretch Transition of Dilute Flexible, Polymers under Ultrahigh Velocity Gradients. J. Chem. Phys. 1974, (60), 5030–5042.

92. Pennings, A.J. Bundle-like Nucleation and Longitudinal Growth of Fibrillar Polymer Crystals from Flowing Solutions. J. Polym. Sci., Polym. Symp. 1977, (59), 55–66.

93. Keller, A., Odell, J.A. The Extensibility of Macromolecules in Solution: A New Focus for Macromolecular Science. Colloid. Polym. Sci. 1985, (263), 181–201.

94. Zwijnenburg, A., Pennings, A.J. Longitudinal Growth of Polymer Crystals from Flowing Solutions. Colloid. Polym. Sci. 1976, (254), 868–881.

95. Mackley, M.R., Sapsford, G.S. Oriented polymer films. European Pat 0230410, assigned to National Research. Develop. Corp. (GB), 1985.

96. Smith, P., Lemstra, P.J., Kalb, B., Pennings, A.J. Ultrahigh-Strength Polyethylene Filaments by Solution Spinning and Hot Drawing. Polym. Bull. 1979, (1), 733–736.

97. Smith, P., Lemstra, P.J., Booij, H.C. Ultradrawing of High-Molecular-Weight Polyethylene Cast from Solution. II. Influence of Initial Polymer Concentration. J. Polym. Sci., Polym. Phys. Ed. 1981, (19), 877–888.

98. P.Smith, P., Lemstra, P.J, Pijpers, J.P.L. Tensile Strength of Highly Oriented Polyethylene. II. Effect of Molecular Weight Distribution. J. Polym. Sci., Polym. Phys. Ed. 1982, (20), 2229–2241.

99. Smith, P., Lemstra, P.J. Ultra-strength Polyethylene Filaments by Solution Spinning/Drawing. Part 2. Makromol. Chem. 1979, (180), 2983–2986.

100. Plummer, C.J.G., Kausch, H.H. Deformation and Entanglement in Semicrystalline Polymers. J. Macromol. Sci., Phys. 1996, (B35), 637–657.

101. Chanzy, H.D., Bonjour, E., Marchessault, R.H. Nascent Structures during the Polymerization of Ethylene. Colloid. Polym. Sci. 1974, (252), 8–14.

102. Tervoort-Engel, Y.M.T., Lemstra, P.J. Morphology of Nascent Ultra-High Molecular Weight Polyethylene reactor Powder: Chain Extended versus Chain Folded Crystals. Polym. Comm. 1991, (32), 343–345.

103. Bieser, J.O., Krupp, S.P., Knickerbocker, E.N. Method of improving melt spinning of linear polyethylene polymers, US Pat.5254299, assigned to Dow Chemical Co., 1993.

104. Barham, P.J., Sadler, D.M. A Neutron Scattering study of the Melting Behaviour of Polyethylene Single Crystals. Polymer 1991, (32), 393–395.

105. Lemstra, P.J., van Aerle, N.A.J.M., Bastiaansen, C.W.M. Chain Extended Polyethylene. Polymer 1987, (19), 85–98.

106. Psarski, M., Piorkowska, E., Galeski, A. Crystallization of Polyethylene from Melt with Lowered Chain Entanglements. Macromolecules 2000, (33), 916–932.

107. Kanamoto, T., Ohama, T., Tanaka, K., Takeda, M., Porter, R.S. Two-Stage Drawing of Ultra-High Molecular Weight Polyethylene reactor Powder. Polymer 1987, (28), 1517–1520.

108. Hirofumi, K., Yoshimu, I., Kazuo, M., Akira, S., Shigeki, Y. Highly oriented polyethylene material, European Pat. 0376423, assigned to Nippon Oil Co., 1990.

109. Rastogi, S., Spoelstra, A.B., Goossens, J.G.P., Lemstra, P.J. Chain Mobility in Polymer Systems: on the Borderline between Solit and Melt. 1. Lamellar Doubling during Annealing of Polyethylene. Macromolecules 1997, (30), 7880–7889.

110. Rastogi, S., Kurelec, L., Lemstra, P.J. Chain Mobility in Polymer Systems: on the Borderline between Solit and Melt. 2. Crystal Size Influence in Phase Transition and Sintering of Ultrahigh Molecular Weight Polyethylene via the Mobile Hexagonal Phase. Macromolecules 1998, (31), 5022–5031.

111. Dijkstra, D.J., Pennings, A.J. Cross-linking of ultra-high strength polyethylene fibres by means of electron beam irradiation. Polym. Bull. 1987, (17), 507–513.

112. de Boer, J., Pennings, A.J. Crosslinking of Ultra-High Strength Polyethylene Fibers by Means of g- Radiation. Polym. Bull. 1981, (5), 317–324.

113. Klein, P.G., Woods, D.W., Ward, I.M. The Effect of Electron Irradiation on the Structure and Mechanical Properties of Highly Drawn Polyethylene Fibers. J. Polym. Sci., Part B: Polym. Phys. 1987, (25), 1359–1379.

114. Hikmet, R., Lemstra, P.J., Keller, A. X-linked Ultra High Strength Polyethylene Fibers. Colloid. Polym. Sci. 1987, (265), 185–193.

115. Lemstra, P.J. Process for preparing polyethylene films having a high tensile strength and a high modulus, GB Patent 2164897, assigned to Stamicarbon, NL, 1986.

116. Capaccio, G., Gibson, A.G., Ward, I.M. Drawing and Hydrostatic Extrusion of Ultra-High Modulus Polymers, Ultra-High Modulus Polymers, Ciferri, A.,Ward, I.A. Eds., Appl. Sci. Pub., London, UK, 1979, 1–76.

117. Taraiya, A.K., Nugent, M., Sweeney, J., Coates, P.D, Ward, I.M. Development of Continuous Die Drawing Production Process for Engineered Polymer Cores for Wire Ropes. Plast. Rub. Comp. 2000, (29), 46–50.

118. Mohanraj, J., Chapleau, N., Ajji, A., Duckett, R.A., Ward, I.M. Fracture Behavior of Die-Drawn Toughened Polypropylene. J. Appl. Polym. Sci. 2003, (88), 1336–1345.

119. Taraiya, A.K., Mirza, M.S., Mohanraj, J., Barton, D.C., Ward, I.M. Production and Properties of Highly Oriented Polyoxymethylene by Die-Drawing. J. Appl. Polym. Sci. 2003, (88), 1268–1278.

120. Berg, E.M., Sun, D.C., Magill, J.H. 3D structure-property relationship in rolltruded polymers: Part I: Mechanical property enhancement in three directions. Polym. Eng. Sci. 1989, (29), 715–721.

121. Berg, E.M., Sun, D.C., Magill, J.H. 3D structure-property relationship in rolltruded polymers: Part II Anisotropic yielding and deformation in triaxially oriented polymers. Polym. Eng. Sci. 1990, (30), 635.

122. Magill, J.H.Rolltrusion processing of polypropylene for property improvement Polypropylene. An A-Z Reference, Karger-Kocsis, J. Ed., Kluwer Acad Pub: Dordrecht, 1999, 728–735.

123. Ciora, R.., Magill, J.H Rolltruded Poly(aryl ether ether ketone) (PEEK) for Membrane Applications. Separ.Sci.Technol. 1997, 32, 899–923.

124. Bartczak, Z., Galeski, A, Morawiec, J., Przygoda, M. Method and appartaus for production of highly oriented polymer shapes, Polish Patent PL-178058 B1 assigned to Centre of Molecular and Macromolecular Studies Polish Academy of Sciences Lodz, 2000.

125. Morawiec, J., Bartczak, Z., Kazmierczak, T., Galeski, A. Rolling of polymeric materials with side constraints. Mater. Sci. Eng. 2001, (A317), 21–27.

126. Bartczak, Z., Morawiec, J., Galeski, A. Structure and properties of isotactic polypropylene oriented by rolling with side constraints. J. Appl. Polym. Sci. 2002, (86), 1413–1425.

127. Bartczak, Z., Morawiec, J., Galeski, A. Deformation of high density polyethylene produced by rolling with side constraints II Mechanical properties of oriented bars. J. Appl. Polym. Sci. 2002, (86), 1405–1412.

6

Microdeformation and Fracture in Semicrystalline Polymers

CHRISTOPHER J.G. PLUMMER

Laboratoire de Technologie des Composites et
Polymères (LTC), Ecole Polytechnique Fédérale
de Lausanne (EPFL)

CONTENTS

I.   INTRODUCTION

Many polymer-based materials whose successful application
depends on their fracture resistance may be considered to be
"nanostructured" in that they show some degree of structural
heterogeneity at the sub-micron level. These include bulk
semicrystalline thermoplastics such as polyethylene (PE) and
isotactic polypropylene (iPP), which self-organize to form crystalline lamellae with typical thicknesses of about 10 nm [1].
The lamellae are generally considered to anchor the chains in
semicrystalline polymers, so that solid-like behavior (revers-
ibility with respect to small deformations) persists at temperatures, T, well above their glass transition temperature, T_g.
PE, in particular, shows excellent ductility and fracture resistance under ambient conditions, even though $T_g \ll 0°C$.

The term "nano-structured" polymer is also commonly
understood to refer to polymers that contain chemically distinct phases in which the heterogeneity has been deliberately
engineered. A particularly high degree of microstructural control in two (or multiple) phase polymers is possible through
self-organization of block copolymers with a well-defined
molar mass, M, which may give rise to a lamellar microstructure on a similar length scale to that of semicrystalline polymers, depending on the ratio of the block lengths [2]. Indeed,
the deformation and toughening behavior of lamellar rub-
bery/rigid block copolymers, which are gaining in importance
commercially, may prove to show interesting parallels with
the behavior of semicrystalline homopolymers.

In what follows, the fracture properties of some practically important semicrystalline polymers will be reviewed.
Both low and high speed fracture behavior will be considered,
because either or both may be critical to performance, depending on the nature of the polymer and its applications. For
example, although the impact behavior of PE is generally not
of as much concern as that of iPP, its slow crack growth (SCG)
behavior is important in applications such as piping, where
it is subject to low-level loading over extended times [3]. Both
impact and SCG often involve crazing, that is, the formation
of fibrillar planar or wedge-shaped deformation zones. Craze

formation, morphology and breakdown will therefore be central to the present discussion, in which emphasis will also be placed on the role of entanglement.

II. MICROMECHANISMS OF FRACTURE IN SEMICRYSTALLINE POLYMERS

Figure 1a shows the micro-structure typically observed in high density polyethylene (HDPE) after processing via standard bulk thermoforming operations, along with representative iPP micro-structures (Figures 1b and c), which will be referred to later. The structure in Figure 1a consists of a network of "dominant" lamellae filled in by "secondary" lamellae, which appear less distinct in the micrograph [1]. Given that conventional processing offers little scope for significant modification of the scale and form of the microstructure, attempts to improve the properties of polyethylene (PE) have focused on using advanced polymerization techniques to tailor the degree of branching and molar-mass distribution. The rationale for this is that, regardless of the details of the subcritical deformation mechanisms, cohesive failure ultimately requires either chain breakage or disentanglement of individual chains from their neighbors (sometimes referred to as "forced reptation" in order to distinguish it from classical reptation via Brownian motion [4]). In an isotropic specimen, it is nevertheless reasonable to postulate an intermediate stage of failure that involves a transformation from an initial state, in which the chains adopt globally random configurations, to an orientated state [5]. Hence, in the case of ductile semicrystalline polymers such as PE, there is a transformation from lamellae characterized by some degree of chain folding (Figure 2a) to an oriented "fringed micelle"-type structure (Figure 2b). (Indeed, it could be argued on the basis of chain packing constraints that the fringed micelle structure is only possible for oriented chains.) In the limit of very long chains, forced reptation may therefore conveniently be described in terms of the "block and tackle" mechanism, originally postulated for amorphous polymers [4,6]. Stress is

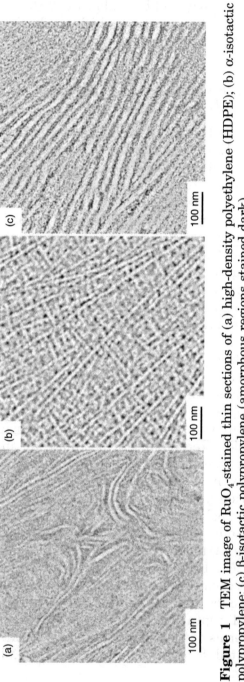

Figure 1   TEM image of RuO$_4$-stained thin sections of (a) high-density polyethylene (HDPE); (b) α-isotactic polypropylene; (c) β-isotactic polypropylene (amorphous regions stained dark).

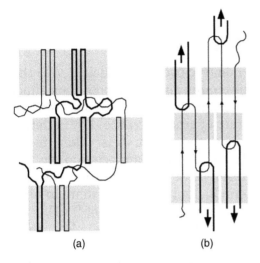

(a)   (b)

Figure 2   Schematic of the initial state of an isotropic semicrystalline polymer before (a) and after (b) ductile necking, illustrating the transformation from a partly chainfolded to a fringed micelle-type arrangement.

transferred to a given chain via its topological interactions with neighboring chains, represented as trapped loops in Figure 2b, and the crystalline regions give rise to a frictional drag that opposes stress relaxation via movement of the chain along its own contour. The effectiveness of anchoring by the crystalline regions (hence the chain mobility and deformation rate) and the length and architecture of the chains then determine the relative extents of chain scission and disentanglement during final failure.

In amorphous polymers, the loops shown in Figure 2b have been identified with entanglements, and the average molar mass associated with each loop is taken to be the molar mass per entanglement, M_e. Because the speed, v, of the chain with respect to its surroundings is nonuniform, for $M \gg M_e$ and affine deformation of the "entanglement points", the mean force $\langle f \rangle$ in a chain during disentanglement by forced reptation has been inferred from the block and tackle model to be

$$\langle f \rangle \approx \frac{\zeta_0 b}{\tau_d} \left(\frac{M}{M_e} \right)^2 \tag{1}$$

where ζ_0 is an effective monomeric friction coefficient, b is the statistical step length of the chains and τ_d is a reptation time, assumed to be fixed in the presence of an imposed deformation rate (for example, a given rate of crack advance) [6]. In deriving equation (1), the force in the chain, f, is taken to be proportional to v locally. The force for chain scission, on the other hand, is relatively insensitive to deformation rate, as discussed in detail elsewhere [3], and for intermediate to high strain speeds, it is estimated to be of the order of 2 nN.

In a semicrystalline polymer, the loop size, M_e^{eff}, may also be identified with the entanglement network if it can be assumed that the entanglement density in a solid semicrystalline polymer reflects that of the melt. Conservation of entanglements is a reasonable assumption for rapid solidification of an isotropic melt, for which the radii of gyration of the chains change little and the topological relationships between the chains are substantially maintained (cf. Figure 2) [7]. Entanglement has been argued to be reduced significantly by crystallization at low rates, as the characteristic times for incorporation of individual chains into the lamellae become comparable with, or exceed the longest melt relaxation times [8–10]. Corresponding decreases in fracture resistance have been observed in semicrystalline polymers that have undergone relatively slow solidification, for example, by isothermal crystallization at low supercooling [8]. However, such conditions are not relevant to the usual industrial processing routes.

Even given an entanglement molar mass close to M_e, the network must remain stable throughout the transformation from a partially chainfolded structure (Figure 2a) to the oriented structure (Figure 2b) for entanglement to be effective in the final stages of fracture. Hence, amorphous sub-chains within the interlamellar regions must be anchored by the crystalline lamellae [11,12]. Assuming approximately equal proportions of amorphous and crystalline material, significant stabilization of the network is expected when M is at least 3 to 4 times greater than the molar mass, M_1, associated with chainfolded crystalline blocks in the lamellae [13,14]. If these blocks consist of an average of 3 adjacent stems (as inferred from classical models of lamellar growth at high undercoolings [15]), then for PE, say, a typical lamellar thickness of about 10 nm implies M_1 to be about 3500 g mol^{-1}. Effective anchoring therefore requires that $M > M_{crit} = 10$–14 kg mol^{-1} in PE. An alternative way of estimating M_{crit} that avoids any assumptions about the degree of chainfolding, is to suppose that effective anchoring only occurs when the root-mean-square end-to-end distance of the chains is of the order of the lamellar long period, which leads to M_{crit} of about 50 kg mol^{-1} [12]. In both cases, the predicted value is very much greater than M_e (about 850 g mol^{-1} in PE [16]). Hence, unlike in amorphous polymers, where $M_c = 2M_e$ defines the threshold for stability of the entanglement network, in semicrystalline polymers it is the scale of the lamellar texture that determines the critical mass for $T > T_g$. Moreover, in regimes of M immediately above M_{crit}, there will remain a significant proportion of amorphous sub-chains associated with free ends ("dangling chains") of molar mass approximately equal to M_1, that cannot contribute to a network. Hence, the effective entanglement molar mass for identical chains of molar mass M may be estimated from

$$M_e^{eff} = \frac{M \cdot M_e}{M - 2M_1} \qquad (2)$$

Figure 3 shows experimental data for iPP specimens with different weight average molar masses, M_w, tested at 1 mm/min in simple tension [17,18], in which significant cohesive strength is observed for M_w down to 100 kg mol^{-1} (Figure 3a). At the same time, however, the natural draw ratio, λ_{max}, in the necked regions of fully ductile specimens shows a sharp increase with decreasing M_w, which may reflect the reduced entanglement constraints predicted by equation (2). By analogy with an amorphous polymer [19],

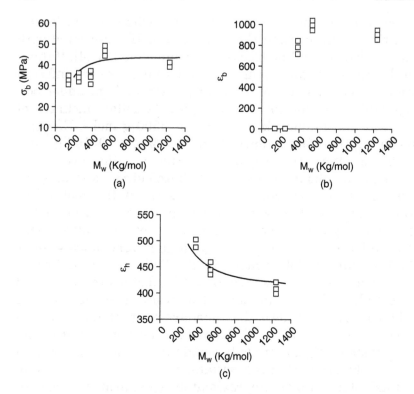

Figure 3 Tensile test data for isotactic polypropylene (iPP) with different weight average molar masses: (a) nominal failure stress; (b) nominal failure strain; (c) nominal strain at the onset of work hardening (ductile specimens).

$$\lambda_{max} \approx \lambda_{max}^{\infty} \sqrt{\frac{M_e^{eff}}{M_e}} = \lambda_{max}^{\infty} \sqrt{\frac{M}{M - 2M_1}} \qquad (3)$$

where λ_{max}^{∞} is the limiting natural draw ratio as $M \to \infty$ (that is, in the absence of chain ends). The results in Figure 3a also indicate a small decrease in the ultimate tensile strength, σ_b, with decreasing M_w. If simple cohesive failure by chain scission is assumed, σ_b, is expected to be proportional to the number of entangled strands crossing unit area of undeformed specimen, which leads to

$$\sigma_b \approx \sigma_b^\infty \sqrt{\frac{M_e}{M_e^{\text{eff}}}} = \sigma_b^\infty \sqrt{1 - \frac{2M_l}{M}} \tag{4}$$

where σ_b^∞ is the limiting ultimate tensile strength as $M \to \infty$. A fit to the observed values in Figure 3a and Figure 3b suggests M_l to be about 55 kg mol^{-1} in each case, which is an order of magnitude higher than the value of 5000 g mol^{-1} estimated for a lamellar thickness of 10 nm. However, the results in Figure 3 are for highly polydisperse specimens, so that equation (2) should be replaced by,

$$\frac{M_e}{M_e^{\text{eff}}} = \int_{M_{\text{crit}}}^{\infty} \phi(M)dM \left(1 - \frac{2M_l \int_{M_{\text{crit}}}^{\infty} \phi(M)dM}{\int_{M_{\text{crit}}}^{\infty} M\phi(M)dM} \right)$$

where $\phi(M)$ is the molar mass distribution, so that if the number average molar mass, $M_n \gg M_{\text{crit}}$, this reduces to

$$\frac{M_e}{M_e^{\text{eff}}} \approx 1 - \frac{2M_l}{M_n} \approx 1 - \frac{2I \cdot M_l}{M_w}$$

The polydispersity, I, is about 5 for the specimens in Figure 3, possibly explaining the discrepancy between the estimated and apparent values of M_l. Even so, the evolution of the ultimate strain (Figure 3c) indicates a ductile-brittle transition to occur at M_w 250 kg mol^{-1}. This is associated with a transition from fully ductile behavior to failure by breakdown of isolated crazes, so that the assumption of a unique failure mechanism implicit in invoking equation (4) is clearly an oversimplification.

Although the above considerations highlight the difficulties in detailed interpretation of data for typical commercial semicrystalline polymers, they are consistent with the general observation that fracture resistance can be improved by increasing M. On the other hand, indefinite increases in M

eventually lead to problems with conventional processing, and reduced crystallinity in semicrystalline polymers. Ultra high molecular weight PE (UHMWPE) cannot therefore be considered a panacea, for example, although its exceptional wear resistance is sufficiently important in certain applications to justify *ad hoc* processing routes [13]. Branching is also expected to improve anchoring of the chains, but high degrees of branching again reduce the degree of crystallinity so that improved toughness is accompanied by a reduction in yield stress and stiffness. Many latest generation high density PEs (HDPEs) destined for pipe applications, in which long term, low-level loading is widely assumed to promote failure by disentanglement, therefore contain a mixture of long, branched "anchor" chains, and shorter linear chains whose role is to maintain the processing characteristics and crystallinity of conventional HDPEs [20].

The importance of disentanglement for long term failure may be inferred from the form of equation (1). Under static conditions, $\langle f \rangle$ is fixed by the external load, so that failure should always occur by disentanglement after sufficiently long times $t = \tau_d$, if ductile yielding does not intervene first. Given that the yield stress, σ_y, is relatively insensitive to rate, this accounts qualitatively for the observed fracture behavior of polymers such as HDPE subject to static loads at temperatures approaching the melting point, T_m, where increased molecular mobility is thought to favor disentanglement (through the effective value of ζ_o, which in this case will be related to chain slip through crystalline regions). Under these conditions, a ductile-brittle transition is observed as the load decreases and failure times increase, beyond which failure times are far shorter than would be expected based on extrapolation from high stress ductile regimes of fracture, as shown in Figure 3. In the ductile failure regimes, the flow rate, $\dot{\varepsilon}$, is usually related to the stress, σ, by an empirical power law. If the failure time $t_f \varepsilon_f / \dot{\varepsilon}$, where ε_f is the failure strain, one obtains

$$\sigma \approx \sigma_o \left(\frac{t_f}{t_{fo}} \right)^{-1/n} \tag{5}$$

with n between 10 and 20 for a wide range of polymers, as illustrated in Figure 4, where n is approximately 20 in the ductile regime [3]. In the brittle regimes of behavior associated with longer failure times, after a certain induction period, SCG is observed. The corresponding crack advance rate is typically given by

$$\dot{a} \approx \dot{a}_o \left(\frac{K}{K_{co}} \right)^m \tag{6}$$

where a is the crack length, $K \propto \sigma \sqrt{a}$ is the stress intensity factor, K_{co} is a scaling parameter and m is between 3 and 5

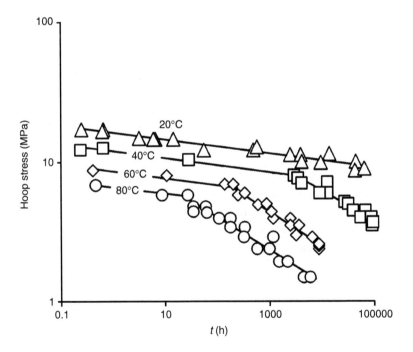

Figure 4 Hoop stress (circumferential stress in a tube subject to internal pressure) vs. time to fail in HDPE pipes at different temperatures. (After Kausch H-H, Polymer Fracture, 2nd Ed. Heidelberg-Berlin: Springer, 1987.)

[21]. For a fixed initial defect size, and assuming $t_f a_f / \dot{a}$ where a_f is some critical amount of crack growth for failure,

$$\sigma \propto \left(\frac{t_f}{t_{fo}} \right)^{-1/m} \tag{7}$$

in static experiments, again as illustrated in Figure 4, where $m \approx 4$ in the brittle failure regime.

Trends toward brittle behavior with decreasing effective deformation rates have also been reported in certain glassy polymers such as polycarbonate (PC), especially close to T_g, where they have again been linked to disentanglement [22,23]. However many commonly available glassy polymers, such as polymethylmethacrylate (PMMA), show fragile behavior over a wide range of strain rates and limited ductility in tension at ambient temperature. This has been attributed to their relatively low entanglement density compared to that in polymers such as PC, which makes it energetically favorable for chains to break at relatively low global stresses, and hence promotes craze formation and craze breakdown [19,24]. The competition between ductile behavior and crazing may also be strongly influenced by the strength of secondary transitions, and their consequences for the local segmental mobility below T_g, which is closely associated with yielding [25]. Thus, although the relative fragility of iPP with respect to PE may be partly due to the lower entanglement density of iPP ($M_e \simeq 5000$ g mol^{-1} [16]), it is also pertinent that T_g in iPP is just below ambient temperature and its melting point T_m is about 170°C, whereas T_g in PE is well below 0°C and T_m is generally between 120 and 140°C. The molecular mobility is therefore inferred to be considerably higher in PE at ambient temperature than in iPP. It follows that the critical resolved shear stress (CRSS) for the most active slip system in PE is reported to be 7.2 MPa, whereas the effective CRSS in iPP is in excess of 20 MPa [26]. The unusual cross-hatched microstructure of the usual α modification of PP (Figure 1b) may also play a role in the observed behavior, since it has been suggested to hinder slip mechanisms with a relatively low CRSS during the early stages of deformation [27,28]. This implies not only a relatively high σ_y, but also a marked yield drop as these slip systems become activated in the latter stages of yielding. The resulting tendency for slip processes to localize may also favor crazing in tension.

A further deformation mode is seen in polyoxymethylene (POM), whose σ_y and degree of crystallinity are both relatively high and for which $M_e \simeq 1600$ g mol^{-1} [16]. In this case, tensile deformation at ambient temperature involves neither ductile drawing, nor fragile rupture, but rather the formation of numerous stable craze-like zones of cavitation [8,29]. The material is able to accommodate relatively large strains (up to 100%) in a nearly reversible manner because the matrix surrounding the cavitated regions is able to recover elastically on unloading. Similar modes of deformation at crack tips render POM relatively tough, in spite of its limited macroscopic ductility. Nevertheless, as the temperature is raised, and the mobility increases, POM shows an increased tendency to neck and, indeed, under slow crack growth conditions at high temperatures ($T > 100°C$), reverts to very similar behavior to that of PE under static loading, that is, a ductile-brittle transition with decreasing load, accompanied by crack tip crazing and craze breakdown [30].

III. CRACK TIP MICRODEFORMATION

To understand the link between macroscopic fracture behavior and microscopic quantities, such as M or the lamellar thickness, with the aim of either developing new strategies for improving fracture properties or understanding and optimizing improvements obtained empirically, it is helpful to observe the microdeformation mechanisms directly in combination with sample preparation techniques that maintain as far as possible the deformation-induced morphology as it appears under load [30]. Figure 5 shows craze microstructures in bulk specimens of HDPE and iPP obtained at moderate strain rates at ambient temperature. In each case, the specimen has been embedded under load with a low viscosity resin, and then stained with RuO_4 prior to thin sectioning, using an ultrami-

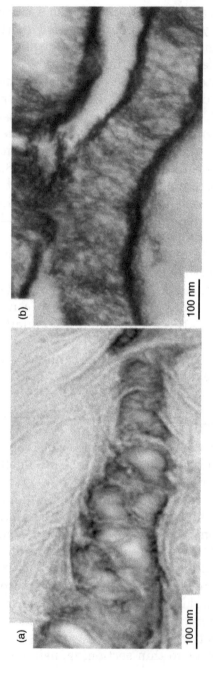

Figure 5 TEM micrographs of crazing in (a) bulk HDPE and (b) bulk α iPP deformed in tension at room temperature (embedded in epoxy and PMMA, respectively, and stained in RuO_4; deformation axis is roughly vertical).

crotome equipped with a diamond knife. HDPE shows a somewhat irregular fibrillar structure, with clear continuity
between the craze fibrils and dominant lamellae visible in the
undeformed material outside the craze. This relationship is
less clear in iPP, but the fibril sizes remain roughly commen-
surate with the lamellar size and spacing. In HDPE such
structures are seen when the degree of plastic constraint is
high, such as at notch tips in bulk specimens, but they may
also be present in unnotched specimens, depending on the
deformation conditions and molecular parameters.

Crazes, or craze-like fibrillar deformation zones are also
characteristic of deformation at the crack tip in HDPE undergoing slow crack growth (SCG) in the brittle regime, as
reflected by Figure 6, which shows part of the crack tip deformation zone in a notched sample subjected to $K = 0.22$
MPam$^{1/2}$ at 80°C [31]. SCG is frequently reported to occur by
a stick slip mechanism such that all or part of the fibrillar
zone at the crack tip breaks down after some characteristic
incubation time. The crack-tip thus advances by a distance
$\Delta a \propto l$, as sketched in Figure 7, and a new fibrillar zone
stabilizes. This behavior is widely accounted for in terms of
a fibril creep failure mechanism and, by implication, disentanglement via forced reptation, discussed earlier in the context of equation (1) [32–34]. If λ is the fibril draw ratio, b^2 is
the effective cross-sectional area of a chain and σ_s is the
traction at the boundary of the fibrillar region, $\langle f \rangle$ is equal
to $\lambda b^2 \sigma_s$. By equating the fibril breakdown time t_c with τ_d in
equation (1), one obtains

$$t_c \approx \frac{\zeta_o}{\lambda b \sigma_s} \left(\frac{M}{M_e^{eff}} \right)^2 \approx \frac{\zeta_o d_o}{\delta \sigma_s b} \left(\frac{M}{M_e^{eff}} \right)^2 \tag{8}$$

where d_o is the initial width of the layer of material that
fibrillates to form the craze, and δ is the crack opening displacement [32]. If σ_s and d_o are uniform and independent of
K, the Dugdale model implies $l \propto K^2$ for $a \gg l$, where l is
the length of the fibrillar region in Figure 7, and $\delta \propto K^2$ [35].
Taking $\dot{a} \propto l / t_c$ it follows that

200 nm

Figure 6 Craze structure in a notched specimen of HDPE subject to an initial applied stress intensity factor of 0.22 MPam$^{1/2}$ at 80°C (embedded in epoxy and stained in RuO$_4$; deformation axis roughly horizontal). (From Plummer CJG, Goldberg A, Ghanem A, Polymer 2001; 42: 9551–9564. With permission.).

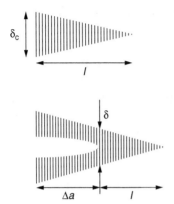

Figure 7 Sketch of the stick-slip mechanism of slow crack growth. (From Plummer CJG. Polymer 2004; 169, 75. With permission.)

$$\dot{a} \propto \frac{\delta b \sigma_s l}{\zeta_o d_o} \left(\frac{M_e^{\mathit{eff}}}{M} \right)^2 \propto \frac{K^4}{M^2} \tag{9}$$

The K^4 dependence predicted by equation (9) is consistent with observation (cf. equation [6]), but the M dependence is difficult to verify since most experimental data pertain to highly polydisperse specimens, and the assumption that $M \gg M_e^{\mathit{eff}}$ is unrealistic in many commercial grades of HDPE. Other questionable assumptions include the independence of d_o and σ_s on deformation conditions, and the linear dependence of the force in a disentangling chain and its displacement rate with respect to its neighbors, particularly since ζ_o is ill-defined for a semicrystalline solid (indeed there is evidence from interfacial studies for a weaker dependence, as will be discussed in the next section [12]).

Alternative approaches to deriving analytical expressions for SCG based on relatively simple models such as that described above include (i) direct measurement of the constitutive behavior of the craze [36–38] and (ii) prediction of the behavior of individual craze fibrils by analogy with macroscopic necks [33]. The results may then be incorporated into finite element simulations of the global response of the crack/cohesive zone [39]. However, although the analogy with plastic necks is arguably justified at relatively high K within the SCG regimes, where fibrillar structures are relatively coarse and their internal texture resembles that of macroscopic necks, the fibrillar structure becomes progressively finer as K decreases, eventually approaching the lamellar thickness (cf. Figure 6). Indeed, at sufficiently low K, accelerated testing of SCG resistant HDPE grades has indicated the mechanism of crack tip deformation to change to one of interlamellar cavitation and crack propagation via lamella cleavage and/or breakdown of interlamellar ligaments, without the formation of a mature craze structure [31]. This results in a mirror-like fracture surface and the deformation in the vicinity of the crack tip is as shown in Figure 8, where regions of interlamellar cavitation associated with coarse and fine lamellar slip are visible, but no large-scale fibrillation [31,40].

At high imposed deformation rates, $\lambda \simeq \lambda_{max}$ in craze fibrils, and, unlike the situation described above for SCG, the onset of crack propagation often occurs at a relatively well defined critical stress intensity, K_c. This suggests the craze breakdown criteria to be insensitive to t. For mode I crack propagation through a (long) single craze under conditions consistent with linear elastic fracture mechanics, and assuming the craze to behave as an orthotropic elastic body with a finite stiffness in the direction parallel to its length (attributed to the presence of "crosstie" fibrils), it has been shown that

$$K_c \approx \left(\frac{DE}{\sigma_s} \right)^{1/2} \sigma_c \approx \left(\frac{DE}{\sigma_s} \right)^{1/2} \frac{f_c}{b^2 \lambda_{max}} \approx \left(\frac{M_b DE}{M_e^{eff} \sigma_s} \right)^{1/2} \frac{f_c}{b^2} \quad (10)$$

where σ_c is the stress to break a craze fibril, D is the fibril spacing and E is the tensile modulus, M_b is the mass of a statistical segment and f_c is the force to break (or disentangle) an entangled strand [41,44]. Equation (10) has been invoked to account quantitatively for crack initiation and failure at high speed in iPP, where a single crack tip craze is observed in tensile tests on compact tensile specimens, and gives a reasonable order of magnitude estimate if chain scission is assumed, that is, $f_c \simeq 2$ nN [45]. Certain parameters required

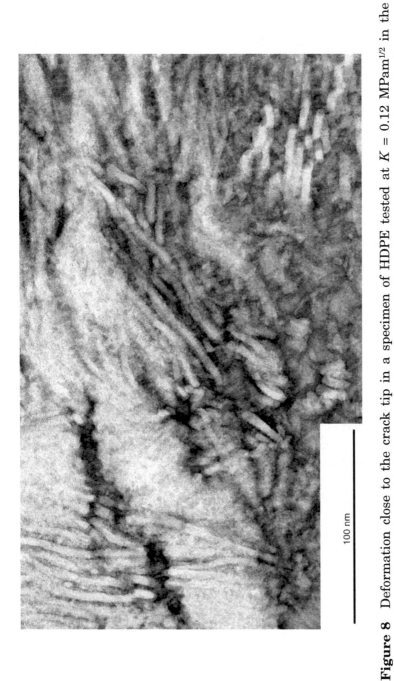

Figure 8 Deformation close to the crack tip in a specimen of HDPE tested at $K = 0.12$ MPam$^{1/2}$ in the presence of a surfactant (embedded in epoxy and stained in RuO$_4$; deformation axis roughly vertical).

in the full analysis, such as the elastic anisotropy of the craze, are nevertheless difficult to estimate, and it is also difficult to vary parameters such as M_e systematically and independently in bulk materials.

Although time dependent failure criteria will not be discussed explicitly in what follows, it is also possible to express K as a function of the crack velocity, $\dot{a}$, by assuming, for example, that

$$K(\dot{a}) \propto f_c(\dot{a}) \tag{11}$$

If breakdown is mediated by disentanglement, then the disentanglement time, τ_d, must be less than equal to the time available for fibril breakdown, $D/\dot{a}$. Hence, by substituting equation (1) into equation (11), one obtains

$$\dot{a} \propto \frac{K}{M^2} \tag{12}$$

and K_c is undefined for crack initiation.

IV. WEDGE TESTING AND INTERFACIAL FAILURE

A convenient way of testing equation (10) is the "wedge test" in which crack advance is achieved by driving a wedge (for example, a razor blade), into a pre-crack, as shown in Figure 9. Provided the physical and geometrical characteristics of the specimen justify the assumption of mode I opening, the critical strain energy release rate, G_c, may be calculated from the crack length ahead of the wedge [46]. K_c is related to G_c through

$$K_c = \sqrt{EG_c} \tag{13}$$

Because the rate of advance of the wedge may be varied systematically, this type of test also provides a straightforward way of testing the rate dependence of the fracture parameters, that is, expressions such as equation (12).

Crack tips are generally characterized by multiple crazing even in brittle bulk polymers so that the detailed assump-

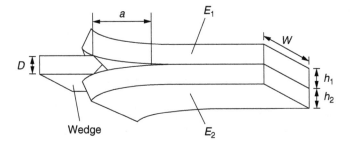

Figure 9 Test geometry used for the wedge test on interfaces.

tions behind equation (10) fail. However, if the crack path along the specimen axis is relatively weak, the single craze criterion may be easier to satisfy. This is often the case in heterogeneous systems such as interfaces between chemically distinct polymers, or partly bonded interfaces between identical polymers, in which Σ, the effective number of entangled strands crossing the unit area of the interface is significantly reduced with respect to its bulk value, $1/(b^2\lambda_{max})$. Indeed, if Σ can be controlled, this type of specimen may be used to investigate the predicted dependence of G_c or K_c on entanglement parameters systematically [47].

In the case of heterogeneous semicrystalline polymer interfaces, Σ has been controlled by using *in situ* chemical reactions to vary the density of covalent bonds straddling the interface [48–51]. Maleic anhydride (MAH) grafted polymers are commonly used for this purpose because MAH reacts readily with a range of functional groups, including the $-NH_2$ terminal groups of polyamides. Thus block or graft copolymers form on heat treatment of interfaces between iPP/iPP-g-MAH blends and polyamide 6 (PA6) [51]. Wedge tests have shown G_c to be proportional to Σ^2 for relatively weak interfaces prepared in this way (G_c up to 100 Jm^{-2}), in which a single crack-tip craze is observed on the iPP/iPP-g-MA side of the interface, consistent with equations (10) and (13) [48–50]. Reasonable agreement has also been obtained between the experimental results and predictions of absolute values of G_c based on the observed craze microstructure at the interface and the assumption of chain scission (G_c is found to be weakly dependent on $\dot{a}$, at least at low $\dot{a}$ [49,50]). However, significant deviations from this simple scaling have also been observed, depending on the nature of the iPP-g-MA anchoring chains, which suggest that the lamellar morphology at the interface plays an important role [48].

Σ may also be varied by varying the thickness of a suitable amphiphilic block copolymer layer deposited at the interface between two incompatible polymers, such that after heat treatment, one block is anchored to one side of the interface and the other block is anchored to the other side of the interface. This technique was first used to investigate interfaces between glassy polymers, for which G_c is again found to be proportional to Σ^2 in the single craze regime [47]. The same approach has been used more recently to investigate interfaces between aPS and PE, with emphasis on the influence of M of the PE compatible block on G_c [12]. This has provided evidence for pull-out, i.e., disentanglement of PE blocks with $M_n < 30$ kg/mol^{-1}. However, scission is observed at higher M_n, and the implied critical value of M for the onset of chain scission is comparable to the simple estimates of M_{crit} in semi-crystalline PE described earlier. Thus disentanglement does not appear to dominate under the conditions studied (constant rate tests at room temperature) as long as anchoring of the entanglement network by the crystalline lamellae remains effective. This provides some direct justification for the earlier assumption of scission in modelling the failure of bulk iPP and iPP-PA6 interfaces [45,48]. Indeed, in glassy polymers, chain lengths equivalent to M_e suffice to induce chain scission at $T \ll T_g$, as has been demonstrated in similar experiments on aPS-poly(2-vinylpyridine) (-PVP) interfaces [52–54]. Systematic investigation of the kinetics of crack propagation along aPS-PE interfaces in the pull-out regime also suggest that $\dot{a} \propto K^{0.4}$, rather than the linear dependence predicted by equation (12), which raises questions as to the general validity of equation (1) for semicrystalline systems [12].

Apart from the role of the lamellar thickness on stabilization of the entanglement network, the specimen microstructure may also influence the qualitative nature of the damage zone, with implications for G_c, given that a more diffuse damage zone is generally expected to lead to a higher crack resistance than for a single craze (owing to increased energy dissipation). iPP is a convenient model system for investigating micro-structural effects, because in addition to the stable α modification (Figure 1b), a metastable β phase appears sporadically in commercial mouldings, and may be induced preferentially using specific nucleation agents. The β phase consists of relatively broad lamellae arranged in twisting parallel stacks (Figure 1c) and is widely associated with improved ductility in iPP. Certainly, in reaction bonded iPP-PA6 interfaces, the presence of β iPP spherulites has been found to result in deviation of crazes and/or crack tips from the interface and nucleation of secondary crazes beyond the interface (in the equatorial or polar regions of the β spherulites) [55]. This is accompanied by relatively diffuse microdeformation locally in the form of either intense lamellar cavitation (Figure 10) or cooperative lamellar shear, depending on the local orientation of the lamellae with respect to the tensile axis. However, the increases in fracture resistance are limited in β-nucleated specimens and the Σ^2 dependence of G_c is maintained, even though the assumption of a single crack tip craze is no longer justified [55–56]. Significantly higher interfacial toughness has been seen in interfaces between rubber modified iPP/iPP-g-MAH and PA6 [56,57]. The crack tip deformation zone in this case consists of rows of cavitated rubber particles, separated by homogeneously necked ligaments of the iPP matrix extending perpendicular to the principal stress axis, with little evidence for crazing (although the layer of iPP immediately adjacent to the PA6 is relatively poor in modifier particles and isolated crazes are observed close to the interface). Nevertheless, immediately adjacent to the crack tip, the whole of the deformation zone forms a continuous highly drawn structure, which may be considered analogous to a single craze, and G_c is once more found to be proportional to Σ^2 [56,57]. This suggests that the general form of equation (10) has relatively wide validity, which underlines the fundamental importance of entanglement in its broadest sense for toughness in semicrystalline polymers.

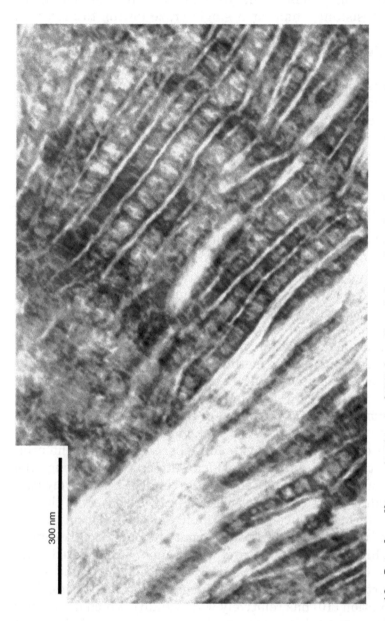

Figure 10 Interlamellar cavitation in βiPP lamellae deformed in the direction perpendicular to their trajectories at ambient temperature (embedded in epoxy and stained in RuO_4).

V. CONCLUSIONS

The role of entanglements in the failure of semicrystalline polymers is widely acknowledged, if only implicitly through reference to concepts such as tie molecules. In the present review, the effects of entanglement have been discussed in terms already commonly employed for amorphous glassy polymers, allowing models developed for these latter to be transposed directly to the semicrystalline case. There are nevertheless considerable difficulties in testing these models with typical commercial semicrystalline thermoplastics, whose molar mass distributions are wide, and, as in the case of latest generation pipe materials, may be convoluted with chemical heterogeneity. Important questions therefore remain open, such as that of the extent to which disentanglement dominates over chain scission as the ultimate fracture mechanism under different conditions. Recent results from wedge tests on heterogeneous interfaces with semicrystalline components have nevertheless provided new insight into the fundamental aspects of fracture in semicrystalline polymers, with practical implications for molecular design in these materials. Although so far limited to a relatively narrow range of experimental conditions, this approach has great potential for parametric fracture studies on semicrystalline polymers.

ACKNOWLEDGMENTS

The author is grateful for the technical support of the Centre Interdisciplinaire de Microscopie Electronique (CIME) of the EPFL, and to H.-H. Kausch, J.-AE. Månson, G. Michler, C. Creton, R. Gensler and many others for their valuable input.

REFERENCES

1. Bassett DC, Principles of Polymer Morphology. Cambridge: Cambridge University Press, 1981.

2. Hamley IW, The Physics of Block Copolymers. Oxford: Oxford University Press, 1998.

3. Kausch H-H, Polymer Fracture, 2nd Ed. Heidelberg-Berlin: Springer, 1987.

4. McLeish TCB, Plummer CJG, Donald AM, Crazing by Disentanglement – Non-diffusive Reptation. Polymer 1989; 30: 1651–1655.

5. Peterlin A, Plastic Deformation of Crystalline Polymers. Polym. Eng. & Sci. 1977; 17: 183–193.

6. Kramer EJ, Berger LL, Fundamental Processes of Craze Growth and Fracture. Adv. Polym. Sci. 1990; 91–92: 1–68.

7. Flory PJ, Yoon DY, Molecular Morphology in Semicrystalline Polymers. Nature 1978; 272: 226–229.

8. Plummer CJG, Menu P, Cudré-Mauroux N, Kausch H-H, The Effect of Crystallization Conditions on the Properties of Polyoxymethylene. J. Appl. Polym. Sci. 1995; 55: 489–500.

9. Plummer CJG, Kausch H-H, Micronecking in Thin Films of Isotactic Polypropylene. Macromol. Chem. Phys. 1996; 197: 2047–2063.

10. Klein J, Ball R, Kinetic and Topological Limits on Melt Crystallization in Polyethylene. Faraday Soc. Discussions 1979; 68: 198–209.

11. Tanzer JD, Crist B, Graessley WW, Chain Dimensions in Plastically Deformed Semicrystalline Polymers. J. Polym. Sci. Part B — Polym. Phys. 1989; 27: 859–874.

12. Benkoski JJ, Flores P, Kramer EJ, Diblock Copolymer Reinforced Interfaces between Amorphous Polystyrene and Semicrystalline Polyethylene. Macromolecules 2003; 36: 3289–3302.

13. Tervoort TA, Visjager J, Smith P, On Abrasive Wear of Polyethylene. Macromolecules 2002; 35: 8467–8471.

14. Tervoort TA, Visjager J, Graf B, Smith P, Melt-Processable Poly(tetrafluoroethylene). Macromolecules 2000; 33: 6460–6465.

15. Hoffman JD, Miller RL, Kinetics of Crystallization from the Melt and Chain Folding in Polyethylene Fractions Revisited: Theory and Experiment. Polymer 1997; 38: 3151–3212.

16. Fetters LJ, Lohse DJ, Graessley WW, Chain Dimensions and Entanglement Spacings in Dense Macromolecular Systems. J. Polym. Sci. Part B – Polym. Phys. 1999; 37: 1023–1033.

17. Gensler R, PhD Thesis, EPFL 1998.

18. Gensler R, Plummer CJG, Kausch H-H, Kramer E, Pauquet J-R, Zweifel H, Thermo-oxidative Degradation of Isotactic Polypropylene at High Temperatures: Phenolic Antioxidants versus HAS. Polym. Degrad. & Stab. 2001; 67: 195–208.

19. Kramer EJ, Microscopic and Molecular Fundamentals of Crazing. Adv. Polym. Sci. 1983; 52–53: 1–56.

20. Böhm LL, Enderle HF, Fleissner M, High Density Polyethylene Pipe Resins. Adv. Mater. 1992; 4: 234–238.

21. Chan MKV, Williams JG, J-Integral Studies of Crack Initiation of a Tough High-Density Polyethylene. Polymer 1983; 24: 234–244.

22. Donald AM, The Effect of Temperature on Crazing Mechanisms in Polystyrene. J. Mater. Sci. 1985; 20: 2630–2638.

23. Plummer CJG, Donald, A.M., Disentanglement and Crazing in Glassy Polymers. Macromolecules 1990; 23: 3929–3937.

24. Donald AM, Kramer, EJ, The Competition between Shear Deformation and Crazing in Glassy Polymers. J. Mater. Sci. 1982; 16: 1871–1879.

25. Wu JH, Xiao CD, Yee AF, Klug CA, Schaefer J, Controlling Molecular Mobility and Ductile-Brittle Transitions of Polycarbonate Copolymers. J. Polym. Sci. Part B - Polym. Phys. 2001; 39: 1730–1740.

26. Galeski A, Strength and Toughness of Semicrystalline Polymer Systems. Prog. Polym. Sci. 2003; 28: 1643–1699.

27. Pluta M, Bartczak Z, Galeski A, Changes in the Morphology and Orientation of Bulk Spherulitic Polypropylene due to Plane-Strain Compression. Polymer 2000; 41: 2271–2288.

28. Aboulfaraj M, G'Sell C, Ulrich B, Dahoun A, *In-situ* Observation of the Plastic Deformation of Polypropylene Spherulites under Uniaxial Tension and Simple Shear in the Scanning Electron Microscope. Polymer 1995; 36: 731–742.

29. Plummer CJG, Béguelin P, Kausch H-H, The Temperature and Strain Rate Dependence of Mechanical Properties in Polyoxymethylene. Polym. Eng. & Sci. 1995; 35: 1300–1312.

30. Plummer CJG, Scaramuzzino P, Kausch H-H, High Temperature Slow Crack Growth in Polyoxymethylene. Polym. Eng. & Sci. 2000; 40: 1306–1317.

31. Plummer CJG, Goldberg A, Ghanem A, Micromechanisms of Slow Crack Growth in Polyethylene under Constant Tensile Loading. Polymer 2001; 42: 9551–9564.

32. Brown N, Lu X, A Fundamental Theory for Slow Crack Growth in Polyethylene. Polymer 1995; 36: 543–548.

33. O'Connell PA, Bonner MJ, Duckett RA, Ward IM, The Relationship Between Slow crack growth and Creep-Behaviour in Polyethylene. Polymer 1995; 36: 2355–2362.

34. Cawood MJ, Channell AD, Capaccio G, Crack Initiation and Fiber Creep in Polyethylene. Polymer 1993; 34: 423–425.

35. Dugdale DS, Yielding of Steel Sheets Containing Slits. J. Mech. Phys. Solid 1960; 8: 100–104.

36. Duan DM, Williams JG, Craze Testing for Tough Polyethylene. J. Mater. Sci. 1998; 33: 625–638.

37. Pandya KC, Williams JG, Cohesive Zone Modelling of Crack Growth in Polymers — Part 1 — Experimental Measurement of Cohesive Law. Plast. Rubber Comp. 2000; 29: 439–446.

38. Pandya KC, Ivankovic A, Williams JG, Measurement of Cohesive Zone Parameters in Tough Polyethylene. Polym. Eng. & Sci. 2000; 40: 1765–1776.

39. Pandya KC, Ivankovic A, Williams JG, Cohesive Zone Modelling of Crack Growth in Polymers — Part 2 — Numerical Simulation of Crack Growth. Plast. Rubber Comp. 2000; 29: 447–452.

40. G'Sell C, Favier V, Giroud T, Hiver JM, Goldberg A, Hellinckx S, Proc. 11th International Conference on Deformation, Yield and Fracture of Polymers, Cambridge, U.K.,10–18 April 2000, p. 73.

41. Brown HR, A Molecular Interpretation of the Toughness of Glassy Polymers. Macromolecules 1991; 24: 2752–2756.

42. Sha Y, Hui CY, Ruina A, Kramer EJ, Detailed Simulation of Craze Fibril Failure at a Crack Tip in a Glassy Polymer. Acta Materiala 1997; 45: 3555–3563.

43. Sha Y, Hui CY, Kramer EJ, Simulation of Craze Failure in a Glassy Polymer: Rate Dependent Drawing and Rate Dependent Failure Models. J. Mater. Sci. 1999; 34: 3695–3707.

44. Sha Y, Hui CY, Ruina A, Kramer EJ, Continuum and Discrete Modelling of Craze Breakdown. Macromolecules 1995; 28: 2450–2459.

45. Gensler R, Plummer CJG, Grein C, Kausch H-H, Influence of the Loading Rate on the Fracture Resistance of Isotactic Polypropylene and Impact Modified Isotactic Polypropylene. Polymer 2000; 41: 3809–3819.

46. Kanninen MF, An Augmented Double Cantilever Beam Model for Studying Crack Propagation and Arrest. Int. J. Fract. 1973; 9: 83–92.

47. Creton C, Kramer EJ, Brown HR, Hui CY, Adhesion and Fracture of Interfaces between Immiscible Polymers: From the Molecular to the Continuum Scale. Adv. Polym. Sci. 2002; 156: 53–136.

48. Plummer CJG, Creton C, Kalb F, Léger L, Structure and Microdeformation of (iPP/iPP-g-MA)-PA6 Reaction Bonded Interfaces. Macromolecules 1998; 31: 6164–6176.

49. Boucher E, Folkers JP, Hervet H, Leger L, Creton C, Effects of formation of Copolymer on the Interfacial Adhesion between Semicrystalline Polymers. Macromolecules 1996; 29: 774–782.

50. Boucher E, Folkers JP, Creton C, Hervet H, Léger L, Enhanced Adhesion between Polypropylene and Polyamide-6: Role of Interfacial Nucleation of the Beta-Crystalline Form of Polypropylene. Macromolecules 1997; 30: 2102–2109.

51. Bideaux J-E, Smith GD, Månson J-AE, Plummer, CJG, Hilborn JG, Fusion Bonding of Maleic Anhydride Grafted PP-PA 6 Blends to PA 6. Polymer 1998; 39: 5953–5948.

52. Washiyama J, Creton C, Kramer EJ, TEM Fracture Studies of Polymer Interfaces. Macromolecules 1992; 25: 4751–4758.

53. Washiyama J, Creton C, Kramer EJ, Xiao F, Hui CY, Optimum Toughening of Homopolymer Interfaces with Block-Copolymers. Macromolecules 1993; 26: 6011–6020.

54. Washiyama J, Kramer EJ, Hui CY, Fracture Mechanisms of Homopolymer Interfaces with Block-Copolymers — Transition from Chain Pull-out to Crazing. Macromolecules 1993; 26: 2928–2934.

55. Plummer CJG, Kausch H-H, Creton C, Kalb F, Léger L, Proc. IUPAC World Polymer Congress — Macro 98, Goldcoast, Brisbane, Australia, 13–17 July 1998.

56. Kalb F, PhD Thesis, Collège de France 1998.

57. Kalb F, Léger L, Plummer CJG, Creton C, Marcus P, Magalhaes A, Molecular Control of Crack Tip Plasticity Mechanisms at a PP-EPDM/PA6 Interface. Macromolecules 2001; 34: 2702–2709.

7

Micromechanical Deformation Mechanisms in Polyolefins: Influence of Polymorphism and Molecular Weight

SVEN HENNING and GOERG H. MICHLER

Institute of Materials Science, Martin-Luther-
University Halle-Wittenberg, Germany

CONTENTS

I.   INTRODUCTION

Polyolefins are a class of thermoplastic polymers that include
the polymerization products of alkenes (polyethylene, polypropylene, etc), the copolymers of ethylene and propylene and
the copolymers of alkenes with vinyl monomers. The aim of
this chapter is to discuss the technically most relevant micromechanical deformation mechanisms in some of these materials: low-density polyethylene (LDPE), linear-low-density
polyethylene (LLDPE), high-density polyethylene (HDPE)
and ultra-high molecular weight polyethylene (UHMWPE)
and the two main crystalline modifications of isotactic
polypropylene. The common feature of these systems is the
semicrystalline morphology, which is controlled by the architecture of the macromolecules (chemical structure, configuration, conformation) and the processing history [1]. The basic
morphological units, i.e., the highly ordered crystalline lamellae that are alternated by amorphous regions, are formed by

self-organization during the crystallization process (see Chapter 1). Their typical dimensions lie within the range of 5 to 25 nm. Hence, these polymers, showing a typical coexistence of two different phases within one chemically homogeneous polymer can be considered as nanostructured materials. Under normal conditions (atmospheric pressure, 23°C, moderate deformation rate), the coexisting phases are a combination of a soft (amorphous) and a relatively stiff (crystalline) component connected by covalently bonded links (tie molecules and entanglements within the amorphous regions) that give rise to a good balance of stiffness and toughness to the polymer material. Such a nanostructure of alternating hard and soft regions seems to be characteristic for "high-end" materials such as lamellar styrene-butadiene block copolymers [2], toughened amorphous polymers, nanocomposites, bone tissue, etc.

The details of the semicrystalline morphology play a dominant role, as they determine the final mechanical properties that are of technical interest. These *nanostructure monitored properties* control the micromechanical mechanisms of deformation and fracture acting when a load is applied to the sample. A variety of micromechanical processes takes place at microscopic, mesoscopic, and nanoscopic scale. Typical microscopic phenomena include micro-yielding, micro-cavitation and orientation mechanisms that are connected with craze- and shear-band formation as well as deformation zones [1].

In the first part of this chapter, results obtained over the last two decades for different polyethylene types will be summarized. In the second part of the chapter, results for a series of iPP samples will be discussed in comparison to the micromechanical mechanisms that are found to be typical for the mechanical properties of polyethylene. Experimental results are focused on scanning electron microscopic (SEM) observations since the procedure for sample preparation is relatively easy and nevertheless effective. Additionally, transmission electron microscopy (TEM) studies were performed in support of the SEM observations and to exclude the misinterpretation of preparation artifacts, gaining insight as well to effects on the nanometer level at regions of special interest. Results will be summarized with respect to the mechanical properties that are of interest for the application of these polymers.

II.   POLYETHYLENE: THE INFLUENCE OF MOLECULAR PARAMETERS ON MORPHOLOGY AND MICROMECHANICAL MECHANISMS

A.   Influence of Chain Architecture: Branching

Polyethylene materials can be roughly classified by the extent, statistical distribution and length of side chains (branches) covalently bound to the linear backbone chain. Branching can be controlled within a wide range by the selection of the thermodynamic conditions in the reactor during the polymerization process, the choice of the catalyst or the addition of higher alkenes as comonomers to the ethylene base monomer. The main types of polyethylenes that are distinguished by the actual chain architecture are low-density polyethylene (LDPE), linear-low-density polyethylene (LLDPE), and high-density polyethylene (HDPE). Usually, ultra-high molecular weight polyethylene (UHMWPE) consisting of very large linear macromolecules is cited as another PE type on its own.

LDPE is characterized by a relatively large amount of branches consisting of shorter and longer chain fragments similar to the backbone macromolecule. Due to that type of branching, the crystallization process is hindered, leading to a low degree of crystallinity and, thereby, a lower density of the material. The melting temperature of the crystalline phase is approximately 105°C, the glass transition temperature of the amorphous portion is approximately –120°C. Morphology and micromechanical mechanisms depend strongly on the molecular weight. The most significant feature is the appearance of banded spherulites.

In contrast, HDPE consists of linear chains. Branches appear only to a small extent, so that crystallization is less disturbed by such chain defects. The linear chain architecture results in higher degrees of crystallinity. As compared to LDPE, the higher content of the crystalline phase yields the higher stiffness of the material. The melting temperature of the more perfect crystalline lamellae is shifted to approximately 130°C.

With respect to the morphology and the deformation behavior, LLDPE combines some typical features of HDPE and LDPE. Very similar to HDPE, there are no long chain branches. In comparison to LDPE, short chain branches are more uniform in their length. The morphology is characterized by the coexistence of long and thicker and short and thinner lamellae, resulting in a broad or bimodal lamellar thickness distribution. Like in LDPE, banded spherulites are formed.

B. Influence of Molecular Weight

The molecular weight has similar influence on the morphology of the different types of polyethylenes. In general, four molecular weight ranges can be defined which are distinguished by their superlamellar structures as well as by length and thickness of the crystalline lamellae and the thickness of the amorphous part [1,3–7]. In the range around $M_w \approx 20,000$ g/mol sheaf-like structures made up of stacks of long lamellae running parallel to each other are typical. The lamellae are densely packed having only a small amount of amorphous material in between. With increasing molecular weight the length of the lamellae is increasing (up to approximately 1 µm), and the lamellar thickness is growing slightly. In the range of $M_w \approx 30,000–100,000$ g/mol there is a transition from sheaf-like structures to spherulites. In contrast to HDPE, LDPE spherulites show a pronounced banding. The thickness of the concentric rings is decreasing with increasing M_w. Lamellar thickness and thickness of the amorphous layer are increasing continuously. For molecular weights $M_w \geq 200,000$ g/mol the spherulitic superstructure of HDPE is replaced by bundles of irregularly distributed lamellae whereas banded spherulites are found in LDPE. The lamellar thickness and the thickness of the amorphous layer are increasing continuously. For ultra-high molecular weights in the region of 1,000,000–2,000,000 g/mol (UHMWPE [10]) randomly distributed short and thick lamellae separated by thick amorphous layers are observed. There is no spherulitic superstructure.

The mechanical behavior of polyethylene also depends strongly on the molecular weight. Below $M_w \approx 50,000$ g/mol the material fails in a brittle manner. For LDPE one observes a pronounced increase of strength and elongation in the range of $M_w \approx 100,000$ g/mol. Above $M_w \approx 100,000$ polyethylenes show a ductile behavior.

For a usual HDPE material with blocks of lamellae in different orientations the typical micromechanical mechanisms include lamellar slip, lamellar rotation, separation and breaking of lamellae, chevron formation, stretching and molecular orientation of the amorphous phase and unfolding or yielding of the crystalline lamellae. These mechanisms are summarized schematically in Table 1 with increasing strain ε for the supermolecular, morphological level, the macromolecular amorphous and crystalline phase.

III. ISOTACTIC POLYPROPYLENE: EXPERIMENTAL

A. Materials and Sample Characteristics

The isotactic polypropylene materials discussed in this chapter are of the α- as well as of the β-type and vary in the molecular weight from $M_w = 230,000$ g/mol up to $M_w = 980,000$ g/mol (Borealis AG, Linz). To obtain a fully β-modified material, one β-nucleated iPP of $M_w \approx 400,000$ g/mol was crystallized from the melt via a multistep crystallization procedure. All samples were prepared by plate pressing at a mold temperature of 200°C. Tensile bars were punched out of the 2 mm thick plates using dumbbell-shaped pierce tools. Two types of tensile bars were produced: standard samples according to DIN and miniaturized dogbone samples that are not following any standard specification.

B. Techniques for the Analysis of Deformation Structures

Tensile testing at room temperature (23°C) was performed using a universal tensile machine at traverse speeds of 1 mm/min and 10 mm/min, respectively. Miniaturized tensile bars were

TABLE 1 Micromechanical Mechanisms in Polyethylenes at Different Stages of Plastic Deformation

Process	Morphology	Amorphous Phase	Crystalline Phase
Original material -Lamellae distributed randomly with respect to straining direction			
$\varepsilon \approx 150\%$ ⟷ -Rotation of lamellae -Collective rotation of lamellar stacks (chevron) -Lamellar slip -Lamellar separation			
$\varepsilon \approx 300\%$ ⟷ -Plastic processes as described above -Breaking of lamellae into microblocks			
$\varepsilon \approx 550\%$ ⟷ -Plastic processes as described above -Unfolding o crystalline lamellae -Chain orientation parallel to straining direction, fibrillation			

Note: The processes described here result in a ductile behavior of the polymeric material. For the example that is illustrated here (cold drawing of a HDPE blown film), no cavitation was observed.

tested using a miniature tensile testing device (MINIMAT materials tester) at a traverse speed of 1 mm/min. For the investigation of structures deformed at lower temperatures, *tensile testing at –5°C* was performed using a universal tensile machine. For tensile testing at –40°C, a miniature tensile device was equipped with a custom-made cooling chamber.

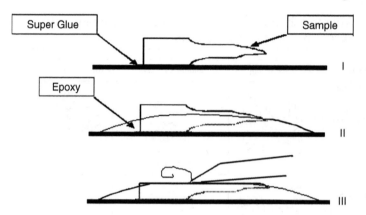

Figure 1 Schematic drawing illustrating the preparation steps for miniaturized tensile bars after deformation, prior to etching and SEM investigation: fixation to a steel plate (I), embedding of the specimen using epoxy resin (II) and microtome cutting using a metal blade (III).

Scanning Electron Microscopy (SEM) was used to record micrographs representing the spherulitic texture and the intraspherulitic lamellar arrangement of the original samples. Moreover, structural changes introduced by plastic deformation during uniaxial testing of miniaturized tensile bars were followed over the complete length of the tensile bar. The single steps of the microtome preparation prior to permanganic etching are shown in Figure 1. As the result of the procedure, a smooth surface representing the center line of the sample is obtained. To reveal the details of the semi-crystalline morphology and the deformation structures, the samples were etched for 20 min at room temperature using a modified permanganic etching agent following the procedure described by Olley et al. [11]. Prior to inspection in the SEM (JEOL JSM 6300), the samples were coated with a gold layer of approximately 12 nm (Edwards sputter coater).

Transmission Electron Microscopy (200 kV TEM, JEOL) was used to image the semicrystalline morphology of ultra thin sections (about 50 to 80 nm thick) ultramicrotomed from the bulk samples before and after deformation. The less

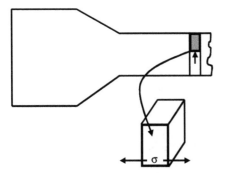

Figure 2 Scheme showing the specimen preparation for TEM inspection. A small block is taken from the region of interest close to the centerline of the deformed tensile bar. The arrow indicates the plane that is presented in the TEM micrographs.

ordered amorphous phase between the highly ordered crystalline lamellae was stained selectively by ruthenium tetroxide (RuO_4) prior to ultramicrotomy yielding appropriate fixation of the material [12]. For the production of the sections at room temperature, a Leica Ultracut Ultramicrotome equipped with a Diatome diamond knife was used. Morphological parameters were derived from TEM micrographs by interactive measurement using ANALY.SIS image processing system. The procedure applied for the *preparation of deformed samples* is depicted in Figure 2. A block of the specimen is taken from the plastically deformed region of interest. After RuO_4 staining and fixation, ultra thin sections are prepared from the block face. By means of this technique, one obtains ultra thin sections from the center of the deformed tensile bar.

IV. ISOTACTIC POLYPROPYLENE: MORPHOLOGY AND MICROMECHANICAL DEFORMATION MECHANISMS

A. Morphology: General Features

The pronounced diversity of isotactic polypropylene (iPP) is also connected with the existence of several crystalline mod-

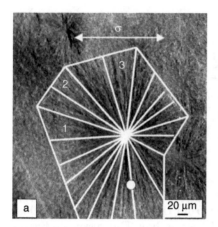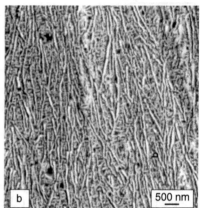

Figure 3 SEM micrographs recorded after permanganic etching, illustrating the typical morphology of α-iPP. (a): The low magnification micrograph shows lamellae radiating from the center of a spherulite. The numbers indicate regions with lamellar orientations parallel (1), diagonal (2) and perpendicular (3) to the strain that will be applied during tensile testing (arrow). The circle marks the region of higher magnification. (b): At high magnifications, the typical cross-hatched arrangement of main lamellae (running top to bottom) and secondary (daughter) lamellae are visible.

ifications. Of particular interest from the technical point of view are the more common α-modification and the more exotic β-modification [13]. Recently, the latter can be produced by the addition of highly efficient β-nucleating agents [14]. It is reported by several authors that the β-iPP has superior mechanical properties compared to the α-form (e.g., enhanced toughness and ductile behavior at high deformation rates) [15–17]. The most significant morphological feature of the α-form of iPP is the so-called "cross-hatched" arrangement of the crystalline lamellae as shown in Figure 3. The main lamellae ("dominant lamellae") are radially growing from an initial site (center of the spherulite, Figure 3a), whereas the "secondary lamellae" are formed by an epitaxial growth onto them, exhibiting a typical angle of 81° (Figure 3b) [18,19].

In contrast to this phenomenon, a stacked, parallel arrangement of bundles of lamellae is found for the β-modified

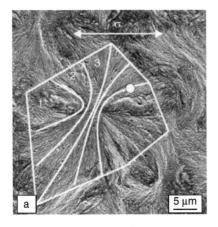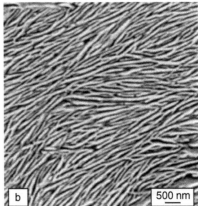

Figure 4 SEM micrographs recorded after permanganic etching, showing the typical morphology of β-iPP. (a): The sheaf-like lamellar arrangement at low magnification. The numbers indicate regions with lamellar orientations parallel (1), diagonal (2) and perpendicular (3) to the strain applied during tensile testing (arrow). The circle marks the region of higher magnification. (b): High magnification micrograph showing bundles of lamellae running parallel to each other and some bifurcations.

material (Figure 4). Here, the lamellae form a more sheaf-like superstructure rather than a radiating, spherulitic growth. The comparison of TEM images obtained for the two main types of iPP is given in Figure 5.

From the SEM and TEM micrographs it becomes clear that the intercrystalline, amorphous regions are distributed more continuously in case of β-iPP. In other words, the latter shows a typical lamellar structure similar to other semicrystalline polymers, especially to the different types of polyethylenes pointed out above. The β-iPP crystalline lamellae are slightly thicker (approximately 15 nm) than the lamellae found in the α-iPP (approximately 10 nm). Due to the nucleating agent added by the producer in case of the β-modified material, the size of the spherulites is different for the two materials. For the α-iPP the average spherulite diameter is approximately 50 μm, while for the β-iPP the spherulite diameter is smaller: approximately 10 μm. From the SEM

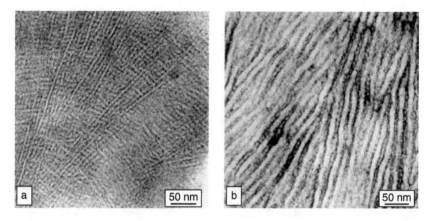

Figure 5 Comparison of lamellar morphologies for the two main iPP types. (a): The lamellae associated to the α modification present a cross-hatched morphology; (b): Lamellae of β-iPP are aligned in a parallel manner forming stacks or bundles. TEM micrographs of ultrathin sections after chemical fixation and staining.

micrographs one can conclude that the β-modified material is "pure" since the whole area of the sample is filled with spherulitic superstructures that can be assigned to the crystalline β-modification.

B. Influence of the Molecular Weight on the Micromechanical Mechanisms

With increasing molecular weight, both α-iPP and β-iPP reveal a characteristic increase in the thickness of lamellae and interlamellar, amorphous layers and in the long period (Table 2). This general tendency is more pronounced for the β-modified material. In order to discuss the influence of molecular weight on the deformation behavior, we have selected two grades of iPP: M_w of 230 kg/mol (iPP_{230}) and 482 kg/mol (iPP_{482}), respectively. The average spherulite size of the undeformed materials varies with M_w. For the iPP with the higher molecular weight (iPP_{482}), the average spherulite diameter is approximately 49 µm, whereas the average diameter of iPP_{230} is approximately 20 µm. It is noteworthy that in the case of

TABLE 2 Characteristics and Morphological Parameters of iPP Samples of Different Molecular Weight

iPP Type, M_w (kg/mol)	Lamellar Thickness [nm]	Thickness Amorphous Layer [nm]	Long Period [nm]	σ_y (MPa)*	ε_b (%)*	σ_b (MPa)*
α-iPP 297,000	7.3	3.4	10.7	36	67	—
α-iPP 565,000	7.2	3.4	10.6	—	—	—
α-iPP 979,000	8.7	4.2	12.9	35	375	41
β-iPP 297,000	6.8	2.3	9.1	31	74	22
β-iPP 565,000	9.1	3.5	12.6	—	—	—
β-iPP 979,000	10.5	4.1	14.6	29	400	44

* Tensile testing at 23°C, traverse speed 10 mm/min

the iPP$_{482}$ one observes single spherulites that can be assigned to the crystalline β-modification scattered randomly within the material formed by α spherulites. This phenomenon is not observed to such an extent for the iPP$_{230}$ sample.

There is a significant difference in the deformation behavior of samples tested in tensile experiments at room temperature (23°C) at the step from 230 to 482 kg/mol, schematically illustrated in Figure 6.

Occasionally, the iPP$_{230}$ material fails in a semiductile manner directly after the yield point in the stress-strain curve is surpassed. The effect of such a premature fracture can be to some extent attributed to defects on the surface (scratches, notches). On the other hand, the iPP$_{582}$ shows a ductile behavior.

Figure 7 illustrates the main differences in the micromechanical mechanisms that are observed for the two materials. The SEM micrograph of iPP$_{230}$ is recorded from a region close to the fracture, representing a local strain of approximately 7% (Figure 7a). Typical crazes running perpendicular to the straining direction are seen. The length of the crazes is in the

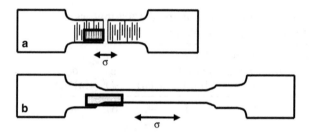

Figure 6 Comparison of the macroscopic deformation observed for samples tested at 23°C with a traverse speed of 10 mm/min for two grades of α-iPP having two different molecular weights. (a): Semiductile fracture and multiple crazing. (b): Ductile behavior showing neck formation and cold drawing.

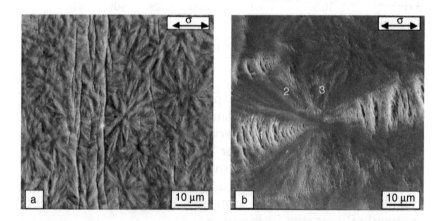

Figure 7 Typical SEM images of deformed structures generated by uniaxial tensile testing at 23°C for two grades of α-iPP having different molecular weight. (a): Crazes close to the plane of fracture, for the 230 kg/mol sample; (b): Craze-like morphology including microvoid formation and fibrillation localized in the polar regions (1) of the spherulite for the 482 kg/mol iPP sample at a strain of about 30%.

range of up to 100 μm, and their growth is not influenced by the morphological units that are involved. In particular, the craze propagation is not influenced by spherulite boundaries or by lamellar orientation with respect to the strain direction.

The SEM micrograph of the deformation structures of iPP_{482} (Figure 7b) corresponds to a local deformation of approximately 30%. Here, the type of deformation strongly depends on the orientation of the lamellae with respect to the straining direction. Intensive craze-like mechanisms including cavitation and fibrillation of the material are observed in the spherulite region 1 (main lamellae oriented parallel to the straining direction). These craze-like mechanisms are limited to a few μm in length. In contrast, no cavitations are observed in the equatorial regions 2 and 3.

The significant differences in the micromechanical mechanisms and the resulting mechanical properties are connected with the decreasing thickness of the amorphous material between lamellae with decreasing molecular weight (Table 2). It can be assumed that the decreasing thickness of the amorphous portion at lower molecular weights is connected with a reduced content of tie molecules or entanglements between lamellae. Cavitations in the amorphous parts cannot be stabilized and local stress concentrations result in longer crazes. Therefore, a decrease in molecular weight initiates a transition from ductile to semiductile or quasibrittle behavior. Such a sharp transition (embrittlement) with decreasing molecular weight and decreasing thickness of the amorphous layers is also known for polyethylene [7].

It is noteworthy that singular β-type spherulites that can be found within the α-iPP exhibit a deformation behavior that is significantly different from the behavior of the surrounding α-type spherulites. An intensive cavitation and fibrillation of the material occurs in the spherulites of the β-iPP even at low molecular weights. Microvoiding is not limited to certain sectors of the spherulite. Similar effects have been reported in the literature [20].

C.   Influence of Crystal Polymorphism on Micromechanical Mechanisms

We have compared the micromechanical processes of the two main polymorphic types of iPP for miniaturized tensile bars tested at 23°C at a traverse speed of 1 mm/min (Figure 8). The

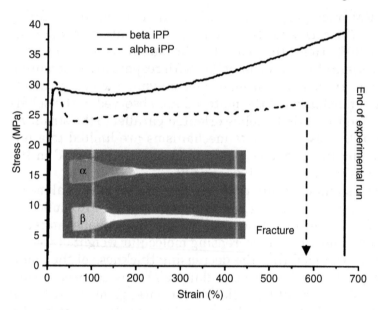

Figure 8 Stress-strain-diagrams recorded at 23°C (traverse speed 1 mm/min) for miniaturized tensile bars made from the two crystalline modifications. Whereas the α-iPP (dotted line) shows a distinct necking, in the β-iPP (solid line) the stress whitening appears homogeneously throughout the length of the tensile bar.

plastic deformation of the tensile bars of both PP types under load proceeds through formation and propagation of two macroscopic necks in opposite directions. The tensile stresses at yield are comparable (approximately 30 MPa in both cases), and the curves for the elastic part before the yield point coincide. Due to the geometrical limitations of the miniature tensile tester, the β-iPP sample was not drawn up to the fracture. The formation of a neck is much more pronounced for the α-iPP. Here, a distinct yield point in the stress-strain curve is observed. In contrast, in case of β-iPP the plastic deformation that is going along with a pronounced stress whitening is distributed more homogeneously over the length of the tensile bar. Accordingly, the yield point in the stress-strain curve is less significant. Since the area under the stress-strain curves can be taken as a measure for the energy that is dissipated

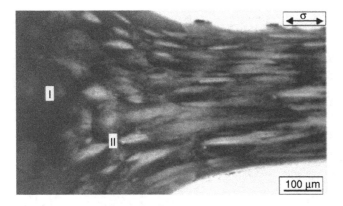

Figure 9 SEM image illustrating the development of a deformed structure across the neck region of a tensile specimen of α-iPP. The regions indicated by the roman capitals I and II, respectively, are shown in more detail in Figures 10 and 11.

by plastic deformation within the sample, the data of Figure 8 confirm that β-iPP is superior to α-iPP.

In Figure 9 the structural changes within the neck region of the α-iPP sample are shown. The formation of microvoids (cavitation that is connected with stress whitening which is observed macroscopically) is limited to polar regions of the spherulites with respect to the straining direction. This observation is comparable to the mechanisms that were described in [21] for α-iPP of similar molecular weight.

Figures 10a and 10b are recorded from region I in Figure 9 at a higher magnification. Here the craze-like nature of the deformation structures becomes visible. Once more, fibrillated microcracks are formed perpendicular to the straining direction. Within the craze-like structures, cavitation and fibrillation occurs. The craze length is limited and does not exceed 2 μm. Within the region 1 (see Figure 3a) the crazes are finely distributed and relatively large in number. The situation is depicted schematically in Figure 10c.

As deformation proceeds (region II in Figure 9 in higher magnification: Figure 11, local strain approximately 100%), the craze-like processes of cavitation and fibrillation are intensified (Figure 11a). In other words, an increasingly larger

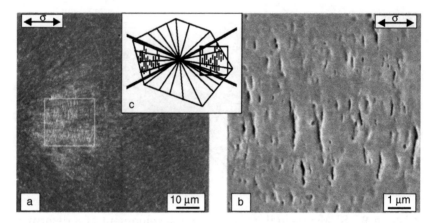

Figure 10   (a, c): SEM image of deformed α-iPP recorded for region I in Figure 9 showing craze-like deformation features localized within the polar region of the spherulite with respect to the straining direction. (b): Multiple formation of short crazes taken at high magnification.

amount of the sample volume is involved in this kind of plastic deformation, including now even the equatorial regions 2 and 3 defined in Figure 3a. The mesh-like structure of the cross-hatched lamellar arrangement is still to be seen (Figure 11b). From this picture it becomes apparent that the deformation process is controlled by the nanoscopic lamellar arrangement. Craze-like deformation is initiated by the interlamellar amorphous material that cavitates as an external stress is applied. The fibrillation that is typical for crazing can be seen. The craze-like processes are limited to the micrometer range by morphological constraints, but they are large in number and occur homogeneously within a large sample volume. Figure 11c summarizes schematically the deformation process.

The micromechanical mechanisms observed for the β-iPP are illustrated in a series of figures representing increasing degrees of local deformation: The first structural changes that appear while a tensile stress is applied are illustrated in Figure 12. The situation can be attributed to a local strain of approximately 20%. The SEM micrograph of Figure 12a shows a collective reorientation of stacks of lamellae that were orig-

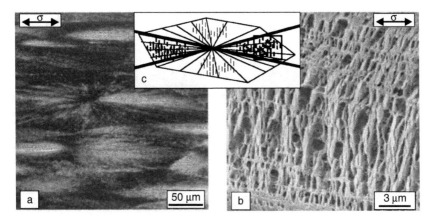

Figure 11 SEM images of deformed α-iPP from region II, Figure 9. (a, c): Multiple craze-like deformation still concentrated within the polar region of the spherulite with respect to the strain direction. (b): Mesh-like texture formed by microvoids; crystalline lamellae and fibrillated material is observed at high magnification.

inally perpendicularly oriented to the strain direction. At this state of deformation, no cavitation is observed. The above phenomenon already described for polyethylene is designed as "chevron structure." A more detailed investigation of the collective twisting was performed using TEM techniques (Figure 12b). After staining, one can observe single lamellae that are partially broken and reoriented (lamellar twisting) into the direction of the applied load. Chevron formation is strictly limited to region 3 as defined in Figure 4a. It is to be noted that such chevron structures have not been found in α-iPP.

The influence of larger deformations leads to the structures shown in Figure 13. The situation that is imaged by SEM methods represents a local strain of approximately 30% (Figure 13a). Now, in addition to chevron formation, a pronounced lamellar separation occurs when the lamellae are oriented in perpendicular direction with respect to the strain direction. Lamellar separation is a process that is initiated within the amorphous phase that shows craze-like mechanisms of cavitation and fibrillation. A similar situation is shown by the TEM micrograph (Figure 13b). After staining,

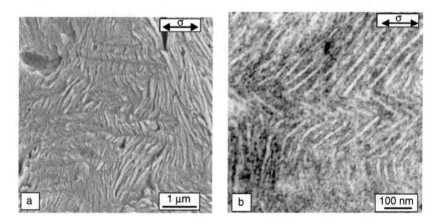

Figure 12 Nanostructure of β-iPP specimen from Figure 8 after plastic deformation. The local strain is approximately 20%. The collective twisting of lamellae gives rise to the formation of so-called chevron structures. (a): SEM micrograph after permanganic etching and (b): TEM micrograph of an ultra-thin section after staining. The strain direction is perpendicular to the lamellar orientation in region 3 as shown in Figure 4a.

the crystalline lamellae can be clearly distinguished from the amorphous material that appears as black regions. Additionally, nano-voids that are not stained by the RuO_4 agent can be seen. The TEM micrograph includes a number of general effects that can be observed in the β-iPP for the initial stages of deformation (strain up to 100%). First, the process of chevron formation is evident. The lamellar reorientation of lamellar stacks is overlaid by multiple lamellar separation. In the latter case, cavitation of the amorphous material results in fibrillation and nano-void formation. Lamellae that are already aligned parallel to the straining direction start to yield. The fracture of single lamellae into mosaic blocks starts in a region that is highlighted by a white circle.

At higher deformations the original spherulitic morphology is increasingly destroyed due to the influence of the plastic deformation mechanisms described above, involving both the amorphous and the crystalline regions of the polymer. Figure 14a illustrates the deformation mechanism for a local strain of

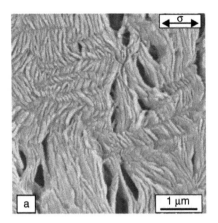
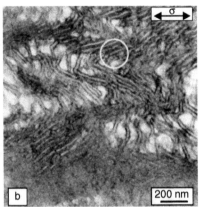

Figure 13 (a): SEM micrograph taken from the same β-iPP spec-imen as in Figure 12, representing a local deformation of approxi-mately 30%. In addition to chevron formation, lamellar separation occurs; (b): TEM micrograph from another sample representing a similar level of deformation. Simultaneously, three main microme-chanical mechanisms are observed: collective twisting of lamellae towards straining direction (chevron formation, reorientation), sep-aration of lamellae that are oriented perpendicular to the straining direction and breaking of lamellae that are oriented parallel to the straining direction. The white circle highlights the fracture of lamel-lae into nano-blocks.

approximately 100%. In this micrograph, the lamellae that are oriented parallel to the straining direction are fractured into smaller fragments. Here again, cavitation and fibrillation can be seen; i.e., the development of a nanoporous structure is not limited to the region 1 (lamellae parallel to the strain direction) but also occurs to a similar extent in the regions 2 and 3 defined above (Figure 4a). In contrast to the results obtained for α-iPP, in the case of β-iPP the whole sample volume contributes to the micromechanical processes that are connected with cavita-tion and fibrillation. The final structure that is obtained by drawing the β-iPP sample up to 670% is given in Figure 14b. The original spherulitic morphology is fully transformed into a fibrillar structure, representing the highly oriented state after cold drawing and strain hardening. In the upper part of

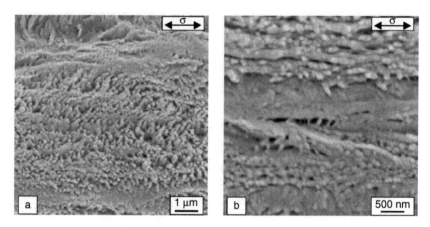

Figure 14 (a): SEM micrograph showing the destruction of the original spherulitic morphology of β-iPP and the intensive microvoid formation at a local strain of approximately 100%. (b): At the end of the deformation process, the original morphology is completely transformed into a microfibrillar structure, representing the highly oriented state as a result of strain hardening (macroscopic strain approximately 670%, see Figure 8).

the SEM image, one clearly distinguishes oriented nano-fibers of 100 to 200 nm in diameter. The porous structure is completely eliminated as the pores close and disappear at high strains, forming nanodefects between the oriented fibers.

The different micromechanical processes that occur depending on the lamellar orientation are schematically summarized in Figure 15. For the initial stages of plastic deformation, the processes are modulated by the mobile amorphous phase. The processes may include (lamellar separation) or exclude (chevron formation) cavitation. Lamellar slip is also controlled by the mobility of the molecules in the amorphous phase. For all of these processes, the lamellar thickness remained constant indicating that the crystalline portion is not affected. In case of α-iPP, the micromechanical mechanisms that are initiated by the amorphous material are hindered by the cross-hatched arrangement of the lamellae. When the crystalline lamellae are aligned parallel to the strain direction, they break into multiple fragments.

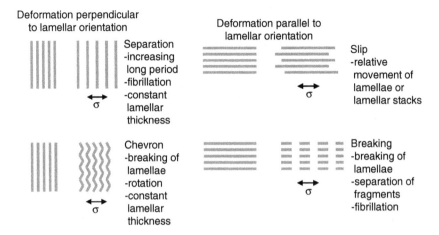

Figure 15 Schematic representation of the main micromechanical mechanisms observed in β-iPP with the typical lamellar morphology after uniaxial tensile deformation at 23°C. The type of deformation process depends on the orientation of the crystalline lamellae with respect to the straining direction.

The nano-mechanical process of lamellar separation is schematically illustrated in Figure 16. Lamellar separation can be interpreted as a craze-like process that is geometrically limited by the lamellar nanostructure. As the amorphous material starts to cavitate, fibrils consisting of oriented macromolecules are formed. The process itself is very similar to crazing in amorphous polymers. In the case of the β-iPP the crazes are highly localized zones of plastic deformation. In contrast to "classical" crazes they are very small ("nanoscopic crazes") though finely distributed, so that a great part of the amorphous material is capable to contribute to plastic deformation.

Lamellar morphologies in the same order of magnitude (10...20 nm) are also formed in some styrene-butadiene block copolymers. The similar arrangement of soft and hard components results in astonishing similarities in the micromechanical mechanisms (Figure 17 in Chapter 10 of this book). In particular, chevron formation as a phenomenon of collective lamellar twisting is observed. At higher deformations, block

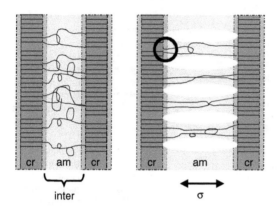

Figure 16 Schematic model of lamellar separation (strain perpendicular to lamellar orientation) showing a pair of lamellae before and after deformation. The highly ordered crystalline phases (cr) are connected by tie molecules and entanglements that are located in the disordered amorphous (am) region. The process of lamellar separation is interpreted as a craze-like mechanism, including microvoid formation and fibrillation that is geometrically limited due to the semicrystalline nanostructure.

copolymers show a so-called thin layer yielding, i.e., the hard component (PS) is drawn up to 300%. In the case of the semicrystalline iPP, high deformation results in a conversion of the folded lamellae into oriented microfibrils. The main difference is that cavitation does not occur in the lamellar block copolymers. The maximum stretching of the hard PS phase of the block copolymer is restricted by the maximum draw ratio of the entanglement network of the soft PB phase. In the semicrystalline iPP, cavitation in the amorphous region allows the higher contribution of the soft phase to macroscopic strain. Theoretically, chain unfolding of the crystalline lamellae may result in an elongation as high as $\lambda_{max} \approx 100$. The pronounced formation of nanoscopic voids in the β-iPP is a precondition for the higher maximum elongation [2].

D.   Influence of the Deformation Temperature

At temperatures below the glass transition temperature (T_g) of the amorphous phase (approximately –5°C) polypropylene

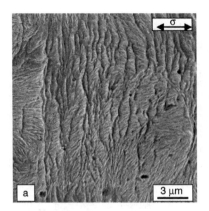 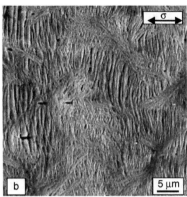

Figure 17 SEM micrographs of β-iPP structures after uniaxial tensile deformation at –5°C (traverse speed of 1 mm/min). (a): Initially a large number of fine crazes is formed. They run perpendicular to the straining direction independently of the actual orientation of the crystalline lamellae that are involved. (b): As deformation proceeds, the number of crazes is growing rapidly throughout the sample volume.

tends to become brittle. Therefore, strategies for toughness enhancement mainly have to overcome the problems connected with low temperature applications. To understand the changes in the mechanical behavior, micromechanical mechanisms were followed for iPP samples of the two main crystalline modifications that were tensile tested at –5°C and –40°C. The traverse speed in both cases was 1 mm/min.

The structural changes observed in β-iPP after tensile deformation at –5°C are shown in Figure 17. The SEM micrograph of Figure 17a depicts a situation representative for a local strain of approximately 20%. A craze-like deformation pattern is observed. The growth of the crazes is not affected by morphological units as there are spherulite boundaries or different lamellar orientations. At a somewhat higher local strain of approximately 40%, more volume becomes involved in the plastic deformation mechanism that is connected with cavitation (Figure 17b).

Figure 18 illustrates high magnification SEM micrographs of the situation described in Figure 17b. For lamellae

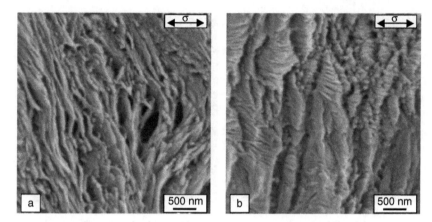

Figure 18 SEM micrographs of the same situation as described in Figure 17b recorded at high magnifications. (a): For lamellae perpendicular to the applied strain, lamellar separation is observed. (b): If lamellae and straining direction are parallel, the crystalline lamellae break up into smaller blocks. In both cases, fibrillated material and microvoids between the edges of the craze-like structures appear. In the regions between the crazes, stacks of lamellae running parallel to the straining direction are visible.

perpendicular to the applied strain (Figure 18a), lamellar separation is observed. Very similar to the structural changes that were generated at 23°C, the lamellar separation process includes cavitation and fibrillation. In other words, it is a craze-like phenomenon that is controlled by the lamellar nanostructure. If the straining direction is parallel to the lamellar orientation (Figure 18b), crystalline lamellae are forced to break up into smaller blocks. In both cases, the fibrillated material and the microvoids appear between the edges of the craze-like structures. Another significant difference with respect to the plastically deformed structures observed at room temperature is the absence of chevron formation.

The material becomes brittle when tested at −40°C, for both polymorphic modifications, α and β, and typical crazes emerge. The craze propagation is not influenced by the spherulitic morphology (different orientations of lamellae within the spherulite, spherulite boundaries). Crazes are running

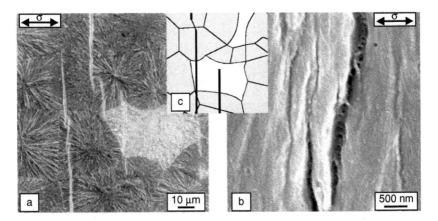

Figure 19 SEM micrographs of α-iPP after uniaxial tensile deformation at −40°C. (a): Only few and relatively coarse crazes can be found. Crazes run perpendicular to the straining direction independently of morphological entities, as there are spherulite boundaries or local orientations of the crystalline lamellae that are involved. (b): High magnification image unveiling a typical feature of crazes, i.e., fine fibrils bridging the two edges of the craze.

perpendicular to the strain direction (Figure 19). Crazes can be up to 100 µm long (Figure 19a) and there is only a small amount of them in the area of the sample directly at the fracture. At higher magnifications (Figure 19b) the typical fibrils bridging the edges of the craze can be seen. The situation is also given as a schematic drawing (Figure 19c).

The crazes that are observed in the β-modified PP are much finer and shorter, and they are higher in number for the same sample area that is close to the fracture region (Figure 20a). This means that even at low temperatures, a larger plastic deformation can take place as crazing is the ruling energy dissipating process. The internal structure of the crazes could not be resolved in the SEM (Figure 20b). The multiple crazing that is observed for the β-iPP (illustrated schematically in Figure 20c) should lead to a higher toughness value. Nevertheless, none of the nanostructure-controlled processes discussed above can be found in the samples deformed at −40°C. The nano-mechanical processes that occur at room

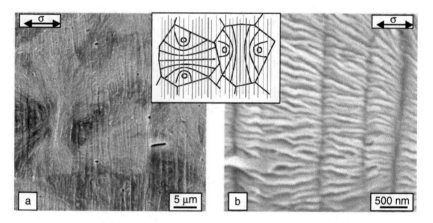

Figure 20   SEM micrographs of β-iPP after uniaxial tensile deformation at –40°C. (a,c): In contrast to α-iPP, the number of crazes observed for the same sample area is significantly higher, and the crazes are finer and shorter. (b): Crazes run perpendicular to the strain direction independently of spherulite boundaries or local orientations of the crystalline lamellae that are involved.

temperature and, to some extent, even at –5°C are initiated by the more mobile amorphous material, and, thus, they are excluded (they are "frozen") as the deformation temperature is well below the T_g value of the amorphous phase.

V. MECHANICAL BEHAVIOR AND MICROMECHANICAL DEFORMATION MECHANISMS IN POLYOLEFINS: COMPARISON OF RESULTS

For polyethylene and polypropylene, *brittle fracture* is observed when the testing temperature (or the service temperature) is below T_g of the amorphous phase ($T_{g,am}$). At temperatures above $T_{g,am}$ brittle behavior is initiated if there are defects within the crystalline phase, lack of interconnecting chains in the amorphous phase (entanglements, tie molecules) or inter-spherulitic defects. In particular, low molecular weight and very high degrees of crystallinity (i.e, thinner amorphous parts) favor brittle fracture. The micromechanical mechanism that characterizes brittle fracture is the occurrence of crazing.

For PE or iPP with a sufficient high molecular weight, *ductile behavior* occurs when the polymer is deformed at temperatures above $T_{g,am}$. Ductile behavior is connected with a significant plastic deformation that can be roughly described as a two-step process. One process includes phenomena that are controlled by the mobile macromolecules of the amorphous phase such as lamellar slip, lamellar rotation and lamellar separation. The other process involves the plastic deformation of the crystalline material [22–24].

Chevron formation is a universal phenomenon observed for different classes of polymers at the initial stages of plastic deformation. This phenomenon is not only found in HDPE, LDPE and the β-iPP discussed in this chapter. Chevron formation is also observed in semicrystalline syndiotactic polystyrene (sPS) deformed at 105°C (close to $T_{g,am}$) and in styrene-butadiene block copolymers consisting of a non-crystalline lamellar hard-soft system (see Chapter 10). This finding supports the conclusion that morphological details at the nanoscopic scale dominate the micromechanical deformation mechanisms and, thereby, the macroscopic mechanical properties of the material.

The correlations found between the macroscopic mechanical deformation types, the micromechanical mechanisms and the deformation conditions observed for the polypropylene types under discussion are summarized in Table 3.

VI. SUMMARY AND CONCLUSIONS

Generally, the micromechanical deformation mechanisms found for isotactic polypropylene are very similar to those described for the different types of polyethylene. The main differences are the following:

- Due to the fact that the glass transition temperature of the amorphous phase ($T_{g,am}$) of iPP is relatively close to room temperature (RT) (i.e., the temperature

TABLE 3 General Types of Mechanical Behavior and Dominating Micromechanical Mechanisms Observed for Different Types of Polyethylenes and Polypropylenes at Different Deformation Temperatures and Deformation Rates (Uniaxial Tensile Testing)*

Mechanical Behavior	Micromechanical Mechanisms	Material Type and Testing Conditions
Brittle σ / ε	• Crazing	• Polyethylenes at T < $T_{g, am}$ • Polyethylenes of $M_w \leq$ 50,000 g/mol at 23°C • **α-iPP of Mw 400,000 g/mol** at −40°C, 1 mm/min
Semiductile with neck formation σ / ε	• Multiple crazing	• Polyethylenes of $M_w \geq$ 50,000 g/mol at 23°C • **α-iPP of $M_w \approx$ 230,000 g/mol** at 23°C • **β-iPP of $M_w \approx$ 400,000 g/mol** at −40°C
Ductile with neck formation and cold drawing σ / ε	• Lamellar separation • Rotation of lamellar stacks towards straining direction • Yielding of the crystalline portion and disintegration of crystalline lamellae	• HDPE, LDPE of $M_w \geq$ 100,000 g/mol at 23°C • **α-iPP of $M_w \geq$ 400,000 g/mol** at 23°C
Cold drawing and strain hardening σ / ε	• Chevron formation (collective twisting of lamellar stacks) • Lamellar separation (nanoscopically limited multiple craze-like deformation) • Yielding of the crystalline portion and disintegration of crystalline lamellae • Formation of microfibrils of highly oriented macromolecules	• HDPE, LDPE of $M_w \geq$ 100,000 g/mol at higher temperature, T < T_m • **β-iPP of $M_w \approx$ 400,000 g/mol** at 23°C, 1mm/min (no distinct neck formation, macroscopically homogeneous deformation)

* Bold letters indicate samples that are discussed in this paper.

of most of the technical applications), the material is more sensitive to external parameters that may have impact on the mobility of the amorphous portion. Lower service temperatures, higher deformation rates (impact), high crystallinity and lower molecular weight (decrease of the density of tie molecules and entanglements) result in a significant embrittlement. The influence of the molecular weight should also be considered with respect to the thermo-oxidative degradation that occurs during processing more rapidly for iPP than for PE.

- The cross-hatched lamellar morphology of the α-iPP is limiting the triggering action of the amorphous portion at initial stages of deformation. Independently of the straining direction with respect to lamellar orientation, the crystalline regions are the ones that yield from the initial stages of the plastic deformation. From this point of view, the deformation mechanism of the common α-form differs from those of other semicrystalline materials, whereas the β-iPP form fits well to the schemes found for PE.

- In a first approximation, the enhanced toughness of β-iPP can be explained in terms of the lamellar nanostructure. The parallel stacking of lamellae without cross-hatching enhances the energy dissipating processes that are connected with the mobility of the amorphous phase, including lamellar rotation, lamellar separation and lamellar slip [25–32]. In particular, at room temperature (RT) chevron formation in iPP is solely observed for β-modification.

- The success of heterogeneous toughening strategies (rubber toughening, nano-particle incorporation, etc.) that are mainly based on the initiation of multiple crazing or cavitation processes is strongly influenced by the above-mentioned effects [17].

ACKNOWLEDGMENTS

This work was funded by the German Science Foundation (SFB 418, Mi 358/8-2) and the Kultusministerium des Landes Sachsen-Anhalt. Financial support from the Deutscher Akademischer Auslandsdienst (DAAD) and from the Alexander von Humboldt Stiftung are acknowledged. A research scholarship from the Max-Buchner-Forschungsstiftung (MBFSt 2280 for S.H.) is thankfully acknowledged. We would like to thank Borealis Aktiengesellschaft, Linz and Prof. J. Karger-Kocsis, Kaiserslautern, for the supply of materials and Mrs. S. Goerlitz for the TEM investigations.

REFERENCES

1. Michler GH. Kunststoff-Mikromechanik: Morphologie, Deformations- und Bruchmechanismen. Hanser, München, 1992.

2. Michler GH, Adhikari R, Henning S. Micromechanical properties in lamellar heterophase polymer systems. J Mat Sci 2004;39:3281–3292.

3. Michler GH. Correlations between molecular weight, morphology and micromechanical deformation processes of polyethylenes. Coll Polym Sci 1992;270:627–638.

4. Michler GH, Gruber K. Elektronenmikroskopische Untersuchungen von Polyethylen niedriger Dichte: II. Einfluss der relativen Molekülmasse auf die Morphologie. Acta Polymerica 1981;32:323–332.

5. Michler GH, Brauer E. Elektronenmikroskopische Untersuchungen von Polyethylenen: V. Einfluss von Verzweigungsgrad und Molekülmasse auf die Morphologie. Acta Polymerica 1983;34:533–545.

6. Fiedler P, Michler GH, Braun D. Einfluss der Molekülmasse auf Morphologie und mechanische Eigenschaften von Polyethylen. Acta Polymerica 1986;37:241–247.

7. Fiedler P, Rätzsch M, Braun D, Michler GH. Einfluss des Knäueldurchmessers auf die Morphologie und die mechanischen Eigenschaften von Polyethylen. Acta Polymerica 1987;38:189–195.

8. Michler GH, Godehardt R. Deformation mechanisms of semicrystalline polymers on the submicron scale. Cryst Res Tech 2000;35:863–875.

9. Godehardt R, Rudolph S, Lebek W, Goerlitz S, Adhikari R, Allert E, Giesemann J, Michler GH. Morphology and micromechanical behaviour of blends of ethylene/1-hexene copolymers. J Macromol Sci-Phys 1999;B38:817–835.

10. Michler GH, Morawietz K. Elektronenmikroskopische Untersuchungen von Polyethylenen: VII. Strukturen von UHMWPE nach Dehnung und verschiedenen thermischen Vorbehandlungen. Acta Polymerica 1991;42:620–627.

11. Olley R, Bassett DC, Hine PJ, Ward IM. Morphology of compacted polyethylene fibers. J Mat Sci 1993;28:1107–1112.

12. Michler GH, Lebek W. (Hrsg.) Ultramikrotomie in der Materialforschung. Hanser, München, 2004.

13. Turner Jones A, Aizlewood JM, Beckett DR. Crystalline forms of isotactic polypropylene. Makromol Chem 1964;75:134–158.

14. Mei-Rong Huang M-R, Li X-G, Fang B-R. Beta-nucleators and beta-crystalline form of isotactic polypropylene. J Appl Polym Sci 1995;56:1323–1337.

15. Tjong SC, Shen JS, Li RKY. Mechanical behavior of injection molded beta-crystalline phase polypropylene. Polym Eng Sci 1996;36:100–105.

16. Fujiyama M. Drawing of beta-crystal nucleator-added PP. Intern Polymer Processing 1999;XIV:75–82.

17. Grein C, Plummer CJG, Kausch H-H, Germain Y, Béguelin Ph. Influence of beta nucleation on the mechanical properties of isotactic polypropylene and rubber modified isotactic polypropylene. Polymer 2002;43:3279–3293.

18. Karger-Kocsis J, Ed., Polypropylene: Structure, Blends and Composites Vol. 1: Structure and Morphology. Chapman & Hall, London, 1995.

19. Phillips RA, Wolkowitcz MD, in: Polypropylene Handbook, Moore EP, Ed., Carl Hanser, Munich, 1996:113ff.

20. Aboulfaraj M, G'Sell C, Ulrich B, Dahoun A. *In situ* observation of the plastic deformation of polypropylene spherulites under uniaxial tension and simple shear in the scanning electron microscope. Polymer 1995;36:731–742.

21. Dijkstra PTS, van Dijk DJ, Huétnik J. A microscopic study of the transition from yielding to crazing in polypropylene. Polym Eng Sci 2002;42:152–160.

22. Baltá-Calleja FJ, Peterlin A. Plastic deformation of polypropylene part 2: The influence of temperature and draw-ratio on the axial long period. J Mat Sci 1969;4:722–729.

23. Peterlin A, Baltá-Calleja FJ. Plastic deformation of polypropylene part III. Small-angle x-ray scattering in the neck region. J Appl Phys 1969;40:4238–4242.

24. Baltá-Calleja FJ, Peterlin A. Plastic deformation of polypropylene IV. Mechanism and Properties. J Macromol Sci - Phys 1970;B4:519–540.

25. Henning S, Michler GH, Ania F, Baltá-Calleja FJ. Microhardness of α- and β-modified isotactic polypropylene at the initial stages of plastic deformation: analysis of micromechanical processes. Coll Polym Sci, submitted.

26. Yoshida T, Fujiwara Y, Asano T. Plastic deformation of oriented lamellae: 3. Drawing behaviour of beta-phase isotactic polypropylene. Polymer 1983;24:925–929.

27. Chu F, Yamaoka T, Kimura Y. Crystal transformation and micropore formation during uniaxial drawing of beta-form polypropylene film. Polymer 1995;36:2523–2530.

28. Chu F, Yamaoka T, Ide H, Kimura Y. Microvoid formation process during the plastic deformation of beta-form polypropylene. Polymer 1994;35:3442–3448.

29. Li JX, Cheung WL. On the deformation mechanisms of beta-polypropylene: 1. Effect of necking on beta-phase PP crystals. Polymer 1998;39:6935–6940.

30. Li JX, Cheung WL, Chan CM. On the deformation mechanisms of beta-polypropylene: 2. Changes of lamellar structure caused by tensile load. Polymer 1999;40:2089–2102.

31. Li JX, Cheung WL, Chan CM. On the deformation mechanisms of beta-polypropylene: 3. Lamella structures after necking and cold drawing. Polymer 1999;40:3641–3656.

32. Varga J. β-Modification of isotactic polypropylene: preparation, structure, processing, properties, and application. J Macromol Sci Phys B 2002; 41(4–6): 1121–1172.

8

Micro-Indentation Studies of Polymers Relating to Nanostructure and Morphology

F.J. BALTÁ-CALLEJA, A. FLORES, and F. ANIA

Instituto de Estructura de la Materia, CSIC, Madrid

CONTENTS

I.   INTRODUCTION

Indentation hardness offers a convenient way to probe the mechanical properties of a polymer surface [1]. The method is based on the local deformation produced on a material surface by a sharp indenter upon application of a given load. Some of the advantages of indentation testing, in relation to other procedures for mechanical characterization, are the possibility of testing the mechanical properties of a device in its original assembly, and the ability to spatially map the surface mechanical properties in the micron or sub-micron range. The latter is of fundamental importance for inhomogeneous polymer systems. Additionally, it has been shown that hardness, H, of homogeneous polymer materials is related to other macroscopic mechanical properties such as the yield stress, σ_y, and the elastic modulus, E [2,3]. The compressive yield stress is shown to correlate with hardness, following the mechanical models of elastoplastic indentation, i.e, tending towards $H \approx 3\sigma_y$ (Tabor's relation for a fully plastic deformation) with decreasing elastic strain (higher E/σ_y ratio).

The method most widely used in determining the hardness of polymers is based on the direct imaging of the residual indentation [1]. Figure 1 shows the impressions left behind by a Vickers indenter on the surface of glassy poly(ethylene terephthalate) (PET). A convenient measure of hardness, H, can be obtained by dividing the peak contact load, P, by the

Figure 1 Impressions left behind by a Vickers indenter on the surface of glassy PET. The material surface on the right-hand side of this figure has been shaded with a thin layer of ink for better visualization of the indentations.

projected area of indentation, A ($H = P/A$). This optical procedure is rather simple and allows for a rapid evaluation of the surface mechanical properties of a polymer material. Applied loads in the interval 0.049 – 1.96 N yield indentation depths in the micron range, which characterizes microhardness.

An alternative procedure for the study of the indentation response of polymers relies on the measurement of the indenter tip penetration as a function of the applied load [4,5]. Depth-sensing indentation devices make use of small loads in the μN range and are capable of producing penetration depths in the sub-micron scale, hence, opening up the possibility of investigating the mechanical properties of thin films and at the near surface of polymers. The main drawback of this technique is that the analysis of the unloading cycle, which is needed to obtain hardness and elastic modulus values, is based on elasticity considerations [6,7]. Hence, its application to polymers encounters great difficulties as a consequence of the visco-elasto-plastic character of these materials [8–10]. So far, there is still not a complete and sound methodology for the derivation of mechanical properties of polymers from depth-sensing data. This is possibly the reason why the literature concerning the mechanical response of polymers by means of depth-sensing instrumentation is more limited than that for metals and ceramics [8–15]. Moreover, most of the references quoted in the present review refer to indentation experiments carried out by means of the imaging method.

Finally, it should be mentioned that the actual trend in hardness testing is to use atomic force microscopy (AFM) operating in force mode to perform indentation tests [10]. However, in addition to the usual uncertainties described in the previous paragraph when dealing with load-depth data, one is also faced with the difficulty of accurately determining the values of the applied load and the proper tip shape area function. Thus, although promising results have already been obtained in some polymeric systems [10,16–20], accurate quantitative measurements are still difficult to achieve.

The microhardness technique has nowadays gained wide-spread application in polymer science. Besides its aforementioned technical advantages, one has to add its extreme sensitivity to detect structural changes occurring in polymer materials [1,21,22]. Behind this ability is the intimate correlation between the mechanisms of local deformation and the polymer basic nanostructural entities. The present review intends to provide a general outlook of the present state of the art on indentation hardness in relation to nanostructure and morphology of glassy and semicrystalline polymers, copolymers, blends and composites.

II. BASIC ASPECTS OF MICRO-INDENTATION: CONTACT GEOMETRY

The mechanical response of a polymer material during an indentation cycle is a contribution of several effects. On loading, time-dependent elastic and plastic strains contribute to the total penetration depth. A zone of plastic deformation extends immediately underneath the indenter toward the sample volume and is surrounded by a larger zone of elastic deformation. As an example, Figure 2 illustrates the stress

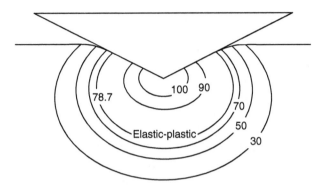

Figure 2 Stress distribution (in MPa) calculated for glassy PET using Finite Element Analysis.

distribution calculated for glassy PET using Finite Element Analysis (10 mN applied force) [23]. A stress level at 78.7 MPa delimits the region where plastic deformation takes place, corresponding to a few times the penetration distance below the indenter tip.

During load release, the instantaneous elastic strains recover and a permanent irreversible plastic deformation remains, the latter one defining the hardness value. In addition, a time dependent contribution during loading (creep) is observed. The creep under the indenter is characterized by a decreasing strain rate that can be described by a law of the form [1,24]:

$$H = H_1 t^{-k} \tag{1}$$

where H_1 is the hardness at a given reference time and k is a parameter (creep constant) which gives the rate at which the material flows under the indenter.

Finally, for low crystallinity and rubbery materials, a long-delayed indentation recovery after load removal (visco-elastic relaxation) is also detected. Thus, hardness should be measured for short loading times (few seconds) and immediately after load removal to minimize creep and visco-elastic relaxation, respectively.

III. STRUCTURE DEVELOPMENT IN POLYMER GLASSES: INFLUENCE OF TEMPERATURE AND TIME OF CRYSTALLIZATION

In the glassy state, hardness involves overcoming the critical stress required to plastically deform the amorphous solid against cohesive intra- and intermolecular forces such as van der Waals interactions, bond forces or internal rotations. Micro-indentation hardness has been shown to provide for direct evidence of changes in the physico-chemical properties of polymer glasses such as those occurring upon physical aging, weathering, UV degradation, γ-irradiation, plasticization, etc. [25–28]. Moreover, the microhardness of glassy polymers has been directly related to the glass transition temperature, T_g [29–31]. Indeed, Figure 3 illustrates the linear relationship found between H and T_g for a variety of polymer materials with dominating single-bonds within the main chain [29]. This correlation is a consequence of the intimate relationship between both, H and T_g, and the cohesive energy density of the amorphous material.

It has also been shown that the microhardness of glassy polymers follows an exponential decrease as a function of temperature. However, H conspicuously decreases at T_g, due to the onset of liquid-like motions, which lead to a much lower cohesive energy density [25,32]. Above T_g, certain glassy polymers are capable of crystallizing. The structure developed will depend upon the crystallization conditions and the microhardness values will accordingly vary.

The ability of the microhardness technique to follow the isothermal crystallization of polymers from the glassy state has been used in a number of studies [33–36]. Figure 4 illustrates the hardness variation with increasing crystallization time, t_c, for PET at different annealing temperatures [33]. The initial rapid hardness increase with increasing crystallization time is associated with the development of primary crystallization, which is connected to the growth of the crystallizing units (spherulites). At the end of primary crystallization, the sample volume is entirely filled up with spherulites and the hardness values level off. During secondary crystallization,

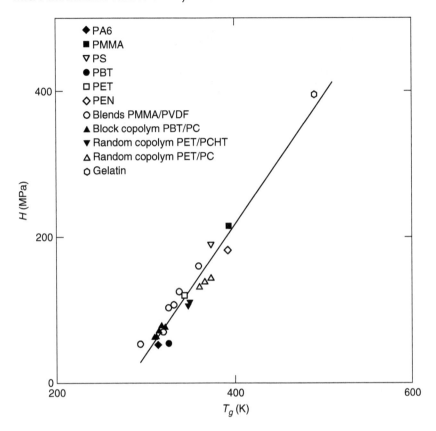

Figure 3 Microhardness linear correlation with T_g for a number of glassy polymers, with dominating single bonds in the main chain.

the following mechanisms can occur: i) formation of new intra-spherulitic lamellar stacks, ii) formation of single crystals within the lamellar stacks or in the interstack regions and iii) increase of the crystal thickness [37]. As will be shown below, one could expect that these structural rearrangements taking place during secondary crystallization should enhance the microhardness of the material. However, in the case of PET, hardness values are shown to remain nearly constant (see Figure 4) — a result that could be related to the occurrence of a rigid amorphous phase, as will be discussed below.

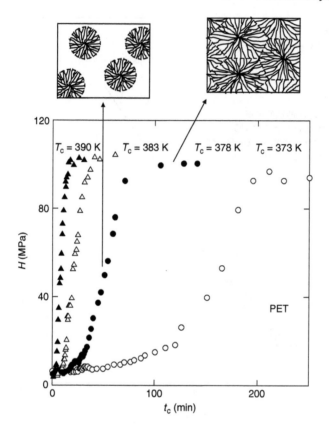

Figure 4 Variation of H as a function of t_c, during cold crystalli-
zation of PET at different temperatures. (From Baltá Calleja FJ,
Flores A. Hardness. In: Kroschwitz J, ed. Encyclopedia of Polymer
Science and Technology, 3rd ed, vol.2. New York: John Wiley & Sons,
Inc., 2003: 678–691. With permission.)

Figure 5 shows the plot of the hardness values of PET
during primary crystallization, measured *in situ* at 117°C and
after cooling at room temperature (RT), as a function of the
volume fraction of spherulites ϕ. The hardness behavior when
the spherulitic growth is not completed can be described following [38]:

$$H = H_{sph}\phi + H_a(1 - \phi) \qquad (2)$$

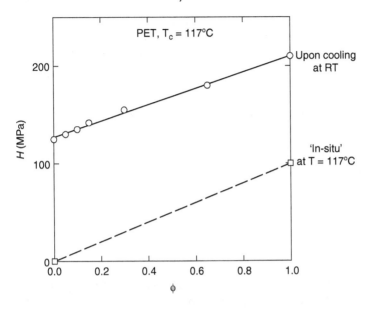

Figure 5 Plot of the microhardness values at room temperature (RT) as a function of the volume fraction of spherulites, during primary crystallization of PET at 117°C. The dashed line represents the H values measured *in situ* at 117°C.

Where H_{sph} and H_a are the hardness values of the spherulitic and the amorphous inter-spherulitic regions respectively. For PET, $H_a \simeq 0$ and $H_{sph} \simeq 100$ MPa when measurements are carried out at 117°C, and $H_a \simeq 120$ MPa and $H_{sph} \simeq 210$ MPa upon cooling to room temperature. Hence, it seems that the hardness of a polymer material during primary crystallization from the glassy state is mainly governed by the volume fraction of crystallizing units, H_{sph} and H_a, remaining constant for a given temperature of measurement. There are other examples in the literature where Equation 2 holds [34,39]. Table 1 collects the H_{sph} and H_a values derived from room temperature microhardness measurements carried out after primary crystallization of various semirigid polyesters.

When spherulitic growth is completed, then $H = H_{sph}$, and the hardness of the material can be expressed following:

TABLE 1   The H_{sph} and H_a Values of Various Semirigid Polyesters, Derived from Micro-indentation Experiments at Room Temperature, Carried Out after Cold Crystallization at T_c

Material	T_c [°C]	H_a [MPa]	H_{sph} [MPa]
Poly(ethylene terephthalate) (PET)	117–240	120	210
Poly(ethylene naphthalene-2,6-dicarboxylate) (PEN)	155–220	185	195–225
Poly(ether ether ketone) (PEEK)	180–340	135	210–250
Thermoplastic polyimide (TPI)	280–330	185	220–280

$$H = H_c\alpha + H_{ar}(1 - \alpha_L)\frac{\alpha}{\alpha_L} + H_a\left(1 - \frac{\alpha}{\alpha_L}\right) \qquad (3)$$

Here H_c is the hardness of the crystals; H_{ar} and H_a are the hardness of the amorphous regions within and between the lamellar stacks respectively; α is the volume degree of crystallinity and α_L is the linear degree of crystallinity, i.e., the degree of crystallinity within the lamellar stacks. Equation 3 is formulated considering the most general case in which the spherulites are constituted by lamellar stacks separated by amorphous interstack regions. It is noticeable that Equation 3 distinguishes between the hardness of the amorphous regions within and outside the lamellar stacks. Indeed, studies carried out in the last decade suggest that a fraction of the amorphous material in semicrystalline systems such as PET, poly(ethylene naphthalene-2,6-dicarboxylate) (PEN) or poly(ether ether ketone) (PEEK), present a hindered mobility [41]. This rigid amorphous fraction material is associated to the regions where several crystalline lamellae are separated by thin amorphous layers. In contrast, the mobile amorphous fraction is associated with thicker regions between the lamellar stacks. Within this context, Equation 3 accounts for the different hardness values that one would expect for the rigid and mobile amorphous phases (H_{ar} and H_a respectively). Otherwise, if $H_{ar} = H_a$, then Equation 3 simplifies to:

$$H = H_c\alpha + H_a(1 - \alpha) \qquad (4)$$

Let us next recall the hardness behavior of PET during secondary crystallization. In Figure 4, the hardness values seem to remain constant with crystallization time and crystallization temperature for this regime. Indeed, room temperature measurements carried out on various PET samples, crystallized at different temperatures for various periods of time (under the fully spherulitic growth regime), reveal a constant microhardness value [38]. However, a parallel increase in the volume degree of crystallinity as a function of crystallization time and temperature is observed. What could be the reason for this apparent contradiction: increasing α and constant H? In the light of the latest investigations favoring the presence of a rigid amorphous phase on PET [41], we suggest that the response to plastic deformation of the amorphous material within the spherulites is not substantially different from that of the crystalline phase, due to the limited mobility of the molecular amorphous chains. Most interesting is the fact that not only the amorphous layers between crystals but also the amorphous interstack regions seem to behave as a rigid amorphous material.

IV. DEPENDENCE OF MICROHARDNESS ON NANOSTRUCTURE OF SEMICRYSTALLINE POLYMERS: MECHANISMS OF DEFORMATION

Equation 4 accounts for the contribution of the amorphous and crystalline phases to the hardness of a semicrystalline polymer. There is now substantial experimental evidence that Equation 4 applies for polymer systems in a variety of morphologies and nanostructures [1,21]. As an example, Figure 6 illustrates the linear variation of H as a function of the degree of crystallinity for isotactic β-polypropylene [42]. The left-hand and right-hand y-axis intercepts represent H_a and H_c respectively. For polymers where $T_g < $ RT, the hardness value of the rubbery amorphous phase has been widely assumed to be negligible with respect to the hardness of the crystals ($H_a \ll H_c$), and Equation 4 simplifies to: $H \simeq H_c\,\alpha$. Recent studies on a series of ethylene-octene copolymers with

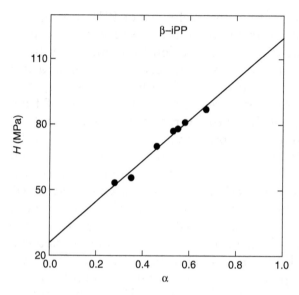

Figure 6 Microhardness linear variation with increasing degrees of crystallinity for isotactic β-polypropylene (β-iPP).

a range of octene content offer values of $H_a \simeq 0.12$ MPa for the highest comonomer (octene) content [43]. Moreover, hardness values for amorphous PET above T_g are shown to lie in the range 3.5–6.5 MPa, with smaller H values as the temperature increases (see Figure 4). A different approach has been proposed by Fakirov et al. [44] to describe the microhardness of semicrystalline polymers comprising amorphous liquid-like regions ($T_g < $ RT). For such purpose, H_a values are estimated from the H-T_g correlation shown in Figure 3, which was initially derived for glassy polymers. The application of the linear H-T_g relationship to T_g values below RT leads to negative H_a values without any physical meaning. However, the idea behind this approach is to introduce a kind of "floating effect" of the solid particles on the soft component [45]. Such correction aims to explain the deviations observed in some experimental H values with respect to the additivity of the hardness of the individual components and/or phases.

The validity of Equation 4 relies on the fact that plastic deformation takes place both in the crystalline and in the amorphous regions. At small applied stresses, plastic yielding under the indenter involves shearing motions of lamellae and lamellar separation. At large stresses (under the indenter tip), lamellar fracture and other mechanisms such as microfibrillization may occur. Recent combined micro-indentation and x-ray synchrotron micro-diffraction experiments in a Nylon 66 and an UHMWPE fiber have provided new stimulating information concerning the mechanisms of deformation during and upon indentation [46]. *Ex situ* experiments in a Nylon 66 single fiber indicate that plastic deformation involves a local perturbation of orientation distribution of the crystal blocks, the degree of crystallinity remaining unchanged. Analogous studies in an UHMWPE fiber reveal, in addition, a partial polymorphic transformation taking place within the indented zone. Finally, simultaneous micro-indentation and x-ray micro-diffraction experiments in an UHMWPE fiber suggest that during the application of the load, the orientation of the crystal blocks splits into two domains, each of them at an angle of 2° with respect to the fiber axis. Retraction of the tip results again in a one-domain orientation of the crystal blocks, however, with a broader crystal orientation distribution with respect to the one present in the non-indented material.

In oriented and isotropic polymeric systems, the material resistance to the applied compressive and shear stresses during an indentation cycle must be related to the cohesive energy density in the amorphous regions and to the intermolecular forces holding the molecules within the crystals (crystal packing). The influence of the crystal thickness on the hardness of the crystalline phase [47] was also soon recognized. On the basis of a heterogeneous deformation model involving the heat dissipated by the generation of a number of shear planes in the lamellar nanostructure (see Figure 7), Baltá Calleja and Kilian derived the following relationship between the crystal hardness, H_c and the crystal thickness, l_c [48]:

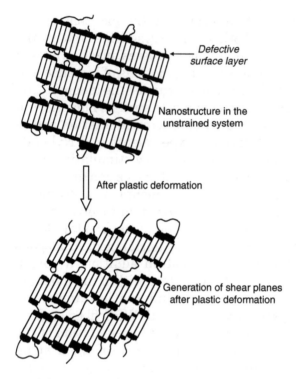

Figure 7   Model of plastic deformation through the generation of a number of shear planes.

$$H_c = \frac{H_c^\infty}{1 + \dfrac{b}{l_c}} \qquad (5)$$

Here, H_c^∞ is the hardness of an infinitely thick crystal and is mainly related to the chain packing density within the crystals. The b-parameter is related to the ratio between the surface free energy of the crystals, σ_e, and the energy required to plastically deform them through a number of shearing planes, Δh ($b = 2\sigma_e/\Delta h$).

Figure 8 shows the plot of $1/H_c$ as a function of $1/l_c$, for various semicrystalline polymers. The y-axis intercept yields

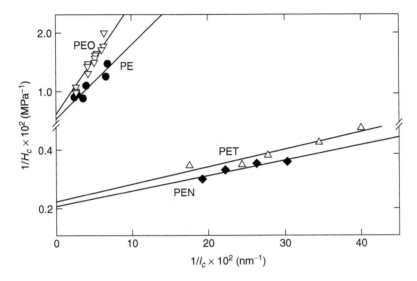

Figure 8 Plot of the H_c^{-1} versus l_c^{-1} for different semicrystalline polymers (see Equation 5).

$1/H_c^{\infty}$. It is worth noting the much higher H_c^{∞} and smaller l_c values for the semirigid polyesters (PET [49] and PEN [39]) in contrast to the flexible polyoleofins (PEO [50] and PE [51]). The fact that H_c^{∞} is higher for PET and PEN ($H_c^{\infty} \simeq 400$ MPa and 550 MPa respectively) than for PEO and PE ($H_c^{\infty} \simeq 150$ MPa and 170 MPa respectively) is related to the stronger intermolecular forces holding the molecular chains within the polyester crystals. The reader is referred to the literature for H_c^{∞} values reported for other polymer systems [34,52–54]. While the influence of crystal size on microhardness has been largely experimentally supported, a recent paper also suggests that this correlation arises from the sensitivity of H to crystal perfection [44].

The similarity of Equation 5 with the Thomson-Gibbs equation, and the fact that polymer materials with high equilibrium melting temperatures, T_m^0, display high H_c^{∞} values, prompted us to seek a correlation between crystal microhardness and melting temperature. Indeed, by combination of Equation 5 with the Thomson-Gibbs equation, one obtains [51]:

$$\frac{1}{H_c} = -\frac{b}{b^* H_c^\infty T_m^0} T_m + \frac{1}{H_c^\infty} \left(1 + \frac{b}{b^*} \right) \qquad (6)$$

where T_m is the melting point of the material and $b^* = 2\sigma_e/\Delta h_f$ (Δh_f is the enthalpy of fusion). Equation 6 substantiates the close relationship between the reciprocal value of the crystal hardness and the melting temperature of the crystals. Moreover, a linear relationship between $1/H_c$ and T_m holds, provided that the b/b^* ratio remains constant. Figure 9 illustrates the variation of $1/H_c$ as a function of T_m for various polymer systems [34,51]. Data for the flexible systems (PEO and PE) show lower T_m and higher $1/H_c$ values, than those for the semirigid polymers (PET and the thermoplastic polyimide, TPI). Also remarkable is the higher slope in Figure 9 for PEO and PE than that for PET and TPI, which can be

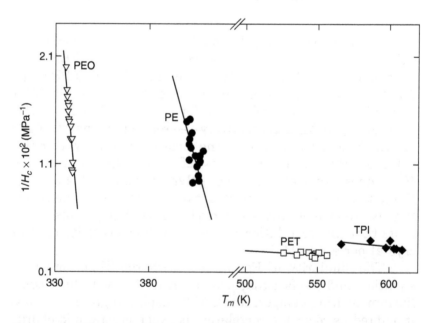

Figure 9 Variation of H_c^{-1} as a function of T_m, for various flexible and semirigid polymers (see Equation 6).

mainly attributed to the higher b/b^* values of the former systems. Since $b/b^* = \Delta h_f / \Delta h$, then, the energy required to plastically deform the crystals is much lower than the melting enthalpy in the case of PE and PEO ($\Delta h_f^{PE} = 45 \, \Delta h^{PE}$ and $\Delta h_f^{PEO} = 95 \, \Delta h^{PEO}$).

V. STUDY OF POLYMORPHISM IN POLYMERS BY MICROHARDNESS

The sensitivity of crystal hardness to chain packing makes micro-indentation hardness of great value for the detection of polymorphic transformations in polymer materials. It has been shown that the amount of α and β phases in iPP can be easily distinguished by means of microhardness measurements [55]. Indeed, the ability of micro-indentation hardness to locally probe the material surface has been used to differentiate the individual α and β spherulites, which are found to coexist in high molecular weight iPP and a random ethylene/propylene copolymer [56]. Moreover, micro-indentation hardness has been applied to the study of the stress-induced polymorphic transition occurring in poly(butylene terephthalate) (PBT) [57], its block copolymers and blends [58] and iPP [59].

The spontaneous solid-state transformation occurring upon aging of isotactic poly(1-butene) (iPBu-1) at room temperature represents a paradigm in which the use of micro-indentation hardness provides valuable information on the kinetics of the polymorphic transformation [60]. When crystallized from the melt, iPBu-1 exhibits the tetragonal form II. The storage of the fresh sample at room temperature eventually leads to the more densely packed hexagonal form I. Figure 10 shows the variation of microhardness with aging time for a high molecular weight iPBu-1 sample ($M_w = 850,000$ g/mol), crystallized at various temperatures. The initial steep increase of microhardness with aging time is associated to the gradual transformation from form II into form I, the latter having larger hardness values, mainly due to a higher H_c^∞ value as a consequence of a denser crystal packing. It is shown that high crystallization temperatures substantially reduce the rate of the II $\rightarrow$ I transformation.

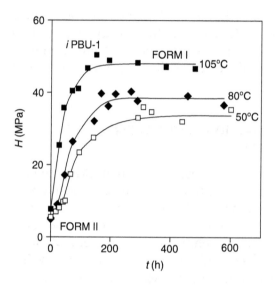

Figure 10 Microhardness values as a function of aging time at room temperature for *i*-PBu1 crystallized at different temperatures. (From Azzurri F, Flores A, Alfonso GC, Baltá Calleja FJ. Macromolecules 2002; 35: 9069–9073. With permission.)

VI. APPLICATION TO POLYMER COMPOSITES

Microhardness has been widely used as a non-destructive method for the evaluation of the mechanical properties of a wide range of polymer composites, which are usually constituted by an organic or inorganic filler embedded in a polymer matrix. The particle or fiber filler is commonly designed to improve the mechanical performance of the material. Microhardness values of the polymer matrix typically increase with the addition of the reinforcing filler, however, the precise rate depends on the specific morphology and nanostructure of each component and the interaction with each other. In addition, indentation hardness offers the unique opportunity to differentiate between the mechanical properties of the individual components and, hence, can provide for direct information on the mechanical properties of the fiber-matrix interphase in a composite material [12,61]. In other cases, as for example in the case of microfibrillar-reinforced composites, microhardness has been successfully used to derive the mechanical properties of the individual microfibrils that are otherwise inaccessible [62].

The application of micro-indentation hardness to the study of inorganic-polymeric composite materials is one of the main issues of great concern in the scientific community, as revealed by a literature search on micro-indentation hardness related papers published in the last five years [63–67]. Among these systems, sol-gel derived organic-inorganic composites are some of the most attractive ones because the inorganic component can be homogeneously dispersed at a molecular level in the polymer matrix through chemical bonding [66,67]. The physical properties of these hybrids can vary from being brittle and hard to rubbery and soft, depending on the inorganic/organic ratio. In addition, a great effort is also being done in the application of micro-indentation hardness to biomedical polymer materials, such as bioactive hydroxy-apatite reinforced polymer composites [68], poly(methyl methacrylate)-based bone mineralizers [69] or ion-implanted polymeric materials [70].

Figure 11 shows the microhardness behavior of polyethylene-fullerene composites, as a function of filler content. This system has been selected as an illustration of the polymer matrix reinforcement via the inclusion of inorganic particles [71]. Indeed, H, for small proportions of the nanofiller, is shown to be an increasing function of fullerene content following:

$$H \approx H_f \phi_{C60} + H_{PE} \tag{7}$$

Where H_f is a constant related to the hardness of the fullerene particles, ϕ_{C60} the fullerene content and H_{PE} the hardness of the polymer matrix. The matrix reinforcement is substantial: H_{PE} rises by 30% when adding about 2.5% of C_{60}, i.e., PE reaches hardness values comparable to those of glassy PET. Furthermore, annealing of the material at 130°C for 10^5 s, leads to H values at 2.5% filler content of 190 MPa, a hardness value which is comparable to that of semicrystalline PET. The substantial hardening of the composite upon annealing is associated to the thickening of the PE crystalline lamellae.

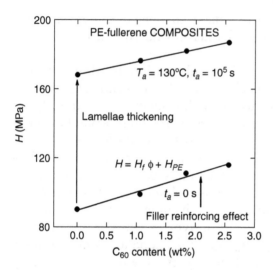

Figure 11 Microhardness behavior with fullerene content, for polyethylene-fullerene dry gel composites. *H* values of the composite films upon annealing at 130°C for 10^5 s are also included.

VII. STRUCTURAL FEATURES OF BLOCK COPOLYMERS: INFLUENCE OF COMPOSITION, STRUCTURE AND PHYSICAL AGING

The microhardness of copolymer systems, including random copolymers of semirigid monomer units, liquid-crystalline copolyesters and thermoplastic elastomers with random or block statistical distribution, has been shown to be well described by a simple additive model of the microhardness of the individual components [72–76]. The microhardness of the latter is controlled by the specific morphology and nanostructure following Equations 4 and 5.

Figure 12 (top) illustrates, as an example, the microhardness variation of a series of PBT/cyclo-aliphatic polycarbonate (PC_c) block copolymers, as function of PC_c content [52]. Quenched samples from the melt display an amorphous morphology for contents of the non-crystallizable component (PC_c) higher than 20 wt%. The microhardness of the freshly

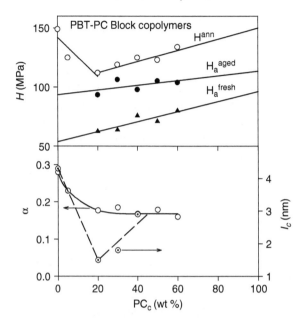

Figure 12 (Top): Variation of microhardness with PC_c content for a series of PBT-PC_c block copolymers: ▲, quenched samples; ●, aged quenched samples; ○, quenched samples annealed at 100 – 130°C. (Bottom): Variation of the degree of crystallinity () and the crystal lamellar thickness (⊙), as a function of PC_c content, for the annealed PBT-PC_c copolymers.

quenched glassy samples (H_a^{fresh}) is shown to linearly increase with increasing PC_c content. The same microhardness additive behavior is evident after storing the samples at room temperature for 6 months (H_a^{aged}). Most interesting is the fact that physical aging induces a hardening effect, due to the increase in the cohesive energy density of the material. An analogous hardness enhancement has also been detected in other aged glassy polymers [25,26].

Figure 12 (top) also shows the hardness values of the quenched PBT/PC_c copolymers after heat treatment at 100 – 130°C (denoted as H^{ann}). In the range of compositions investigated, all the samples develop crystallinity after annealing, with the PC_c segments taking part of the amorphous regions and the PBT segments forming the crystalline phase. H values increase with respect to H_a^{fresh}, due to the reinforcing contribution of the developing crystalline domains (see Equation 4). The microhardness of the semicrystalline copolymers is shown to initially decrease with increasing PC_c content, up to 20 wt%. For higher PC_c contents, a linear microhardness increase with increasing PC_c wt% is observed. The deviation of the microhardness behavior of the crystallized copolymers with respect to the additivity law of the individual homopolymers can be attributed to two factors: i) the change in the degree of crystallinity of the copolymers with respect to the PBT homopolymer and ii) variations in the crystalline lamellar thickness with PC_c content. Indeed, one can see from Figure 12 (bottom) that the degree of crystallinity does not linearly decrease with increasing PC_c, deviating to lower values for PC_c contents < 30 wt% and to higher ones for PC_c wt% > 40. Moreover, the crystal thickness displays a minimum value at a PC_c content of 20 wt% (see Figure 12, bottom), hence, resembling the dependence of H with the amount of PC_c.

From the H and α data in Figure 12, one can derive the H_c values of the PBT crystalline phase within the copolymer system, making use of the microhardness additivity law (using $H_a^{PCc} = 96$ MPa), together with Equation 4 ($H_a^{PBT} = 54$ MPa). Using a similar representation as that shown in Figure 8, from Equation 5, one can calculate the H_c value for PBT, which is found to be 370 MPa, i.e., only slightly smaller than the H_c value reported for PET (400 MPa).

VIII. MICROHARDNESS: MORPHOLOGY CORRELATIONS IN BLENDS OF GLASSY POLYMERS

The use of micro-indentation hardness for the study of polymer multicomponent blends has been demonstrated to yield valuable information with respect to the chemical composition of the blends. This is of fundamental importance in the case of polyester blends, where trans-esterification processes may lead to copolymer segments with a random or block character [76,77]. It has also been shown that H of a polymer blend behaves as an additive property of the two independent components, both for blends of semicrystalline materials [78] and for blends of noncrystallizable polymers [79]. Deviations from the microhardness additivity law can be accounted for on the basis of variations of the individual morphologies and nano-structures with respect to those of the homopolymers [80,81].

More recently, the microhardness variation of star block styrene-butadiene-styrene copolymer (ST2)/polystyrene blends has been investigated as a function of styrene composition [82]. The peculiarity of these blends is that their constituents are both amorphous, and yet they can display a morphology, which resembles a lamellar semicrystalline system (see Figure 13a). Alternated styrene and butadiene components form a layer-like morphology (see Chapter 3). Most interesting is the fact that the lamellar periodicity can be detected by means of small angle x-ray scattering. The pure ST2 copolymer exhibits a polystyrene layer periodicity of about 20 nm, as shown in Figure 13b. By adding increasing amounts of the polystyrene homopolymer (hPS) to the star copolymer, the PS layer thickness distribution broadens and shifts to higher values (see Figures 13a and b). Figure 14 shows the microhardness measurements carried out on the ST2/hPS blends. The x-axis represents the total PS content within the blend, i.e., the amount of PS within the star copolymer in addition to the hPS content in the blend. A significant deviation of H from the additivity law is apparent. Most interesting is the observation that the smaller the layer thickness is, the larger is the deviation of H from the microhardness additivity law. This behavior resembles the microhardness variation with crystal lamellar thickness, observed for semicrystalline polymers. This analogy has prompted us to represent the H values of the blends as a function of the reciprocal lamellar thickness of polystyrene, $1/D_{PS}$ (see Figure 15). Here, pure polystyrene is assumed to display an infinite lamellar thickness. Surprisingly, the data points fit into a straight line, suggesting that the lamellar amorphous block copolymer systems may be regarded, indeed, as an analogue to the semicrystalline systems. Hence, an equivalent equation to that introduced for the microhardness of polymer crystal lamellae

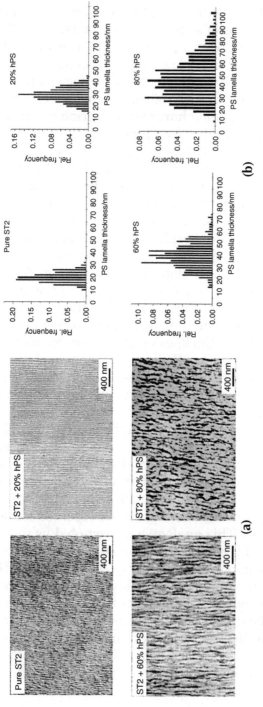

Figure 13 (a): Scanning electron micrographs of star block styrene-butadiene-styrene copolymer/polystyrene blends with different polystyrene content. (b): Lamellar thickness distributions for the PS component within the amorphous blends shown in Figure 13a. (Reprinted from Baltá Calleja FJ, Cagiao ME, Adhikari R, Michler GH. Polymer 2004; 45, 247–254. Copyright 2004, with permission from Elsevier.)

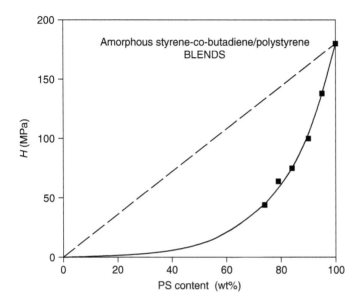

Figure 14 Variation of microhardness as a function of PS content, for the star block styrene-butadiene-styrene copolymer/polystyrene blends. The dashed line follows the additive law of the individual components (styrene and butadiene). (Reprinted from Baltá Calleja FJ, Cagiao ME, Adhikari R, Michler GH. Polymer 2004; 45, 247–254. Copyright 2004, with permission from Elsevier.)

(see Equation 5) could be written for the star copolymer/hPS blends:

$$H = \frac{H^{PS}}{1 + \dfrac{K}{D_{PS}}} \tag{8}$$

where K is now a constant similar to the b-parameter in Equation 5. In summary, the increasing tendency of the H values to approach the additivity law with added hPS content can be explained as a result of the thickness increase of the PS layers that are similarly organized as the stacks of crystalline lamellae.

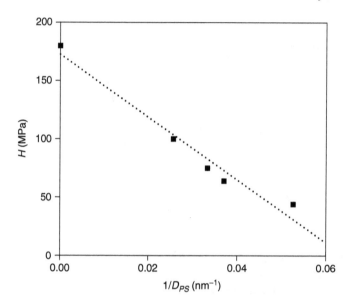

Figure 15 *H* values as a function of the reciprocal lamellar thickness of polystyrene, for the amorphous blends of Figures 13 and 14. (Reprinted from Baltá Calleja FJ, Cagiao ME, Adhikari R, Michler GH. Polymer 2004; 45, 247–254. Copyright 2004, with permission from Elsevier.)

IX. OUTLOOK

Micro-indentation has been shown to give valuable information on the local deformation produced by a sharp indenter on a polymer surface at the micron and sub-micron range. One may expect that the microhardness technique can lead to future developments in various fields.

Although physical aging of glassy polymers has been the object of investigation during the last decade, the detailed molecular mechanisms are still open to debate [83]. Recent studies reveal the role of the thermal history on the precrystalline embrionic order of amorphous polymers (PET) and on the resulting physical aging as revealed by micro-indentation [36,84].

In semicrystalline polymers, as discussed above, *H* can be described on the basis of a composite consisting of hard lamellae intercalated by compliant disordered layers. A generalized Tabor relationship between H and σ_y accounting for the dependence of crystal thickness, crystallinity and temperature explains the correlation of mechanical properties with microstructure [3]. The correlation found between H_c and T_m (Equation 6) opens up new routes for the detailed understanding of the deformation mechanism of polymer crystals in the light of thermodynamical parameters [51]. We also expect that this technique will be of great value in the detection and analysis of micromechanical processes of other polymorphic transitions such as, for example, the recently reported stress-induced α-β transition in iPP during the initial stages of deformation [85].

Microhardness offers, in addition, future possibilities for the mechanical characterization of specific parts of processed polymers on a micron scale (prepared by extrusion, injection molding, etc.) [86] or in blends of polyolefins with recycled polymer components [87]. Polymer microlayering is a new approach to prepare polymer architectures for material improvement [88]. Micro-indentation experiments on multilayered PET/PC films produced by continuous layer multiplying co-extrusion, having layer thicknesses in the micron and submicron range, have been recently performed [89]. The deformation mechanisms under the indenter reveal in this case the small influence of the interphase on the microhardness value.

Work is now in progress to examine the micro-indentation hardness of nanocomposites based on multiblock polyester elastomers and carbon nanotubes [90]. These composites represent a new class of reinforced utility materials with improved physical properties [91]. Micro-indentation hardness can also significantly play an important role in the coming years concerning the characterization of biomedical composites, such as bone replacements, using hydroxy-apatite reinforced polymer matrices [92]. The localized nature of the micro-indentation test will allow the retrieval of information as regards the heterogeneity of composites that is often not available with other analytical techniques.

With the increasing tendency nowadays to use renewable resources, the possibility to employ natural polymers for replacement of synthetic ones is lately developing [93]. For instance, the micro-indentation of injection-molded starch has been investigated in relation to the water sorption mechanism [94]. Thermally treated silk membranes represent another example of a natural material that has been studied using this technique. However, silk membranes exhibit hardness values around 400 MPa, which are notably larger than those of common polymers, opening up new application possibilities [95].

As pointed out above, several reports have appeared lately on the mechanical properties of polymer-, blend- and composite-surfaces at sub-micron resolution [8,11,13,15]. New data on the influence of molecular weight and molecular weight distribution on the microhardness at the near surface of glassy polymers have been recently reported [96]. As indicated in the introduction, nanoindentation techniques using AFM for probing the mechanical behavior of the top few tens of nanometers of bulk and thin polymer film specimens open up future possibilities to investigate the elastic and plastic flow properties of the near-surface region of polymer materials [16–20].

ACKNOWLEDGEMENTS

The authors acknowledge the Ministerio de Educación y Ciencia (Grant No F1S2004-01331), MCYT, Spain, for the generous support of this investigation.

REFERENCES

1. Baltá Calleja FJ, Fakirov S. Microhardness of polymers. Cambridge, UK: Cambridge University Press, 2000.

2. Benavente R, Pérez E, Quijada R. Effect of comonomer content on the mechanical parameters and microhardness values in poly(ethylene-co-1-octadecene) synthesized by a metallocene catalyst. J Polym Sci: B: Polym Phys 2001; 39: 277–285.

3. Flores A, Baltá Calleja FJ, Attenburrow GE, Bassett DC. Microhardness studies of chain-extended PE: III. Correlation to yield stress and elastic modulus. Polymer 2000; 41: 5431–5435.

4. Pethica JB, Hutchings R, Oliver WC. Hardness measurements at penetration depths as small as 20 nm. Phil Mag A 1983; 48 (4): 593–606.

5. Pollock HM. Nanoindentation. In: ASM Handbook, vol 18: Friction Lubrication, and Wear Technology. Materials Park, Ohio: ASM International, 1992.

6. Doerner MF, Nix WD. A method for interpreting the data from depth-sensing indentation instruments. J Mater Res 1986; 1(4): 601–609.

7. Oliver WC, Pharr GM. An improved technique for determining hardness and elastic modulus using load and displacement sensing indentation experiments. J Mater Res 1992; 7(6): 1564–1583.

8. Flores A, Baltá Calleja FJ. Mechanical properties of poly(ethylene terephthalate) at the near surface from depth-sensing experiments. Phil Mag A 1998; 78(6): 1283–1297.

9. Hochstetter G, Jiménez A, Loubet JL. Strain-rate effects on hardness of glassy polymers in the nanoscale range. Comparison between quasi-static and continuous stiffness measurements. J Macromol Sci Phys 1999; B38: 681–692.

10. VanLandingham MR, Villarubia JS, Guthrie WF, Meyers GF. Nanoindentation of polymers: an overview. Macromol Symp 2001; 167: 15–43.

11. Briscoe BJ, Sebastian KS, Sinha SK. Application of the compliance method to micro-hardness measurements of organic polymers. Phil Mag A 1996; 74(5): 1159–1169.

12. Hodzic A, Stachurscki ZH, Kim JK. Nano-indentation of polymer-glass interfaces. Part I. Experimental and mechanical analysis. Polymer 2000; 41: 6895–6905.

13. Briscoe BJ, Akram A, Adams MJ, Johnson SA, Gorman DM. The influence of solvent quality on the mechanical properties of thin cast isotactic poly(methyl methacrylate) coatings. J Mater Sci 2002; 37: 4929–4936.

14. Beake BD, Leggett GJ. Nanoindentation and nanoscratch testing of uniaxially and biaxially drawn poly(ethylene terephthalate) film. Polymer 2002; 43: 319–327.

15. Gilbert JL, Cumber J, Butterfield A. Surface micromechanics of ultrahigh molecular weight polyethylene: micro-indentation testing, crosslinking and material behavior. J Biomed Mater Res 2002; 61(2): 270–281.

16. Amitay-Sadovsky E, Cohen SR, Wagner HD. Anisotropic nanoindentation of transcrystalline polypropylene by scanning force microscope using blade-like tips. Appl Phys Lett 1999; 74: 2966–2968.

17. Han Yanchun, Schmitt S, Friedrich K. Nanoscale indentation and scratch of short carbon fiber reinforced PEEK/PTFE composite blend by atomic force microscopy lithography. Appl Comp Mater 1999; 6: 1–18.

18. Mareanukroh M, Eby RK, Scavuzzo RJ, Hamed GR, Preuschen J. Use of atomic force microscope as a nanoindenter to characterize elastomers. Rubber Chem Tech 2000; 73: 912–925.

19. Bischel MS, Vanlandingham MR, Eduljee RF, Gillespie JW Jr., Schultz JM. On the use of nanoscale indentation with AFM in the identification of phases in blends of linear low density polyethylene and high density polyethylene. J Mater Sci 2000; 35: 221–228.

20. Gao SL, Mäder E. Characterisation of interphase nanoscale property variations in glass fibre reinforced polypropylene and epoxy resin composites. Composites A 2002; 33: 559–576.

21. Baltá Calleja FJ. Microhardness studies of polymers and their transitions. Trends Polym Sci 1994; 2(12): 419–425.

22. Baltá Calleja FJ, Flores A. Hardness. In: Kroschwitz J, ed. Encyclopedia of Polymer Science and Technology, 3rd ed, vol.2. New York: John Wiley & Sons, 2003: 678–691.

23. Rikards R, Flores A, Ania F, Kushnevski V, Baltá Calleja FJ. Numerical-experimental method for the identification of plastic properties of polymers from microhardness tests. Comp Mater Sci 1998; 11: 233–244.

24. Baltá Calleja FJ, Flores A, Ania F, Bassett DC. Microhardness studies of chain-extended PE: II. Creep behavior and temperature dependence. J Mater Sci 2000; 35: 1315–1319.

25. Ania F, Martínez-Salazar J, Baltá Calleja FJ. Physical ageing and glass transition in amorphous polymers as revealed by microhardness. J Mater Sci 1989; 24: 2934–2938.

26. Rueda DR, Varkalis A, Viksne A, Baltá Calleja FJ, Zachmann HG. Physical aging in poly(ethylene naphthalene-2,6-dicarboxylate) in relation to sorbed water, enthalpic relaxation, and mechanical properties. J Polym Sci: B: Polym Phys 1995; 33: 1653–1661.

27. Kusy RP, Whitley JQ, Kalachandra S. Mechanical properties and interrelationships of poly(methyl methacrylate) following hydration over saturated salts. Polymer 2001; 42: 2585–2595.

28. Katare R, Bajpai R, Datt SC. Microhardness testing to detect radiation-induced crosslinking in polystyrene: poly(methyl methacrylate) polyblends. Polym Test 1994; 13(2): 107–112.

29. Fakirov S, Baltá Calleja FJ, Krumova M. On the relationship between microhardness and glass transition temperature of some amorphous polymers. J Polym Sci: B: Polym Phys 1999; 37: 1413–1419.

30. Baltá Calleja FJ, Privalko EG, Fainleib AM, Shantalii TA, Privalko VP. Structure-microhardness relationship in semi-interpenetrating polymer networks. J Macromol Sci Phys 2000; B39(2): 131–141.

31. Scrivani T, Benavente R, Pérez E, Pereña JM. Stress-strain behavior, microhardness, and dynamic mechanical properties of a series of ethylene-norbornene copolymers. Macromol Chem Phys 2001; 202: 2547–2553.

32. Cerrada ML, de la Fuente JL, Madruga EL, Fernández-García M. Preparation of poly(tert-butyl acrylate-g-styrene) as precursors of amphiphilic graft copolymers: 2. Relaxation processes and mechanical behavior. Polymer 2002; 43: 2803–2810.

33. Baltá Calleja FJ, Santa Cruz C, Asano T. Physical transitions and crystallization phenomena in poly(ethylene terephthalate) studied by microhardness. J Polym Sci: B: Polym Phys 1993; 31: 557–565.

34. Cagiao ME, Connor M, Baltá Calleja FJ, Seferis JC. Structure development in a thermoplastic polyimide. Cold crystallization as revealed by microhardness. Polym J 1999; 31(9): 739–746.

35. Kajaks J, Flores A, García Gutiérrez MC, Rueda DR, Baltá Calleja FJ. Crystallization kinetics of poly(ethylene naphthalene-2,6-dicarboxylate) as revealed by microhardness. Polymer 2000; 41: 7769–7772.

36. Kiflie Z, Piccarolo S, Brucato V, Baltá Calleja FJ. Role of thermal history on quiescent cold crystallization of PET. Polymer 2002; 43: 4487–4493.

37. Zachmann HG, Wutz C. Studies of the mechanism of crystallization by means of WAXS and SAXS employing synchrotron radiation. In: Dosière M, ed. Crystallization of Polymers. The Netherlands: Kluwer Academic Publishers, 1993: 403–414.

38. Santa Cruz C, Baltá Calleja FJ, Zachmann HG, Stribeck N, Asano T. Relating microhardness of poly(ethylene terephthalate) to microstructure. J Polym Sci: B: Polym Phys 1991; 29: 819–824.

39. Rueda DR, Viksne A, Malers L, Baltá Calleja FJ. Influence of morphology on the microhardness of poly(ethylene naphthalene-2,6-dicarboxylate). Macromol Chem Phys 1994; 195: 3869–3876.

40. Deslandes Y, Alva Rosa E, Brisse F, Meneghini T. Correlation of microhardness and morphology of poly(ether-ether-ketone) films. J Mater Sci 1991; 26: 2769–2777.

41. Wunderlich B. Reversible crystallization and the rigid-amorphous phase in semicrystalline macromolecules. Prog Polym Sci 2003; 28: 383–450.

42. Martínez-Salazar J, García Tijero JM, Baltá Calleja FJ. Microstructural changes in polyethylene-polypropylene blends as revealed by microhardness. J Mater Sci 1988; 23: 862–866.

43. Flores A, Mathot V, Michler GH, Baltá Calleja FJ. Microhardness study of amorphous ethylene-1-octene copolymers (unpublished results).

44. Fakirov S, Krumova M, Rueda DR. Microhardness model studies on branched polyethylene. Polymer 2000; 41: 3047–3056.

45. Fakirov S. Private communication.

46. Gourrier A, García Gutiérrez MC, Riekel C. Combined microindentation and synchrotron radiation microdiffration applied to polymers. Macromolecules 2002; 35: 8072–8077.

47. Baltá Calleja FJ. Dependence of micro-indentation hardness on the superstructure of polyethylene. Colloid Polym Sci 1976; 254: 258–266.

48. Baltá Calleja FJ, Kilian HG. A novel concept in describing elastic and plastic properties of semicrystalline polymers: polyethylene. Colloid Polym Sci 1985; 263: 697–707.

49. Baltá Calleja FJ, Santa Cruz C, Chen D, Zachmann HG. Influence of composition and molecular structure on the microhardness of liquid crystalline copolymers. Polymer 1991; 32(12): 2252–2257.

50. Baltá Calleja FJ, Santa Cruz C. Novel aspects of the microstructure of poly(ethylene oxide) as revealed by microhardness: influence of chain ends. Acta Polym 1996; 47: 303–309.

51. Flores A, Baltá Calleja FJ, Bassett DC. Microhardness studies of chain-extended PE: I. Correlations to microstructure. J Polym Sci: B: Polym Phys 1999; 37: 3151–3158.

52. Giri L, Roslaniec Z, Ezquerra TA, Baltá Calleja FJ. Microstructure and mechanical properties of PBT-PC$_c$ block copolymers: Influence of composition, structure, and physical ageing. J Macromol Sci Phys 1997; B36(3): 335–343.

53. Flores A, Aurrekoetxea J, Gensler R, Kausch HH, Baltá Calleja FJ. Microhardness-structure correlation of iPP/EPR blends: Influence of molecular weight and EPR particle content. Colloid Polym Sci 1998; 276: 786–793.

54. Azzurri F, Flores A, Alfonso GC, Sics I, Hsiao BS, Baltá Calleja FJ. Polymorphism of isotactic polybutene-1 as revealed by micro-indentation hardness. Part II: Correlations to microstructure. Polymer 2003; 44: 1641–1645.

55. Baltá Calleja FJ, Martínez-Salazar J, Asano T. Phase changes in isotactic polypropylene measured by microhardness. J Mater Sci Lett 1988; 7: 165–166.

56. Seidler S, Koch T. Determination of local mechanical properties of α and β-PP by means of microhardness measurements. J Macromol Sci Phys 2002; B41(4–6): 851–861.

57. Fakirov S, Boneva D, Baltá Calleja FJ, Krumova M, Apostolov AA. Microhardness under strain. Part I. Effect of stress-induced polymorphic transition of poly(butylene terephthalate) on microhardness. J Mater Sci 1998; 17: 453–457.

58. Boneva D, Baltá Calleja FJ, Fakirov S, Apostolov AA, Krumova M. Microhardness under strain. Part III. Microhardness behavior during stress-induced polymorphic transition in blends of poly(butylene terephthalate) and its block copolymers. J Appl Polym Sci 1998; 69: 2271–2276.

59. Krumova M, Karger-Kocsis J, Baltá Calleja FJ, Fakirov S. Strain-induced β-α polymorphic transition in iPP as revealed by microhardness. J Mater Sci 1999; 34(10): 2371–2375.

60. Azzurri F, Flores A, Alfonso GC, Baltá Calleja FJ. Polymorphism of isotactic polybutene-1 as revealed by micro-indentation hardness. Part I: Kinetics of the transformation. Macromolecules 2002; 35(24): 9069–9073.

61. Amitay-Sadovski E, Wagner HD. Hardness and Young's modulus of transcrystalline polypropylene by Vickers and Knoop micro-indentation. J Polym Sci: B: Polym Phys 1999; 37(6): 523–530.

62. Krumova M, Fakirov S, Baltá Calleja FJ, Evstatiev M. Structure development in PET/PA6 microfibrillar-reinforced composites as revealed by microhardness. J Mater Sci 1998; 33: 2857–2868.

63. Martínez-Burgos JM, Benavente R, Pérez E, Cerrada ML. Effect of short glass fiber on structure and viscoelastic behavior of olefinic polymers synthesized with metallocene catalyst. J Polym Sci: B: Polym Phys 2003; 41(11): 1244–1255.

64. Mamunya YP, Privalko EG, Lebedev EV, Privalko VP, Baltá Calleja FJ. Structure-dependent conductivity and microhardness of metal-filled PVC composites. Macromol Symp 2001; 169: 297–306.

65. Krumova M, Klingshirn C, Haupert F, Friedrich K. Microhardness studies on functionally graded polymer composites. Comp Sci Tech 2001; 61(4): 557–563.

66. Makishima A, Mackenzie JD. Hardness equations for ormosils. J Sol-Gel Sci Tech 2000; 19: 627–630.

67. Perrin FX, Nguyen V, Vernet JL. Mechanical properties of polycrylic-titania hybrids. Microhardness studies. Polymer 2002; 43: 6159–6167.

68. Sousa RA, Reis RL, Cunha AM, Bevis MJ. Structure development and interfacial interactions in high density polyethylene/hydroxyapatite (HDPE/HA) composites molded with preferred orientation. J Appl Polym Sci 2002; 86(11): 2873–2886.

69. Fini M, Giavaresi G, Aldini NN, Torricelli P, Botter R, Beruto D, Giardino R. A bone substitute composed of polymethylmethylmethacrylate and α-tricalcium phosphate: results in terms of osteoblast function and bone tissue formation. Biomaterials 2002; 23(23): 4523–4531.

70. Turos A, Jagielski J, Piatkowska A, Bielinski D, Dariusz S, Ludomir M, Nabil K. Ion beam modification of surface properties of polyethylene. Vacuum 2003; 70(2–3): 201–206.

71. Baltá Calleja FJ, Giri L, Asano T, Mieno T, Sakurai A, Ohnuma M, Sawatari C. Structure and mechanical properties of polyethylene-fullerene composites. J Mater Sci 1996; 31: 5153–5157.

72. Santa Cruz C, Baltá Calleja FJ, Zachmann HG, Chen D. Mechanical properties and structure of glassy and semicrystalline random copolymers of poly(ethylene terephthalate) and poly(ethylene naphthalene-2,6-dicarboxilate). J Mater Sci 1992; 27: 2161–2164.

73. Flores A, Ania F, Baltá Calleja FJ. Novel aspects of microstructure of liquid crystalline copolyesters as studied by microhardness: influence of composition and temperature. Polymer 1997; 38(21): 5447–5453.

74. Benavente R, Pereña JM, Pérez E, Bello A, Ribeiro MR, Portela MF. Dynamic mechanical relaxations and microhardness indentations of styrene-ethylene copolymers obtained with heterogeneous catalysts. Eur Polym J 2000; 36(5): 879–887.

75. Flores A, Pietkiewicz D, Stribeck N, Roslaniec Z, Baltá Calleja FJ. Structural features of random polyester-amide copolymers as revealed by x-ray scattering and micro-indentation hardness. Macromolecules 2001; 34(23): 8094–8100.

76. Minkova L, Peneva Y. Microhardness of PET-based liquid crystalline copolyesters: Influence of the microstructure. Polymer 2003; 44: 6483–6488.

77. Connor MT, García Gutiérrez MC, Rueda DR, Baltá Calleja FJ. Cold crystallization studies of PET/PEN blends as revealed by microhardness. J Mater Sci 1997; 32: 5615–5620.

78. Martínez-Salazar J, Baltá Calleja FJ. Mechanical model of polyethylene blends as revealed by microhardness. J Mater Sci Lett 1985; 4: 324–326.

79. Martínez-Salazar J, Canalda Cámara JC, Baltá Calleja FJ. Mechanical study of poly(vinylidene fluoride)-poly(methyl methacrylate) amorphous blends. J Mater Sci 1991; 26: 2579–2582.

80. Baltá Calleja FJ, Santa Cruz C, Sawatari C, Asano T. New aspects of the microstructure of PE/iPP gel blends as revealed by microhardness: influence of composition. Macromolecules 1990; 23: 5352–5355.

81. Yordanov Hr, Minkova L. Microhardness and thermal stability of compatibilized LDPE/PA6 blends. Eur Polym J 2003; 39: 951–958.

82. Baltá Calleja FJ, Cagiao ME, Adhikari R, Michler GH. Relating microhardness to morphology in styrene/butadiene block copolymer/polystyrene blends. Polymer 2004; 45, 247–254.

83. Pethrick RA. Physical aging — an old problem revisited. Trends Polym Sci 1993; 1: 226–227.

84. Rueda DR, García Gutiérrez MC, Baltá Calleja FJ, Piccarolo S. Order in the amorphous state of poly(ethylene terephthalate) as revealed by microhardness: creep behavior and physical aging. Int J Polym Mater 2002; 51: 897–908.

85. Boyanova M, Baltá Calleja FJ, Fakirov S. New aspects of the β - α polymorphic transition in plastically deformed isotactic polypropylene studied by micro-indentation hardness. J Mater Sci (submitted)

86. Boyanova M, Baltá Calleja FJ, Fakirov S, Kuehnert I, Mennig G. Influence of processing conditions on the weld line in doubly injection-molded glassy polymers: micro-indentation hardness study. Adv Polym Tech. In press.

87. Berdjane K, Berdjane Z, Rueda DR, Bénachour D, Baltá Calleja FJ. Microhardness of ternary blends of polyolefins with recycled polymer components. J Appl Polym Sci 2003; 89: 2046–2050.

88. Baer E, Kerns J, Hiltner A. Processing and properties of polymer microlayered systems. In: Cunha A, Fakirov S, eds. Structure Development during Polymer Processing. Dordrecht: Kluwer Academic Publishers, 2000: 327–344.

89. Puente Orench I, Ania F, Baer E, Hiltner A, Bernal T, Baltá Calleja FJ. Basic aspects of micro-indentation in multi-layered PET/PC films. Phil Mag A. 2004; 84(18): 1841–1852.

90. Mina F, Broza G, Schulte K, Ania F, Baltá Calleja FJ. Microhardness studies of polymer nanocomposites based on carbon nanotubes (unpublished work).

91. Roslaniec Z, Broza G, Schulte K. Nanocomposites based on multiblock polyester elastomers and carbon nanotubes. Composites Interfaces 2003; 10(1): 95–102.

92. Hench, LL. Medical materials for the next millennium. Mat Res Soc Bull 1999; 24(5): 13–19.

93. Poutanen K, Forssell P. Modification of starch properties with plasticizers. Trends Polym Sci 1996; 4(4): 128–132.

94. Ania F, Dunkel M, Bayer RK, Baltá Calleja FJ. Microhardness and water sorption in injection-molded starch. J Appl Polym Sci 2002; 85: 1246–1252.

95. Puente Orench I, Putthanarat S, Baltá Calleja FJ, Eby RK, Stone M. Ultra-micro-indentation at the surface of silk membranes. Polymer. 2004; 45: 2041–2044.

96. Baltá Calleja FJ, Flores A, Michler GH. Micro-indentation studies at the near surface of glassy polymers: Influence of molecular weight. J Appl Polym Sci. 2004; 93: 1951–1956.

9

Micromechanics of Particle-Modified Semicrystalline Polymers: Influence of Anisotropy Due to Transcrystallinity and/or Flow

J. A. W. VAN DOMMELEN and H. E. H. MEIJER

Eindhoven University of Technology

CONTENTS

I.   INTRODUCTION

Semicrystalline polymeric materials are widely used in a range of engineering applications. Despite many advantages such as low cost and weight, their application is limited by some unfavorable mechanical properties, such as brittle response in plain or notched applications. Toughening can be enhanced by rubber blending, and a criterion proposed by Wu

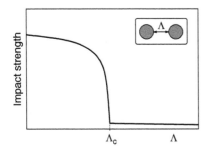

Figure 1 Impact toughness of nylon/rubber blends vs. the average surface-to-surface interparticle ligament thickness. Redrawn from S. Wu. Polymer, 26:1855–1863, 1985.

[1] states that a sharp brittle-to-tough transition occurs for, e.g., nylon/rubber blends when the average interparticle matrix ligament thickness Λ is reduced below the critical value $\Lambda_c = 0.3$ μm, as is schematically depicted in Figure 1. The critical value was shown to be independent of the rubber volume fraction and the particle size. Similar critical interparticle distances were reported in rubber-modified polyethylene [2] and poly(ethylene terephthalate) [3–5]. The explanation offered by Wu for this transition addressed the mutual interaction of particle-disturbed stress fields, enhancing matrix yielding. Ramsteiner and Heckmann [6] concluded that the energy-dissipating deformation mode for rubber-modified nylon is shear yielding. Borggreve et al. [7] confirmed the existence of a critical interparticle distance for the brittle-to-tough transition, however questioned the physical explanation by Wu. A modified theory was proposed, in which the critical ligament thickness corresponds to a local plane strain-to-plane stress transition in the matrix [8, 9]. Based on numerical investigations, Fukui et al. [10] and Dijkstra and Ten Bolscher [11] attributed the toughening effect to extensive shear yielding due to the interaction of stress fields. However, since the stress field theory is only effective for changes in geometrical ratios, it can be concluded that stress field interaction is incapable of explaining an absolute length scale such as a critical interparticle distance. Clearly, a more sound explanation for the critical interparticle distance was needed.

A.   Toughening Mechanism

A physical explanation of the absolute length parameter was offered by Muratoğlu et al. [12,13], who recognized the brittle-to-tough transition as a true material characteristic. It was attributed to thin layers of transcrystallized material, with a reduced plastic resistance, appearing in the microstructural morphology of particle-modified semicrystalline materials. Effectively, the crystallization behavior of the matrix is influenced by the particle/matrix interface, leading to a layer of parallel crystalline lamellae [2,12–20], with the crystalline planes having the lowest plastic resistance parallel to the interface. It was experimentally established that these transcrystalline layers have a well-defined thickness of approximately $\Lambda_c/2$. When the average matrix ligament thickness Λ is below the critical value Λ_c, the preferentially oriented material percolates through the system, bridging between the dispersed particles, as is depicted in Figure 2(a). Situations with $\Lambda \gg \Lambda_c$ are represented in Figure 2(b) and (c) and interesting in this respect is Figure 2(c) that represents the same volume fraction as in Figure 2(a), only with larger particle size. The system represented in Figure 2 consists of (i) rubber particles

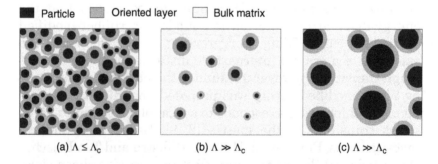

(a) $\Lambda \le \Lambda_c$ (b) $\Lambda \gg \Lambda_c$ (c) $\Lambda \gg \Lambda_c$

Figure 2   Transcrystallized layers around second-phase particles for (a) material with a decreased plastic resistance percolating through the blend, enhancing the toughness (adopted from [13] and [2]), and for materials with a brittle response, with (b) a smaller volume fraction and (c) the same volume fraction as in (a), but larger particles.

with a low modulus, (ii) a preferentially oriented anisotropic matrix material, enveloping the particles and (iii) the bulk matrix material with a randomly oriented structure and consequently isotropic material properties. According to the toughening mechanism postulated by Muratoğlu et al. [13], after cavitation of the rubber particles, the regions with a lowered yield resistance promote large plastic deformation and thereby improve the toughness. Toughness indeed can be defined as a delocalizing strain mechanism [21]. Tzika et al. [22] used a micromechanical numerical model, with a staggered array of particles, to study the influence of preferentially oriented anisotropic layers, modeled with anisotropic Hill plasticity, on the deformation mechanisms under high triaxiality conditions. They observed plastic deformation in the matrix to occur diagonally away from the particles (i.e., in the matrix material between particles, parallel to the interfaces) for $\Lambda \leq \Lambda_c$. The anisotropic matrix material was found to act as a nonstretching shell around the (cavitated) particles, leading to extensive shear yielding.

Bartczak et al. [2,17] generalized the Wu criterion to high density polyethylene (HDPE) and showed the critical interparticle distance ($\Lambda_c = 0.6$ μm for HDPE) to be an intrinsic property of the matrix material, thereby opening the possibility of using mineral fillers for the toughening of semicrystalline polymers, the advantage of which would be an improved modulus of the blend, as schematically indicated in Figure 3. They argued that debonding of hard filler particles could be an alternative for the cavitation of the rubbery phase. However, the Bartczak et al. results showed a distinct effect of processing conditions on the toughness obtained. The importance of process conditions was demonstrated by Schrauwen et al. [23–25], who found toughness to be dominated by flow-induced effects. By using calcium carbonate filler particles in a nylon-6 matrix, Wilbrink et al. [26] did not obtain the tough response of nylon/rubber blends, as was reported by Muratoğlu et al. [13,27], and attributed this to the development of triaxial stresses. A four times increase of the Izod impact energy was obtained by Thio et al. [28] by incorporation of calcium carbonate particles in polypropylene,

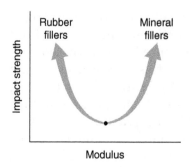

Figure 3 Influence of soft (rubber) vs. hard (mineral) particles on the mechanical properties. (Schematically drawn after Z. Bartczak, A. S. Argon, R. E. Cohen, and M. Weinberg. Polymer, 40:2347–2365, 1999.)

reportedly resulting from combined mechanisms of crack deflection and local plastic deformation of interparticle ligaments. Similar results have been obtained by Zuiderduin et al. [29]. Again, it can be concluded that more refined modeling is needed to explain the experimental findings and to suggest routes to design tough polymers.

B.   Modeling Strategy

Rigorous modeling of toughness enhancement of semicrystalline polymers should in the above context be based on describing the mechanisms schematically shown in Figure 4. It is based on the hypothesis that local anisotropy, induced by a specific microstructure, which results from preferred crystallization of polymeric material, leads to macroscopically tough behavior. The potential validity of this hypothesis is examined by methods of micromechanical modeling. Crystallization behavior itself is left out of consideration (because it is outside the scope of this study), and the starting point is an assumed microstructure of a particle-modified system.

The deformation of polymeric materials, and thus also their either brittle or tough responses, are the result of the interplay of various effects and mechanisms at different levels, such as for example [30] chain scission, microyielding,

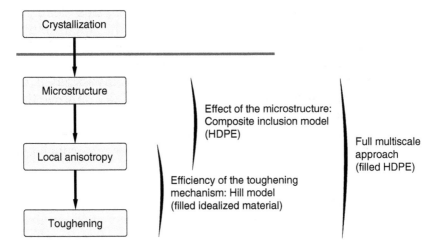

Figure 4   Modeling strategy to explain toughness.

microcavitation, crazing, shear band formation, crack initiation and propagation, and fracture. For semicrystalline materials, also phenomena such as interlamellar slip and
intralamellar deformation mechanisms such as crystallographic slip, twinning, and stress-induced martensitic transformation play a role [31–33]. A quantitative prediction of
toughness would require a coupled and detailed modeling of
the various deformation mechanisms and criteria for the different failure modes, which is at present still not feasible.
Therefore, merely the influence of the microstructure on the
qualitative individual occurrence of some of these phenomena,
or the conditions that may induce them, is investigated.

   First, a micromechanical investigation of the potential of
local anisotropy around dispersed particles for enhancing the
material properties is presented. Calculations are performed
on an idealized semicrystalline material, blended with rubber
particles, which are assumed to be cavitated and are thus, for
simplicity, represented as voids. The system contains a scale
parameter, which is the ratio of the average distance between
particles and a critical distance. This length parameter is
represented in the calculations by the relative thickness of
the transcrystalline layer that results from the surface-nucle-

ated crystallization process. The simulations on this idealized polymeric material are used to investigate the applicability of various types of representative volume elements (RVE) for soft and hard particle-modified semicrystalline materials. Furthermore, with these calculations, it is investigated in detail (i) whether local anisotropy can potentially improve the toughness of rubber-modified systems; (ii) what is the occurring effect of anisotropy that may improve the toughness; and (iii) what type of anisotropy (i.e., mesoscopic morphology) would be required.

To anticipate the conclusions of our calculations, it is pointed out that the simulations, with both axisymmetric and multiparticle RVEs, show that local anisotropy of the matrix material around the particles can effectively replace localization by dispersed shear yielding and change the occurring hydrostatic stresses, potentially leading to toughened material behavior. However, to achieve these improvements, a morphology must be pursued with a radially oriented structure around the dispersed particles and it should, moreover, provide a sufficiently large amount of anisotropy. Furthermore, the possibility of using mineral fillers, rather than low-modulus rubber, for the toughening of semicrystalline polymers is evaluated. The presence of hard, easily debonding, particles is found to negatively affect the anisotropy-induced toughening mechanism, although some improvements remain, in accordance with the existing experimental evidence.

Now in more detail: to investigate whether the above-mentioned requirements can be achieved by a transcrystallized microstructure, a micromechanically based numerical model for the elasto-viscoplastic deformation and texture evolution of semicrystalline polymers has been developed. For actual particle-modified polymeric systems, a distinction between three different scales is made, as is schematically depicted in Figure 5. The constitutive properties of the material are identified at the microscopic scale. At this scale, the individual crystallographic lamellae and amorphous layers determine the local material response. At the mesoscopic scale, an aggregate of individual phases is formed, which can be a spherulite or a sheaflike aggregate of preferentially ori-

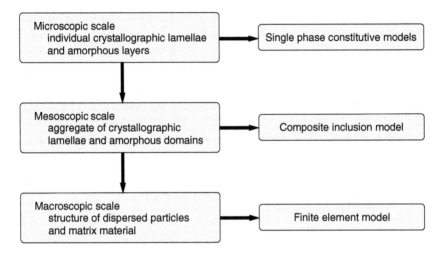

Figure 5 Different scales which can be identified in particle-toughened semicrystalline polymeric systems.

ented material. The local inclusion-averaged deformation and stress fields are related to the mesoscopic fields by a polycrystalline aggregate model. The effect of a transcrystallized structure of matrix material versus randomly oriented material on both mesoscopic and microscopic results is examined. A limited effect of the preferential orientations is observed. Further improved properties are obtained for a partly flow-induced microstructure, if loaded in the appropriate direction.

II. SOFT FILLERS: CAN LOCAL ANISOTROPY INDUCE TOUGHNESS?[1]

The potential of plastic anisotropy for enhancing the toughness of semicrystalline polymeric materials is investigated. Calcu-

[1]Partly reprinted from J. A. W. van Dommelen, W. A. M. Brekelmans, and F. P. T. Baaijens. Mechanics of Materials, 35:845–863, 2003. With permission from Elsevier.

lations are performed on idealized semicrystalline materials, blended with rubber particles, which are assumed to be cavitated and are represented by spherical voids. The system contains a length scale parameter, which is the ratio of the average distance between voids and a critical distance, and is represented by the absolute thickness of the transcrystalline anisotropic layer around the voids. We will start the analysis using a simple Hill anisotropic plasticity constitutive model, that is, subsequently, applied to a micromechanical RVE.

A.   Constitutive Model of the Anisotropic Bulk Polymer

Our idealized polymeric material is modeled by isotropic elasticity (characterized by a Young's modulus E^m and a Poisson's ratio v^m) combined with classical [35] anisotropic plasticity. During yield, the anisotropic Hill yield criterion is applied:

$$F(\sigma_{22} - \sigma_{33})^2 + G(\sigma_{33} - \sigma_{11})^2 + H(\sigma_{11} - \sigma_{22})^2 + 2L\sigma_{23}^2 +$$
$$2M\sigma_{13}^2 + 2N\sigma_{12}^2 = \sigma_y^2 \tag{1}$$

where σ_{ij} are the stress components with respect to a local material vector basis, and the anisotropic constants F, G, H, L, M and N are given by:

$$F = \frac{1}{2}\left(\frac{1}{R_{22}^2} + \frac{1}{R_{33}^2} - \frac{1}{R_{11}^2}\right); \quad G = \frac{1}{2}\left(\frac{1}{R_{11}^2} + \frac{1}{R_{33}^2} - \frac{1}{R_{22}^2}\right) \tag{2}$$

$$H = \frac{1}{2}\left(\frac{1}{R_{11}^2} + \frac{1}{R_{22}^2} - \frac{1}{R_{33}^2}\right); \quad L = \frac{3}{2R_{23}^2}; \quad M = \frac{3}{2R_{13}^2}; \tag{3}$$

$$N = \frac{3}{2R_{12}^2}$$

The constants R_{11}, R_{22} and R_{33} are the ratios of the actual tensile yield strength values of the anisotropic material, relative to the actual virtual bulk tensile yield strength, σ_y. The constants R_{12}, R_{13}, and R_{23} are the ratios of the yield strength values in shear to the shear yield strength τ_y of the virtual bulk material, with $\tau_y = \sigma_y / \sqrt{3}$. The Hill yield criterion was previously used for anisotropic polymeric material by Kobayashi and Nagasawa [36] and Tzika et al. [22]. Here, a linear dependency of the yield strength σ_y on the effective plastic deformation measure $\tilde{\varepsilon}_p$ and a power law dependency of σ_y on the corresponding rate $\dot{\tilde{\varepsilon}}_p$ are assumed for the polymeric material:

$$\sigma_y = \sigma_{y0} \left\{ h\tilde{\varepsilon}_p + q^{1/n} \left[1 + \left(\frac{\dot{\tilde{\varepsilon}}_p}{q\dot{\gamma}_0} \right)^2 \right]^{1/2n} \right\} \tag{4}$$

where σ_{y0} is the reference yield strength, h is the linear hardening parameter and n is the stress exponent of the strain rate. A rate-independent contribution is introduced for strain rate values which are considerably smaller than the reference strain rate $\dot{\gamma}_0$, controlled by the parameter q. The plastic strain measure $\tilde{\varepsilon}_p$ and the corresponding rate are, for anisotropic plasticity, assumed to be given by:

$$\tilde{\varepsilon}_p = \int_0^t \dot{\tilde{\varepsilon}}_p dt; \quad \dot{\tilde{\varepsilon}}_p = \frac{\sigma : \dot{\varepsilon}_p}{\sigma_y} \tag{5}$$

where σ is the Cauchy stress tensor and $\dot{\varepsilon}_p$ is the plastic rate of deformation tensor. The material parameters that are used here are summarized in Table 1. As mentioned earlier, Equations (1)–(3), with the anisotropic strength ratios R_{ij}, are applied in a local coordinate system. The transcrystallized material around the voids is assumed to have a reduced plastic resistance with respect to the local 12 and 13 shear com-

TABLE 1   Material Parameters for a Fictitious Polymer Matrix

E^m [GPa]	v^m	σ_{y0} [MPa]	h	$\dot{\gamma}_0$ [s^{-1}]	n	q	R_{11}	R_{22}	R_{33}	R_{12}	R_{13}	R_{23}
1	0.45	25	0.6	10^{-3}	9	10^{-2}	1	1	1	$1/\zeta$	$1/\zeta$	1

ponents (at the void/matrix interface, the 1-direction is defined to be perpendicular to the interface), and the reduction is controlled by only one adjustable parameter ζ.

B.   Micromechanical Models

The heterogeneous particle-filled system is described by a finite element model of a representative volume element that generally and intrinsically has a three-dimensional structure. Full three-dimensional numerical analyses, however, are computationally extremely demanding and, therefore, first two different simplified models are used. For an adequate representation of the triaxial stress state around a void, an axisymmetric RVE is used, as suggested by Socrate and Boyce [37] and Tzika et al. [22]. However, because of the regular void stacking associated with such RVEs, important effects such as sequential yielding of the matrix material between the different voids [38] cannot be accounted for. In order to capture the essentially irregular nature of a system of dispersed voids, also a plane strain RVE is used, which is incapable of capturing the triaxial stress effects. A comparison of the simplified two-dimensional models with three-dimensional calculations is presented in Van Dommelen et al. [34].

1.   The First Model: An Axisymmetric RVE, the SA Model

An axisymmetric RVE model of a staggered array of voids (referred to as the *SA model*) is considered which resembles a body centered tetragonal stacking of voids. A schematic representation of the unit cell, with $L_0 = R_0$, is shown in Figure 6. The axis of rotational symmetry, as well as the loading direction are horizontal. The RVE is subjected to anti-symmetry conditions (with respect to point M) along the outer radius, which were introduced by Tvergaard [39,40]. Axial compatibility along the radial boundary Γ_{34} is written as

$$u_z(z_0\big|_M - \eta) + u_z(z_0\big|_M + \eta) = 2u_z\big|_M \qquad (6)$$

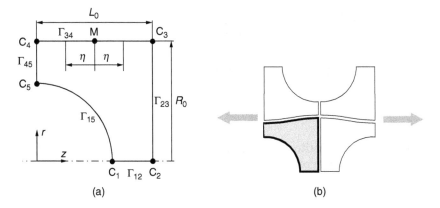

Figure 6 (a) Schematic visualization of an axisymmetric RVE model of a staggered array of voids [22,37] and (b) its position with respect to three neighboring RVEs in a deformed state.

The combined cross-sectional area of neighboring cells is assumed to remain constant along the axial coordinate:

$$[R_0 + u_r(z_0|_M - \eta)]^2 + [R_0 + u_r(z_0|_M + \eta)]^2 = 2[R_0 + u_r|_M]^2 \quad (7)$$

Symmetry conditions along the right and left boundaries are written as

$$u_z|_{\Gamma_{23}} = u_z|_{C_2} \quad (8)$$

and

$$u_z|_{\Gamma_{45}} = u_z|_{C_5} \quad (9)$$

respectively. Since the axis of rotational symmetry coincides with boundary Γ_{12}, the following condition is imposed on this boundary:

$$u_r|_{\Gamma_{12}} = 0 \quad (10)$$

The axisymmetric RVE is subjected to tension at a macroscopically constant strain rate:

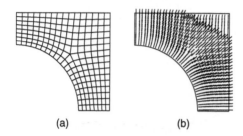

(a)   (b)

Figure 7   (a) Finite element mesh, and (b) local material orientations for the axisymmetric SA model.

$$u_z\Big|_{C_2} - u_z\Big|_{C_5} = L_0[\exp(\dot{\varepsilon}t) - 1] \qquad (11)$$

where the deformation rate $\dot{\varepsilon}$ is set equal to the material reference shear rate $\dot{\gamma}_0$ in Equation (4). The finite element mesh of the axisymmetric SA model, with void fraction $f = 0.2$, is visualized in Figure 7(a). In each integration point of the 196 four-noded bilinear elements, a local coordinate system is generated. The local 1-directions are taken perpendicular to the closest particle/matrix interface, taking into account the periodicity of the structure, and are shown in figure 7(b).

2.   The Second Model: A Multiparticle Plane Strain RVE, the ID Model

To account for the irregular nature of particle-dispersed systems, a plane strain RVE with irregularly dispersed voids (referred to as the *ID model*) is used. In Figure 8, a schematic illustration of this RVE is shown, as well as its arrangement with respect to the neighboring RVEs. The periodicity assumption requires full compatibility of each opposite boundary pair. The corresponding kinematic and natural boundary tyings [41] for related points on opposite boundaries are given by:

$$\boldsymbol{u}\Big|_{\Gamma_{34}} - \boldsymbol{u}\Big|_{C_4} = \boldsymbol{u}\Big|_{\Gamma_{12}} - \boldsymbol{u}\Big|_{C_1} \qquad (12)$$

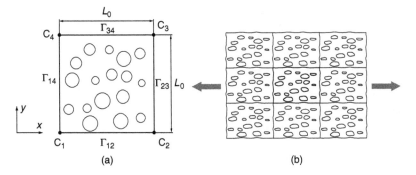

Figure 8 (a) Schematic visualization of a multiparticle plane strain RVE [41] and (b) its position with respect to neighboring RVEs in a deformed state.

$$\boldsymbol{u}\big|_{\Gamma_{14}} - \boldsymbol{u}\big|_{C_1} = \boldsymbol{u}\big|_{\Gamma_{23}} - \boldsymbol{u}\big|_{C_2} \tag{13}$$

$$\boldsymbol{\sigma}\cdot\boldsymbol{n}\big|_{\Gamma_{12}} = -\boldsymbol{\sigma}\cdot\boldsymbol{n}\big|_{\Gamma_{34}} \tag{14}$$

$$\boldsymbol{\sigma}\cdot\boldsymbol{n}\big|_{\Gamma_{14}} = -\boldsymbol{\sigma}\cdot\boldsymbol{n}\big|_{\Gamma_{23}} \tag{15}$$

where $\boldsymbol{n}$ denotes the outward unit normal of the boundary. A tensile loading condition in x-direction is prescribed:

$$u_x\big|_{C_2} - u_x\big|_{C_1} = L_0[\exp(\dot{\varepsilon}t) - 1] \tag{16}$$

where $\dot{\varepsilon}$ is set equal to the reference strain rate $\dot{\gamma}_0$ of the material. Furthermore, rotations are prevented by the following condition for the vertices C_1 and C_2:

$$u_y\big|_{C_1} = u_y\big|_{C_2} \tag{17}$$

The relative displacements of C_4 are unspecified and follow from the analysis, whereas the displacements of C_3 are tied to the other vertices.

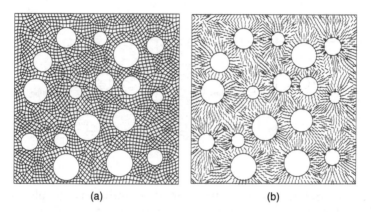

| (a) | (b) |

Figure 9 (a) Finite element mesh and (b) local material orientations for the multiparticle plane strain RVE.

A structure with 20 vol% irregularly dispersed voids is generated using a procedure from Hall [42] and Smit et al. [38]. In order to obtain initially straight boundaries, no void is allowed to cross a boundary. The mesh with 2622 four-noded bilinear plane strain elements is shown in Figure 9(a). A local orientation field is generated by taking the local 1-direction perpendicular to the closest void/matrix interface, taking into account the periodicity of the structure, and is shown in Figure 9(b).

C.   Results

The constitutive behavior and micromechanical models just discussed were implemented in the finite element package ABAQUS [43] to study the potential of local anisotropy enveloping dispersed voids for the toughening of semicrystalline polymers. The potential of toughening by local anisotropy is investigated by using different amounts of anisotropy in both the SA and the ID models.

Isotropic material, with $\zeta = 1$, is used as a reference case for large-scale, randomly crystallized material. For the small-scale transcrystallized material, several levels of anisotropy are considered, ranging from $\zeta = 1.5$ to $\zeta = 5$. In Figure 10,

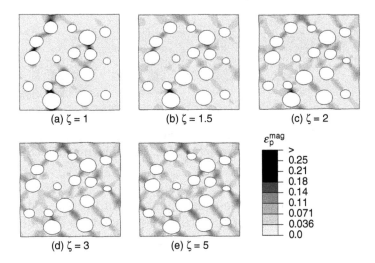

(a) $\zeta = 1$ (b) $\zeta = 1.5$ (c) $\zeta = 2$

(d) $\zeta = 3$ (e) $\zeta = 5$

ε_p^{mag}

> 0.25
0.21
0.18
0.14
0.11
0.071
0.036
0.0

Figure 10 The influence of radially oriented anisotropy on the magnitude of plastic deformation, ε_p^{mag}, for the irregular plane strain model, at $\dot{\varepsilon}t = 0.05$.

the magnitude of the plastic deformation is given for the irregular plane strain model. For fully isotropic material, the deformation is strongly localized in a specific path through the microstructure, determined by the irregular void arrangement. For increasing anisotropy, increasingly dispersed shear yielding is observed, which is highly favorable to enhance the toughness (equivalent to delocalization of strain). Maximum ε_p^{mag}-values are reduced, and are located both in the matrix material, away from the void surface and at the void/matrix interface for the transcrystallized material. In the largely anisotropic material, double shear bands can be observed at each side of a void, positioned at the inclined 30°–50° off-polar regions, whereas in the isotropic material mostly single shear bands at the void equators are found. The shear yielding mechanism becomes truly effective for anisotropy ratios $\zeta = R_{11}/R_{12}$ above the value of 3.

Although not shown here, it is mentioned that for all levels of anisotropy, in the ID model, the highest tensile tri-axial stresses are found at the void equator regions, which is

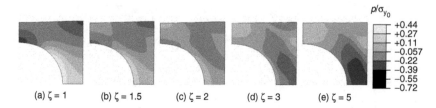

Figure 11 The influence of radially oriented anisotropy on the normalized hydrostatic pressure, p/σ_{y_0}, for the axisymmetric SA model, at $\dot{\varepsilon}t = 0.05$.

not in agreement with three-dimensional simulations [34]. More realistic predictions of the triaxial stress state around dispersed voids are obtained with the axisymmetric SA model, as represented in Figure 11. In the isotropic material, the maximum tensile triaxial stresses are found to occur at the void equators. For small anisotropy levels, these negative hydrostatic pressures are reduced. At higher levels of anisotropy, however, the tensile pressures again increase with increasing ζ, but the maximum values are now found in the matrix material near the void pole. The SA model is in much better agreement with the three-dimensional situation [34] than the ID model. Despite this, the peak values of the tensile hydrostatic stress are significantly lower for the axisymmetric SA model than for the full three-dimensional model.

The distinct effect of anisotropy on the triaxial stress field is also reflected in the maximum in-plane principal stress, σ_{max}, as is shown in Figure 12. For the isotropic matrix material, the locations of large equivalent stress and negative hydrostatic stress coincide, leading to a relatively high maximum in-plane principal stress, concentrated at the void equators, and the void-bridging ligaments. As the anisotropy parameter ζ is increased, the values of σ_{max} are considerably reduced and the location of the largest σ_{max} changes to the void poles for $\zeta = 5$.

D.   Failure Mechanisms

Based on the investigations with both the ID and the SA model, two distinct effects of local, radially oriented, anisot-

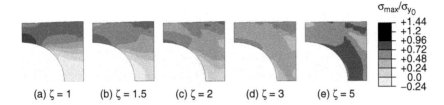

$\sigma_{max}/\sigma_{y_0}$

+1.44
+1.2
+0.96
+0.72
+0.48
+0.24
0.0
−0.24

(a) $\zeta = 1$ (b) $\zeta = 1.5$ (c) $\zeta = 2$ (d) $\zeta = 3$ (e) $\zeta = 5$

Figure 12 The influence of radially oriented anisotropy on the normalized maximum in-plane principal stress, $\sigma_{max} / \sigma_{y_0}$, for the axisymmetric SA model, at $\dot{\varepsilon}t = 0.05$.

ropy can be observed. The influence on the triaxial stress field is a change of the position of maximum tensile values. Under high tensile triaxial stress, craze-like features, such as interlamellar separation and voiding of amorphous regions [44–47], may be initiated in the matrix material, and upon extension and coalescence of cavities, crazes are formed. Although crazing leads to brittle behavior, the crazing process itself may under certain conditions also lead to some plasticity via a multiple crazing process [48,49].

In the simulations, the highest negative hydrostatic pressures are found at the void equators for the large-scale, isotropic material. For the small-scale anisotropic material, however, the large tensile pressures are shifted towards the void poles. Therefore, the initiation of crazing may, for the voided system with radially oriented anisotropy, be expected to occur at the particle poles, rather than in the equator region. The growth of initiated crazes is likely to occur along planes which are perpendicular to the direction of maximum principal stress [45,46,50]. In Figure 13, the direction of the maximum in-plane principal stress is given for the SA model, for both isotropic ($\zeta = 1$) and largely anisotropic ($\zeta = 5$) material. For both situations, in the region of large tensile triaxial stress, the maximum in-plane principal stresses are relatively large, and directed parallel with the void interface. Therefore, craze growth is expected to occur perpendicular to the interface. Moreover, the large principal stresses at the particle equators of the isotropic material will lead to brittle fracture

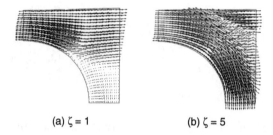

(a) ζ = 1   (b) ζ = 5

Figure 13 The influence of radially oriented anisotropy on the direction and magnitude of the maximum in-plane principal stress, σ_{max}, for the SA model, at $\dot{\varepsilon}t$ = 0.05.

of the matrix material given the direction of the craze growth, perpendicular to the loading direction. The decrease of maximum in-plane principal stresses in the anisotropic material and the shift of the directions of an eventual craze growth, makes this system less susceptible to brittle fracture. It is noted that in the current simulations, no failure and crazing initiation/growth criteria have been used (as for example in [51–53]). But still the conclusions based upon the analysis this far can be schematically summarized, see Figure 14. The actual mechanism that will occur for a specific matrix material will depend on the values of the brittle fracture strength, the resistance against craze initiation and growth and the yield strength of the material [54]. In isotropic materials,

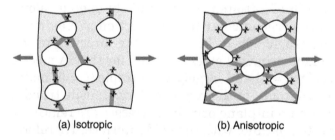

(a) Isotropic   (b) Anisotropic

Figure 14 Schematic illustration of the effect of radially oriented anisotropy on the occurrence of matrix shearing and triaxial stresses.

possible crazes are initiated at the particle equators, and grow transversely to the macroscopic tensile direction. The plastic deformation is localized in a few bands, located in crazing regions. For this large-scale system, the crazes may act as precursors to cracks, and ultimately failure. For the small-scale situation, with radially oriented anisotropic material around the cavitated particles, maximum tensile triaxial stresses are predominantly found in zones of limited plastic deformation. Possible crazes are initiated at the particle poles, and grow in the direction of macroscopic loading. In this situation, crazing may become a mechanism of energy-absorbing inelastic deformation. Additionally, extensive matrix yielding, which is another beneficial energy-absorbing mechanism, occurs in noncrazing regions. Therefore, by changing the nature of matrix crazing, reducing principal stresses and inducing extensive matrix shearing, a local, radially oriented anisotropy, with sufficiently reduced shear strengths, may be a highly efficient method for the toughening of semicrystalline materials.

E.   Conclusions

The effect of matrix materials with a reduced yield strength in the local shear directions around well-dispersed voids has been investigated by numerical simulations of idealized systems. The local principal anisotropy directions were assumed to be radially oriented around second-phase soft particles. The fictitious polymeric material was modeled in the context of anisotropic Hill plasticity, where the yield strength has been taken to depend on an effective plastic deformation measure, and its time derivative. Two extreme size scales were investigated; the largest scale with completely isotropic material properties and the smallest scale having a fully percolated anisotropic structure.

The three-dimensional structure of the voided material was simplified to two different micromechanical models. The irregular distribution of voids is captured by a multiparticle plane strain RVE. The irregular nature of this RVE is essential in capturing the effects of local anisotropy on the mechanics of plastic deformation. For the large scale, plastic deformation localizes in a specific path through the matrix material, inducing macroscopically brittle behavior, whereas for the small-scale configuration, a heterogeneous field of void-bridging shear bands was found throughout the entire domain, which could lead to a macroscopically toughened behavior. The extent of shear yielding increases with increasing anisotropy. The localization of deformation vanishes due to a large reduction of local shear yield strengths. However, the ID model is incapable of capturing the distinct effects of local anisotropy on the triaxial stress state. A better representation thereof is obtained with the axisymmetric SA model, where a regular distribution of voids is assumed. The most striking effect of local anisotropy on the hydrostatic stress field is a shift of the highest tensile triaxial stresses from the void equator region (where the surface normals are perpendicular to the loading direction) to the void polar region (the surface area where the normals are aligned with the loading direction). Consequently, the maximum principal stresses at the void equator are considerably reduced.

The calculations confirm that the mechanism as proposed by Muratoğlu et al. [13] could indeed lead to toughened material behavior. The presence of an absolute length scale is related to the thickness of a layer of anisotropic matrix material enveloping the dispersed voids. Required for toughening by this mechanism is then (i) a structure of well-dispersed voided particles with an average surface-to-surface interparticle distance which is smaller than the critical length parameter of the matrix material; (ii) locally anisotropic material with the principal 1-direction radially oriented with respect to the nearest void surface; and (iii) a sufficiently reduced shear yield strength in the local 12- and 13-directions (with $\zeta = R_{11}/R_{12}$ at least of the order of 3).

It is noted that although the large potential of local anisotropy for toughening of semicrystalline polymers was shown, the origin of this anisotropy has not been addressed here. The material was merely assumed to be oriented with the principal 1-direction of anisotropy toward the nearest void and having a finite anisotropic layer thickness. The origin of these layers is attributed to a preferred crystallization at the particle/matrix interface by Muratoğlu et al. [13] and Bartczak et al. [2,17]. The consequences of such a morphology for local anisotropy will be addressed in Section IV.

III. HARD PARTICLES: CAN THEY ALSO ACT AS TOUGHNESS MODIFIERS?[1]

A physically based mechanism for the toughening of semicrystalline polymeric materials due to the dispersion of particles originates from the presence of a layer of anisotropic transcrystallized material enveloping the particles, and was proposed for nylon by Muratoğlu et al. [13]. Bartczak et al. [2,17] generalized this mechanism to other semicrystalline materials (like high-density polyethylene) and showed the critical interparticle distance to be an intrinsic property of the matrix material, thereby opening the possibility of using mineral fillers instead of rubber particles for the toughening of semicrystalline polymers, the advantage of which would be an improved modulus of the blend. They argued that debonding of hard filler particles could be an alternative for the cavitation of the rubbery phase.

In Section II, an idealized, polymeric matrix material was modeled by anisotropic Hill plasticity, and various representative volume elements were used to describe the system containing dispersed voids. It was shown that a local plastic anisotropy of matrix material around the voids can effectively replace localization by dispersed shear yielding and change the occurring hydrostatic stresses, potentially leading to toughened material behavior. However, to achieve these improvements, a morphology should be pursued that has a radially oriented structure around the dispersed voids and provides a sufficiently large amount of anisotropy.

---

[1]Partly reprinted from J. A. W. van Dommelen, W. A. M. Brekelmans, and F. P. T. Baaijens. Computational Materials Science, 27:480–492, 2003. With permission from Elsevier.

In this section, the consequence of using hard mineral particles for the toughening of semicrystalline polymers is investigated. For this purpose, again the anisotropic Hill model is used. The system contains a scale parameter, which is the ratio of the average distance between particles and a critical distance. The value of this parameter is represented in the calculations by the relative thickness of an anisotropic layer around the particles. Large- and small-scale configurations are modeled again by entirely isotropic or anisotropic matrix material, respectively.

A.   Model Description

The influence of mineral (i.e., hard) filler particles in semi-crystalline polymeric material is investigated, and particularly the effect of these fillers on the mechanism of toughening by locally induced anisotropy. As a reference situation, voided matrix material will be used. A distinction is made between fully bonded particles, for which a tied particle/matrix interface is used, and debonding particles. For the latter, a contact algorithm [43] with a relatively low maximum tensile strength $\sigma^i / \sigma_{y_0} = 0.4$ is used to describe the particle/matrix interaction. The effect of relatively stiff rubber particles is presented in Van Dommelen et al. [55].

For the yield behavior, again the anisotropic Hill yield criterion [35] is used, with a rate-dependent and hardening yield stress, as discussed in Section II. The mineral filler particles are modeled as linearly elastic, with Young's modulus $E^p = 80$ GPa and Poisson's ratio $\nu^p = 0.3$. Particle-modified material is again described by a finite element model of a representative volume element (RVE). The particle-modified system, having a three-dimensional nature, is simplified to a two-dimensional RVE, for which two different approaches are used, as discussed in Section II.

B.   Results

In Figure 15, the magnitudes of plastic deformation as obtained by the ID model are shown for systems containing

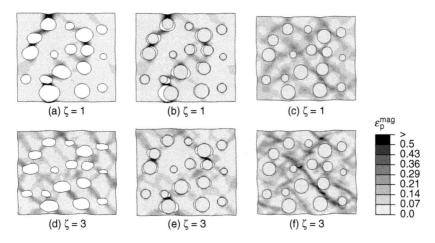

$$\varepsilon_p^{mag}$$

(a) $\zeta = 1$ (b) $\zeta = 1$ (c) $\zeta = 1$

(d) $\zeta = 3$ (e) $\zeta = 3$ (f) $\zeta = 3$

> 0.5
0.43
0.36
0.29
0.21
0.14
0.07
0.0

Figure 15 The magnitude of plastic deformation, ε_p^{mag}, for the ID model with (a), (d) voids, (b), (e) easily debonding hard particles, and (c), (f) fully bonded hard particles, at $\dot{\varepsilon}t = 0.1$.

both debonded and fully bonded hard filler particles, for both isotropic ($\zeta = 1$, i.e., large scale) and anisotropic ($\zeta = 3$, i.e., small scale) matrix material, respectively. For the voided isotropic matrix material (Figure 15(a)), the macroscopic contraction in the free direction is small, corresponding to the growth of voids, due to stretching of relatively thin ligaments. Therefore, for this matrix material, the inclusion of easily debonding hard particle fillers has no significant effect on the deformation observed, as can be seen by comparison with Figure 15(b), where the interface strength σ^i is negligibly small. For radially oriented anisotropic voided material, a dispersed mode of massive shear yielding is observed, with double shear bands at each side of a particle. As a result of matrix shearing, for the voided anisotropic system, however, the voids become smaller in the macroscopically free direction, see Figure 15(d). Consequently, the presence of hard mineral fillers interferes with the mechanism of matrix shearing, as can be observed in Figure 15(e). Therefore, although there is some effect of anisotropy, the mechanism of toughening by locally induced anisotropy is expected to be considerably less

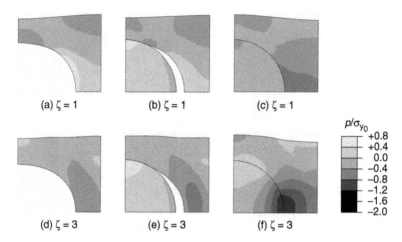

(a) $\zeta = 1$ (b) $\zeta = 1$ (c) $\zeta = 1$

(d) $\zeta = 3$ (e) $\zeta = 3$ (f) $\zeta = 3$

p/σ_{y0}

+0.8
+0.4
0.0
-0.4
-0.8
-1.2
-1.6
-2.0

Figure 16 The normalized hydrostatic pressure, p/σ_{y0}, for the SA model, with (a), (d) voids, (b), (e) easily debonding hard particles, and (c), (f) fully bonded hard particles, at $\dot{\varepsilon}t = 0.1$.

efficient for non-adhering hard fillers than for low modulus rubber particles.

For materials filled with well-bonded stiff particles, which is shown in Figures 15(c) and (f), massive shear yielding is found for both isotropic and anisotropic matrix behavior. However, bonding has negative effects concerning craze initiation and growth direction. In Figure 16, the effect of hard filler particles on the normalized hydrostatic pressure, p/σ_{y0}, as predicted by the SA model, is displayed for (large scale) isotropic and (small scale) anisotropic matrix material. For voided systems, the effect of anisotropy on the triaxial stress field is a change of the position of maximum tensile values. The highest negative (tensile) hydrostatic pressures are found at the particle equators for the isotropic material (Figure 16(a)). For the anisotropic material, however, large tensile pressures are found in the polar regions (Figure 16(d)). Therefore, the initiation of crazing may, for the voided system with radially oriented anisotropy, be expected to occur at the particle poles, rather than in the equator region. For easily debonding hard particles, a similar effect of local anisotropy on the tensile triaxial stresses is observed, with an increase

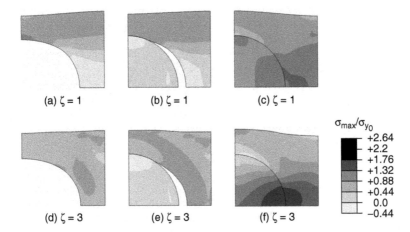

(a) $\zeta = 1$ (b) $\zeta = 1$ (c) $\zeta = 1$

(d) $\zeta = 3$ (e) $\zeta = 3$ (f) $\zeta = 3$

σ_{max}/σ_{y0}

+2.64
+2.2
+1.76
+1.32
+0.88
+0.44
0.0
−0.44

Figure 17 The normalized maximum in-plane principal stress, $\sigma_{max} / \sigma_{y_0}$, for the SA model, with (a), (d) voids, (b), (e) easily debonding hard particles, and (c), (f) fully bonded hard particles, at $\dot{\varepsilon}t = 0.1$.

of the peak value for the anisotropic situation. The well-bonded configurations both show peak tensile triaxial stresses at the poles. The growth of initiated crazes is likely to occur along planes which are perpendicular to the direction of the maximum principal stress. In Figure 17, the normalized maximum in-plane principal stress, $\sigma_{max} / \sigma_{y_0}$ is depicted for the SA model, for both isotropic ($\zeta = 1$) and anisotropic ($\zeta = 3$) material, with either a void, or easily debonding or adhering hard particles. Moreover, in Figure 18, the direction of the maximum in-plane principal stress is given for the systems containing a hard particle. For both non-adhering situations, the maximum in-plane principal stresses in the region of large tensile triaxial stress are parallel with the particle/matrix interface. Therefore, craze growth is expected to occur perpendicular to the interface, i.e., perpendicular to the loading direction for the isotropic material and parallel to the loading direction for the anisotropic system. However, for the well-bonded systems, which did show advantageous shear yielding for both the isotropic and the anisotropic configuration, the maximum principal stresses in the polar region (where the largest tensile triaxial stresses are observed) are directed

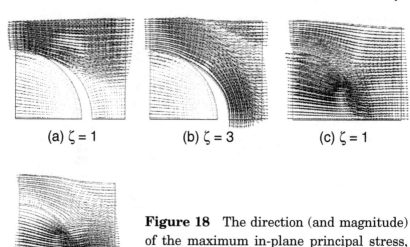

(a) $\zeta = 1$ (b) $\zeta = 3$ (c) $\zeta = 1$

(d) $\zeta = 3$

Figure 18 The direction (and magnitude) of the maximum in-plane principal stress, σ_{max}, for (a, b) easily debonding hard fillers, and (c, d) well-bonded particles, at $\dot{\varepsilon}t = 0.1$.

approximately in the loading direction. Consequently, for these systems craze growth or microcracking may be expected to occur perpendicular to the loading direction, thereby possibly leading to macroscopic failure. At any rate, for the systems containing well-bonded hard particles, an isotropic matrix seems to be favorable over locally anisotropic material.

C. Conclusions

Fictitious, idealized, polymeric matrix materials were modeled by anisotropic Hill plasticity in Section II, where the distinct effect of local plastic anisotropy of matrix material around the voids was shown, viz. an effective replacement of localization by dispersed shear yielding and a change of the occurring hydrostatic stresses, potentially leading to toughened material behavior. In this section, a similar modeling approach was used to investigate the influence of mineral filler particles on this toughening mechanism.

The use of mineral filler particles for toughening of polymeric materials requires debonding in order to prevent excessive tensile hydrostatic stresses. These debonded hard particles show a relocation of tensile triaxial stresses to the particle polar

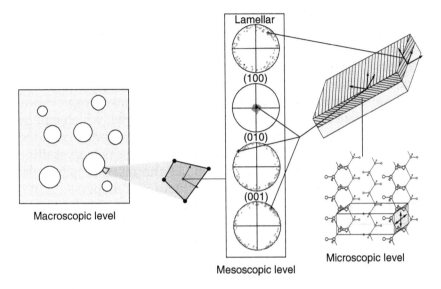

Figure 19   The various levels involved in the multiscale model.

areas by local anisotropy, similar to anisotropic voided systems, with the maximum principal stresses directed such that crazes or microcracks are expected parallel to the loading direction. However, the anisotropy-induced shear yielding mechanism is affected by the presence of stiff inclusions.

IV. FULL MULTISCALE MODELING: WHAT DID WE LEARN?[1]

To investigate the possibility that a particular microstructure satisfies the requirements for sufficient anisotropy, a full multiscale micromechanically based numerical model for the elasto-viscoplastic deformation and texture evolution of semicrystalline polymers was developed [57] and is used to simulate the behavior of particle-modified high density

[1]Partly reprinted from J. A.W. van Dommelen,W. A. M. Brekelmans, and F. P. T. Baaijens. Journal of Materials Science, 38:4393–4405, 2003.

polyethylene (HDPE). For the analysis of these systems, a distinction between three different scales is made, as schematically depicted in Figure 19. The constitutive properties of the material components are characterized at the microscopic scale. At this level, the individual crystallographic lamellae and amorphous layers are identified. At the mesoscopic scale, an aggregate of individual phases is considered, which can be a spherulite or a sheaflike aggregate of preferentially oriented material. To bridge between those scales, the polycrystalline composite inclusion model is used. At the macroscopic scale, for particle-modified materials, a structure of dispersed particles and matrix material can be identified. At this level, the system is represented by a finite element model using various representative volume elements, as suggested by the Hill-type simulations of Section II. A bridge to the mesoscopic level is obtained by using an aggregate of composite inclusions as a representative microstructural unit in each integration point. The effect of transcrystallized orientations of matrix material versus randomly oriented material on both mesoscopic and microscopic results is investigated, as well as a hypothesized microstructure, which may be the result of process conditions.

A.   Microscopic Scale: Material Models

The anisotropic constitutive behavior of intraspherulitic material is modeled by an aggregate of elasto-viscoplastic two-phase composite inclusions [58,59]. Each inclusion consists of a crystalline and an amorphous phase. In this section, the constitutive models of the constituent phases are shortly discussed. For a more elaborate presentation of these models, see Van Dommelen et al. [57].

1.   Crystalline Phase

The crystalline domain of polymeric material consists of regularly ordered molecular chains. The crystal structure results in (i) anisotropic elastic behavior where the elastic properties are given with respect to the crystallographic directions, and (ii) plastic deformation governed primarily by crystallographic slip on a limited number of slip planes [32,60]. Moreover, plastic deformation may result from mechanical twinning or stress-induced martensitic phase transformations [32,33,61,62]. Since crystallographic slip is assumed to be of most importance for polymeric materials, in the modeling process the latter two mechanisms are left out of consideration.

The elastic component of the deformation in the crystalline phase is characterized by an anisotropic fourth-order elastic modulus tensor. The anisotropic elastic properties are coupled to the crystallographic directions. For the viscoplastic behavior of the crystalline phase, a rate-dependent crystal plasticity model is used. In this model, the plastic velocity gradient of the crystalline lamella, consisting of a single crystal, is composed of the contributions of all physically distinct slip systems, being 8 for high density polyethylene (HDPE), with a lowest slip resistance of 8 MPa. The shear rate of each slip system is assumed to be related to the resolved shear stress according to a viscoplastic power law.

2.   Amorphous Phase

The amorphous phase of semicrystalline polymeric material consists of an assembly of disordered macromolecules, which are morphologically constrained by the neighboring crystalline lamellae. Plastic deformation in these domains occurs by a thermally activated rotation of segments. At room temperature, the amorphous phase of HDPE, which is the material of interest in this work, is in the rubbery regime, with the glass transition temperature near $-70°C$. The deformation in this regime is characterized by a limited strain rate-sensitivity and a strong entropic hardening at large deformations.

The initial elastic resistance of the rubbery amorphous phase is well below the elastic resistance of the crystalline domain. Consequently, elastic deformations can be considerably large and are modeled by a generalized neo-Hookean relationship. A relatively strain rate-insensitive viscoplastic power law relation between an effective shear strain rate and the effective shear stress [58] is used in conjunction with a back stress tensor for which the Arruda-Boyce eight-chain network model of rubber elasticity [63] is used.

B.   Mesoscopic Scale: Composite
          Inclusion Model

The mechanical behavior at the mesoscopic level is modeled
by an aggregate of layered two-phase composite inclusions as
was first proposed by Lee et al. [58,59] for rigid/viscoplastic
material behavior. Each separate composite consists of a crystalline lamella which is mechanically coupled to its corresponding amorphous layer, as is shown in the upper right part
of Figure 19. The stress and deformation fields within each
phase are assumed to be piecewise homogeneous; however,
they may differ between the two coupled phases. It is assumed
that the crystalline and amorphous components remain fully
mechanically coupled. Interface compatibility within the composite inclusion and traction equilibrium, across the interface,
are enforced. To relate the volume-averaged mechanical
behavior of each composite inclusion to the imposed boundary
conditions for an aggregate of inclusions, a hybrid local–global
interaction law is used. This class of hybrid-inclusion models
was introduced by Lee et al. [58,59] for rigid/viscoplastic composite inclusions. A more detailed description of the composite
inclusion model is presented elsewhere [57]. Moreover, an
application of this model to study the intraspherulitic deformation of polyethylene is presented in Van Dommelen et al.
[64]. Some aspects of the finite element implementation in
ABAQUS [43] are given in Van Dommelen et al. [65].

The anisotropy of preferentially oriented intraspherulitic
material, as predicted by the composite inclusion model, is
investigated. A mesoscopic aggregate of composite inclusions,
represented by a set of crystallographic orientations and corresponding lamellar orientations, is subjected to constant
strain rate uniaxial tension in the three principal directions
e_i of the material coordinate system. Consider the right polar
decomposition $\bar{F} = \bar{R} \cdot \bar{U}$ of the volume-averaged deformation
gradient, where $\bar{R}$ is the mesoscopic rotation tensor, and $\bar{U}$
is the corresponding right stretch tensor. Then, the following
conditions are imposed:

$$\bar{R} = I; \quad \bar{U}_{ii} = \lambda(t); \quad i = 1 \vee i = 2 \vee i = 3 \qquad (18)$$

with

$$\lambda(t) = \exp(\dot{\varepsilon}t) \tag{19}$$

where $\dot{\varepsilon}$ is set equal to the material reference shear rate $\dot{\gamma}_0$. Furthermore, the components of the Cauchy stress tensor should satisfy:

$$\bar{\sigma}_{jj} = \bar{\sigma}_{12} = \bar{\sigma}_{13} = \bar{\sigma}_{23} = 0; \quad j \in \{1,2,3 \mid j \neq i\} \tag{20}$$

In another test case, pure shear deformation is applied by prescribing one of the basic shear components ij of the (symmetric) right stretch tensor:

$$\bar{R} = I; \quad \bar{U}_{ij} = \gamma(t); \quad ij = 12 \vee ij = 13 \vee ij = 23 \tag{21}$$

with

$$\gamma(t) = \frac{1}{2}\sqrt{3}\,\dot{\gamma}_0 t \tag{22}$$

and

$$\bar{\sigma}_{11} = \bar{\sigma}_{22} = \bar{\sigma}_{33} = \bar{\sigma}_{kl} = 0; \quad kl \in \{12,13,23 \mid kl \neq ij\} \tag{23}$$

1.   Anisotropy of Preferentially Oriented Material

The full multiscale model is used to examine the effect of the microstructural morphology on the mechanics of particle-modified systems [56]. First, the anisotropy, at the mesoscopic level, of (microscopically) preferentially oriented material, as predicted by the composite inclusion model, is investigated.

Randomly oriented material: The local spherulitic structure of melt-crystallized HDPE is represented by an aggregate of 125 composite inclusions with randomly generated initial orientations of the crystallographic phases, having an orthorhombic lattice. Experimental studies of melt-crystallized polyethylene show that lamellar surfaces are of the $\{h0l\}$-type, where the angle between the chain direction c and the lamellar normal n varies between 20° and 40° [66,67]. Gautam et al. [68] found by molecular simulations the $\{201\}$ planes to provide the lowest amorphous/crystalline interface energy. For randomly oriented material, the initial angle between c_0 and n_0^{I} is set at 35°, corresponding to the {201} planes. Since the distribution of the crystallographic orientations is random, the mechanical behavior of this aggregate will be quasi isotropic.

Transcrystallized material: For thin HDPE films, crystallized on rubber and calcium carbonate substrates, Bartczak et al. [18] found a sheaflike morphology of the lamellae, which were oriented preferentially edge-on against the substrate, resulting from a surface-induced crystallization. The (100) planes, containing the two crystallographic slip systems with the lowest slip resistance, (100)[001] and (100)[010], were found to be directed preferentially parallel to the plane of the film with a random orientation of the molecular chains within this plane. The lamellar normals were either parallel to the plane of the substrate or somewhat tilted with respect to the plane. However, twisting of lamellae was found to be substantially reduced. The preferred crystallographic planes for the crystalline/amorphous interface remain unclear for this morphology. It can be assumed that these planes are still of the {h0l}-type. The observation by Bartczak et al. [18] that crystal growth is unidirectional with little divergence sideways, and the reduction of lamellar twisting, suggest a smaller angle between the crystallographic chain direction c and the lamellar normal n^{I} than observed in randomly crystallized material. Here, the plane of the crystalline/amorphous interface is assumed to be of the {102}-type, corresponding to an initial chain tilt angle of 9.7°. Besides the crystallographic and lamellar orientations, all microscopic material properties are assumed to be identical to the properties of the randomly crystallized material. A set of crystallographic orientations is generated with the (100) poles preferentially aligned in the direction of the normal of the substrate, with a certain random deviation from the substrate normal direction. Furthermore, a random rotation around this normal direction is applied. Therefore, the mechanical properties at the mesoscopic scale can be expected to be transversely isotropic with the (fiber) symmetry direction corresponding to the substrate normal direction. All differences in mechanical response in the 22 and

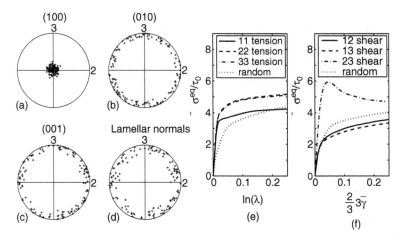

Figure 20 Equal area projection pole figures representing (a)–(c) the principal crystallographic lattice directions, and (d) the lamellar normals of a set of transcrystallized orientations and (e), (f) the normalized equivalent mesoscopic stress $\bar{\sigma}^{eq} / \tau_0$, vs. the imposed deformation for tension and shear, respectively, in the basic material directions as predicted by the composite inclusion model.

33 tensile directions and the 12 and 13 shear directions will be of statistical origin. After a set of crystallographic orientations has been generated, the lamellar normals are obtained as described above. In Figure 20(a)–(d), the orientations of a set of 125 composite inclusions are displayed. The view direction of the equal area pole figures is the substrate normal direction. This direction is the preferred direction for the (100) poles. In Figure 20(e) and (f), the corresponding mesoscopic stress-strain response of the composite inclusion model is shown in the basic tensile and shear directions. The equivalent stresses are normalized by the lowest slip resistance $\tau_0 = 8$ MPa. Additionally, the response of a randomly oriented aggregate is shown also. With respect to the randomly oriented aggregate, the equivalent stresses in the transverse tensile directions are increased. The reduction of the 12 and 13 shear resistances is related to these tensile 22 and 33 resistances. The ratio of transverse tensile and 12/13 shear resistances at the onset of yielding is of the order of 2. At higher strains,

this ratio decreases to approximately 1.5. Simulations with an anisotropic Hill plasticity model (Section II) showed that for the effectiveness of the toughening mechanism under investigation, a larger amount of anisotropy would be necessary ($\zeta \geq 3$). A sharper texture, however, does not increase the R_{22}/R_{12} anisotropy ratio.

C.   Macroscopic Models

For particle-modified materials, a structure of dispersed particles and matrix material can be identified. The system is described by a finite element model of a representative volume element (RVE). The blended system, having a three-dimensional nature, is simplified to a two-dimensional RVE, for which two different approaches are used (see Section II). In order to capture the important effects of the essentially irregular nature of particle-dispersed systems, a plane strain RVE with randomly dispersed particles is used, where the cavitated particles are represented by voids. A similar structure as was used in the Hill-type simulations in Section II is adopted, containing 20 vol% irregularly dispersed voids. The mesh with 565 four-noded bilinear reduced integration plane strain elements is shown in Figure 21(a). An orientation field is generated by taking the local 1-directions perpendicular to the closest void/matrix interface, respecting the periodicity of the structure, as shown in Figure 21(b).

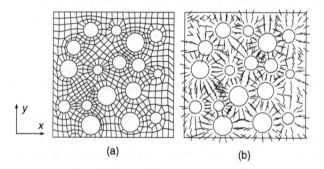

(a)   (b)

Figure 21   (a) Finite element mesh of multiparticle plane strain RVE and (b) local material orientations.

For the representation of the triaxial stress state around a particle, an axisymmetric RVE model of a staggered array of particles is considered. The finite element mesh of the SA model, with 20 vol% voids, was visualized in Figure 7(a). In each integration point of the 196 reduced integration four-noded bilinear elements, a local coordinate system is generated, such that the local 1-directions are again perpendicular to the closest void surface.

D.   Effect of Transcrystallized Anisotropy on Toughness

Now, the full multiscale model will be used to investigate the effect of a transcrystallized orientation on the deformation of particle-modified systems. Both RVE models, as described in the previous section, are applied, with in each integration point either an aggregate of randomly generated orientations or a (unique) set of orientations with a similar distribution as in Figure 20. For the latter situation, the local (fiber) symmetry directions correspond to the 1-directions as described in Section IV.C. In each integration point, 64 composite inclusions per aggregate are used.

In Figure 22, for the plane strain ID model, the obtained fields of the magnitude of plastic deformation, $\overline{\varepsilon}_p^{mag} = \sqrt{\frac{2}{3} \overline{\varepsilon}_p : \overline{\varepsilon}_p}$,

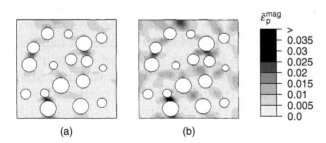

(a)   (b)

$\overline{\varepsilon}_p^{mag}$

> 0.035
0.03
0.025
0.02
0.015
0.01
0.005
0.0

Figure 22 The influence of radially oriented anisotropy on the magnitude of the plastic deformation, $\overline{\varepsilon}_p^{mag}$, for the ID model, at $\dot{\varepsilon}t = 0.025$, with (a) randomly generated initial orientations, and (b) transcrystallized preferential orientations.

are shown for $\dot{\varepsilon}t$ = 0.025. For the large scale RVE, containing randomly oriented, and thus quasi isotropic, matrix material, the plastic deformation is localized in particle-bridging paths, percolating through the matrix, approximately perpendicular to the loading direction (Figure 22(a)). The small scale RVE, having transcrystallized orientations, shows more widespread localized plastic deformation, also with shear bands in relatively thick interparticle ligaments, in the 30° to 50° direction with respect to the particle poles (the term *pole* refers to the location where the particle/matrix interface normals are aligned with the loading direction, Figure 22(b)). In the relatively thin ligaments, still localized deformation is observed. In Figures 23 and 24, the magnitude of the plastic deformation is shown for $\dot{\varepsilon}t$ = 0.05, as well as some selected microscopic texture evolutions and deformation quantities in two integration points, for random and transcrystallized initial orientations, respectively. For both situations, most plastic deformation is concentrated in relatively thin interparticle ligaments. The presence of a layer of preferentially crystallized material with significant thickness around cavitated rubber particles does have some effect on the mechanism of matrix shear yielding. This effect is, however, limited due to the relatively small level of anisotropy in the material.

In the pole figures showing the evolution of crystallographic and morphological texture, the initial orientation of each composite inclusion is represented by a dot. The arrow connects it with the corresponding final orientation, which is located at the arrowhead. In the pole figures showing microscopic deformation quantities, the location of each dot denotes the initial orientation of the lamellar normal of an inclusion and its gray intensity represents the value of the indicated quantity for the inclusion. To enrich the information shown in the latter pole figures, the mirror location of each pole with respect to the central point of the plot is also given. The view direction is the macroscopic out-of-plane direction. The term *intralamellar* deformation is employed for the magnitude of the deformation of the crystalline phase. For the amorphous deformation, a distinction is made between so-called *inter-*

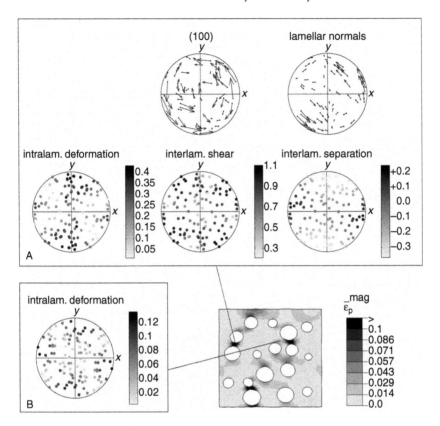

Figure 23 The magnitude of the plastic deformation, $\bar{\varepsilon}_p^{\text{mag}}$, and selected microscopic results for the ID model, at $\dot{\varepsilon}t = 0.05$, with randomly generated initial orientations.

lamellar shear and *interlamellar separation*. Let $\boldsymbol{y}^{a^i}$ be a material vector in the amorphous phase of inclusion i, with $\boldsymbol{y}_0^{a^i} = \boldsymbol{n}_0^{I^i}$. Then, interlamellar shear is assumed to be represented by the angle (in radians) between the convected material vector, $\boldsymbol{y}^{a^i} = \boldsymbol{F}^{a^i} \cdot \boldsymbol{n}_0^{I^i}$, and the current lamellar normal, $\boldsymbol{n}^{I^i}$. Lamellar separation is represented by $\ln(\lambda_{nn}^{a^i})$, with $\lambda_{nn}^{a^i} = \boldsymbol{n}^{I^i} \cdot \boldsymbol{y}^{a^i}$.

The integration point indicated by A in Figures 23 and 24, represents, for the initially randomly oriented material, a material point in the highly localized zone. Since this integration point is located in the equatorial area (the *equator* is

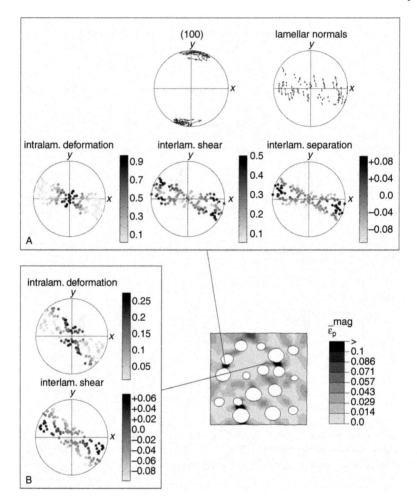

Figure 24 The magnitude of the plastic deformation, $\bar{\varepsilon}_p^{mag}$, and selected microscropic results for the ID model, at $\dot{\varepsilon}t = 0.05$, with transcrystallized initial orientations.

defined as the area where the particle/matrix interface normal is perpendicular to the loading direction), the local 1-direction is almost perpendicular to the global loading direction. The microscopic results for this point show moderate crystallographic deformation, mainly for inclusions with their lamellar normals close to the local 1-direction. The (100) poles, which represent the planes containing the two most easily activated slip systems, migrate towards a direction which is approximately 40° away from the local 1-direction. The lamellar normals are moving towards the same direction, with the largest activity for lamellar poles initially far from the target direction. Amorphous deformations are relatively large, with interlamellar shear predominantly in inclusions with their crystalline/amorphous interface approximately 45° inclined with the loading direction. Interlamellar separation is found predominantly in inclusions with their interface normals perpendicular to the local 1-direction. For the RVE with transcrystallized orientations (Figure 24), deformation is still localized in the ligament containing integration point A. In this point, the maximum intralamellar (crystallographic) deformation has increased with respect to the quasi-isotropic material, whereas both maximum interlamellar shear and separation have decreased. Crystallographic deformation is concentrated in inclusions with their lamellar normals perpendicular to the loading direction. Also in integration point B, the maximum intralamellar deformations are approximately doubled, whereas the magnitude of the interlamellar deformations is comparable to the isotropic situation.

In Figures 25 and 26, the normalized hydrostatic pressure $\bar{p}/\tau_0$, as well as some selected microscopic texture evolutions and deformation quantities in two integration points, are shown for the SA model, with random and transcrystallized initial orientations, respectively. The region of peak tensile triaxial stresses is located in the matrix material near the polar region for the preferentially oriented material, rather than in the equator area, as is observed for the randomly oriented material. In the equator region, the hydrostatic pressures remain negative; however, the absolute values are reduced with respect to the quasi-isotropic material. In Figure 27, the direction and the magnitude of the normalized maximum in-plane principal stress, $\bar{\sigma}_{max}/\tau_0$, are shown for the SA model. For the large scale, quasi-isotropic material, the maximum principal stresses are found to be negligible in the polar region, whereas for the small scale configuration, with transcrystallized orientations, also in this

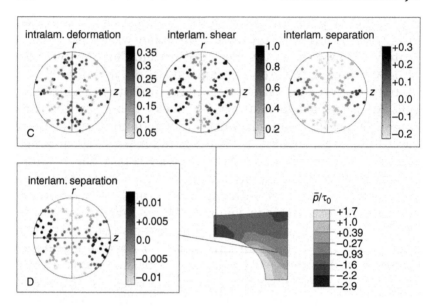

Figure 25 The normalized hydrostatic pressure, $\bar{p}/\tau_0$, and selected microscopic results for the SA model, at $\dot{\varepsilon}t = 0.1$, with randomly generated initial orientations.

region, maximum principal stresses are significant. In the equator region, maximum values are slightly increased with respect to the principal stresses in the isotropic material.

For the initially randomly oriented configuration, in integration point C, which is located in the equatorial region, the intralamellar deformations are relatively small, and are found predominantly for inclusions with their lamellar normals either perpendicular or parallel to the loading direction. In the remaining inclusions, interlamellar shear is considerably large and a significant amount of interlamellar separation is found for inclusions with their lamellar normals aligned with the loading direction. For these inclusions, the preferred direction of possible craze growth, perpendicular to the direction of the maximum principal stress [50], is parallel to the crystalline/amorphous interface. For the material with transcrystallized initial orientations, the maximum

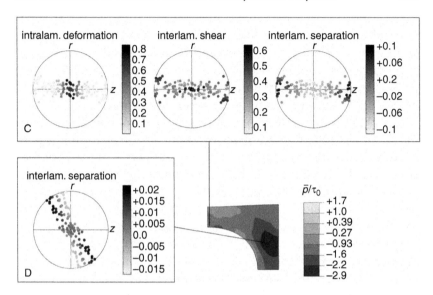

Figure 26 The normalized hydrostatic pressure, $\bar{p}/\tau_0$, and selected microscopic results for the SA model, at $\dot{\varepsilon}t = 0.1$, with transcrystallized initial orientations.

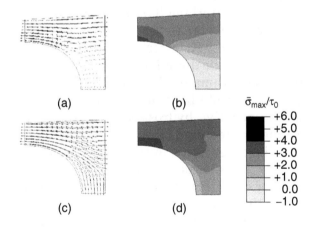

Figure 27 The direction and the magnitude of the normalized maximum in-plane principal stress, $\bar{\sigma}_{max}/\tau_0$, for the SA model, at $\dot{\varepsilon}t = 0.1$, with (a)–(b) randomly generated orientations, (c)–(d) transcrystallized preferential orientations.

intralamellar deformations are more than doubled with respect to the randomly oriented material for inclusion with their lamellar normal perpendicular to the loading direction. Interlamellar deformations are considerably reduced. For integration point D, which is located in a high tensile triaxial stress area, lamellar separations remain small.

In polymeric materials, the principal mechanisms leading to deformation and fracture [46] are shear yielding of matrix material, voiding and the occurrence of craze-like features [45,47] under triaxial stress conditions, and brittle fracture of the matrix by chain scission, induced by high tensile principal stresses. Whether or not the material will show toughened behavior will depend on which of these phenomena will predominantly occur. Massive shear yielding, with energy-absorbing inelastic deformation, has a beneficial effect on toughening. However, for the transcrystallized orientation currently considered, the increase of matrix shear yielding (replacing strain localization) is limited. Intralamellar deformation is favored over interlamellar deformation. In the quasi-isotropic material, craze-like features may be initiated in the equator region, where the peak tensile triaxial stresses are maximal, and significant interlamellar separation occurs. Since they will grow perpendicular to the direction of maximum principal stress, the growth direction will be transversely to the macroscopic tensile direction. These crazes may act as precursors to cracks, ultimately leading to failure. For the material having transcrystallized preferential orientations, in this region the negative hydrostatic pressures and interlamellar separation are reduced, diminishing the chance of craze initiation. On the other hand, in the polar region, for this material, relatively large tensile triaxial stresses are found, possibly initiating voids. However, interlamellar separation remains small at this location. In this area, the growth direction of possible crazes will be in the direction of macroscopic loading, and crazing may become an energy-absorbing mechanism. Therefore, transcrystallized orientations may lead to some degree of toughening, however, the effect is limited by the relatively small amount of anisotropy.

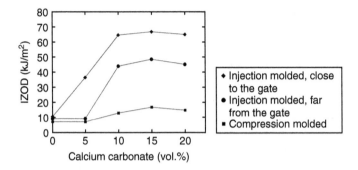

Figure 28 Influence of processing conditions on toughness [25].

E.   Influence of Processing Conditions

In the foregoing discussion, a fully radially oriented, transcrystallized, microstructure was shown to have a beneficial, but limited effect on the mechanics of deformation in particle-modified systems. This was due to the relatively small reduction in yield strength in the local 12 and 13 shear directions. A further decrease of these shear yield strengths would increase matrix shear yielding. Alternatively, an increase of the local 22 tensile yield strength would have a similar effect, since concurrently it reduces the strengths in shear.

A small flow-related crystallographic orientation was found by Bartczak et al. [17] in HDPE with calcium carbonate fillers. The importance of process conditions was demonstrated by Schrauwen et al. [23–25], who found toughness to be dominated by flow-induced effects, see Figure 28. Compression molded specimens, having a transcrystallized microstructure, showed only a minor increase in impact toughness. However, a substantial increase in toughness was found for injection molded samples, indicating the presence of flow-induced effects. Moreover, the actual toughness was found to depend on the location of the Izod specimen in the mold, confirming Bartczak's original results. A row structure of polyethylene lamellae was found in extruded alternating high density polyethylene and polystyrene thin layers by Pan et al. [69]. The long axes of the lamellae, which are the crystallographic

b-axes, were oriented in the plane of the layers and perpendicular to the extrusion direction. The a-directions were found to be predominantly normal to the layer surface and lamellar surface normals were aligned with the direction of flow. Moreover, only partial twisting was observed. In thicker layers, an unoriented structure was observed, similar to bulk polyethylene, with the corresponding lamellar twisting.

The effect of a hypothetical microstructure, with preferential orientations that may be the result of an influence of the process conditions on the anisotropic crystallization of matrix material, is now investigated for a voided macrostructure. The transcrystallized preferential orientations as used previously, were axisymmetric with respect to the local 1-direction, i.e., within the plane of the particle/matrix interface, the orientation was assumed to be random. Here, an additional preferential orientation of the molecular chains and the lamellar normals in the local 2-direction is assumed. The lamellar row structure obtained thereby may be the result of an influence of the flow on the crystallization behavior, and resembles the structure that was reported by Pan et al. [69]. Again, the crystallographic {102} planes are assumed to constitute the crystalline/amorphous interface, with an initial tilt angle of 9.7°. In Figure 29, a generated set of 125 composite inclusion orientations of this type are displayed. The stress-strain behavior obtained when this aggregate of composite inclusions is subjected to tension and shear in the basic material directions, is shown in Figures 29(e) and (f). Because of the lack of a fiber symmetry in this material, large differences between the 22 and 33 tensile loading configurations are found. With respect to the transcrystallized orientations (Figure 20), the yield strength in the 22-direction has considerably increased, whereas the 33 yield strength is reduced. At $\ln(\lambda) = 0.05$ and $2/3\sqrt{3}\,\gamma = 0.05$, the ratio of the 22 tensile and 12 shear yield strength is 3.4. The ratio of the 22 tensile resistance and the 12 shear resistance of transcrystallized material is 3.0. When material with this microstructural morphology would, in a particle-dispersed system, be oriented appropriately with respect to the loading conditions, an addi-

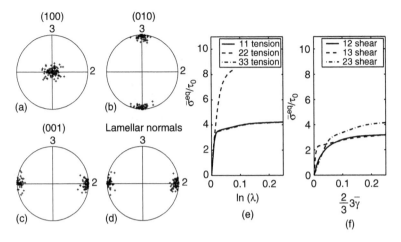

Figure 29 Equal area projection pole figures representing (a)–(c) the principal crystallographic lattice directions, and (d) the lamellar normals of an anisotropic set of orientations with an assumed influence of processing conditions and (e), (f) the normalized equivalent mesoscopic stress $\bar{\sigma}^{eq} / \tau_0$, vs. the imposed deformation for tension and shear, respectively, in the basic material directions as predicted by the composite inclusion model.

tional beneficial effect on the amount of matrix shear yielding may be expected, based on the simple analyses of Section II.

A microstructure of matrix material around well-dispersed voided particles is hypothesized that consists of lamellar crystals that are nucleated at the particle/matrix interface. An influence of processing conditions is assumed for the matrix material in the equatorial regions (with respect to the flow direction). This hypothetical morphology is realized by assigning aggregates of composite inclusions with crystallographic and morphological orientations similar to the orientation set in Figure 29 to specific elements of the finite element meshes which were previously used for the ID model and the SA model. These elements are located in the equatorial areas with respect to the flow direction. For the remaining elements, again transcrystallized orientations are assumed (similar to Figure 20). In Figure 30, the assigned flow-influenced areas are shown for both models, for either flow in the

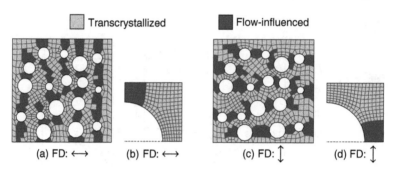

(a) FD: ↔   (b) FD: ↔   (c) FD: ↕   (d) FD: ↕

Figure 30 Assumed influence of flow on crystallization, with (a)–(b) flow in the loading direction and (c)–(d) flow perpendicular to the loading direction. The flow direction (FD) is indicated by the arrows.

loading direction or flow perpendicular to the loading direction. In each integration point, the local 1-directions are assumed to be radially oriented with respect to the nearest particle, as was previously used for the transcrystallized situation. In Figure 31, the effect of this microstructure on the obtained field of plastic deformation is shown for the ID model at $\dot{\varepsilon}t = 0.035$. In this figure also, the fully transcrystallized situation is represented. When the macroscopic loading is applied perpendicular to the flow direction, no significant

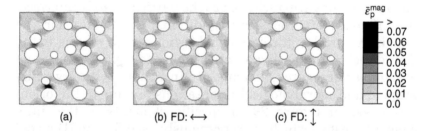

(a)   (b) FD: ↔   (c) FD: ↕

Figure 31 The influence of (a) transcrystallized orientations and (b), (c) flow-influenced orientations on the magnitude of the plastic deformation, $\bar{\varepsilon}_p^{mag}$, for the ID model, at $\dot{\varepsilon}t = 0.035$, with (b) flow in the loading direction, (c) flow perpendicular to the loading direction. The flow direction (FD) is indicated by the arrows.

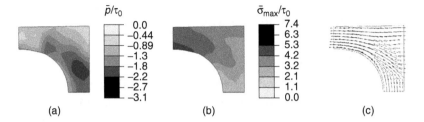

Figure 32 (a) The normalized hydrostatic pressure, $\bar{p}/\tau_0$ and (b, c) the magnitude and the direction of the normalized maximum in-plane principal stress, $\bar{\sigma}_{max}/\tau_0$, for the SA model, at $\dot{\varepsilon}t = 0.075$, with the influence of flow in the loading direction.

effect of the flow-influenced orientations can be observed, compared to fully transcrystallized material. However, when the RVE is loaded in the direction of the flow, the plastic deformation is no longer localized in the relatively thin inter-particle ligaments, but occurs predominantly in the matrix material, away from the particle surfaces and at the particle surface at an inclined location.

In Figure 32(a), the normalized hydrostatic pressure field is represented for the SA model, loaded in the flow direction. Results for the material loaded perpendicular to the flow direction are not shown because of the similarity with the fully transcrystallized situation. As for the fully transcrystallized situation, the area of peak tensile hydrostatic stress is relocated from the equatorial region for the quasi-isotropic material to the matrix material near the particle pole. Figure 32(b) and (c) show the normalized maximum in-plane principal stress for the SA model with the influence of flow in the loading direction. The largest maximum principal stresses are again observed in the equatorial area.

Thus, the hypothesized microstructure with local material orientations that results from an influence of process conditions on crystallization increases the toughening effects, if loaded in the appropriate direction with respect to the flow direction. Then, localization of deformation is replaced by dispersed shear yielding. Although a hypothesized micro-structure is used, this example demonstrates the possible importance of the processing conditions for the particle-toughening of semicrystalline material.

F.   Conclusions

A physically based mechanism for the toughening of semicrystalline polymeric materials by the dispersion of particles relies on the presence of a layer of anisotropic transcrystallized material around the particles [2,13,17]. A multiscale model was used to investigate the effect of preferentially oriented matrix material in HDPE blended with rubber particles, which were assumed to be cavitated. The particle-dispersed system was described by both a plane strain RVE model with irregularly dispersed voids and by an axisymmetric RVE model with an assumed regular stacking of voids, which were loaded in constant strain rate tension. In each integration point of the finite element model, an aggregate of composite inclusions was used as a representative microstructural element that provides the constitutive behavior of the material at the mesoscopic level. Constitutive properties were assigned at the microstructural level to the amorphous and the crystalline domains. Besides these properties, the mesoscopic constitutive behavior was affected by the crystallographic and lamellar orientations of the composite inclusions. By using preferential initial orientations, a mesoscopically anisotropic constitutive behavior was obtained.

Simulations on voided polymeric material with a large average interparticle matrix ligament thickness, having quasi-isotropic constitutive behavior at a mesoscopic level, showed a strongly localized deformation, along a path through the matrix, perpendicular to the loading direction. Large tensile triaxial stresses were found in the equator region (with respect to the loading direction) near the particles. In this area, interlamellar separations were relatively large. A particle-modified system having a relatively small average interparticle matrix ligament thickness has been realized by using initially preferentially oriented lamellae, with the crystallographic (100) planes approximately parallel to the void/matrix interface. For this system, a more dispersed field of plastic deformation was found, induced by a small relative reduction of the shear yield strength. Moreover, a relocation of the tensile triaxial stresses in the polar region, where deformation by interlamellar separation remains small, was observed, diminishing the likeliness of initiation and growth of critical craze-like features in the amorphous domains. These phenomena could indeed lead to some degree of toughening of the particle-modified material if the interparticle distance is small. However, with the level of anisotropy as predicted by the composite inclusion model, the effects of these locally preferential orientations remain limited. Simulations on idealized polymeric materials, modeled by anisotropic Hill plasticity (Section II), showed a much larger potential of local anisotropy for toughening of particle-dispersed semicrystalline materials, if the amount of anisotropy would be sufficiently large ($\zeta > 3$).

The level, and thereby the effect of local anisotropy was found to be improved if an additional hypothetical possibly flow-induced row structure of the transcrystallized lamellae was assumed in certain regions. When loaded in the direction of the macroscopic flow, plastic deformation was no longer localized in relatively thin interparticle ligaments, but was dispersed through the matrix. This massive shear yielding, in combination with the effects on craze-initiating conditions, has a further beneficial influence on the behavior of this material, however, only if loaded in the appropriate direction. Although the employed microstructure was hypothetical, the calculations demonstrated the important role of processing conditions in particle-toughening of semi-crystalline polymers.

V. DISCUSSION

The hypothesis that local anisotropy in particle-modified polymeric systems may lead to macroscopically toughened behavior was investigated by numerical methods. The anisotropy is assumed to be induced by a specific microstructure, which results from preferred crystallization of the polymeric matrix material. The role of the particles in this mechanism is (i) to create a microstructure with anisotropic constitutive behavior during crystallization, and additionally (ii) to provoke local stress concentrations during loading, thereby inducing extensive matrix shear yielding, meanwhile (iii) decreasing or relocating critical triaxial stresses that induce crazes. The validity of this hypothesis has been investigated by micromechanical modeling methods. The precise crystallization behavior was left out of consideration, and the starting point was an estimated microstructure for the particle-modified system.

The effect of a matrix material with a reduced yield strength in the local shear directions around well-dispersed voids was investigated by finite element simulations for idealized systems. The fictitious polymeric material was modeled in the context of anisotropic Hill plasticity. The three-dimensional structure of the voided material was simplified by using two different micromechanical models. The calculations confirmed that the mechanism as proposed by Muratoğlu et al. [13] could indeed lead to toughened material behavior. Required for toughening by this mechanism is apparently (i) a structure of well-dispersed voided particles; (ii) locally anisotropic material, radially oriented with respect to the nearest void surface; and (iii) sufficiently reduced shear yield strengths. If these requirements are satisfied, local plastic anisotropy of matrix material around the voids can effectively replace localization by dispersed shear yielding and can relocate the occurring tensile triaxial stresses from the equator to the particle poles, potentially leading to toughened material behavior.

A similar modeling approach was used to investigate the influence of mineral filler particles on this toughening mechanism. These systems require debonding in order to prevent excessive negative hydrostatic pressures. These debonded hard particles show a relocation of tensile triaxial stresses to the particle polar areas by local anisotropy, similar to anisotropic voided systems, with the maximum principal stresses directed such that crazes or microcracks are expected to propagate parallel to the loading direction. Moreover, the anisotropy-induced shear yielding mechanism is affected by the presence of stiff inclusions. Although some effect of the anisotropy was observed, the mechanism of toughening by local anisotropy is considered to be less effective for non-adhering hard particles, which have the advantage of increasing the blend modulus, than for low stiffness rubber fillers.

Thereafter, to investigate whether the requirements for the toughening mechanism can be achieved by a transcrystallized microstructure, a micromechanically based numerical model for the elasto-viscoplastic deformation and texture evolution of semicrystalline polymers was developed and used to simulate the behavior of particle-modified high density polyethylene (HDPE). For these blended polymeric systems, a distinction between three different scales has been made. The constitutive properties of the distinguishable material components were characterized at the microscopic scale. At the mesoscopic scale, an aggregate of individual phases was considered. To bridge between those scales, a composite inclusion model has been formulated. The model is based on (a simplified representation of) the underlying morphology and deformation mechanisms of this material. As a representative microstructural element, a two-phase layered composite inclusion has been used, with a lamellar structure as is commonly observed in semicrystalline polymers. Both the crystalline and the amorphous phase are represented in the composite inclusion model and are mechanically coupled at the interface. For both phases, micromechanically based constitutive models have been used. The local inclusion-averaged deformation and stress fields are related to the mesoscopic fields of the aggregate using an interaction model. In the model, the effect of transcrystallization on microscopic properties other than orientation, such as crystallinity and amorphous and crystalline constitutive behavior (e.g. as a result of the lamellar thickness), are not accounted for.

A full multiscale model was used to investigate the effect of preferentially oriented matrix material in HDPE blended with rubber particles, which were represented by voids. In this model, the structure-property relationship is addressed at various levels: (i) the arrangement of chain segments, represented in the constitutive behavior of the individual phases; (ii) the arrangement of the lamellae, influencing the response of the polymeric matrix material; and (iii) the particle-modified macroscopic structure. Simulations on voided polymeric material with a large average interparticle matrix ligament thickness, having quasi-isotropic constitutive behavior at the mesoscopic level, showed strongly localized deformation. Maximum tensile triaxial stresses were found in the equator regions near the particles. A particle-modified system having a relatively small average interparticle matrix ligament thickness was realized by using initially preferentially oriented lamellae, with the crystallographic (100) planes approximately parallel to the void/matrix interface. For this system, a more dispersed field of plastic deformation was found, induced by a relative reduction of the shear yield strength. Moreover a relocation of the peak tensile triaxial stresses to the polar region was observed, diminishing the initiation and growth of critical craze-like features in the amorphous domains. These phenomena could indeed lead to some degree of toughening of the particle-modified material when the interparticle distance is small. However, with the level of anisotropy as predicted by the composite inclusion model, the effects of these locally preferred orientations remained limited. The simulations on idealized polymeric materials, modeled by anisotropic Hill plasticity, showed a substantially increased potential capacity of local anisotropy for toughening of particle-dispersed semicrystalline materials, if the amount of anisotropy was sufficiently large.

The level, and thereby the effect of local anisotropy was made more pronounced by assuming a hypothetical additional row structure of the transcrystallized lamellae in certain regions, which represents the result of the processing history. When loaded in the direction of the macroscopic flow, plastic deformation was no longer localized in relatively thin interparticle ligaments, but was largely distributed through the matrix. This shear yielding, in combination with the effects on craze-initiating conditions, has a further beneficial effect on the mechanical behavior of this material, however only when loaded in the appropriate direction.

Based on these simulations, Figure 3, which was the starting point of this work, illustrating the potential of rubber and mineral fillers for improving mechanical properties by

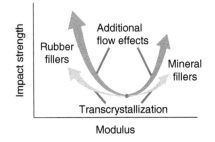

Figure 33 The influence of voids (cavitated rubber) vs. hard (mineral) particles in combination with microstructure-induced anisotropy on the mechanical properties.

the hypothesized mechanism, can be further refined. This refinement is displayed in Figure 33. Whereas in the reference (large scale) isotropic system, tensile triaxial stresses were found in the particle equator areas, for all anisotropic (small scale) systems, maximum negative pressures were observed in the polar area. For these systems, craze-like events are expected to propagate in the loading direction. For rubber-filled systems, transcrystallized layers had a limited effect on matrix shearing, whereas the presence of row-structured material more efficiently changed the mode of deformation to dispersed shear yielding. However, the mechanism of toughening by local anisotropy was concluded to be less effective for non-adhering hard particles.

ACKNOWLEDGMENTS

This research was funded by the Dutch Polymer Institute. The authors gratefully acknowledge Prof. D. M. Parks and Prof. M. C. Boyce of the Massachusetts Institute of Technology for their stimulating and helpful discussions concerning the micromechanical modeling of semicrystalline polymers and Dr. B. A. G. Schrauwen for the valuable contributions to this research.

REFERENCES

1. S. Wu. Phase structure and adhesion in polymer blends: a criterion for rubber toughening. Polymer, 26:1855–1863, 1985.

2. Z. Bartczak, A. S. Argon, R. E. Cohen, and M. Weinberg. Toughness mechanism in semi-crystalline polymer blends: I. High-density polyethylene toughened with rubbers. Polymer, 40:2331–2346, 1999.

3. A. Sánchez-Solís, M. R. Estrada, J. Cruz, and O. Manero. On the properties and processing of polyethylene terephthalate/styrene-butadiene rubber blend. Polymer Engineering and Science, 40:1216–1225, 2000.

4. W. Loyens and G. Groeninckx. Ultimate mechanical properties of rubber toughened semicrystalline PET at room temperature. Polymer, 43:5679–5691, 2002.

5. W. Loyens and G. Groeninckx. Rubber toughened semicrystalline PET: influence of the matrix properties and test temperature. Polymer, 44:123–136, 2003.

6. F. Ramsteiner and W. Heckmann. Mode of deformation in rubber-modified polyamide. Polymer Communications, 200:199–200, 1985.

7. R. J. M. Borggreve, R. J. Gaymans, J. Schuijer, and J. F. Ingen Housz. Brittle-tough transition in nylon-rubber blends: effect of rubber concentration and particle size. Polymer, 28:1489–1496, 1987.

8. S. Wu. A generalized criterion for rubber toughening: the critical matrix ligament thickness. Journal of Applied Polymer Science, 35:549–561, 1988.

9. A. Margolina and S. Wu. Percolation model for brittle-tough transition in nylon/rubber blends. Polymer, 29:2170–2173, 1988.

10. T. Fukui, Y. Kikuchi, and T. Inoue. Elastic-plastic analysis of the toughening mechanism in rubber-modified nylon: matrix yielding and cavitation. Polymer, 32:2367–2371, 1991.

11. K. Dijkstra and G. H. Ten Bolscher. Nylon-6/rubber blends. Part iii. Stresses in and around rubber particles and cavities in a nylon matrix. Journal of Materials Science, 29:4286–4293, 1994.

12. O. K. Muratolğu, A. S. Argon, and R. E. Cohen. Crystalline morphology of polyamide-6 near planar surfaces. Polymer, 36:2143–2152, 1995.

13. O. K. Muratolğu, A. S. Argon, and R. E. Cohen. Toughening mechanism of rubber-modified polyamides. Polymer, 36:921–930, 1995.

14. V. P. Chacko, F. E. Karasz, R. J. Farris, and E. L. Thomas. Morphology of CaCO$_3$-filled polyethylenes. Journal of Polymer Science: Part B: Polymer Physics, 20:2177–2195, 1982.

15. F. Rybnikář. Interactions in the system polyethylene–solid filler. Journal of Macromolecular Science – Physics, B19:1–11, 1981.

16. F. Rybnikář. Orientation in composite of polypropylene and talc. Journal of Applied Polymer Science, 38:1479–1490, 1989.

17. Z. Bartczak, A. S. Argon, R. E. Cohen, and M. Weinberg. Toughness mechanism in semi-crystalline polymer blends: II. High-density polyethylene toughened with calcium carbonate filler particles. Polymer, 40:2347–2365, 1999.

18. Z. Bartczak, A. S. Argon, R. E. Cohen, and T. Kowalewski. The morphology and orientation of polyethylene in films of sub-micron thickness crystallized in contact with calcite and rubber substrates. Polymer, 40:2367–2380, 1999.

19. G.-M. Kim, D.-H. Lee, B. Hoffmann, J. Kressler, and G. Stöppelmann. Influence of nanofillers on the deformation process in layered silicate/polyamide-12 nanocomposites. Polymer, 42:1095–1100, 2001.

20. P. H. Nam, P. Maiti, M. Okamoto, T. Kotaka, N. Hasegawa, and A. Usuki. A hierarchical structure and properties of intercalated polypropylene/clay nanocomposites. Polymer, 42:9633–9640, 2001.

21. H. E. H. Meijer and L. E. Govaert. Multi-scale analysis of mechanical properties of amorphous polymer systems. Macromolecular Chemistry and Physics, 204:274–288, 2003.

22. P. A. Tzika, M. C. Boyce, and D. M. Parks. Micromechanics of deformation in particle-toughened polyamides. Journal of the Mechanics and Physics of Solids, 48:1893–1930, 2000.

23. B. A. G. Schrauwen, L. E. Govaert, and H. E. H. Meijer. Toughness of high-density polyethylene with hard filler particles. In Proceedings of the Seventeenth Annual Meeting of the Polymer Processing Society, Montreal, Canada, 2001.

24. B. A. G. Schrauwen, L. E. Govaert, G. W. M. Peters, and H. E. H. Meijer. The influence of flow induced crystallization on the impact toughness of high-density polyethylene. In Proceedings of the International Conference on Flow Induced Crystallization of Polymers, Salerno, Italy, 2001.

25. B. A. G. Schrauwen, L. E. Govaert, G. W. M. Peters, and H. E. H. Meijer. The influence of flow-induced crystallization on the impact toughness of high-density polyethylene. Macromolecular Symposia, 185:89–102, 2002.

26. M. W. L. Wilbrink, A. S. Argon, R. E. Cohen, and M. Weinberg. Toughenability of nylon-6 with $CaCO_3$ filler particles: new findings and general principles. Polymer, 42:10155–10180, 2001.

27. O. K. Muratolğu, A. S. Argon, R. E. Cohen, and M. Weinberg. Microstructural processes of fracture of rubber-modified polyamides. Polymer, 36:4771–4786, 1995.

28. Y. S. Thio, A. S. Argon, R. E. Cohen, and M. Weinberg. Toughening of isotactic polypropylene with $CaCO_3$ particles. Polymer, 43:3661–3674, 2002.

29. W. C. J. Zuiderduin, C. Westzaan, J. Huétink, and R. J. Gaymans. Toughening of polypropylene with calcium carbonate particles. Polymer, 44:261–275, 2003.

30. G. H. Michler. Micromechanics of polymers. Journal of Macromolecular Science – Physics, B38:787–802, 1999.

31. J. Petermann and H. Ebener. On the micromechanics of plastic deformation in semicrystalline polymers. Journal of Macromolecular Science – Physics, B38:837–846, 1999.

32. C. G'Sell and A. Dahoun. Evolution of microstructure in semicrystalline polymers under large plastic deformation. Materials Science and Engineering A, 175:183–199, 1994.

33. L. Lin and A. S. Argon. Structure and plastic deformation of polyethylene. Journal of Materials Science, 29:294–323, 1994.

34. J. A. W. van Dommelen, W. A. M. Brekelmans, and F. P. T. Baaijens. Micromechanical modeling of particle-toughening of polymers by locally induced anisotropy. Mechanics of Materials, 35:845–863, 2003.

35. R. Hill. The Mathematical Theory of Plasticity. Oxford University Press, London, 1950.

36. K. Kobayashi and T. Nagasawa. Mechanical properties of polyethylene crystals. II. Deformation process of spherulite. Journal of Polymer Science: Part C, 15:163–183, 1966.

37. S. Socrate and M. C. Boyce. Micromechanics of toughened polycarbonate. Journal of the Mechanics and Physics of Solids, 48:233–275, 2000.

38. R. J. M. Smit, W. A. M. Brekelmans, and H. E. H. Meijer. Prediction of the large-strain mechanical response of heterogeneous polymer systems: local and global deformation behaviour of a representative volume element of voided polycarbonate. Journal of the Mechanics and Physics of Solids, 47:201–221, 1999.

39. V. Tvergaard. Effect of void size difference on growth and cavitation instabilities. Journal of the Mechanics and Physics of Solids, 44:1237–1253, 1996.

40. V. Tvergaard. Interaction of very small voids with larger voids. International Journal of Solids and Structures, 30:3989–4000, 1998.

41. R. J. M. Smit, W. A. M. Brekelmans, and H. E. H. Meijer. Prediction of the mechanical behavior of non-linear heterogeneous systems by multi-level finite element modeling. Computer Methods in Applied Mechanics and Engineering, 155:181–192, 1998.

42. R. Hall. Computer modelling of rubber-toughened plastics: random placement of monosized core-shell particles in a polymer matrix and interparticle distance calculations. Journal of Materials Science, 26:5631–5636, 1991.

43. HKS. ABAQUS/Standard User's Manual, version 6.2. Hibbitt, Karlsson & Sorensen, Inc., Pawtucket, RI, 2001.

44. K. Friedrich. Crazes and shear bands in semi-crystalline thermoplastics, volume 52/53 of Advances in Polymer Science, pages 225–274. Springer-Verlag, Berlin, 1983.

45. I. Narisawa and M. Ishikawa. Crazing in semicrystalline thermoplastics, volume 91/92 of Advances in Polymer Science. Springer-Verlag, Berlin, 1990, 353–391.

46. H.-H. Kausch, R. Gensler, Ch. Grein, C. J. G. Plummer, and P. Scaramuzzino. Crazing in semicrystalline thermoplastics. Journal of Macromolecular Science – Physics, B38:803–815, 1999.

47. G. H. Michler and R. Godehardt. Deformation mechanisms of semicrystalline polymers on the submicron scale. Crystal Research and Technology, 35:863–875, 2000.

48. A. S. Argon, R. E. Cohen, O. S. Gebizlioglu, and C. E. Schwier. Crazing in block copolymers and blends, volume 52/53 of Advances in Polymer Science. Springer-Verlag, Berlin, 1983, 275–334.

49. A. S. Argon, R. E. Cohen, and T. M. Mower. Mechanisms of toughening brittle polymers. Materials Science and Engineering A, 176:79–90, 1994.

50. E. J. Kramer. Microscopic and molecular fundamentals of crazing, volume 52/53 of Advances in Polymer Science. Springer-Verlag, Berlin, 1983, 1–56.

51. M. G. A. Tijssens, E. Van der Giessen, and L. J. Sluys. Modeling of crazing using a cohesive surface methodology. Mechanics of Materials, 32:19–35, 2000.

52. R. Estevez, M. G. A. Tijssens, and E. van der Giessen. Modeling of the competition between shear yielding and crazing in glassy polymers. Journal of the Mechanics and Physics of Solids, 48:2585–2617, 2000.

53. S. Socrate, M. C. Boyce, and A. Lazzeri. A micromechanical model for multiple crazing in high impact polystyrene. Mechanics of Materials, 33:155–175, 2001.

54. J. Bicerano and J. T. Seitz. Molecular origins of toughness in polymers. Polymer Toughening. Marcel Dekker, New York, 1996, 1–59.

55. J. A. W. van Dommelen, W. A. M. Brekelmans, and F. P. T. Baaijens. A numerical investigation of the potential of rubber and mineral particles for toughening of semicrystalline polymers. Computational Materials Science, 27:480–492, 2003.

56. J. A. W. van Dommelen, W. A. M. Brekelmans, and F. P. T. Baaijens. Multiscale modeling of particle-modified polyethylene. Journal of Materials Science, 38:4393–4405, 2003.

57. J. A. W. van Dommelen, D. M. Parks, M. C. Boyce, W. A. M. Brekelmans, and F. P. T. Baaijens. Micromechanical modeling of the elasto-viscoplastic behavior of semi-crystalline polymers. Journal of the Mechanics and Physics of Solids, 51:519–541, 2003.

58. B. J. Lee, D. M. Parks, and S. Ahzi. Micromechanical modeling of large plastic deformation and texture evolution in semi-crystalline polymers. Journal of the Mechanics and Physics of Solids, 41:1651–1687, 1993.

59. B. J. Lee, A. S. Argon, D. M. Parks, S. Ahzi, and Z. Bartczak. Simulation of large strain plastic deformation and texture evolution in high density polyethylene. Polymer, 34:3555–3575, 1993.

60. A. S. Argon. Morphological mechanisms and kinetics of large-strain plastic deformation and evolution of texture in semi-crystalline polymers. Journal of Computer-Aided Materials Design, 4:75–98, 1997.

61. F. C. Frank, V. B. Gupta, and I. M. Ward. The effect of mechanical twinning on the tensile modulus of polyethylene. Philosophical Magazine, 21:1127–1145, 1970.

62. R. J. Young and P. B. Bowden. Twinning and martensitic transformations in oriented high-density polyethylene. The Philosophical Magazine, 29:1061–1073, 1974.

63. E. M. Arruda and M. C. Boyce. A three-dimensional constitutive model for the large stretch behavior of rubber elastic material. Journal of the Mechanics and Physics of Solids, 41:389–412, 1993.

64. J. A. W. van Dommelen, D. M. Parks, M. C. Boyce, W. A. M. Brekelmans, and F. P. T. Baaijens. Micromechanical modeling of intraspherulitic deformation of semicrystalline polymers. Polymer, 44:6089–6101, 2003.

65. J. A. W. van Dommelen, D. M. Parks, M. C. Boyce, W. A. M. Brekelmans, and F. P. T. Baaijens. Multi-scale modeling of the constitutive behavior of semi-crystalline polymers. In Proceedings of the European Conference on Computational Mechanics, Cracow, Poland, 2001.

66. A. Keller and S. Sawada. On the interior morphology of bulk polyethylene. Die Makromolekulare Chemie, 74:190–221, 1964.

67. D. C. Bassett and A. M. Hodge. On the morphology of melt-crystallized polyethylene I. Lamellar profiles. Proceedings of the Royal Society of London A, 377:25–37, 1981.

68. S. Gautam, S. Balijepalli, and G. C. Rutledge. Molecular simulations of the interlamellar phase in polymers: effect of chain tilt. Macromolecules, 33:9136–9145, 2000.

69. S. J. Pan, J. Im, M. J. Hill, A. Keller, A. Hiltner, and E. Baer. Structure of ultrathin polyethylene layers in multilayer films. Journal of Polymer Science: Part B: Polymer Physics, 28:1105–1119, 1990.

10

Micromechanical Mechanisms of Toughness Enhancement in Nanostructured Amorphous and Semicrystalline Polymers

GOERG H. MICHLER

Institute of Materials Science, Martin-Luther-
University Halle-Wittenberg, Germany

CONTENTS

I. INTRODUCTION

In nearly all applications of polymers, the mechanical behavior cannot be ignored. Important mechanical properties include stiffness, strength, elongation at break, and, as an average property, toughness or fracture toughness. Within this context, the term "toughness" denotes the absorption of mechanical energy during a deformation that ends up in fracture. The aim of improvement or modification of polymeric materials is often to develop a material with high toughness and a large plastic elongation of break, whilst retaining a high level of other desirable properties such as stiffness and strength. These are opposed demands, and the usual technique of manufacturing of high-impact polymers — the rubber toughening — has the disadvantage of a pronounced decrease in strength and stiffness due to the rubber content [1]. Additionally, in many applications of polymers, a good balance of the mechanical properties with other properties (e.g., transparency, electric properties, flame retardancy) and a good processability is demanded. With our present knowledge, a combination of such different properties is impossible to realize in a homogeneous polymer on the basis of new monomers but only in a heterogeneous one by modifying the polymer structure on a meso-, micro- and sub-micrometer scale. There is a large variety of macromolecular and supermolecular structures (the morphology), but not all of them are of equal importance for property improvements. Only certain details of the structure are decisive for the mechanical behavior, and these details are called *property-determining structures* [2]. Often, structural defects exist that are responsible for premature failure of materials. A better understanding of such important structures or defects and their role in influencing mechanical properties is a task of particular scientific and economic importance and a key to modify and to improve the properties of polymeric materials.

Structure, morphology and mechanical properties are linked by the micromechanical processes of deformation and fracture [2,3]. A detailed knowledge about structure-property correlations and their underlying micromechanical mechanisms enables criteria to be defined for modifying polymer structure and morphology and for producing polymers with defined improved or new properties, known as *microstructural construction of polymers* [4].

Micromechanical processes are usually highly localized, and several different techniques have been applied for their study. Besides spectroscopic and scattering techniques, the techniques of electron and atomic force microscopy are particularly useful for direct determination of micromechanical properties of polymers [2,3,5]. Using special micro-tensile devices for the electron and atomic force microscopes, *in situ* deformation tests can be performed. Since these techniques also allow the study of morphological details, they enable us to investigate structure-property correlations in a very direct way. A brief overview of these techniques is given in Section II.

It has been well known for many years that thermoplastics can be toughened by adding 5 to 25% of a suitable rubber [1]. The process of rubber toughening is of major importance to the plastics industry, and it has proved so effective that this technology has been extended to almost all of the commercial glassy thermoplastics, including polystyrene (PS), poly(styrene-acrylonitrile) copolymer (SAN), polymethylmethacrylate (PMMA) and polyvinylchloride (PVC). It has also been applied to several thermosetting resins and to semicrystalline polymers, notably polypropylene (PP) and polyamides (PA). Recently, some new techniques and micromechanical mechanisms were found as alternative toughening mechanisms. An overview of some of these mechanisms is given in Section III.

Some additional micromechanical mechanisms for toughness enhancement are discussed in detail in this chapter, including toughening effects due to core-shell particles in amorphous and semicrystalline polymers with mechanisms of *core flattening* and *low-temperature toughness*, the so-called *thin-layer yielding* mechanism in nanostructured block copolymers and effects of *multiple crazing* in multi- and nanolayered polymer systems.

II. TECHNIQUES TO STUDY TOUGHNESS MECHANISMS

A. Materials and Sample Preparation

In this chapter several amorphous and semicrystalline heterogeneous polymer blends, block copolymers and composites are used, which are described and characterized in detail in connection with the results. The investigated samples are commonly used in the form of injection molded plates or extrudates.

B. Study of Morphology

Every study of micromechanical mechanisms must be coupled with a detailed investigation of the morphology. For polymer studies several techniques of electron microscopy (EM) can be employed, including conventional transmission electron microscopy (TEM), high-resolution electron microscopy (HREM), high-voltage electron microscopy (HVEM), analytical electron microscopy (AEM), scanning electron microscopy (SEM), environmental scanning electron microscopy (ESEM), and atomic force microscopy (AFM). In general two preparation techniques and investigation methods have been applied [2,6,7]:

1. The preparation of special surfaces from internal parts of the samples, such as brittle fracture surfaces or smooth and selectively etched surfaces. These surfaces are investigated directly in the SEM or AFM.
2. The preparation of semi-thin or ultra-thin sections by ultramicrotomy, generally after special fixation and staining procedures or at lower temperatures by cryoultramicrotomy. Investigations of the usually $0.05 - 1.0$ µm thick sections are carried out by TEM, HVEM, or AFM.

C. Study of Micromechanical Properties

Depending on type, morphology and the main micromechanical properties of the polymeric material of interest, several different methods have been applied, which enable the investigation of micromechanical processes over a wide range of temperatures and magnifications often together with the morphology, interfaces, and others.

1. The easiest way is to study fracture surfaces of bulk samples by SEM or ESEM (microfractography), yielding information about the processes of crack initiation and crack propagation up to final fracture. The influence of structural heterogeneities ("defects") on the initiation, as well as on the propagation of cracks can be studied.

2. Deformation of bulk material (e.g., stress-elongation test, bending test) is followed by inspection of the changes at the surfaces by SEM, ESEM or AFM. Changes inside the bulk material are studied by producing flat surfaces and SEM investigation (occasionally after selective etching) or by preparing ultra-thin sections using an ultramicrotome (occasionally after chemical staining) and by TEM investigation.

3. Deformation of semi-thick samples, thin films or ultramicrotomed thin sections in a special micro-tensile or bending device and direct investigation of the deformed samples by SEM, ESEM, AFM, HVEM, or TEM. Some of the tensile or bending devices enable one to perform deformation tests at lower or higher temperatures (from −180°C to +200°C). Further details of the different microscopic techniques are given in Referencs [2] and [5].

D.   Determination of Mechanical Properties

Tensile testing of dog-bone shaped tensile bars was performed using a universal tensile machine at a cross head speed of usually 50 mm/min and at room temperature (23°C). Some of the samples were tested using miniaturized specimens by means of a Minimat miniature materials testing device (Polymer Laboratories, UK) at a cross-head speed of 1 mm/min at room temperature.

III. MECHANISMS OF TOUGHNESS ENHANCEMENT

A. General Aspects and Overview

There are different mechanisms to enhance the toughness of polymers at room temperature or at lower temperatures. Some of them are broadly applied in the plastics industry, others are theoretical possibilities. Toughness enhancement with retaining the other mechanical properties such as stiffness and strength is possible only in a heterogeneous polymeric material by modifying the polymer morphology in such a way as to promote a large number of local energy-absorbing plastic deformation processes on a micro- and sub-micrometer scale. Therefore, all of the toughening mechanisms, which are discussed in this and in the next chapter, appear in two- or multiphase polymeric materials with a heterogeneous morphology, including nanostructured amorphous and semicrystalline polymers.

The different toughening mechanisms can be summarized as follows:

- Rubber particle toughening, or in general, modifier particle toughening with the details
  - Three-stage mechanism of particle toughening
  - Homogeneous and core-shell particles
  - Particles due to phase separation
- Rubber network yielding
- Inclusion yielding
- Particle-filled systems (composites)
  - Usual inorganic particles
  - Nanoparticles
- Self-reinforcement
- Phase transformation toughness
- Thin-layer yielding

B. Rubber Particle Toughening

Rubber-toughened thermoplastics were first manufactured in the late 1940s and have since been studied very extensively. Fracture toughness can be increased by up to one order of magnitude by adding a small amount (usually 5 to 25%) of a suitable rubber or elastomer to the thermoplastics. This effect was initially utilized in PS and SAN by grafting to butadiene rubber, yielding high-impact polystyrene (HIPS) and acrylonitrile-butadiene-styrene copolymer (ABS), respectively. The addition of rubber particles promotes energy absorption through the initiation of local yielding. This deformation mechanism involves three important stages (the so-called *three-stage mechanism of toughening* [1,2]) (Figure 1):

1. Elastic deformation, resulting in the generation of stress concentrations around the rubber particles, and (in some cases) cavitation in the rubber particles
2. Plastic strain softening, characterized by local yielding of the matrix, through multiple crazing (fibrillated or homogeneous crazes), extensive shear yielding, or some combination of both
3. Strain hardening of the yield zone, a process to which stretching of the rubber phase to very high strains makes a significant contribution, especially when the rubber content is high; in specimens containing sharp notches or cracks, rubber particles cause crack tip blunting and consequently crack stop, preventing a premature crack propagation and fracture

The dominant mechanism of deformation in the matrix varies not only with the chemical structure and composition of the matrix material, but also with the test temperature, the strain rate, and the morphology, shape, and size of the rubber particles [1,2,8–10]. Formation of fibrillated crazes and fibrillation of the rubber membranes of the "salami" particles is typical of high impact polystyrene (HIPS). The formation of fibrillated matrix crazes in combination with dilatational shear bands, where the voids are confined to the rubber phase, is typical of many grades of toughened PSAN or ABS blends. Craze-like dilatational shear bands are also formed in toughened grades of PVC, PMMA, PP, and epoxy resins. Elastomeric particles can be distributed in the matrix in the form of homogeneous or heterogeneous particles. In systems with

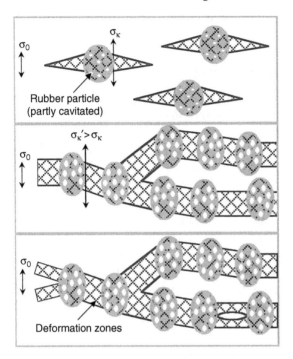

Figure 1 Schematic illustration of the three-stage mechanism of rubber particle toughening (σ_0 applied stress, σ_k stress concentration at a single rubber particle, σ'_k enlarged stress concentration due to stress field superposition).

"salami" or "core-shell" particles, hard polymer sub-inclusions occasionally can be deformed by a *mechanism of core flattening* as a new additional mechanism of energy absorption (see Section IV.A).

The chemical structure and composition of the matrix material determines not only the type of the local yield zones, but also the critical parameters for toughening. In amorphous polymers with the dominant formation of fibrillated crazes, the *particle diameter D* is of primary importance. In some other amorphous and in semicrystalline polymers with the dominant formation of dilatational shear bands or intense shear yielding the critical parameter is the *interparticle distance A* (the thickness of matrix strands between the particles)

— for details see the literature cited above and Chapters 9, 11, and 12.

C.   Rubber Network Toughening

It is often assumed that rubber toughening is synonymous with the addition of rubber particles. However, there is an alternative and also very effective approach to rubber toughening of amorphous polymers, using *rubber networks*, which has been known for some decades [2,11–13]. This method involves small particles of a thermoplastic being embedded in the rubber network to form a honeycomb structure with very thin layers of rubber separating the thermoplastic particles. A blend might consist, for example, of a rubbery ethylene-vinylacetate copolymer (EVAc) or a nitrile-butadiene rubber (NBR) matrix encapsulating primary particles of PVC which are about 0.5 to 1 µm in diameter. In the example of Figure 2, a network of a rubbery EVAc phase contains small particles of PVC. Since there are very tiny network layers, the rubber content is usually kept below 10 vol%; that is, the PVC content reaches more than 90%.

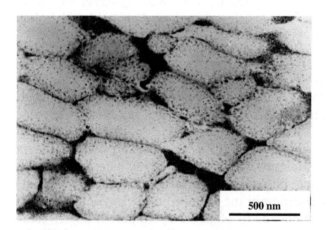

500 nm

Figure 2   Rubber-toughened PVC with a network structure: a network of a rubbery phase (EVAc, dark appearance due to selective staining) contains many small particles of PVC (bright); (stained ultrathin section, TEM).

When the rubbery network is stretched, the PVC particles start to yield and absorb energy. In the micrographs of
Figure 3, the two phases are clearly visible without any chemical staining due to stretching of the sample in the microscope.
Owing to the lower density and the predominant deformation
of the rubber phase, the rubber network appears bright, and
the PVC particles appear dark (effect of "straining induced
contrast enhancement" [2]). During loading, the following
deformation steps appear:

1. At the beginning of deformation, the weaker rubber
   phase is stretched, building up a three-dimensional
   stress state everywhere in the network (Figure 3a).
2. The rubber phase transfers stresses from one PVC
   particle to the next, high enough to reach the yield
   stress of PVC, the particles start to deform plastically (Figure 3b and c); the yielding of the numerous
   PVC particles absorbs the main part of the total
   fracture energy.
3. By rupture of parts of the rubber network, microvoids
   appear, yielding an intense fibrillation of the network
   and an additional plastic yielding of PVC particles
   (Figure 3d).

One critical parameter of this mechanism is the thickness of
the rubber network layers, which must be in the range of a
few tens of nanometers [2,11,12]. Only such a thin-walled
network can create a three-dimensional (hydrostatic) stress
state high enough to reach the yield stress of PVC. A disadvantage of these systems and the reason why they are not
used in the plastics industry is the sensitivity during processing with destruction of the network and its transformation
into a fine rubber particle distribution (Figure 4). After this
phase separation the toughness enhancement is drastically
reduced because of the relatively low rubber content and the
small particles [2].

Other examples of such a type of blends are a polyethylene (PE) or a polybutadiene (PB) network containing embedded polystyrene (PS) particles [2,14]. However, in this case
the PS particles do not yield as whole particles (as in the case

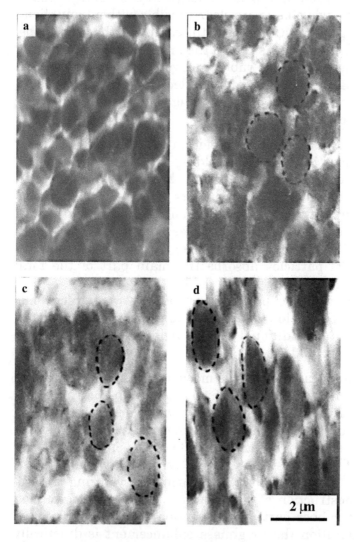

Figure 3 Rubber-toughened PVC with a network structure of rubber (EVAc, bright appearance) around PVC particles (dark): a, b, c
and d show increasing elongation, deformation direction is vertical;
(1000 kV HVEM micrographs, *in situ* – deformation test).

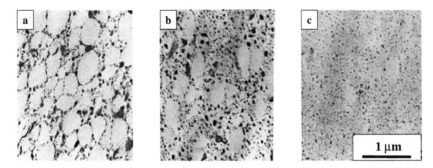

Figure 4 Destruction of the network structure in PVC/EVAc blends due to processing with increasing blending time: a) starting network structure, b) beginning of structure destruction, c) after destruction with phase separation (small rubber particles in PVC matrix). (Selectively stained ultra-thin sections, TEM.)

of PVC), but at best via a surface drawing mechanism [2] (very similar to the core flattening mechanism in the core-shell particles, described in Section IV.A.3).

D.   Inclusion Yielding

A deformation mechanism similar to that observed in rubber network blends is *inclusion yielding* or *hard particle yielding*, in which stiff thermoplastic particles are distributed in a matrix with a slightly lower yield stress, e.g., PSAN particles embedded in a polycarbonate (PC) matrix [15,16]. Under load, stresses transferred to rigid inclusions via the softer matrix can exceed the yield stress of the inclusions and, therefore, the particles are forced to deform plastically and absorb energy.

Both types of mechanisms — rubber network yielding and inclusion yielding — act in systems with stiffer polymeric particles in a rubbery network or a softer matrix. The difference lies in volume content of the stiff particles, which is higher in the network structure, where the very thin rubber layers of the network do not really form a matrix. It is often not believed that stiffer particles can be plastically deformed due to a softer surrounding, since the usual, well-known case is the opposite situation (yielding of softer particles in a stiffer matrix). However, experiments on these types of blends demonstrate that the stresses transferred to rigid inclusions via a softer matrix can exceed the yield stress of the inclusions. Of course, an optimum morphology and a strong adhesion (interfacial strength) are necessary for these effects.

E.   Particle-Filled Polymers (Composites)

Inorganic filler particles are usually added to thermoplastics to enhance the stiffness and — if there is a strong interfacial strength — also to increase the yield stress. In case of low or missing interfacial strength (adhesion), debonding and cavitation appears during loading. The microvoids around the filler particles act as stress concentrators (like the elastomeric particles in rubber particle toughening) and can, therefore, initiate local yielding processes. In dependence on particle size and particle distribution several cases often appear (Figure 5):

    a. The basic effect is debonding/cavitation and local stress concentration.

    b. Large particles create large voids with the disadvantage of void coalescence and production of cracks of overcritical lengths.

    c. Agglomerates of small particles can rupture, producing sharp cracks.

    d. Small, homogeneously distributed particles initiate local yielding between the particles/microvoids.

Only the case d) gives rise to an increased toughness. Particle-filled PE and PP systems are examples of this mechanism [2,17–19] and Figure 6 illustrates this effect in a polypropylene (PP) matrix modified with small filler particles (Al_2O_3, average particle diameter about 1 µm, filler content 10 wt%). The lower as well as the higher magnification show the effects of debonding and void formation around the filler particles, the elongation of microvoids, and the plastic shear deformation of the matrix strands between the particles/voids. The local deformation reaches several hundred percent and can be larger than the average elongation at break of the unmodified matrix material. It is interesting to note that there is a remarkable similarity

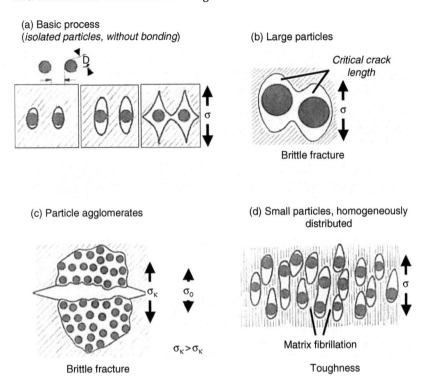

Figure 5 Schematics of particle-filled polymers and different cases of local processes, depending on particle size and distribution (a, b, c, d — see text).

of the deformation mechanisms in these particle-filled PPs and in rubber toughened PP, revealing that particle-filling in principle can create not only stiffer but also tougher materials.

Preconditions of this toughening mechanism are particles, which are small enough (to prevent crack initiation at the microvoids around the particles) with a narrow size distribution (to prevent premature crack formation at the largest particles), with a homogeneous distribution and good spatial separation and with optimum distances between the particles in the matrix (optimum critical interparticle distance) [2,18,19]. Therefore, there is only an optimum "window" in particle size D or interparticle distance A and in adhesion for tough filled systems (see Figure 7).

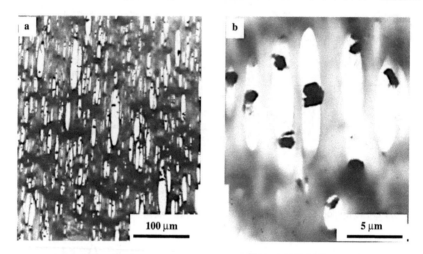

Figure 6 TEM micrographs showing the deformation of particle-filled PP (10 wt% of Al$_2$O$_3$ filler particles): debonding and microvoid formation around the particles, elongation of microvoids, plastic stretching of the matrix strands between the particles/voids; a) lower and b) higher magnification (deformed in a 1000 kV HVEM; deformation direction vertical).

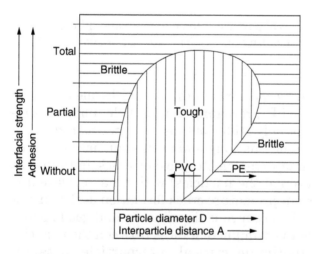

Figure 7 Influence of particle size D/interparticle distance A and interfacial strength/adhesion on mechanical behavior; there is only a small "window" for toughness; the borders between tough and brittle fields depend on matrix behavior.

The application of nanometer particles (nano-composites) should give additional possibilities to enhance the toughness. An overview about morphology and properties of nanotube and nanofiber-reinforced polymers is given in Chapter 14.

F.   Self-Reinforcement and Compaction

It is well known that strong orientation of the macromolecules increases stiffness and strength of thermoplastic polymers, e.g., PP and PE. To maintain strong molecular orientation, relaxation processes during cooling must be suppressed due to a special process of fast solidification [20] or the preparation only of thin films or fibers. Using thin films or fibers, bulk material with good toughness at a high level of stiffness and strength can be produced with special compaction techniques [21,22]. Such a procedure and the resulting effects are discussed in detail in Chapter 16.

G.   Phase Transformation

Following the concept of toughening of ceramic materials due to the phase transformation mechanism (high toughened ceramics), a theoretical estimation shows that a stress-induced transformation from one crystalline phase into another one could absorb energy (e.g., transition from β- to α-crystals in iPP) [23]. However, up to now, there has been no experimental confirmation of this mechanism in polymers.

H.   Thin Layer Yielding Mechanisms

In lamellae-forming styrene/butadiene star block copolymers, an unexpected large homogeneous plastic deformation of PS lamellae up to 200 to 300% was found instead of the usual brittle fracture following craze formation. This new toughening mechanism based on a sub-micromechanical or a nano-mechanical effect was called a *thin layer yielding mechanism* and is discussed in detail in Section IV.C.

Modified thin layer mechanisms with toughness enhancement were also found in lamellae-forming block copolymer blends as well as in multilayered systems and are described in Sections IV.C and IV.D.

IV.   RESULTS AND DISCUSSION

A.   Amorphous Polymers Modified with Core-Shell Particles

1.   Materials Used

Very fine and well-dispersed morphologies can be obtained by blending the matrix polymer with core-shell particles. The particles used contain a hard core of cross-linked PMMA with a diameter of about 180 nm, a rubbery shell consisting of poly(butyl acrylate-co-styrene) (PBA) about 40 nm thick, and an outer thin-grafted PMMA shell for good adhesion between particles and matrix [24]. The overall diameter of the particles is in the order of 260 nm (Figure 8, left). These preformed particles were melt compounded with

- A PMMA matrix with volume contents of core-shell particles from 4 to 35%
- A PSAN matrix with 20 vol% particles (a styrene-acrylonitrile-acrylate copolymer [ASA] with a morphology similar to ABS)

From injection molded plates, ultra-thin sections were prepared for morphology investigations and semi-thin sections for studying micromechanical properties.

2.   Toughening Mechanism in General

The characteristic deformation structures of core-shell particles are shown in the schematic drawing on the left-hand side and in an electron micrograph from a deformed thin section of ASA on the right-hand side of Figure 8. In the modifier particles, a cavitation mechanism begins in the PBA shell, with subsequent fibrillation of the rubber. Fibrils separated by microvoids around the cores are connected to the matrix and, therefore, the fibrillated shells with the internal PMMA

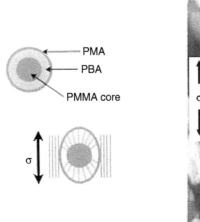

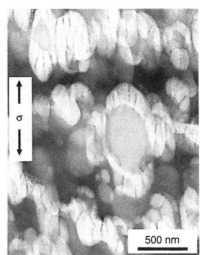

Figure 8 Left: Scheme of a core-shell particle undeformed and deformed. Right: TEM micrograph of a deformed sample of ASA (SAN/PBA blend) (arrow shows the deformation direction). (From Michler GH, Bucknall CB. Plastics, Rubber and Composites 2001;30:110–115. With permission.)

cores resemble spiders. The matrix material is plastically deformed between the cavitated and elongated particles. The matrix material (here SAN) deforms mainly in the form of homogeneous shear deformation zones and a small number of short and relatively thin fibrillated crazes.

Figure 9 shows deformation structures in a rubber-toughened PMMA. The lower magnification (micrograph in Figure 9a) shows multiple crazes between the particles. Crazes are initiated at the particles and propagate into the matrix perpendicular to the tensile direction. The higher magnification (Figure 9b) reveals also void formation, stretching and fibrillation in the rubbery shell. Cavitation in the particles produces stress concentration at and between particles, initiating plastic deformation of the adjacent matrix strands. With increasing particle volume content, a transition from crazing to shear yielding occurs [25].

 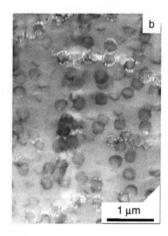

Figure 9 Craze formation in rubber-toughened PMMA at cavitated core-shell particles, particle volume content 7.5%: (a) lower magnification; (b) higher magnification (HVEM micrographs, deformation direction vertical).

3. Advantages of Core-Shell Particles

The rigid cores in the core-shell particles have an indirect effect upon toughness, because they increase the volume fraction of particles (similar to the salami particles in HIPS). The small size with a very narrow size distribution and the homogeneous spatial distribution in the matrix allows for the manufacture of transparent toughened polymers as a second advantage.

An additional advantage of this type of particle over homogeneous rubber particles lies in the stabilization of cavities by the core-shell structure: the cavities in the PBA shells are limited in their size by the fibrils connecting the matrix to the rigid PMMA cores, and these fibrils also clearly support substantial stresses. This useful difference between homogeneous rubber particles and core-shell particles is illustrated in Figure 10. In both cases, the single mechanism with cavitation inside the particles, stress concentration at the surface of the particles, and initiation of plastic yielding of adjacent matrix parts is very similar. However, if we consider superposition effects between the particles, differences become vis-

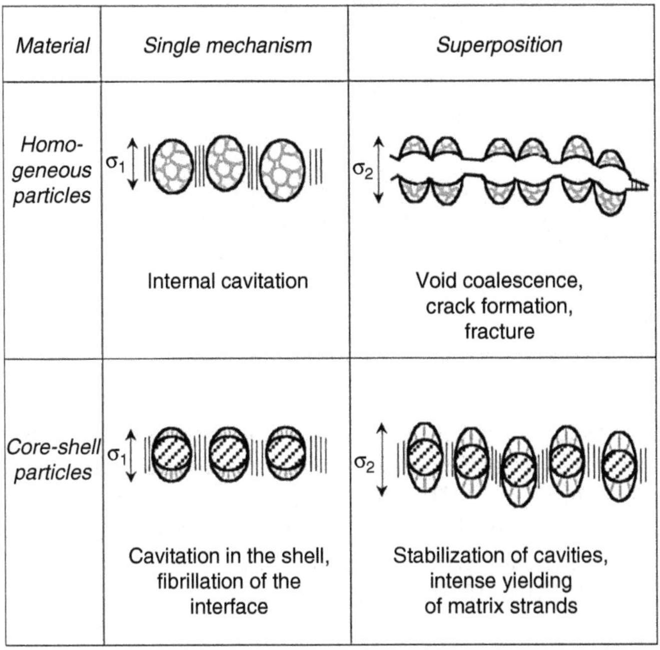

Material	Single mechanism	Superposition
Homo-geneous particles	Internal cavitation	Void coalescence, crack formation, fracture
Core-shell particles	Cavitation in the shell, fibrillation of the interface	Stabilization of cavities, intense yielding of matrix strands

Figure 10 Comparison of the action of role played by homogeneous particles in contrast to core-shell ones in rubber-toughened polymers concerning the single mechanism acting on particles and the superposition effect between particles. (From Michler GH. J Macromol Sci: Phys 1999;38:787–802. With permission.)

ible: If there are closely connected homogeneous rubber particles, the voids inside the particles can coalesce, leading to larger voids, crack formation, and, consequently, to a premature fracture. Using core-shell particles, the individual microvoids are stabilized in their size, and void coalescence is prevented, with the result that an additional intense plastic yielding of the matrix is possible.

Usually, the cores of the modifier particles are not deformed, as is visible in the micrographs of Figures 8 and 9. This is usually accepted since it hardly seems possible to deform a glassy polymer core inside a rubbery shell. However, there are results that show that a plastic yielding of hard PS sub-inclusions in salami particles of HIPS should be possible

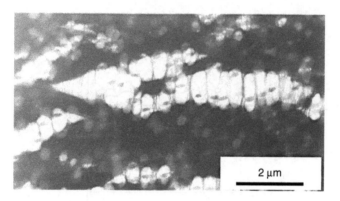

Figure 11 ASA blend with dilatational bands, which are formed by core-shell particles extended in deformation direction (vertical) and crazes between the particles; note cavitation and fibrillation of rubber phase and flattening of PMMA cores. (1 μm semi-thin micro-tomed section stretched to high strains at room temperature, 1000 kV HVEM). (From Michler GH, Bucknall CB. Plastics, Rubber and Composites 2001;30:110–115. With permission.)

[26] (see also Section III.D). One example of this behavior in an ASA blend containing acrylic core-shell particles is shown in Figure 11. The micrograph shows a pattern of light-colored dilatational bands formed as a result of crazing in the PSAN matrix and cavitation in the modifier particles. Most of these particles feature dark, lens-shaped PMMA cores. Both the fibrils and the minor axis of the "lenses" are oriented parallel to the tensile direction, indicating that as the deformation progresses to high tensile strains, PMMA is drawn from the "poles" of the cores, which were originally spherical [24,26].

This process of large strain plastic deformation of glassy cores of the modifier particles can be regarded as a new toughening mechanism (so-called *core flattening*), which is potentially capable of contributing significantly to energy absorption during the deformation and fracture of the blend [26,27]. Figure 12 is a schematic diagram, illustrating the morphology of a cavitated core-shell particle immediately before fibrils of glassy polymer are drawn from the polar regions of the core. That drawing process is almost identical

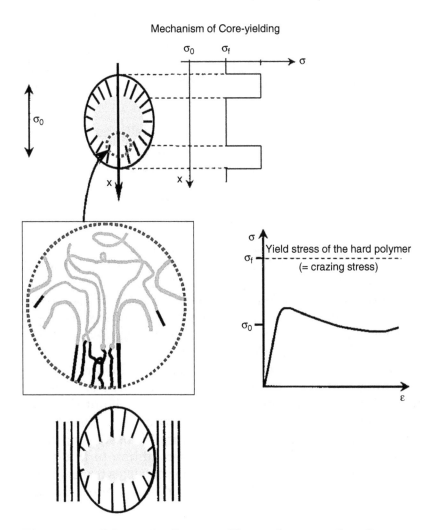

Figure 12 Schematic diagram illustrating transfer of stresses from matrix via rubber fibrils to rigid polymer core; true stresses in the rubbery shell are high enough to pull out fibrils from core, as in craze formation (σ_0 applied stress; σ_f higher stress in fibrils, comparable with yield stress (or crazing stress) of the hard polymer). (From Michler GH, Bucknall CB. Plastics, Rubber and Composites 2001;30:110–115. With permission.)

to the drawing of fibrils from the walls of a craze [2,28] and, therefore, requires substantial stresses, probably in the range of 50 to 80 MPa at room temperature, to be applied to the solid core. As these stresses must be in equilibrium with the surrounding matrix, the true stresses in the rubber fibrils must be in a corresponding size. This is quite realistic for two reasons:

- Due to the void content in the shell, the stresses in the fibrils are increased accordingly.
- Natural rubber in the bulk state can reach engineering stresses (load/original area) of 30 MPa and extension ratios of 10 or higher, which means that the true stress in the fully stretched state exceeds 300 MPa.

Therefore, the yield (crazing) stress of the hard polymer particles (PS inclusions in the salami particles of HIPS or PMMA particles in the core-shell particles of ASA) can be reached, and fibrils are pulled from the surface of the particles (there is no homogeneous yielding of the whole particle as discussed in the mechanism's *rubber network toughening* (Section III.C) and *inclusion yielding* (Section III.D)).

In view of the large stresses and strains involved in core flattening and the relatively high volume fractions of the glassy inclusions (occupying about 50 and 80% of the total content of the core-shell and salami particles) in typical thermoplastic blends, there is a possibility of dissipating large amounts of energy in this way. A precondition to this effect is that the rubbery shell of the modifier particles forms fibrils, which show a remarkable strain-hardening effect as discussed above.

B.   Low Temperature Toughness in Semicrystalline Thermoplastics

1.   Materials Used

Core-shell particles show advantageous effects also in toughening of semicrystalline polymers. To illustrate these effects, two different types of modified polypropylene (PP) were chosen. One was a commercial PP homopolymer modified by melt compounding with 20 vol% of ethylene/propylene/diene terpolymer (EPDM). The other was a "reactor blend," in which the PP is modified with rubber particles consisting of an ethylene/propylene rubber (EPR) shell surrounding a rigid semicrystalline PE core. The core-shell particles have average diameters of about 300 nm [29]. Both grades of PP were supplied by PCD Polymers, Linz (now Borealis).

2. Deformation Mechanism

The toughness of semicrystalline polymers such as polyamides (PA) and PP can be increased similar to the amorphous polymers by the adding of relatively small amounts of rubber particles such as EPDM or EPR. In these modified polymers, the main energy absorption mechanism under tensile loading at room temperature is shear deformation of the matrix (mechanism of *multiple shear deformation*) [3]. Usually, toughened PP shows a decrease in toughness with decreasing temperature. If the temperature is below the glass transition temperature of the modifier particles (e.g., EPDM), the particles can no longer act as rubbery stress concentrators, the initiation of plastic deformation is lost and the materials break in a brittle manner. However, a low temperature toughness is demanded for many practical applications.

The use of core-shell particles is favorable in comparison with homogeneous rubber particles in two respects. One is similar to the case referred to above (cf. Figure 10, effect of preventing void coalescence). The other is demonstrated in Figure 13 with a sample of a PP blend containing EPR core-shell particles: A thin section was strained *in situ* on the stage of a 200 kV-TEM at the very low temperature of −100°C. The EP copolymer shell is cavitated and elongated in the load direction with the formation of coarser fibrils (the EP shells became thinner and, therefore, appear bright in the micrograph). The cavities in the shell act as stress concentrators and initiate crazes in the adjacent matrix. It is known that PP also well below its glass transition temperature T_g can deform via crazes [2,27]. This mechanism

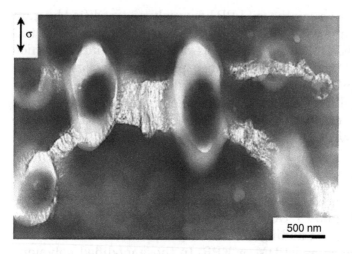

Figure 13 PP modified with EPR core-shell particles; thin section deformed at −100°C, showing cavitated and deformed core-shell rubber particles and crazes in PP matrix, revealing an effect of low temperature toughness (deformed and investigated in a 200 kV-TEM). (From Michler GH, Bucknall CB. Plastics, Rubber and Composites 2001;30:110–115. With permission.)

reminds one of the toughening mechanism in high-impact polystyrene (HIPS) or ABS. It is an effect far below the glass transition of the rubber phase, and it demonstrates that toughening is not limited by T_g of the modifier particles and that low-temperature toughness of PP and other semicrystalline polymers is not a problem of principle [3,26]. The essential structural precondition is cavitation of particles and stabilization of the formed cavities to avoid their growth into cracks, as it can be realized in the core-shell particles. There is a similarity between formation of microvoids and fibrillation of the rubber shell of the core-shell particles below the glass transition temperature and the processes of craze formation in polymers well below T_g (e.g., crazing in PS, PP, connected also with processes of microcavitation and fibrillation [2]). In both cases the polymeric material is deformed plastically below T_g on a sub-micrometer level (thicknesses of fibrils are in the range of 10 nm).

C.   Block Copolymers

Block copolymers allow the control of structure on the nano-meter scale and enable the combination of good mechanical properties with other properties such as transparency, recy-clability, etc. Due to the connectivity of the component chains and their inherent chemical incompatibility, the dissimilar blocks prefer to segregate, which gives rise to a rich variety of ordered structures called microphase-separated structures. In addition to the variation of the overall composition of block copolymers, the morphology can be adjusted by changing the processing conditions (processing temperature, cooling rate, etc. or solution casting using different solvents), molecular architecture, using more than two types of monomers, etc. (see Chapter 3).

1.   Materials Used and Sample Preparation

Lamellae-forming asymmetric styrene/butadiene block copol-ymers (number average molecular weight ~ 100,000 g/mol and total styrene volume content of 0.74) were investigated. A linear (LN) and a star-shaped (ST) copolymer were used. Unlike diblock copolymers, due to presence of asymmetric architecture and tapered transition, these block copolymers reveal lamellar morphology in spite of large polystyrene con-tent. Due to modified molecular architecture, the glass tran-sition temperature of the soft butadiene phase was significantly increased (–55°C for LN and –78°C for ST) with respect to that of pure polybutadiene (–98°C). The samples were prepared by injection molding (mass temperature 250° and mold temperature 45°C) and solution casting using tolu-ene as solvent. The materials were supplied by the BASF Aktiengesellschaft, Ludwigshafen. (For more details see [30,31] and Chapter 3.)

2.   Deformation Mechanism and Effect of Thin Layer Yielding

It is known that the mechanical and micromechanical prop-erties of block copolymer systems with lamellar morphology

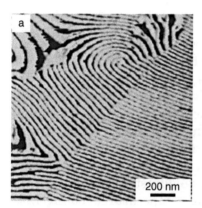 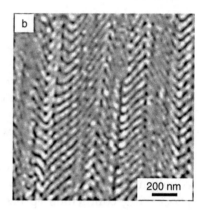

Figure 14 AFM phase images showing the morphology of a solution cast lamellar SBS triblock copolymer: (a) before deformation, and (b) after deformation in perpendicular direction to lamellae arrangement in vertical direction (cryo-microtomed surface imaged in tapping mode). (From Michler GH, Adhikari R, Henning S. J Mater Sci 2004;39:3281–3292. With permission.)

are strongly affected by the loading direction relative to the orientation of the microphase-separated structures.

If the material is loaded perpendicular or at an angle to the lamellar orientation direction, the lamellae are folded in a fish-bone-like arrangement (*chevron morphology*). The morphology of a lamellar SBS triblock copolymer (LN) film cast from solution is shown before and after deformation in AFM phase images, Figure 14. After deformation, the regions where the lamellae were initially perpendicular to the strain direction turn into so-called *chevron morphology* or fish-bone morphology (zig-zag pattern). Formation of *chevron morphology* under mechanical loading was recently reported in oriented lamellar block copolymer samples subjected to tensile deformation perpendicular to the lamellar alignment [32].

A different picture appears if the samples are loaded in such a way that the lamellae are parallel to the deformation direction. Figure 15 shows the morphology of an SBS star block copolymer (ST) before and after tensile deformation along the injection direction (or lamellar orientation direc-

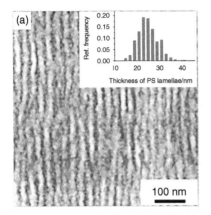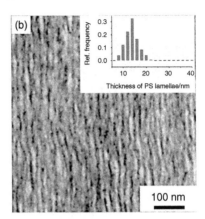

Figure 15 TEM images showing the morphology of an injection molded lamellar SBS triblock copolymer; insets show frequency distribution of thickness of PS lamellae: (a) before deformation, and (b) after deformation in parallel direction to lamellae arrangement (thin sections selectively stained by OsO_4). (From Michler GH, Adhikari R, Henning S. J Mater Sci 2004;39:3281–3292. With permission.)

tion). In the undeformed sample, the thickness of the PS lamellae and the lamellar spacing lies in the range of 20 nm and 42 nm, respectively (Figure 15a). The tensile deformation parallel to the injection direction (i.e., the lamellar orientation direction) leads to an extreme plastic drawing of both PS and PB lamellae (Figure 15b). In the deformed sample the thickness of the PS lamellae and the lamellar spacing have been reduced to about half of their values before deformation. It is worth mentioning that the lamellae were stretched to a very high degree without any cavitation or microvoid formation. In contrast to the diblock copolymers, where the deformation localization in the craze-like zones is the principal deformation mechanism [33,34], no local deformation bands were observed.

A summary of the deformation processes illustrated in Figure 14 and Figure 15 are collected in a TEM image of a solution cast star block copolymer (ST) in Figure 16. If the deformation direction is parallel to the local lamella orienta-

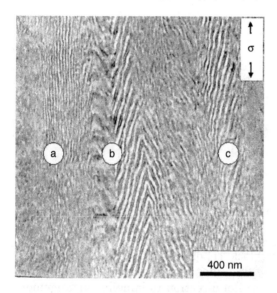

Figure 16 TEM micrograph showing different deformation structures (a, b, c) in a lamellar star block copolymer (ST) prepared by
solution casting (strain direction is shown by an arrow; OsO_4 stained
ultra-thin section). (From Michler GH, Adhikari R, Henning S. Macromol Symp 2004;214:47–71. With permission.)

tion (region **a**) or perpendicular or oblique to the local lamellar
orientation (regions **b** and **c** in Figure 16), different mechanisms may act simultaneously or consecutively. Firstly, a shift
of adjacent lamellae (gliding process) occurs. Thereafter, the
lamellae yield (region **a**) or break into smaller domains, rotating towards the loading direction, forming chevron-like or fir-
tree like morphologies (regions **b** and **c** in Figure 16; see also
scheme in Figure 17a).

It should be noted that the thickness of PS lamellae
(about 20 nm) remains practically unchanged in the chevron-
folded structures after deformation. However, the lamellar
long period is increased due to widening of the PB lamellae
in the folded "hinges." Moreover, the thickness reduction of
the PB layers during the parallel deformation was found to
be more pronounced. These observations indicate that the
rubbery phase reacts earlier towards the applied stress, which

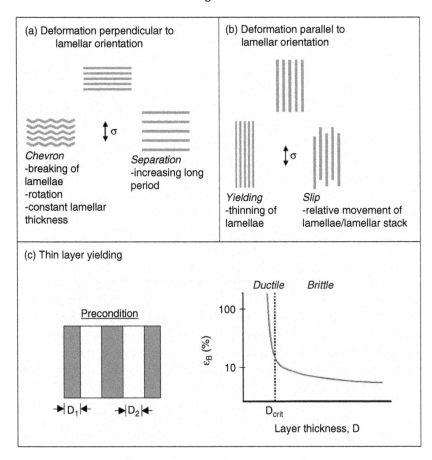

Figure 17 Scheme of micromechanical processes observed in lamellar systems investigated: (a) lamellae perpendicular to strain direction; (b) lamellae parallel to strain direction; and (c) scheme showing the principle of the thin layer yielding mechanism.

was further supported by Fourier transform infrared (FTIR) spectroscopy results [35].

If the deformation direction is parallel to the lamellar orientation, a homogeneous plastic yielding of both PS and PB lamellae is detectable (region **a** in Figure 16 and scheme in Figure 17b). Apparently there is no change in the phase morphology besides a significant reduction of the layer thicknesses and the long period. The lamellae are strongly aligned towards the strain direction, and the well-defined lamellar structure is even partly destroyed. The large plastic deformation of the glassy PS lamellae under tensile strain found in the star block copolymer studied is in line with the results reported earlier by Kawai et al., who observed a cooperative, drawing, shearing and kinking of the block copolymer micro-domains [36]. This homogeneous plastic deformation even without forming any localized deformation zones endows the PS with a ductile property, whereas it is otherwise brittle in the bulk state. This mechanism of homogeneous plastic deformation of PS lamellae (thin PS layers) together with adjacent PB lamellae can be described by a new deformation mechanism called *thin layer yielding* [37], which is schematically represented in Figure 17c. This new toughening mechanism based on an effect on nanometer scale yields high impact and transparent polymers. The effect appears if the thickness of the PS layers lies below a critical thickness (D_{crit}). This critical thickness is comparable to the maximum craze fibril thickness in polystyrene homopolymer, i.e., in the range of 20 nm. The difference between the craze fibril yielding and the yielding of the PS lamellae lies in the fact that the craze fibrils stretch between microvoids while the PS lamellae undergo yielding between PB lamellae. Thus, in this case, the PB lamellae act similar to microvoids in PS crazes and do not hinder the plastic deformation of the glassy polystyrene layers.

From the reduction of the lamellar spacing and the thickness of the PS lamellae in the undeformed and deformed samples, a local deformation of about 300% can be estimated. This is the same order of magnitude as the craze fibrils stretching in the crazes of PS (craze fibril extension ratio, $\lambda \approx 4$ [2]). This confirms that the yielding of the lamellae in the star block copolymer is analogous to the drawing of craze fibrils in polystyrene homopolymer.

The large homogeneous plastic deformation of the PS lamellae at room temperature under tensile loading conditions is limited to lamella thickness below a critical value of about 20 nm. The deformation mechanism changes from this homogeneous drawing of the lamellae to the formation of the usual local craze-like zones, if the average thickness of the PS lamellae increases to about 30 nm (see Section IV.D). Additional micromechanical details in block copolymers with lamellar morphologies are discussed in Reference [38].

3.   Influence of Processing Conditions and
        Deformation Temperature

Variation in processing conditions can have a dramatic impact on the morphology and mechanical behavior of block copolymers. Figure 18 plots the stress-strain curves of the linear block copolymer (LN) processed by three different methods: solution casting, compression molding and injection molding. The compression molded and injected samples met the requirements of standard test (ISO 527) while miniaturized tensile bars (50 mm long, 0.5 mm thick) were prepared from solution cast films. Therefore, the comparison of properties is rather qualitative.

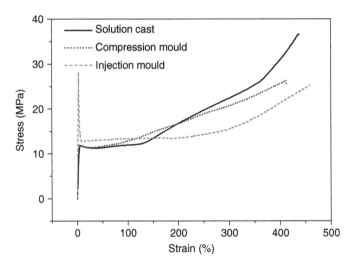

Figure 18 Stress-strain curves of the lamellar block copolymer (LN) samples prepared by different methods; injection molded samples were strained parallel to the injection direction. (From Adhikari R, Godehardt R, Lebek W, Goerlitz S, Michler GH, Knoll K. Macromol Symp 2004;214:173–196. With permission.)

What one can immediately notice in Figure 18 is the significantly high yield stress (ca. 29 MPa) of the injection molded samples. The yield points of solution cast films and compression molded lie nearly at the same low level (12 MPa). The high yield stress of the injection molded bars can be attributed to the orientation of lamellae along the flow direction (loading is parallel to the injection direction, i.e., parallel to the lamellae orientation direction). The yield point defines the start of the yielding of the PS lamellae as in Figure 15 (thin layer yielding mechanism). Loading is parallel to the lamellar orientation, and, therefore, load is covered by the hard (PS) lamellae. Due to comparable thickness of soft and hard lamellae, the stress covered by the PS lamellae is twice that of the applied stress, that means $2 \times \sigma_y \sim 2 \times 30 = 60$ MPa and this correlates roughly with the yield stress of bulk PS [2].

It was shown that the solution cast and compression molded samples reveal "polygranular" structures, which deform by cooperative shearing, breaking and yielding of lamellae (as in Figure 16). The first step of shearing and interlamellar gliding appears by chain movements in the soft (PB) lamellae at low stresses. After the second step of breaking, twisting and chevron formation of the lamellae, the third step of lamellae yielding starts with an increasing stress.

Another noticeable difference is in the tensile strength (maximum stress at break). The tensile strength decreases in the following order: solution cast film → compression mold → injection mold. It is well known that the tensile strength in styrene/diene block copolymers is a direct function of extent of phase separation [39]. Thus, decreasing tendency of tensile strength in the sequence noted above can be correlated to the decreasing degree of phase separation. The structures in solution cast films are closest to the equilibrium ones (clearly separated lamellae, which ensure the largest tensile stress) while the injection mold has the smallest chance of forming the well phase-separated morphology owing to the rapid cooling of highly sheared melt. Details on the influence of processing conditions on deformation mechanisms of styrene/butadiene block copolymers can be found elsewhere [40].

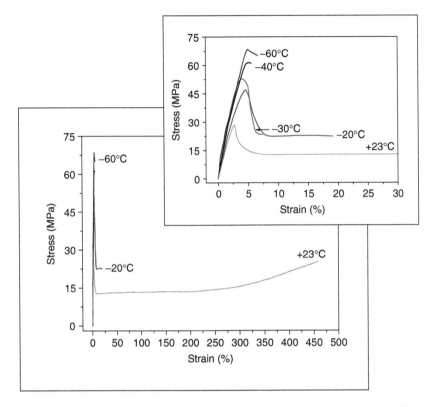

Figure 19 Stress-strain curves of the injection molded lamellar block copolymer (LN), loaded at different temperatures as indicated; strain rate 50 mm/min; initial part of the curves are magnified in the inset. (From Adhikari R, Godehardt R, Lebek W, Goerlitz S, Michler GH, Knoll K. Macromol Symp 2004;214:173–196. With permission.)

Representative stress-strain curves of the linear block copolymer (LN) measured at different temperatures are plotted in Figure 19. Deformation behavior was examined from room temperature down to the temperature close to the glass transition temperature of the soft phase, $T_{g\text{-PB}}$ (details in [40]). With decreasing temperature, as expected, the strain at break decreases while the yield stress (a measure of strength) and Young's modulus (a measure of stiffness) increases. Above −40°C, a ductile failure occurs, which is accompanied by necking of the tensile specimens. A drastic reduction in strain at break takes place on lowering the temperatures (e.g., from about 450% at +23°C to about 20% at –30°C). At temperatures below –40°C the block copolymer shows a brittle behavior.

4.   Comparison with Other Lamellae-Forming Polymers

The typical lamellae-forming polymers are the semicrystalline polymers with amorphous interlamellar layers. Therefore, it is interesting to compare the micromechanical behavior of the block copolymers discussed here with semicrystalline polymers (e.g., PE, iPP — Chapter 17). In PE and iPP, the crystalline lamellae have thicknesses of about 20 nm and long spacings in the range of 40 nm. This is the same size of order of the lamellar dimension in block copolymers. We can define a morphology of hard (crystalline phase, PS) lamellae and soft (amorphous phase, PB) lamellae in between. The comparable lamellar arrangement and nanostructures in these two entirely different classes of materials (PP and PE versus block copolymers), result in very similar micromechanisms [41,42].

Depending on the loading direction relative to the lamellar alignment, different micromechanisms exist as shown schematically in Figure 17(a) and (b). When loaded perpendicular to the lamellar orientation direction, similar deformation structures appear with lamellae separation, breaking and twisting of hard lamellae yielding to the formation of chevron structures. In HDPE and also in β-iPP, cavitation and fibrillation in the amorphous phase occur at larger deformation (Figure 17(a)).

When loaded in parallel direction to the lamellae orientation in the semicrystalline polymers, an interlamellar gliding mechanism appears with plastic yielding of the amorphous layers between the lamellae and separation of the lamellae into smaller crystalline fragments with chain unfolding finally leading to the formation of microfibrils at very high deformation. In lamellar SBS block copolymers, interlamellar gliding is followed by plastic yielding of the butadiene and styrene lamellae corresponding to the *thin layer yielding* mechanism (Figure 17(b)).

The high plastic deformation of the soft (amorphous or rubbery) regions is the precondition of these micromechanisms leading to a high ductility of the sample. The high molecular mobility of the soft layers enhances the local interlamellar shearing (if the lamellae are parallel to the loading direction) and lamellar separation and chevron formation (if the load acts perpendicular to the lamellar orientation direction). Since the amorphous regions in the semicrystalline polymers contain a number of defects (such as chain ends), they are more prone to cavitation leading to fibrillated crazes in the amorphous phase. The absence of microvoids in the soft phase of the SBS block copolymers is a most striking difference. The butadiene phase, covalently bonded to the styrene chains (no molecular defects), allows cavitation only after chain scission.

The micromechanical behavior of lamellae-forming heterophase polymers (based on semicrystalline polymers and amorphous block copolymers) is dictated by the nature and arrangement of these structures and shows two basic stages. The initial stage is characterized by a plastic deformation of the soft phase (rubbery or amorphous) with a reorganization of the hard (glassy or crystalline) lamellae. The second stage is determined by the alignment of the hard phase towards the deformation direction and their plastic yielding. In this way, the knowledge gained using a set of polymers may be used to understand the deformation mechanisms of another set of polymers with comparable phase morphology.

D. Block Copolymer–PS blends

1. Materials Used

For the detailed analysis of deformation behavior of the lamellar systems with increasing thickness of the lamellae, blends of a lamellar star block copolymer (ST) and polystyrene homopolymer (hPS; number average molecular weight ca. 80,000 g/mol) having wide molecular weight distribution were studied by preparing the samples by injection molding (details on materials used in Reference [43]).

2.   Deformation Mechanisms

With increasing molecular weight of the added PS (M_{PS}), the
compatibility with the corresponding PS block of the block
copolymer decreases (Chapter 3). If M_{PS} is higher than that
of the corresponding block, then the macrophase-separation
predominates under equilibrium conditions. However, the
morphology of block copolymer/PS blends can be strongly
influenced by processing conditions. Due to the high speed
processing and rapid cooling conditions during the molding
process, there is not enough time for the components to form
the structures that may be expected under equilibrium conditions.

Figure 20 (top) presents TEM micrographs of some of
the injection molded blends. Since the thickness of the structures doesn't exceed 100 nm, the blends may be regarded as
being microphase separated. This is the reason why the
blends were nearly as transparent as the pure block copolymer itself. The microphase-separated structures are oriented
in the direction of shear force (i.e., along the injection direction) in the pure block copolymer as well as in the blends.
The thickness of the PS lamellae (or PS strands) increases
continuously with increasing hPS concentration in the blends,
whereas, the thickness of the PB lamellae doesn't change
significantly. This suggests that the hPS chains are accommodated at the PS block lamellae of the pure star copolymer.
At higher PS content, the PB lamellae appear as elongated
worms dispersed in the PS matrix. Details on the morphology
and micromechanical behavior of these blends may be found
in [43].

The deformation tests of the investigated blends show a
drastic reduction in elongation at break at hPS concentration
of 20 wt% and more (see Figure 20 at the bottom). It was
found that this change is associated with average PS lamella
thickness and a transition in deformation mechanism from
homogeneous plastic flow to the formation of craze-like deformation zones. The transition in deformation mechanism from
ductile plastic yielding of PS lamellae (thin layer yielding
mechanism) to crazing was found to take place when the

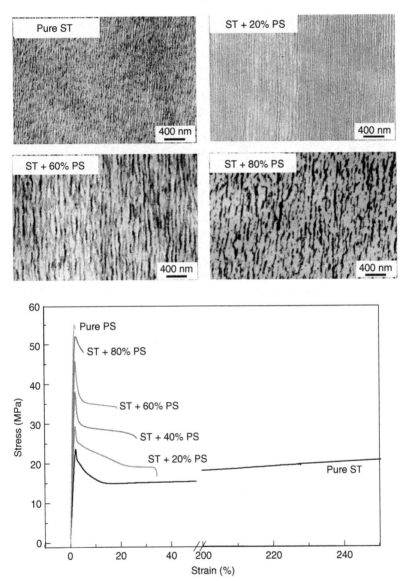

Figure 20 Top: representative TEM images showing the morphology of injection molded samples of star block copolymer/hPS blends; injection direction vertical; the blend composition is indicated. Bottom: stress-strain curves of these blends determined by tensile testing at 23°C. (Reprinted from Baltá Calleja et al. Polymer 2004;45(1):247–254. Copyright 2004, with permission from Elsevier.)

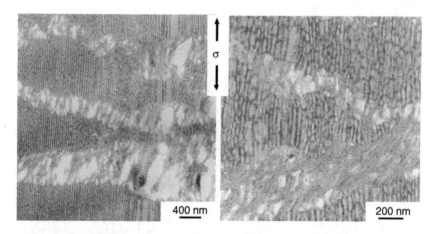

Figure 21 Lower (left) and higher (right) magnification of TEM micrographs showing the deformation structures in an injection molded blend of 80 wt% block copolymer and 20 wt% hPS, deformation direction is shown by an arrow (samples are stained by OsO_4).

average thickness of the PS lamellae reaches a value of ca. 30 nm. These results are strong evidence of the proposed *thin layer yielding* mechanism.

The craze-like deformation zones in Figure 21 are characterized by a high degree of stretching of the lamellae accompanied by void formation. Locally, an elongation ratio of $\lambda >$ 4 may be estimated. However, the macroscopically achievable elongation of this sample is relatively low (only about 34%) due to the extreme localization of the deformation zones. With increasing PS content (and, therefore, increasing PS lamellae thickness) the number of crazes decreases, yielding to the observed reduction in elongation at break.

A blend with a higher PS content is shown in a TEM image in Figure 22 and illustrates how the thickness of the glassy layers affects the deformation micromechanism. The sample, a blend of 20 wt% of a linear block copolymer and 80 wt% of hPS (i.e., a total PS content of $\approx$ 95%), was prepared by extrusion with PS layers aligned along the flow direction. Here, the PS layer thickness (average thickness ~350 nm) clearly exceeds the critical value for the "thin layer yielding"

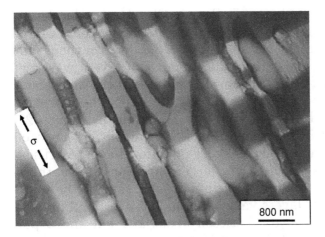

Figure 22 TEM image of a blend consisting of an SBS triblock copolymer and standard polystyrene (20/80 by weight) after deformation; the crazes are exclusively localized in the PS layers; deformation direction is shown (sample prepared by extrusion, staining by OsO_4). (From Michler GH, Adhikari R, Henning S. Macromol Symp 2004;214:47–71. With permission.)

mechanism with the result of the only appearance of usual fibrillated crazes inside.

E. Multilayered Systems

1. Materials and Sample Preparation

During the last decades the study of mechanical properties of polymers in the form of very thin films and layers became a field of increasing interest in polymer research. Micro- and nanolayering is an attractive approach for modifying the properties of the component polymers [44–47].

Here, two macroscopically ductile polymers, PC and PET, were used to prepare a new kind of multilayered tapes. PC (polycarbonate) from the Dow Chemical Company, Midland, USA, and PET (polyethylene terephthalate) from M&G Cleartuff had both molecular weights of about 30,000 g/mol. Microlayers of PET/PC having compositions of 20/80, 50/50, and 80/20, respectively, were coextruded with the two-component microlayer system of E. Baer in Cleveland — for details see Reference [48] and Chapter 15. Both components can be coextruded as uniform laminates with hundreds or thousands of alternating layers. Here, the number of layers was kept constant at 1024. It is assumed that besides the amorphous PC also PET is in an amorphous state under the extrusion conditions used. The morphology is visible in Figure 23, revealing that in all compositions the microlayers are well ordered and oriented in parallel direction to the extrusion direction. However, there are larger variations in the thicknesses between different parts of the same sample: the thicknesses of the PC layers vary between 100 nm and 2 µm and that of the PET layers between about 100 nm and 3 µm. Additional details of the morphology depending on composition and number of layers are described in References [49,50] and systems discussed in Chapter 15.

2.   Deformation Mechanisms

Figure 24 presents HVEM micrographs of the PET/PC samples with compositions 20/80, 50/50, and 80/20, respectively. The visibility of the different layers in these lower magnifications is not so good as in Figure 23. Increase in the PET layer thicknesses results in a pronounced transformation in the deformation structure. While 20/80 PET/PC shows a coexistence of deformation zones and shear bands similar to the behavior of pure PC, the opposite composition 80/20 PET/PC deforms only with deformation zones like amorphous PET [49]. The sample with 50/50 shows deformation zones and shear bands across numerous PET and PC layers without stopping at the interfaces. That means that the deformation mechanism of the major component determines the behavior of the composite. Such cooperative microdeformation mechanisms are only possible if the interfacial strength (adhesion) between the two components is high enough to ensure stress transfer. The good adhesion is also proved by a very seldom observed delamination during deformation studies of the samples.

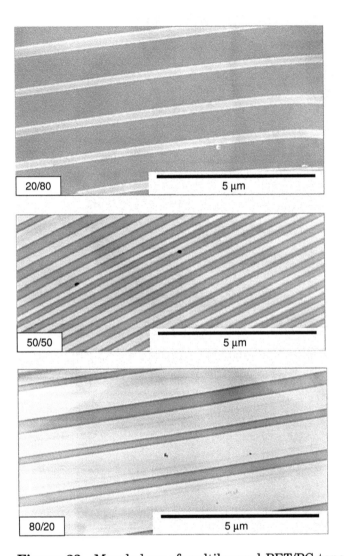

Figure 23 Morphology of multilayered PET/PC tapes having different composition (1 μm thick sections, stained with RuO$_4$ [PC appears dark], 1000 kV HVEM). (From Ivankova EM, Michler GH, Hiltner A, Baer E. Macromol Mater Eng 2004;289:787–792. With permission.)

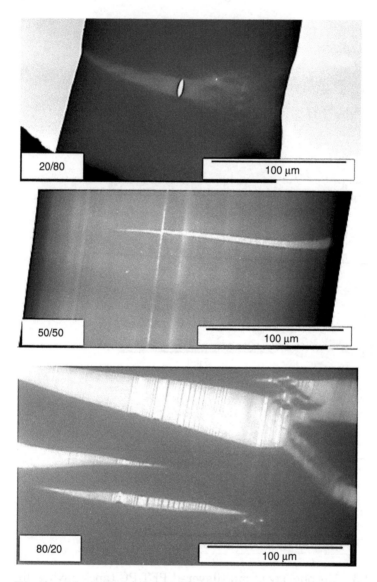

Figure 24 Deformation-induced structures in PET/PC multilayered tapes of different compositions (deformed and investigated in a
HVEM; deformation direction vertical). (From Ivankova EM, Michler
GH, Hiltner A, Baer E. Macromol Mater Eng 2004;289:787–792. With
permission.)

Annealing of the samples before mechanical stretching causes significant changes in the deformation behavior. Figure 25 reveals transitions in the different samples. In the 20/80 PET/PC sample deformation zones dominate instead of the coexistence of deformation zones and shear bands in the unannealed sample (such as in pure PC due to annealing [51]). In the annealed 80/20 PET/PC sample shear bands dominate instead of the only existence of deformation bands in the sample without annealing (in accordance with the transition in pure PET due to annealing). Therefore, a transition appears from deformation zones in 20/80 PET/PC to an intense shear band formation with increasing PET [43]. In a similar manner as in the previous case (before annealing, Figure 24), the micromechanical deformation mechanism of the major component determines the mechanism of the whole composite. The effect of annealing in the 80/20 PET/PC sample with transition from deformation bands to shear bands should be connected with an increase of the degree of crystallinity and of toughness of the PET. However, TEM inspection of the annealed PET layers didn't reveal clear lamellae. It seems that the crystallization process in very thin PET layers is constrained. This correlates with SAXS and WAXS synchrotron irradiation measurements, which revealed a crucial reduction of the degree of crystallinity of the PET phase, if the layer thicknesses were below 10 µm [52]. This effect of hindered crystallization in confined systems could act as an additional effect to modify mechanical properties of multi- and nanolayered polymer systems.

Another example of multilayered systems are coextruded PS/PMMA multilayers, which also reveal an increasing toughness with increasing number of layers, corresponding with decreasing layer thicknesses [53].

V. CONCLUSIONS/OUTLOOK

There are several different mechanisms of toughness enhancement of polymers. The usual *rubber toughening* of amorphous and semicrystalline polymers is of major impor-

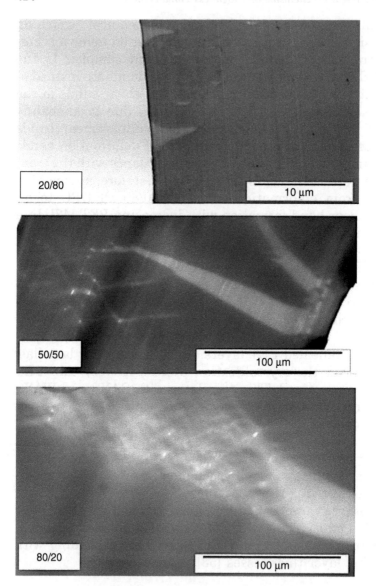

Figure 25 Deformation induced structures in PET/PC multilayered tapes of different compositions after annealing (deformed and investigated in a HVEM). (From Ivankova EM, Michler GH, Hiltner A, Baer E. Macromol Mater Eng 2004;289:787–792. With permission.)

tance to the plastics industry. The basic effects are stress concentration at the particles and initiation of plastic yielding at and between the particles. In the past, enhancement of the toughening effect appeared mainly via optimization of the rubber particle size or the inter-particle distance and less via an improvement of the yielding behavior of the matrix itself. However, the results of micromechanical behavior of semi-crystalline polymers (HDPE, iPP) with that of lamellae forming amorphous SBS block copolymers reveal several similarities, which could be used to improve the yielding ability particularly of the semicrystalline matrix strands between the modifier particles. Aspects of interest in this sense are:

1. The softer interlamellar phase is decisive for the starting processes of deformation in the form of twisting, reorganization and chevron formation of the crystalline lamellae (if deformation direction is perpendicular to the lamellar arrangement) or for the process of interlamellar yielding (if deformation direction is parallel to the lamellae orientation).
2. A controlled cavitation inside the amorphous interlamellar phase can improve the lamellar orientation processes.
3. Intense plastic yielding of the crystalline lamellae themselves are assisted by breaking of lamellae into shorter crystalline blocks and transformation into microfibrils.

However, there are some other mechanisms which can be used to enhance the toughness. In the mechanisms of *rubber network toughening* and *inclusion yielding* hard, semi-ductile polymer particles (about and below 1 μm in diameter) are initiated to plastic deformation (in the form of a homogeneous yielding). If we are going to a sub-micrometer level, in the mechanism of *core flattening*, hard, brittle polymer particles such as from PS and PMMA also can be plastically deformed via a surface drawing or craze-like process inside core-shell particles.

In multilayered or multilamellar polymer systems, a modification of the deformation processes, which are known from the corresponding bulk materials, occurs in the form of

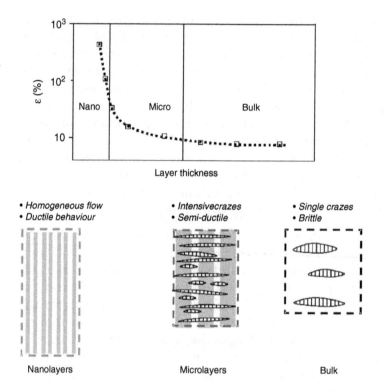

Figure 26 Schematics of changes in the deformation mechanism showing the transition from bulk material to multi- and nano-layered systems and the corresponding increase in elongation at break or toughness.

multiple formation of crazes, deformation or shear bands and contributes to an enhanced toughness. If the lamellae or layers fall below critical thicknesses in the range of a few 10 nm, a new yield mechanism arises, the *thin layer yielding*, that means that usually brittle polymeric materials can be plastically deformed up to several 100%. In Figure 26 it is schematically shown that with transition from bulk polymers to multilayered and nanolayered systems, changes in the micro- and nanomechanical mechanisms arise with a transition (e.g., for PS) in the formation of single crazes to numerous crazes and finally to the thin-layer yielding effect and with a corresponding increase in elongation at break or toughness. This new nanomechanical mechanism has been studied in detail in SBS block copolymers and blends with PS, but it should be valid also for other nanostructured brittle polymers. The significance of this mechanism from a practical point of view may be found in the fact that it can be used as an alternative toughening mechanism for transparent, brittle amorphous and semicrystalline polymers.

ACKNOWLEDGMENTS

The author thanks Borealis AG, Linz for supplying the PP samples, Röhm GmbH, Darmstadt for PMMA materials, and BASF Aktiengesellschaft for the block copolymers. I am indebted to Dr. R. Adhikari, Dr. E. Ivankova, Dr. G.-M. Kim, and Mrs. S. Goerlitz for EM and AFM investigations and the Deutsche Forschungsgemeinschaft (DFG) and the Bundesministerium für Bildung und Forschung (BMBF) for financial support.

REFERENCES

1. Bucknall CB. Toughened Plastics. London: Applied Science Publishers, 1977.

2. Michler GH. Kunststoff-Mikromechanik: Morphologie, Deformations und Bruchmechanismen. München: Carl Hanser Verlag, 1992.

3. Michler GH. Micromechanics of polymers. J Macromol Sci: Phys 1999;38:787–802.

4. Michler GH. Microstructural construction of polymers with improved mechanical properties. Polym Adv Technol 1998;9:812–822.

5. Michler GH. Electron microscopic techniques for direct investigation of micromechanical mechanisms in polymers. Trends Polym Sci 1995;3:124–131.

6. Michler GH. Electron microscopy in polymer science. Appl Spectroscopy 1993;28:327–384.

7. Michler GH, Lebek W, eds. Ultramikrotomie in der Material-forschung. München: Hanser-Verlag, 2004.

8. Michler GH, Starke JU. Investigation of micromechanical and failure mechanisms of toughened thermoplastics using electron microscopy. In: Riew CK, Kinloch AJ, eds. Toughened Plastics. Washington DC: American Chemical Society, 1996: 251–277.

9. Bucknall CB. Rubber toughening. In: Haward RN, Young RJ, eds. The Physics of Glassy Polymers, 2nd Ed. London: Chapman and Hall, 1997:363–412.

10. Bucknall CB. Deformation mechanisms in rubber toughened polymers. In: Paul DR, Bucknall CB, eds. Polymer Blends, Vol. 2. New York: Wiley and Sons, 2000: 83–117.

11. Michler GH, Gruber K. Beiträge zum Bruchverhalten von schlagzähen Zweiphasenpolymeren. Plaste u. Kautschuk 1976;23:346–351.

12. Michler GH. Determination of the morphology and mechanical microprocesses in polymer combinations by electron microscopy. Polymer 1986;27:323–328.

13. Liu ZH, Zhang XD, Zhu XG, Li KY, Wang ZN, Choy CL. Effect of morphology on the brittle ductile transition of polymer blends: 4. Influence of the rubber particle spatial distribution in poly(vinyl chloride)-nitrile rubber blends. Polymer 1998;39:5035–5046.

14. Borsig E, Fiedlerova A, Michler GH. Structure and properties of an interpenetrating polymer network-like system consisting of polystyrene-polyethylene. Polymer 1996;37:3959–3963.

15. Kolařik J, Lednicky F, Locati G, Fambri L. Ultimate properties of polycarbonate blends: effects of inclusion plastic deformation and of matrix phase continuity. Polym Eng Sci 1997;37:128–137.

16. Kelnar I, Stephan M, Jakisch L, Fortelny I. Ternary reactive blends of nylon-6 matrix with dispersed rigid brittle polymer and elastomer. J Appl Polym Sci 1999;74:1404–1411.

17. Bartczak Z, Argon AS, Cohen RE, Weinberg M. Toughening mechanism in semicrystalline polymer blends II. High density polyethylene toughened with calcium carbonate filler particles. Polymer 1999;40:2347–2365.

18. Kim G-M, Michler GH. Micromechanical deformation processes in toughened and particle filled semicrystalline polymers. Polymer 1998;39:5689–5697; Polymer 1998;39:5699–5703.

19. Lazzeri A, Thio YS, Cohen RE. Volume strain measurements on $CaCO_3$/polypropylene particulate composites: the effect of particle size. J Appl Polym Sci 2004;91:925–935.

20. Ehrenstein GW, Martin Cl. Zum Spritzgießen von eigenverstärktem Polyethylen. Kunststoffe 1985;75:105–110.

21. Jordan ND, Bassett DC, Olley RH, Hine PJ, Ward IM. The hot compaction behaviour of woven oriented polypropylene fibres and tapes. II. Morphology of cloths before and after compaction. Polymer 2003;44:1133–1144.

22. Bjekovic R. Monocomposite Schichtwerkstoffe auf Basis von Polypropylen. Düsseldorf: VDI Verlag, 2003.

23. Karger-Kocsis J. How does phase transformation toughening work in semicrystalline polymers? Polymer Eng Sci 1996;36:203–210.

24. Starke J-U, Godehardt R, Michler GH, Bucknall CB. Mechanism of cavitation over a range of temperatures in rubber-toughened PSAN modified with three-stage core-shell particles. J Mater Sci 1997;32:1855–1860.

25. Laatsch J, Kim G-M, Michler GH, Arndt T, Süfke T. Investigation of the micromechanical deformation behaviour of transparent toughened poly(methyl-methacrylate) modified with core-shell particles. Polym Adv Technol 1998;9:716–720.

26. Michler GH, Bucknall CB. New toughening mechanisms in rubber modified polymers. Plastics, Rubber and Composites 2001;30:110–115.

27. Michler GH. *In situ* characterization of deformation processes in polymers. J Macromol Sci – Phy 2001;40:277–296.

28. Kausch HH, ed. Crazing in Polymers II. Berlin: Springer-Verlag, 1990.

29. Kim G-M, Michler GH, Gahleitner M, Fiebig J. Relationship between morphology and micromechanical toughening mechanisms in modified polypropylenes. J Appl Polym Sci 1996;60:1391–1403.

30. Knoll K, Nießner N. Styrolux and Styroflex — from transparent high impact polystyrene to new thermoplastic elastomers. Macromol Symp 1998;132:231–243.

31. Adhikari R, Godehardt R, Lebek W, Weidisch R, Michler GH, Knoll K. Correlation between morphology and mechanical properties of styrene/butadiene block copolymers: a scanning force microscopy study. J Macromol Sci: Phys 2001;40:833–847.

32. Cohen Y, Albalak RJ, Dair BJ, Capel MS, Thomas EL. Deformation of oriented lamellar block copolymer films. Macromolecules 2000;33:6502–6516.

33. Schwier CE, Argon AS, Cohen RE. Crazing in polystyrene-polybutadiene diblock copolymer containing cylindrical polybutadiene domains. Polymer 1985;26:1985–1993.

34. Weidisch R, Michler GH. Correlation between phase behaviour, mechanical properties and deformation mechanisms in weakly segregated block copolymers. In: Baltá Calleja FJ, Roslaniec Z, eds. Block Copolymers. New York: Marcel Dekker, 2000:215–249.

35. Huy TA, Adhikari R, Michler GH. Deformation behaviour of styrene-block-butadiene-block-styrene triblock copolymers having different morphologies. Polymer 2003;44:1247–1257.

36. Fujimora M, Hashimoto T, Kawai H. Structural change accompanied by plastic-to-rubber transition of SBS block copolymers. Rubber Chem Technol 1978;51:215–224.

37. Michler GH, Adhikari R, Lebek W, Goerlitz S, Weidisch R, Knoll K. Morphology and micromechanical deformation behaviour of styrene/butadiene block copolymers: I. Toughening mechanism in asymmetric star block copolymers. J Appl Polym Sci 2002;85:683–700.

38. Adhikari R, Michler GH, Goerlitz S, Knoll K. Morphology and micromechanical behavior of SB-block copolymers: III. Star block copolymer/PS-homopolymer blends. J Appl Polym Sci 2004, in press.

39. Holden G. Understanding Thermoplastic Elastomers. Munich: Carl Hanser Verlag, 2000.

40. Adhikari R, Godehardt R, Lebek W, Goerlitz S, Michler GH, Knoll K. Morphology and micromechanical behaviour of SBS block copolymer systems. Macromol Symp 2004;214:173–196.

41. Michler GH, Adhikari R, Henning S. Toughness enhancement of nanostructured amorphous and semicrystalline polymers. Macromol Symp 2004;214:47–71.

42. Michler GH, Adhikari R, Henning S. Micromechanical properties in lamellar heterophase polymer systems. J Mater Sci 2004;39:3281–3292.

43. Ivankova E, Adhikari R, Michler GH, Weidisch R, Knoll K. Investigation of micromechanical deformation behaviour of styrene-butadiene star block copolymer/polystyrene blends using high voltage electron microscopy. J Polym Sci Polym Phys 2003;41:1157–1167.

44. van der Sanden MCM, Buijs LGC, de Bie FO, Meijer HEH. Deformation and toughness of polymeric systems: 5. A critical examination of multilayered structures. Polymer 1993;35:2783–2792.

45. Ebeling T, Hiltner A, Baer E. Delamination failure mechanisms in microlayers of polycarbonate and poly(styrene-co-acrylonitrile). J Appl Polym Sci 1998;68:793–806.

46. Kerns J, Hsieh A, Hiltner A, Baer E. Mechanical behavior of polymer microlayers. Macromol Symp 1999;147:15–25.

47. Lin CH, Yang ACM. Super-plastic behavior of the brittle polymer film in multilayer systems. J Mater Sci 2000;35:4231–4242.

48. Mueller C, Kerns J, Ebeling T, Nazarenko S, Hiltner A, Baer E. Microlayer coextrusion: processing and applications. In: Coates PD, ed. Polymer Process Engineering 97. Cambridge: University Press, 1997:137–157.

49. Ivankova EM, Michler GH, Hiltner A, Baer E. Micromechanical processes in PET/PC multilayered tapes: high voltage electron microscopy investigations. Macromol Mater Eng 2004;289:787–792.

50. Adhikari R, Lebek W, Godehardt R, Henning S, Michler GH, Baer E, Hiltner A. Investigating morphology and deformation behaviour of multilayered PC/PET composites. Polym Adv Technol 2005, in press.

51. Starke J-U, Schulze G, Michler GH. Craze formation in amorphous polymers in relation to the flow and main transition. Acta Polymerica 1997;48:92–99.

52. Puente I, Ania F, Baltá Calleja FJ, Funari SS, Baer E, Hiltner A, Bernal T. Confined crystallization in PET/PC micro- and nanolayers: influence of layer thickness. Scientific report to HASYLAB synchrotron irradiation laboratory. Hamburg: DESY, 2002.

53. Ivankova EM, Krumova M, Michler GH, Koets PP. Morphology and toughness behaviour of coextruded PS/PMMA multilayers. Colloid Polym Sci 2004;282:203–208.

Part III

Mechanical Properties Improvement
and Fracture Behavior

11

Structure-Property Relationship in Rubber Modified Amorphous Thermoplastic Polymers

W. HECKMANN, G.E. MCKEE, and F. RAMSTEINER

BASF Aktiengesellschaft, Polymer Research
Laboratory, Ludwigshafen, Germany

CONTENTS

I. INTRODUCTION

The main aim of the rubber modification of thermoplastic homopolymers is to improve their toughness. There are various methods to increase the toughness of amorphous homopolymers, e.g., by copolymerization and by the incorporation of a second phase like other thermoplastics, inorganic materials, very small voids and spherical rubber particles.[1] The last method is the most widely used, just the same there are still some questions open for discussion. In this chapter mainly the influence of rubber particles on the increase of toughness will be considered in more detail. The other mechanical properties are of course also influenced by a rubber particle modification, but their changes are in most cases not the primary object of material development for an intended application. Young's modulus, e.g., decreases on rubber modification due to the soft rubber phase as described by the constituent equation developed for filled polymers. Accordingly, the modulus is mainly given by the modulus of the matrix and the rubber concentration; however it is independent of the particle size. The tensile strength is also reduced due to the introduction of stress intensities at the rubber particles and the substitution of the stiff matrix by soft rubber,

---

[1]An overview is given in Chapter 10

but this reduction happens in a more complicated way. Clearly the strength of the matrix, which also depends on other parameters such as its molecular weight or plasticizer contents, influences significantly toughness, but these molecular aspects of the matrix will not be considered here. Also the deformation behavior and peculiarities of the deformation structure of the pure homopolymer are only regarded if they are important for understanding the deformation behavior in the rubber modified blend. Fracture strain increases on rubber modification, which ultimately is responsible for the higher toughness of the material. Since the main intention of rubber modification is to toughen brittle polymers, this chapter will address mainly the influence of the structure and properties of rubber particles on the toughness of an amorphous polymer matrix. Toughness is defined nowadays in two principally different ways in polymer testing. A relatively modern characterization of toughness follows the methods of fracture mechanics. Toughness in this framework means mainly the deformation energy dissipated up to the beginning of failure. In the standard testing, which we will consider in this chapter, toughness regards the whole energy dissipation up to fracture — it is the fracture energy. These values are given e.g., by impact strength in the three-point bending test. Both characterizations must of course not correspond to each other because the energy for the beginning of failure is different from the fracture energy containing the complete deformation and fracture of the matrix and rubber particles.

In the case of spherical soft rubber particles, the energy-dissipating deformation processes mainly inherent to the matrix can be intensified and stabilized. The main deformation processes in polymers are given by shearing and crazing. In the compression mode of deformation, the volume constant shear processes dominate and therefore the polymer is tough. In tensile tests, however, with an additional dilatational component of the strain tensor, there is a competition between cavitation processes and volume constant shearing. The choice between both deformation mechanisms depends on details of the matrix's chemistry [1] and the rubber phase [2]. If the molecules in the host matrix are flexible under the test conditions, shearing dominates like in polycarbonate (PC) and polyvinylchloride (PVC) at room temperature. These types of polymers show below their glass transition temperature a secondary relaxation process, which indicates some segmental mobility of their backbone chain between this lower temperature and the glass transition temperature. Parts of the molecules can slide past each other. In consequence, these materials are semi-ductile already below their glass transition temperature. On the other hand, stiff polymers like polystyrene (PS), poly(styrene-co-acrylonitrile) (PSAN) and poly(styrene/diphenylethylene) without this type of secondary relaxation process below their glass transition temperatures deform at room temperature preferentially by crazing. This deformation process leads to failure, i.e., the polymers are brittle due to the concomitant stress intensities by cavitation and chain scission, if this energy dissipating craze mechanism is not stabilized, e.g., by rubber particles. PMMA with a beginning mobility of the side groups slightly above room temperature and in direct succession also of parts of the backbone chain is at room temperature in the intermediate range in this classification.

Generally, crazing tends to prevail at low temperatures and/or at very high tensile deformation rates, when the molecules have no time to rearrange under the stress field. In contrast to this brittleness-causing situation, if enough time for possible rearrangement is given above the secondary relaxation temperatures or above the glass transition and/or at low deformation rates, the polymers tend to shear. In accordance with this trend, the stiff rubber modified syndiotactic polystyrene (sPS) and atactic polystyrene (aPS) deform by crazing at room temperature but by shearing above the glass transition temperature [2] either in the melt as in aPS or in the bulk of the semicrystalline sPS. For demonstration, Figure 1a shows the TEM micrographs of crazes in rubber modified sPS after deformation at room temperature. In Figure 1b the deformation morphology of the same material, deformed at 110°C, shows in the transmission electron microscope the herringbone pattern caused by shearing of crystalline lamellae.

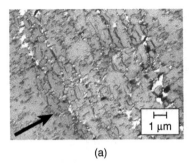

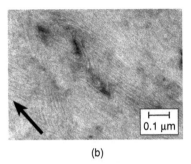

(a)   (b)

Figure 1 Transmission electron micrographs of rubber modified syndiotactic PS (arrow indicates tensile direction). (a) Crazes after deformation at room temperature; (b) herringbone patterns caused by shearing of crystalline lamellae at 110°C.

It is to be expected that highly flexible polymer chains form a higher density of entanglements than the stiffer ones. On the basis of this assumption, Wu [3] tried to correlate quantitatively stiffness and entanglement density with the transition from crazing to shearing in different polymer types with partial success. According to this classification, polymers with a critical entanglement density above 0.15 mmol/cc should deform by shearing and below this critical level by crazing. However the flexibility of the backbone chains in thermoplastic polymers seems more likely to be the main parameter for this craze/yield transition because when cooling down the semi-tough polymers below the temperature of their secondary relaxation process, they become brittle although the entanglement density does not change at this temperature. Therefore Wu additionally included in his model a characteristic ratio parameter, which stands for the mobility of the polymer chain.

The rubber particles in modified polymers act in manifold ways to improve toughness in polymers:

- They initiate the deformation process in the host polymer and multiply it
- They stabilize crack propagation by bridging the developing cracks and they blunt locally the crack tip

- They can reduce the detrimental dilational stress
  field in the matrix by internal voiding, dependent on
  their volume concentration and particle size they
  limit the free crack length in the host matrix

The details of the effectiveness of the individual rubber particles in toughening depend mainly on their structure and the
deformation process in the host polymer. The interdependence
between the structure of the rubber particles and the mechanical properties of polymers, as studied by micromechanics,
helps to understand the macromechanics of polymers. This
chapter will therefore focus on the influence of the structure
and properties of the rubber particles in enhancing the toughness of thermoplastic polymers.

II. INFLUENCE OF THE STRUCTURE AND PROPERTIES OF RUBBER PARTICLES ON THE TOUGHNESS OF AMORPHOUS POLYMERS

A.   Glass Transition Temperature of the Rubbery Phase

Generally, independent of the type of deformation process,
whether crazing or shearing, the rubbery phase must have
its glass transition temperature below the test temperature
to be effective: first to generate around the particle, due to its
lower stiffness, a local stress field in the stiffer host matrix
to initiate there the deformation processes; and second, to
reduce additional uncontrolled stress intensities elsewhere in
the matrix by internal cavitation or stretching under the
external deformation strain. Figure 2 shows as an example
the logarithmic decrement of ASA (acrylnitrile-styrene acrylate) measured by the torsion pendulum method (ISO DIN
6721/3) and additionally the temperature dependence of the
notched impact strength and impact strength for this polymer
measured in Charpy mode [4].

The toughness improves significantly near the glass transition temperature of the rubbery phase which consists in
ASA of polybutylacrylate rubber particles grafted with SAN

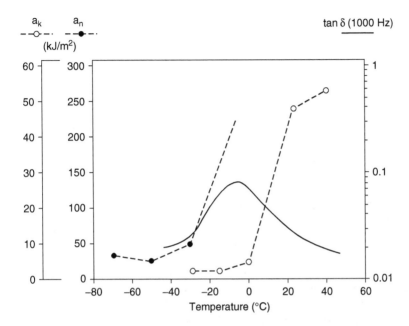

Figure 2 Correlation of tan δ (torsion pendulum test) with the notched (a_k) and unnotched (a_n) impact strength for different test temperatures in ASA.

in a PSAN [poly(styrene-acrylonitrile)] matrix. Generally impact strength improves already in the lower part of the region of the glass transition, the more severe notched impact strength in the upper part. Acrylic polymers for the rubbery phase are used for PSAN when impact resistant outdoor applications are of primary interest. If brittle/ductile transitions at much lower temperatures are required, then silicone rubber with its much lower glass transition temperature near –110°C can be used. In the standard poly(acrylonitrile-butadiene-styrene) copolymer (ABS) and HIPS (high impact polystyrene) polybutadiene is used as the rubbery phase with a glass transition temperature near –80°C. Figure 3 shows by means of the temperature dependence of the notched impact resistance the brittle/ductile transitions for the latter two rubber types in PSAN [5]. In both cases the rubber concentration is 20% and the brittle/ductile transition correlates

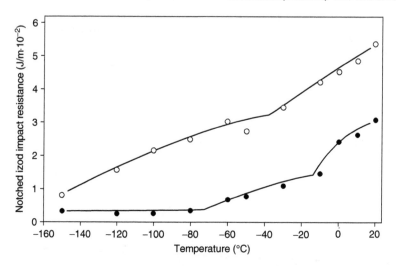

Figure 3 Brittle/ductile transition for polysiloxane (o) and poly-butadiene (●) rubbers using the notched impact resistances of their blends in PSAN. (According to Saam JC, Mettler CM, Falender JR, Dill TJ. J Appl Polym Sci 1979; 24:187–199.)

quite well with the region of the glass transition temperature of the rubbery phase. Polybutadiene is the most used rubber for polymer modification.

B. Grafting of the Rubber Particles

1. Adhesion

From the beginning of the development of rubber-modified polymers, it has become clear that the rubber particles must be adequately grafted to make their surface material compatible with the surrounding matrix. If this compatibility is present, then the molecules at the interface can interdiffuse for good cohesion. The effective part of this grafting covers the surface of the rubber particles like a shell. Such a situation is shown in Figure 4 for HIPS [6] where the grafted PS-shell around the butadiene rubber particle was made visible by dissolving the matrix polystyrene and the included PS with MEK/acetone directly on the TEM grid. The covalently bonded

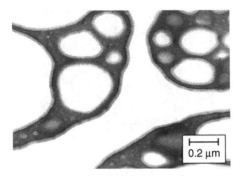

Figure 4 Transmission electron micrograph of the graft shell on the polybutadiene particles in HIPS.

grafted PS molecules cannot, however, be extracted and so the grafted PS-shell becomes visible. The grafted PS-molecules give rise to a good adhesion between the PBu rubber particles and the PS matrix, thus stopping the rubber particles from debonding from the matrix during tensile deformation. Figure 5 shows the fracture surfaces of two rubber modified polystyrenes. In one of the blends, pure homopolybutadien particles were mixed into PS (Figure 5a). At the fracture surface, these particles can be detected very well, because the crack propagated around the particles. There is no bonding between the matrix and the particles, because these rubber particles are not grafted. The view is quite different on the fracture surface of HIPS in which the rubber particles are grafted with PS (Figure 5b). In this material, the rubber particles are bonded to the matrix by the grafted shell. Consequently the crack has propagated preferentially straight forward through the rubber particles. This difference in the fracture surfaces is also reflected in the impact strength of these two materials. In the case of the pure mixture, the notched impact strength is 14 kJ/m², whereas for the HIPS with grafted particles, 50 kJ/m² was measured. The debonded rubber particles cannot contribute to the fracture energy by stretching, voiding and bridging the gap of the propagating crack. At best they can initiate some crazes.

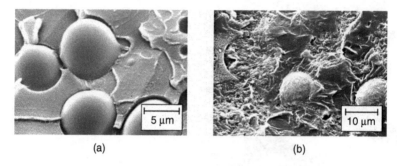

(a)   (b)

Figure 5 Fracture surface of rubber modified PS. (a) Ungrafted polybutadiene rubber; (b) polybutadiene grafted with polystyrene.

In ABS, the polybutadiende (PBu)-rubber particles are grafted with SAN or another polymer, which is compatible with PSAN. The strong adherence between rubber particles and matrix is essential in crazing materials as described above for HIPS. Otherwise, not only crazes but also cavities are formed at the interface, with the consequence that detrimental stress and crack initiation intensifies and the modified polymer remains brittle due to these extra cracks.

2.   Agglomeration

In addition to bonding, grafting is also important for the dispersion of the rubber particles. This aspect is relevant for crazing as well as for volume constant shearing processes. Very large agglomerates are ineffective in toughening. Figure 6 shows the distribution of the same rubber particles in styrene polymeric host copolymers with different AN contents. The rubber particles were polymerized in emulsion. They consist of 58% cross-linked polybutylacrylate (PBA) grafted with 40% styrene and acrylonitrile copolymer (SAN) which consists of 75% styrene and 25% AN (S75/AN25). The primary particles have a diameter of about 80 nm. For these experiments, a grafted rubber concentration of 50% was chosen. Figure 6a shows the distribution of these agglomerated rubber particles in pure polystyrene. The ultra-thin sections for the electron microscopy were taken from deformed injection molded spec-

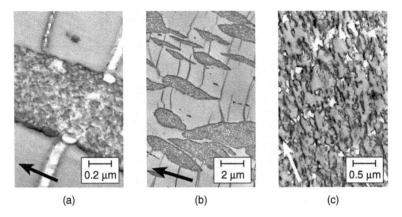

(a)   (b)   (c)

Figure 6 Transmission electron micrographs of blends of a core PBA-shell (PSAN/75:25) modifier in PS (a and b) and in PSAN 75:25 (c). Arrow indicates tensile direction.

imens. The rubber particles have agglomerated significantly to big units, because the PS matrix is not compatible with the composition of the grafted polymer. There is a pronounced mismatch between the composition of the SAN shell surrounding the rubber particles and the PS matrix. At lower magnification in Figure 6b, long crazes are shown which have developed during deformation and which run from one rubber particle agglomerate to another. Figure 6c shows a TEM image of the distribution of the rubber particles after deformation in a matrix consisting of S75/AN25, which has the same composition as the grafted shell. In this mixture the rubber particles are well dispersed in the matrix, not as isolated particles but as much smaller agglomerated units than in PS. After deformation nearly no crazes but some voids have formed in this polymer. The material with this high concentration of individual particles, which are well dispersed, has deformed preferentially by shearing. This can be deduced from the rubbery units being elongated in the tensile direction in this originally compression molded specimen. Some of the voids are arranged lateral to the deformation elongated rubber particles, that means perpendicular to the tensile direction. Between the cavitated particles, the ligament of matrix

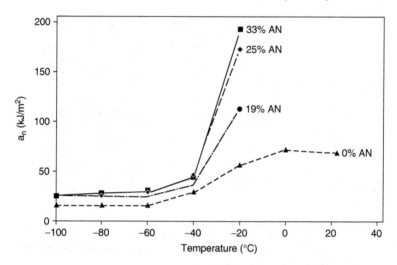

Figure 7 Impact resistance of blends of a core (PBA)/shell (75S + 25AN) impact modifier in different PSAN matrices.

material must have been stretched like fibrils in the standard crazes. Therefore this deformation pattern can be named "macro crazes." The impact strengths of the injection molded specimens of this set of polymers are plotted in Figure 7 as a function of the temperature. Toughness increases as expected for this polybutylacrylate rubber modified polymer only beyond the glass transition temperature of the rubber phase near −40°C. In the pure PS matrix, the PBA rubber particles have only a modest influence on the impact strength. Due to their agglomeration, the number of the rubber particle agglomerate units is too low and their size too big. In the case where the matrix and the grafted material have the same composition, that is to say 75% styrene and 25% AN (S75/AN25), but not necessarily the same sequence of the individual monomers in the grafted and the host material, impact strength at −20°C is as expected higher than for the less matched matrix with 71% styrene and 19% AN. The polymer matrix with 33% AN content (S67/AN33) is obviously still sufficiently compatible with the S75/AN25-type graft shell to afford high impact strength. At room temperature,

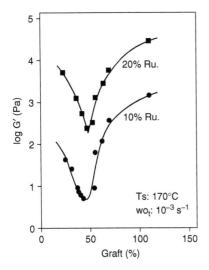

Figure 8 Storage modulus of molten ABS containing 10% and 20% rubber (Ru) as a function of the degree of grafting. (According to Aoki Y. Macromolecules 1987; 20:2208–2213. With permission.)

the rubber-modified polymers with 25% and 33% grafted AN portions in the matrix do not break totally in this test in contrast to the polymer with the incompatible PS matrix.

For good adhesion and dispersion, the grafted molecules must be not only compatible with the host polymer, but the graft shell of the rubber particles must also be sufficiently well developed, i.e., in amount and coverage of the particle surface. Aoki [7] found by dynamic viscoelastic measurements on molten ABS that an optimum dispersion of the rubber particles with a diameter of 170 nm is achieved with a degree of grafting of about 45% (ratio of weight of the grafted polymer to weight of the rubber). The results are shown in Figure 8, where the storage modulus of the molten ABS is plotted as a function of the grafting degree for 10% and 20% rubber. These measurements were made in the melt at 170°C. It is postulated that the elastic modulus at long test times and low frequencies is a minimum when the particles are well dispersed and minimum interaction between the particles exists. This assumption was verified by electron microscopy. At low grafting degrees, the particles agglomerate because of poor compatibility; in solutions the corresponding particle agglomeration would be called flocculation. At too high grafting degrees attraction effects are postulated due to the interaction of the thick grafted shells called depletion effects. At approximately 50% graft, interaction is minimized due to best dispersion. The grafting degree however cannot be the only parameter to characterize agglomeration, since it is not trivial whether there are, at a given grafting degree, many short grafted polymer chains which cover the surface or only a few very long ones. The graft density, which is given by the number of grafted polymers per unit surface area has also to be taken into consideration. Chang and Nemeth [8] tried to separate these effects in more detail. They found that for a given graft level agglomeration increases by increasing graft molecular weight, lowering particle size and increasing compositional mismatch between the graft shell and the matrix polymers of the particle. Figure 9 shows from their measurements the number of the rubber particles with the original diameter of 0.19 μm in an agglomerate as a function of the graft molecular weight for three different graft levels (GL). Clearly agglomeration of particles is reduced by shorter graft molecules and higher graft levels. Ahn [9] later showed for ABS that with increased grafting, an increase of the amount of PSAN inclusions in the polymer particles also occurs. Thus grafting modifies not only the surface of the rubber particles but also their size and morphology.

Breuer, Haaf and Stabenow [10] showed that with PVC blended with an ABS modifier, the degree of particle agglomeration is reduced with an increasing degree of grafting in the range of 25% to 60%. With increasing dispersion of the rubber particles, with a diameter of about 80 nm, the notched impact strength was reduced from 26 to 5 kJ/m². The specimens had been compression molded at 160°C. Thus in PVC as in PSAN, some agglomeration of small rubber particles as a consequence of less grafting is advantageous for higher toughness.

In addition to grafting, the flow conditions of the melt during processing for preparing household items can also

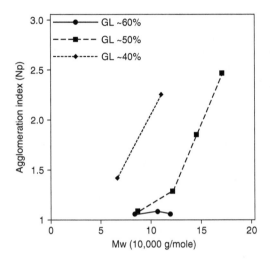

Figure 9 Dependence of the number of particles in an agglomerate in ABS as a function of the molecular weight of the graft chains for three degrees of grafting. (According to Chang MCO, Nemeth RL. J Appl Polym Sci 1996; 61:1003–1010. With permission.)

influence the agglomeration of especially small rubber particles. Figure 10 shows transmission electron micrographs of an ASA type, which was injection molded at three different temperatures. At the low injection temperature of 220°C the particles were finely dispersed. At the slightly higher temperature of 250°C some agglomeration with rubber free interspace has occurred and at 280°C the agglomeration process has progressed to form larger units. The corresponding notched impact strength of these materials is plotted in Figure 11 as a function of the injection molding temperature (curve a in Figure 11). Obviously some agglomeration seems to be advantageous for optimum toughness of ASA. Finely dispersed or highly agglomerated units with larger rubber free regions are less effective in toughening. If the sample, which was injection moulded at 220°C and having a fine dispersion of the rubber particles, is annealed for 30 min at 190°C to give the particles in the melt time to rearrange further, then the notched impact strength improves dramatically, as can be seen from the diagram in Figure 11 (curve

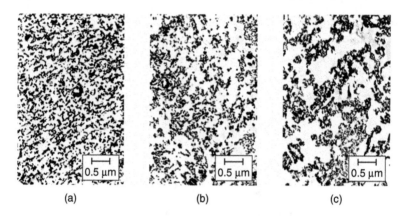

(a)   (b)   (c)

Figure 10 Transmission electron micrographs showing the influence of the molding temperature for ASA on the degree of agglomeration of the impact modifier particles (a) 220°C (b) 250°C (c) 280°C.

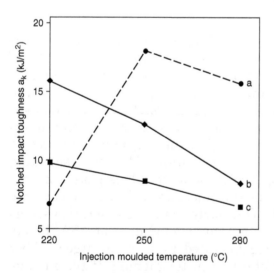

Figure 11 Notched impact strength of ASA. (a) Influence of injection molding temperature after injection molding; (b) after annealing at 190°C for 30 min; (c) after annealing at 220°C for 30 min.

b), whereas in the already slightly agglomerated products after injection molding at 250 and 280°C the toughness decreases by further agglomeration beyond the optimum level. Annealing at the still higher temperature of 220°C reduces (curve c in Figure 11) toughness in all three at different temperatures injection molded specimens compared to the annealing step at 190°C. It is likely that agglomeration proceeded during the annealing time at 220°C. Details are given in the original paper [11].

Breuer, Haaf and Stabenow [10] blended PVC with an ABS modifier with a graft degree of 25% at different temperatures. In contrast to the former example with ASA, in PVC the dispersion of the rubber particles was finest at the higher temperature (Figure 12). Since a suspension PVC was used for these blends it is believed that the PVC microparticles of the polymerization process only melt at the high temperature of 185°C and therefore only at this high mixing temperature

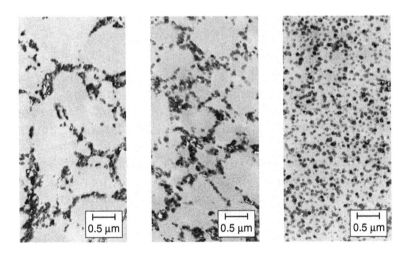

Figure 12 Transmission electron micrographs of PVC blended with ABS modifiers (degree of grafting 25%) mixed at various temperatures. (According to Breuer H, Haaf F, Stabenow J. J Macromol Sci Phys 1977; B14:387–417). (a) 140°C, a_k = 42 kJ/m^2; (b) 160°C, a_k = 26kJ/m^2; (c) 185°C, a_k = 8kJ/m^2.

a fine dispersion of the particles is attainable, with the consequence of low toughness. At lower temperatures the PVC microparticles had not properly melted and do not become homogeneous, and so the particles were arranged around the original suspension particles. As in ASA the highest impact strength is not attained with the finest dispersion of the rubber particles but when some agglomeration is present. Thus with the pronounced network structure of the rubber particle arrangement, notched Charpy impact strength is 42 kJ/m², while in the material with the highly dispersed rubber particles after milling at 185°C only 8 kJ/m² is reported.

In conclusion, every modification of rubber particles by grafting influences the adhesion, agglomeration, morphology and in consequence of course also the mechanical properties of the polymers. Therefore grafting and processing must be optimized to fit the virgin homopolymer when rubber-modified polymers are to be developed.

C. Particle Size of the Rubber Particles

To effectively toughen homopolymers by rubber modification, the rubber particles must be sufficiently large to multiply in their stress field the deformation processes inherent to the polymer matrix. Thus the deformation process starts in rubber-modified polymers at many sites in the material, namely at the rubber particles, instead of at a few inherent natural defects which would lead to overloading and failure. Additionally, the rubber particles should also be able to reduce the subsequent crack propagation (compare with the "three stage mechanism," Figure 1, Chapter 10).

1. Pseudo-Ductile Homopolymers

In pseudo-ductile matrices, shear deformation prevails when tested between the temperature of its secondary relaxation and the glass transition. PVC and PC are typical representatives of amorphous polymers for this type. Thus above the glass transition temperature all semicrystalline polymers like

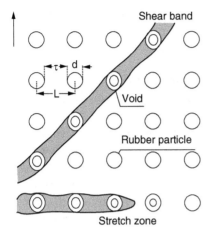

Figure 13 Scheme of a stretch zone and a shear band in rubber-modified polymer.

sPS (Figure 1) deform by shearing, if they are not highly cross-linked. In semicrystalline polymers, the crystals stiffen the materials beyond their glass transition temperature, whereas amorphous polymers become fluid. The shear processes are mainly initiated or facilitated between the rubber particles. To be most effective the distance in the matrix between the rubber particles should be reduced, so that the constraint of the matrix between the particles is reduced and instead of the original plane strain a plane stress field is locally built up, which is ideal for shearing. If the rubber particles cavitate additionally during the deformation, the constraint is further reduced by the adjacent voids and the reduction of the dilatation field, and consequently ductility is increased. The distance τ (Figure 13) between the surfaces of two adjacent rubber particles with the diameter d in a cubic arrangement with the side length L is given by

$$\tau = L - d$$

The volume concentration Φ of the particles is given by

$$\Phi = 4\pi(d/2)^3/3L^3$$

Combining both equations gives for the distance between the particle surfaces

$$\tau = d[\pi/(6\Phi)^{1/3} - 1)]$$

In this simple assessment, the concentration of the rubber particles is assumed to be low enough so that they do not touch each other. As expected, at a given rubber concentration the distance between the surfaces of the particles decreases with decreasing particle size, or for a given diameter the concentration must be increased to reduce distance between the surfaces of the particles. According to the experimental data in the literature [3] the critical surface-surface distance τ, where the brittle ductile transition occurs for a given material, appears to be constant. The critical distance probably only depends on the deformation rate and the temperature. Thus if the matrix ligament between the rubber particles is thinned to a certain critical thickness, the polymer can shear, which entails toughness. Therefore small rubber particles at high concentrations should be applied in toughening semi-ductile matrices.

In Figure 13 the progress of the plastic stretching deformation zone is shown schematically as observed for rubber modified PC [12]. In this case voids formed in the rubber particles, which facilitate stretching because the constraint in the material is additionally reduced. The deformation zone started at a notch and grew perpendicular to the tensile direction. In some cases the rubber-modified pseudo-plastic polymers also deform along shear planes oblique to the tensile direction as shown, e.g., in Figure 14 for PVC shear bands as seen in an optical microscope image. The pattern of light scattering, which shows the preferential oblique shear bands is inserted in Figure 14. This situation is shown schematically in Figure 13. Obviously it depends on yield and shear conditions near the rubber particles as to which way the material deforms in detail by shearing. For the beneficial voiding within rubber particles, they should be bonded to the matrix, otherwise the material cavitates near the interphase. Such large voids in or near the matrix have detrimental effects on toughness, due to additional stress intensities in the matrix.

Figure 14 Optical micrograph of shear bands in PVC with small angle light scatter pattern inserted. Arrow indicates tensile direction.

In contrast the voids inside the rubber particles are blunted and therefore lead to less crack initiation.

Besides the voids within the rubber particles in pseudo-ductile materials, the rubber particles have mainly the function to thin the pseudo-ductile material to make it able to shear easily. It has been revealed by experiments that it is still better for high toughness if the small rubber particles show some agglomeration in a network-like structure than to be finely dispersed (Figure 12). Within these areas of higher particle concentration, i.e., the agglomerates, shear is still further facilitated because at a given rubber concentration, the inter-particle distance is locally further reduced. And from these sources of increased shearing, the deformation zones grow into the adjacent material with a low particle concentration, if this is not too brittle. Therefore a network formed from agglomerated particles is especially advantageous and an optimum agglomeration for toughness can be expected; if the particles are less agglomerated than at the optimum, normal shearing is observed. If the agglomeration has developed too far, the region between the rubber-rich agglomerates is too brittle and then there are no rubber particles to continue stable shearing. The unnotched impact strengths and the corresponding transmission electron micrographs of the PVC blended with chlorinated polyethylene were reported by Haaf, Breuer, Echte, Schmitt and Stabenow [13]. The same modified PVC was milled at temperatures between 150 to 190°C. Very similar to Figure 12, the authors found that with increasing processing temperature, the rubber phase passes from an agglomerated state through a network structure to a very fine dispersion. Impact strength shows a maximum for the network structure at 170°C processing temperature.

In practice semi-ductile polymers such as plain PC do not need rubber modification because they are sufficiently tough at room temperature. But if these polymers are also used at lower temperatures or if they are processed to thicker items with plain strain, then detrimental cavitation in the pure material can be avoided by rubber modification. At low temperature shearing is facilitated between the rubber particles. In the case of thick specimens, voiding within the rubber particles reduces dilatational stress and supports the transition from plane strain to plane stress between the rubber particles thus favoring shearing and impeding fracture.

2.   Brittle Homopolymers

In brittle homopolymers as in PS or PSAN without a secondary relaxation process of the backbone chain, crazing is the dominating deformation process. Within the crazes numerous detrimental voids are formed between the fibrils. The aim of rubber modification of this kind of polymer is to start these crazes at the large number of particles, where the stress intensity is increased. Thus the few crazes originating at natural defects in the neat polymer, with the risk of premature overstressing of one of them leading to failure, are exchanged by many crazes in the rubber-modified material. The whole material is extensively crazed and fibrillated and with the stretching of the fibrils in the crazes, energy is dissipated. Additionally these rubber particles blunt the tips of the crazes and are also stretched to bridge the gap of the craze. To initiate crazes, the matrix adjacent to the rubber particles must be opened by their stress field. Since the ratio between the extension of the stress field around a rubber particle and the particle diameter is constant, larger particles with widely extended stress fields are needed for crazing to open the matrix network. Small particles are ineffective when their stress field only stretches some segments of the host molecule. The size of the particles should be large compared to the length of the matrix molecules between their entanglements.

Turley and Keskkula [14] polymerized in a batch process styrene with 7% diene rubber in different ways to change the particle size while keeping the rubber content constant. Thus salami-type rubber particles were produced with diameters ranging from about 3 μm to less than 1 μm. The occluded portion of polystyrene was higher in the larger particles than in the small ones. The rubber phase volume (occluded PS and rubber) is higher in the larger particles. In qualitative agreement with this model regarding influence of particle size, the notched Izod impact of HIPS with smaller particles is reduced in comparison to HIPS with larger particles by 50%, the elongation even up to an order of magnitude.

Figure 15 shows the transmission electron micrograph of PS modified with 50% small particles. These rubber particles have a diameter of 80 nm and consist of a 50% PS core and a 50% PS/PBu shell (1:1). After deformation, zones of crazes with voided particles are only observed perpendicular to the tensile direction. Between these deformation zones, undeformed regions prevail. Thus toughness was low because craze initiation at small particles is impeded and not the whole of the material is included into the deformation process. The volume change of this specimen during deformation amounts to 95% in the compression molded specimens and 80% in the injection molded ones. Thus most of the deformation is performed via voiding in rubber particles and crazes within these zones. Shear deformation in this HIPS can be neglected. Figure 16 shows a transmission electron micrograph of a mixture of small and large particles in HIPS which was deformed only 2.6%. Clearly the first crazes (arrows in

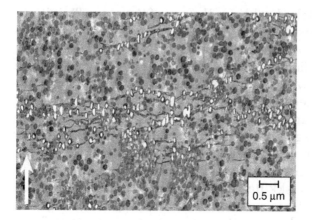

Figure 15 Transmission electron micrograph of PS modified with 80 nm PBu particles after deformation. Arrow indicates tensile direction.

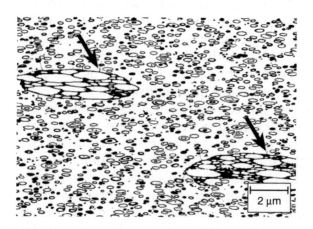

Figure 16 Transmission electron micrograph of PS modified with large and small particles and deformed 2.5%. The arrows indicate crazes starting at the large particles.

Figure 16) were initiated only at the larger salami-type rubber particles, especially near the equator, where the maximum tensile stress is generated. This observation is in agreement with experiments by Okamoto et al. [15], who observed initial crazing only at larger particles in their mixture of small and large particles at a constant rubber concentration near to 20%.

In some cases synergetic effects were reported when mixing small and larger particles in PS. The increase in toughness by the addition of large particles to small ones can be clearly understood on the basis of this explanation: by adding large particles to small ones, the number of craze initiation points is increased and facilitated, owing to the large particles. Addition of small particles to large ones decreases the free craze length between the large particles if the rubber concentration is kept constant. Wrotecki and Charentenay [16] explained the synergistic effect by assuming that crazes are also initiated at the small particles in the field of the large ones and that crazes are only stopped by large particles. Hobbs [17] assumes that the probability of craze initiation is independent of the particle size and that only large particles are effective in terminating crazes. Okamoto et al. [15] postulate that crazes are induced from the large particle's surface at the initial stage of loading. As these crazes grow, minute crazes from the small rubber particles in the vicinity of the large particles are induced, and then they overlap with the crazes growing from the large particles. Consequently long extended crazes starting from the surface of the larger particle are stabilized by the smaller ones in the stress field of bimodal rubber particles. According to the experiments, it can be regarded for certain that the smaller particles are less effective in craze initiation than the larger ones. It is likely that with a bimodal distribution of the rubber particles, the small particles can guide the propagating longer crazes, which are initiated at the larger particles. This thus ensures that the maximum craze length between the large particles does not quickly exceed the critical length, which leads to failure. The extent of the synergistic effect of the bimodal distribution of the rubber particles may depend on the difference in their size.

With the primary intention to increase its glass transition temperature, styrene was copolymerized with diphenylethylene (S/DPE) [18]. Due to the two phenyl groups per monomer unit this material is stiffer than PS and conse-

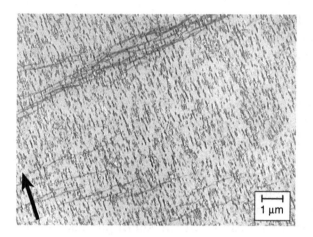

Figure 17 Transmission electron micrograph of S-DPE copolymer
(85:15) modified with PBA-g-S core-shell particles with a diameter
of 120 nm after deformation. Arrow indicates tensile direction.

quently more brittle, thus extensive toughening can only be
expected by intensifying crazing at large particles. If for
rubber modification of S/DPE rubber particles produced by
the emulsion process with diameters of only about 120 nm
are used, then they are too small for extensive crazing (Figure
17). They consist of cross-linked poly(butyl acrylate) (PBA)
particles which are grafted with polystyrene (60:40 wt%).
35% of these particles were blended with a P(S/DPE) copolymer having 15 wt% DPE units, with which the polystyrene
graft shell is compatible. After injection molding, the particles in the blend are highly dispersed and individually oriented in the direction of the injection molding. With these
small particles only some zones with crazes are observed
after the deformation, as is shown in the TEM images of
Figure 17. Energy dissipation processes are only restricted
to a few regions. The notched impact strength of this copolymer with 15% DPE is with 2.3 kJ/m^2, only three times
higher than without modification. When 36% of larger particles consisting of polystyrene-hydrogenated polybutadiene-
polystyrene block copolymer are used instead of the PBA
particles, particles of about 1μm are formed and the notched

Figure 18 Transmission electron micrograph of a deformed S-DPE copolymer modified with a PS-hydrogenated PBu-PS block copolymer after deformation. Arrow indicates tensile direction.

impact strength increases to 19.5 kJ/m². As can be seen in Figure 18 on impact the energy has been dissipated by voids in the rubber particles and crazes in the matrix, which have been formed throughout the material. In all these modified PS polymers only crazing and no shear bands are observed, independent of the particle size.

The situation can be different for ABS with its AN modified PS-matrix. Figure 19 shows transmission electron micrographs of two ABS samples with different sized rubber particles. LABS 321 (Figure 19a) is an ABS, produced in solution with 12% rubber in the form of salami-like particles with a diameter of up to 3 μm. In the rubber particles PSAN is occluded. LABS 312 (Figure 19b) has 15% rubber, the particle diameter is however 0.5 μm smaller. In Figure 20 the damping (tan δ) is plotted for both materials in vibration at 1000 Hz (ISO DIN 6721/3) and the notched impact strength as a function of the temperature. 1000 Hz was chosen for the vibration frequency to make the linear elastic measurement time comparable to impact with fracture times near to 1 μs. Impact strength increases, as expected, only beyond the glass transition temperature of the rubbery phase of the rubber

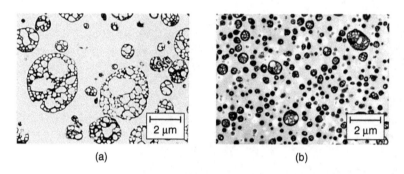

(a)   (b)

Figure 19 Transmission electron micrographs of ABS with average particle size (a) 3μm (LABS 321) and (b) 0.5 μm (LABS 312).

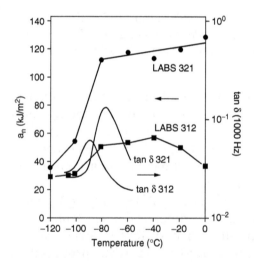

Figure 20 Tan δ and unnotched impact strength (a_n) as a function of temperature for LABS 321 and LABS 312.

particles, where they are sufficiently soft to form stress intensities. As can be observed, LABS 321 with the larger particles has a 300% higher impact strength than the LABS 312 with the smaller particles, in spite of having less polybutadiene rubber. The rubber particles in LABS 321, with the higher amount of occluded PSAN, have a higher glass transition temperature than the more compact polybutadiene phase in

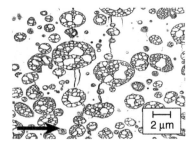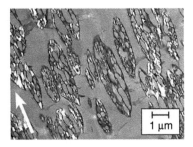

Figure 21 Transmission electron micrographs of LABS 321 after (a) deformation to 20% of yield strain, (b) deformation to yield strain. Arrow indicates tensile direction.

the smaller particles of LABS 312. Obviously the stiff matrix material adjacent to the thin rubbery strands in the large particles restricts the mobility of the rubbery material, so the glass transition temperature is higher. More decisive for toughness is, however, the particle size, the larger particles leading to higher toughness, as predicted by the model prescribed before. Figure 21 shows transmission electron micrographs of LABS 321. When deformed only up to 20% of the yield strain, crazes are observed mainly between the larger particles and are generated near the equator of these rubber particles. When deformed up to the yield strain, numerous crazes with voids inside the rubber particles have developed. The particles are elongated in the tensile direction. The larger particles are clearly advantageous for toughening. Wu [3] studied the relationship between the entanglement density in the host material and the optimum particle size for crazing. The results are summarized in Table 1, where the optimum particle sizes are given for three brittle amorphous materials together with their entanglement densities.

Regardless of the scatter of the data, it is without any doubt that the low entanglement density in PS needs larger rubber particles for optimum toughness than PSAN with its higher entanglement density. For PMMA, with the highest entanglement density, small particles bring the highest toughness. Thus the conception that with increasing chain length between the entanglements, larger rubber particles are

TABLE 1 Characteristic Data for Rubber-Modified Polymers with Brittle Matrix

	PS	PSAN	PMMA
Entanglement density mmol/cm^3	0.056	0.093	0.127
Optimum rubber particle diameter, μm	2.5	0.75	0.25

Source: Wu S. Polymer International 1992; 29: 229–247. With permission.

needed for stretching the entanglement network to form crazes is in accordance with these experiments. Additionally an optimum size is to be expected, because as mentioned before, small particles cannot initiate crazes. On the other hand if the particles are too large, the distance between the rubber particles is too long and consequently the length of the crazes can become over critical for the host material, leading to premature fracture. In Figure 22 this concept is schematically shown, where the effectiveness of craze formation is plotted against particle size for host polymers with high and low entanglement densities. Michler [19, 31] considered the individual relevant influences in more detail.

Agglomeration of small particles to larger units, e.g., as is shown in Figure 6a for a PS matrix, can also be helpful for

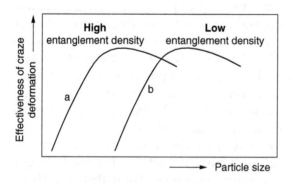

Figure 22 Schematic plot of effectiveness of crazing for toughness as a function of rubber particle size for host polymers having high and low entanglement densities.

TABLE 2   Fracture Energy [mJ] on Impact of 10×4 mm^2 Notched Specimens of ASA3 after Injection Molding and after Annealing at 180°C for 30 min

	Injection molded	Annealed at 180°C for 30 min
Notch depth [mm]	Fracture energy, mJ	Fracture energy, mJ
4 mm	65	155
5 mm	55	240

Source: Ramsteiner F, McKee GE, Heckmann W, Fischer W, Fischer M. Acta Polymer 1997; 48:553–561.

promoting crazing in polymers modified with small particles. This agglomeration can be generally controlled by grafting and flow behavior in the melt. This principle works and is applied in many materials. However in the case of Figure 6a with PS as the host matrix, the agglomeration is too extensive. The free craze length between the agglomerates is too long and ends in premature fracture leading to the measured low brittleness, probably additionally intensified in this case by the missing adhesion between the matrix and the rubber particles.

At higher concentrations of small rubber particles the interparticle distance can become so short that a transition in the deformation mechanism from crazing to shearing is observed, even in the brittle polymer, PSAN. This situation is demonstrated for a rubber modified PSAN in the following section. The ASA3 product [4] consists of a PSAN matrix with 50% core-shell particles composed of polybutylacrylate and grafted with PSAN. The particle diameter is 80 nm. This type of ASA embrittles as expected below the glass transition temperature of its rubber particles as shown in Figure 2 for a similar ASA sample by means of the notched and unnotched impact strength. The fracture energies for notched specimens of ASA3 are given in Table 2.

The specimens were 4 mm thick and notched along the 10 mm long side. They were injection molded and some of them were annealed afterwards in a mold to save their shape. Testing was performed in the Charpy mode at 23°C.

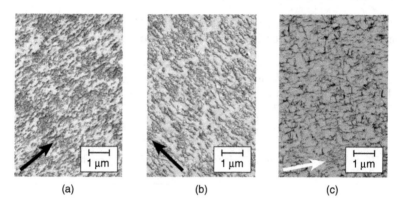

(a)   (b)   (c)

Figure 23 Transmission electron micrographs of ASA after deformation; (a) injection molded, notch depth 3.65 mm (b) as with (a) but annealed 30 min at 180°C before testing, notch depth 3.4 mm, (c) as (b) but with a notch depth of 5.26 mm. Arrows indicate tensile direction.

From Table 2 it is seen that the injection molded specimens are less tough than those annealed at 180°C after injection molding. TEM images of sheets cut parallel to the deformation direction are shown in Figure 23. Although the small rubber particles have agglomerated to large units, where the initiation of crazes should be possible, the deformation in this polymer, which is highly filled with rubber particles, occurs obviously by shearing. No crazes are observed (Figure 23a — only injection molded) in contrast to the LABS with its large particles in Figure 21. This material ASA3 with its high rubber concentration has obviously deformed not by crazes but by shearing because the interparticle distance has become sufficiently short. The specimens, which were annealed before the impact test, are tougher than the injection molded ones. That this prehistory of annealing the polymer has led to deformation by shearing without crazing must be concluded from the TEM image (Figure 23b) of the specimens with a 3.4 mm long initial notch. The structure of the agglomerates of the particles has not changed significantly by annealing in the melt at 180°C. Obviously the influence of the preprocessing of the specimens affects the matrix properties.

It is likely that the already high orientation in the injection molded specimens allows less further orientation of the molecules by stretching during the deformation than in the unoriented matrix, where the matrix molecules are able to undergo more easily additional stretching on impact. Therefore the dissipation of energy by the deformation processes is increased by annealing and consequently also toughness. If the residual cross-section is thin because the notches of the specimen are deeper, then some single crazes are seen (Figure 23c), which bridge the rubber-free region between the agglomerates. The toughness has enormously increased by annealing. At a notch depth below 4 mm, the products are less tough because specimens with the thicker residual cross-section ahead of the initial notch break already near the peak stress of the deformation curve [4]. The transition in toughness at about 4 mm deep notched specimens is mainly caused by the failure strain, where the thinner residual cross-sections at longer notch depth allow higher deformation beyond the peak stress [4]. These results taken from fracture mechanics testing are helpful to understand this size-dependent mechanical behavior better.

Figure 10 shows that, on processing ASA with small rubber particles, the extent of agglomeration can be regulated. This dependence influences the impact strength. As in the pseudo-plastic PVC, ASA with many small rubber particles with the tendency to initiate shearing, has an optimum degree of agglomeration for maximum toughness. A perfect dispersion of small rubber particles, as well as too much agglomeration is detrimental for toughness.

3. Semi-Brittle Homopolymers

The molecules in amorphous PMMA become partly mobile near room temperature as indicated by the secondary relaxation process at 50°C which is caused by the beginning mobility of the side groups. The glass transition temperature is only about 80°C higher. Thus it is not surprising that at room temperature or at slightly higher temperatures and at slow deformation rates, shearing and stretching dominate,

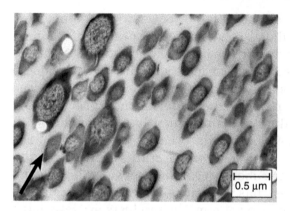

Figure 24 Transmission electron micrograph of PMMA modified with particles of 300 nm and 600 nm after deformation. Arrow indicates tensile direction.

whereas at lower temperatures and/or higher deformation rates crazes are observed. For increasing shearing, small particles at high concentrations are helpful; for crazing, larger particles are needed. According to the literature the optimum particle diameter in PMMA for crazing is about 200 nm [3,20]. In the TEM image reproduced in Figure 24 the deformation structure is shown after impact in a PMMA, which was modified with 42% rubber particles: 33.6% had a diameter of about 300 nm, the rest 600 nm. The rubber particles have a core and two-shell structure. The core consists of PMMA which was grafted with butylacrylate/PS in the first step and with MMA/ethylacrylate (EA) in the outer one. To increase the contrast of the rubbery phase, the specimen was stained after impact with RuO_4 to make visible the BA/S of the first shell. In addition the sample was stained with OsO_4 to make the crazes visible. The outer shell with EA is not stained by this method. As documented in Figure 24 the deformation is mainly by shearing and stretching, only very few crazes are discernable. Furthermore some particles have voided in the first BA/S shell near the poles of the rubber particles. The impact strength of this and the other materials of this set of experiments with various particle sizes are plotted in Figure

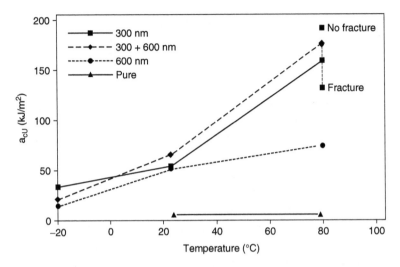

Figure 25 Impact strength as a function of test temperature for PMMA modified with rubber particles having different diameters.

25 as a function of the temperature. Rubber modification increases the impact strength significantly. This finding is especially expected above room temperature where shearing prevails. Larger particles in this temperature range are less effective than the smaller ones, when shearing between the particles dominates. Schirrer et al. [21] studied a PMMA modified with 40% core-shell particles over a wide range of temperatures and deformation rates. Volume change and light scattering during deformation were used as indicators for deformation mechanisms. The diameter of the rubber particles was about 200 nm, their structure consisted of a PMMA core, a cross-linked rubbery shell (butylacrylate-styrene) and an outer PMMA shell.

The authors summarized their results, which are in agreement with the above-described model, in the following way: compared to pure PMMA, rubber modification increases toughness and at very low strain rates or high temperatures the volume of the specimens remains constant and the material transparent. Deformation is entirely due to plastic deformation by shearing without voiding. At intermediate strain

rates and temperatures light scattering indicates particle cavitation in bands inclined to the direction of impact. Micro
shear bands propagate in the matrix from the rubber particles, accompanied by some voiding inside the rubber particles,
until the shear bands reach another particle which cavitates
in turn. No volume change was evident during the deformation process within the experimental limit of less than 1%.
However light absorption due to particle voiding increases
approximately linearly with strain. Obviously voids were created in the rubber particles without changing the overall
volume of the specimen. The damage is described by the
authors as "ordered," i.e., a macroscopically homogeneous triggering of progressive voiding and micro yielding. At high
strain rates or low temperatures, disordered cavitation, i.e.,
rubber tearing and matrix plasticity, was observed. TEM
experiments in combination with volume measurements performed by Béguelin, Plummer and Kausch [22] cast even more
light on the deformation process. These authors used two
types of spherical rubber particles, one type consisted of a soft
core and a grafted PMMA shell, the second type had a glassy
core with an elastomeric shell, which was grafted with
PMMA, to ensure adequate adhesion to the matrix. Both
particles had diameters near to 200 nm. In agreement with
the other measurements the authors thought it likely that in
tensile tests at low speeds and/or high temperatures (60°C),
both crazing and cavitation are apparently suppressed in
specimens with 40% rubber particles. Especially the sample
deformed at 60°C showed little stress whitening as would be
indicated by voiding. Stress whitening is a reliable indication
for any voiding due to the inherent light scattering. Some
crazes were visible in samples strained at room temperature,
however extensive crazes were only present in those specimens which were deformed at higher speed and/or samples
with a low volume faction of rubber particles (20%).

Summing up, when PMMA is modified with particles
optimized for crazing, crazes are only relevant at temperatures near or below room temperatures and at high deformation rates. At higher volume concentrations of the rubber
particles (40%) and/or at lower deformation rates and/or at

higher test temperatures, the shear process either in form of shear bands or stretching is observed, as for the pseudo-ductile materials. In both cases the deformation of the matrix, either by shear processes or fibrillation in crazes, toughen rubber-modified PMMA, in some cases supported by cavitation in the rubber particles.

D. Structure of the Rubber Particles

For the initiation of stress fields in rubber-modified polymers, mainly three types of rubber particles are used. In HIPS and solution ABS salami particles are preferred. These particles contain much occluded matrix so that the particles are sufficiently large for initiating crazing while the rubber content is low, limiting the decrease in Young´s modulus. The crazes are initiated near the equator region of the rubber particles perpendicular to the tensile stress direction, as shown in Figure 26. In this region the normal stress component of the stress tensor is highest. It has been described in the preceding section, that if only a small concentration of very large particles instead of a larger number of middle range particles are present in the blend, then the efficiency of the particles for

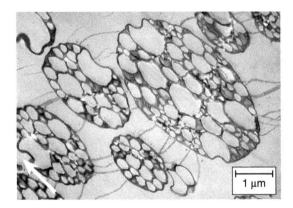

1 µm

Figure 26 Transmission electron micrograph of HIPS with salami particles and crazes after deformation. Arrow indicates tensile direction.

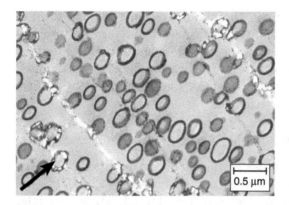

Figure 27 Transmission electron micrograph of PS modified with core-shell particles after deformation. Arrow indicates tensile direction.

toughening is reduced. The reason is attributed to the large distances between the larger particles in combination with the over critical long crazes, which are formed on deformation, combined with crack instability. Michler [23] further suggests, that with a few large particles the number of crazes which can be initiated is lower than when a large number of middle sized particles are present.

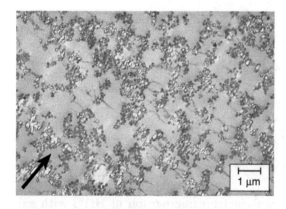

Figure 28 Transmission electron micrograph of ABS containing agglomerates of small particles and crazes perpendicular to the tensile direction after deformation. Arrow indicates tensile direction.

Core-shell particles (Figure 27) are used especially in transparent polymers. In this type of rubber particle the core is very often formed from the matrix material and is covered with a thin rubbery shell, which is grafted with an outer second shell. If in these particles the shell thickness is small compared to the wavelength of light, then light scattering is reduced and the modified polymers maintain some transparency. The TEM image in Figure 27 reveals that in PS modified with core-shell particles the crazes propagate preferentially along the poles of the particles, and it is also likely that they are initiated at these stress poles where the dilatational strain component of the stress field for supporting voiding is highest. Voiding is also observed or even fibrillation in the rubbery phase between the core and the outer grafted shell [23]. (Compare Figures 8 to 10 in Chapter 10.) This craze structure was also reported for rubber-modified PMMA [22].

In the emulsion polymerization process small particles (Figure 28) are generated with mainly a rubbery core and a grafted surface. These small particles are blended with polymers at high concentrations for initiating shearing in semiductile polymers like PVC or PC. At high concentrations in ABS shearing is also the predominant method of energy dissipation. Although the diameters of these rubber particles are only approximately 100 nm, crazes can be initiated if some agglomeration is present. Figure 28 shows as an example crazes perpendicular to the tensile direction between agglomerated particles in ABS. The advantages of this type of small particles are:

- The possible transition from crazing to shearing if less stress whitening is demanded
- Better weatherability because small destroyed particles on the surface of an item is less disturbing than larger ones
- Better gloss, which results from the small particles size

E.   Voiding of Rubber Particles

It was shown in the preceding parts of this chapter, that voiding in the rubber particles very often goes hand in hand with toughening. This voiding was thought to be important not only to reduce the dilatational strain in the matrix but also to facilitate the deformation processes. Opening of crazes, e.g., is simplified by highly stretchable non-debonding rubber particles in contrast to stiff ones. The voids themselves in the particles are blunted in a first step by the surrounding rubbery material and therefore do not generate detrimental stress intensities. Especially Ayre and Bucknall [24] postulated that crazing is mainly initiated by a preceding voiding process in the rubber particles. Due to the concomitant energy release rate and an energy barrier, voiding in small particles is more difficult than in large ones. Therefore crazing at small particles is more difficult than at large particles. To summarize, each measure, which facilitates cavitation in rubber particles, should contribute to increased toughness and all steps that obstruct voiding will be less effective in toughening the material. To prove this model, voiding was impeded on the one hand by cross-linking the rubbery phase and on the other voiding was promoted by addition of oil to the rubber particles to ease the nucleation of voids. These two last points are discussed in more detail in the following sections.

1.   Cross-Linking

Rubber particles must be at least slightly cross-linked, otherwise the rubber phase loses its individual particular structure in processing and is transferred, e.g., to an interpenetrating network. These blends can also be tough, but their commercial production and the deformation mechanisms are different from modification with rubber particles and therefore not the topic of this chapter. Now it is known from literature that HIPS and ABS can embrittle if they are processed too long at too high of temperatures, which leads to thermal cross-linking of the rubbery phase. Embrittlement is also observed if the rubbery phase in the particles is cross-linked chemically or by radiation [25,26]. According to Bergmann and Gerberding, cited in [27], decreasing notched impact strength in ABS was observed with increasing cross-linking (Figure 29). The authors changed the degree of cross-linking by annealing at

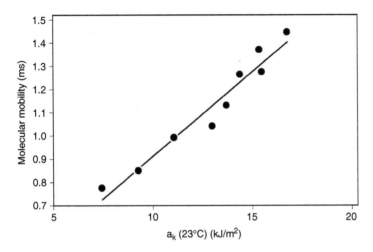

Figure 29 Increase of notched impact strength with increasing molecular mobility in the rubbery phase of ABS as determined by NMR relaxation measurements. (From Ramsteiner F, Heckmann W, McKee GE, Breulmann M. Polymer 2002; 43:5995–6003.)

high temperatures for different times and characterized the cross-linking by measuring the NMR relaxation times for the rubber phase. The reduction in relaxation times reflects reduced mobility of the molecules in the rubbery phase, resulting from the increased cross-link density and consequently higher strength. Thus the observed embrittlement might be caused by the concomitant resistance to voiding.

Figures 30 and 31 show the dynamic mechanical properties of a HIPS with a polybutadiene phase and of an ABS with a copoly(butadiene-butylacrylate) rubber phase in forced torsional mode at 1Hz. In both materials the glass transition temperature of the rubber phase is shifted by 10°C to higher values after annealing for 30 min at 280°C. This increase in the glass transition temperature is caused by an increased thermal cross-linking, which leads to a corresponding shift of the brittle/ductile transition in rubber-modified materials. However as long as the test temperature is higher than the rubber glass transition temperature and consequently the cross-linked rubber particles do not have the modulus of the

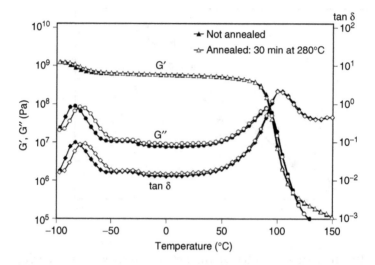

Figure 30 Dynamic mechanical properties of annealed and non-annealed HIPS measured using the torsional vibration test (1 Hz).

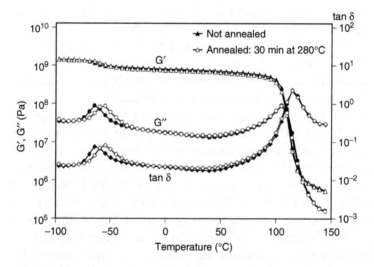

Figure 31 Dynamic mechanical properties of annealed and non-annealed ABS with a copoly(butadiene-butylacrylate) rubber phase, measured using the torsional vibration test (1Hz).

matrix, crazes should be generated anyway in the stress field of the still soft particles, leading to toughening. The embrittlement by the annealing process reduced the impact strength of this HIPS from 72 to 58 kJ/m² and of the ABS from 154 to 109 kJ/m². Reduced fracture strain and not a drop in the matrix strength was responsible for the embrittlement as demonstrated by the force deflection diagrams recorded during impact strength testing [27]. Figure 32 shows this reduced strain to fracture for the ABS type of Figure 31. Transmission electron micrographs confirmed that a possible change in the agglomeration structure of the particles was not responsible for this embrittlement. Figure 33 shows the TEM images of the virgin and annealed HIPS after fracture. Whereas in the virgin material numerous voids and crazes have developed during impact (Figure 33a), these deformation structures are formed much less in the annealed material (Figure 33b). Obviously the crazes cannot be formed in the abundant way after thermal cross-linking, because the cohesion energy of the rubbery phase has become too strong for voiding.

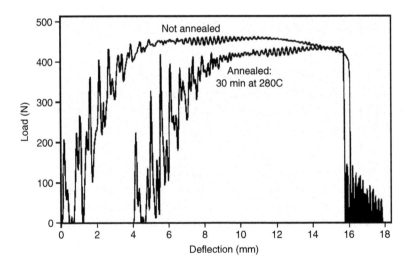

Figure 32 Force deflection diagrams recorded during impact strength testing for a non-annealed and an annealed ABS with a copoly(butadiene-butylacrylate) rubber phase.

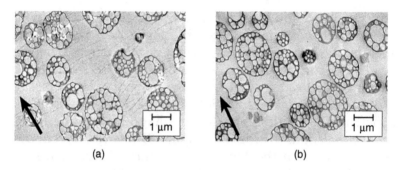

(a)   (b)

Figure 33 Transmission electron micrographs of (a) non-annealed HIPS after deformation and (b) after annealing followed by deformation. Arrows indicate tensile direction.

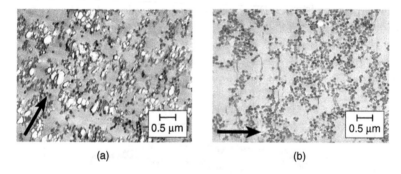

(a)   (b)

Figure 34 Transmission electron micrographs of a (a) deformed ABS with a copoly(butadiene-butylacrylate) rubber phase and (b) after annealing followed by deformation. Arrows indicate tensile direction.

Figure 34 shows the TEM images of the deformed ABS. In the non-annealed material (Figure 34a) much voiding has occurred during impact. Perpendicular to the tensile direction "macro crazes" in the form of lateral arranged voids have formed. Voiding facilitates this shearing and stretching in the matrix ligaments between the rubber particles. In the annealed ABS sample, however (Figure 34b), cavitation is drastically reduced and only some conventional crazes have appeared, which bridge the larger regions between the agglomerates of the rubber particles. Obviously thermal cross-linking of the rubber particles by annealing has not only increased agglomeration but has also reduced the formation of voids and "macro crazes" and consequently the energy dissipation process. In conclusion, cross-linking of the rubbery phase in HIPS and ABS embrittles the materials which results from a reduction of the ease voiding, which initiates crazing.

2.   Addition of Oil

Since Morbitzer et al. [28] showed some years ago that the addition of small amounts of oil can influence the mechanical properties of ABS, this phenomenon has been studied in more detail. With the intention of facilitating voiding within the rubber particles, 5% of silicone oil was added to an ABS, which contained particles with a PBu rubber core grafted with PSAN. Dynamic mechanical measurements (Figure 35) reveal no influence of the silicone oil on the glass transition temperature of the PSAN-matrix. However silicone oil broadens the

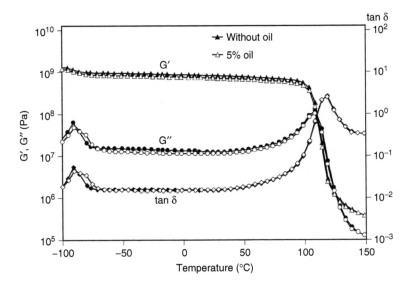

Figure 35 Dynamic mechanical testing of ABS with (5%) and without silicone oil.

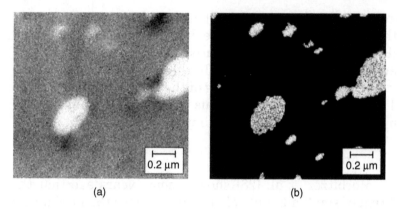

(a)   (b)

Figure 36 ABS (2.7% rubber) modified with 0.43% silicone oil. (a)
TEM bright field image (unstained); (b) corresponding Si elemental
distribution map.

glass transition temperature of the rubber phase to higher
temperatures. DSC measurement has shown that the silicon
oil crystallizes at −90°C, leading to a stiffening of the rubber
particles in this low temperature range. These measurements
show that silicone oil is at least partly soluble in the rubber
phase of ABS. The SEM images with back-scattered electrons
reveal the presence of abundant silicone oil droplets in the
matrix. To concentrate the oil mainly in the rubber particles,
only 0.43% oil was added to ABS. To decide where the silicone
oil is located, a Si elemental distribution map by means of
electron spectroscopic imaging (ESI) was recorded. As can be
clearly seen in Figure 36 [27] the silicone oil has quantitatively concentrated in the rubber particles and not in the
interface or in the matrix. Thus plasticizing effects are not to
be expected. Therefore it is likely that any improvement of
toughness by small amounts of silicone oil is mainly caused
by a promotion of deformation processes at rubber particles.
Bucknall et al. [29] claimed that if ABS samples are cooled
down to liquid nitrogen temperatures and then warmed up
again, several voids appear inside the rubber particles in the
silicone oil modified ABS, in contrast to the unmodified version. Therefore silicone oil facilitates voiding inside the rubber

particles simply on the basis of the different expansion coefficients of the rubbery and matrix phases. The relaxation of the rubbery phase by voiding is besides crystallization effects the reason for the broadening of the glass transition temperature (Figure 35) to higher temperatures. This follows since the relaxed rubber has a lower free volume than the particles still under stress. Thus adding silicone oil is likely to act as a nucleus for easier voiding and in consequence improve toughness. The addition of only 0.025% silicone oil to ABS increased impact strength from 18 to 23 kJ/m^2. The toughening of rubber modified PMMA due to micro-crack formation inside rubber particles containing a small amount of silicone oil was recently reported by Yamashita and Nabeshima [30].

Still open to question at the moment is, whether

- Crazes are initiated after voiding, as Bucknall postulates
- Or the precursor of a craze in the matrix at a rubber particle can only easily open to a real craze if the adjacent rubber particle cavitates to compensate for the local volume change
- Or both deformation structures develop simultaneously

III. SUMMARY

It is well known in polymer development that blending rubber particles into homopolymers improves their toughness. The interdependence of the structure of the rubber particles and the mechanical properties of the unmodified polymers determines fundamentally the mechanical behavior of the rubber-modified polymers. The rubber modification results in an enhancement of the deformation processes, which are given in polymers by crazing or shearing in the matrix. It is likely that voiding itself can be regarded as an unimportant energy-dissipating process compared to the extensive deformation of the matrix. To be effective, the glass transition temperature of the rubbery phase in the particles must be lower than the test temperature in order to initiate additional deformation processes in the stress field of the rubber particles. The surface of the rubber particles should be compatible with the matrix to guarantee good adhesion between the two. Further, to obtain the optimal particle size, a controlled agglomeration on processing may be helpful, which in turn can be controlled by the grafting step. The size of the rubber particles must be adjusted to the deformation mechanism in the matrix. Shear processes require preferentially many small particles with short interparticle distances. Crazing is initiated preferentially at larger particles. At low temperatures and high deformation processes, crazing is preferred; at high temperatures and low deformation processes shearing prevails. It depends on the brittleness of the homopolymer at the test temperature as to whether it is possible to initiate by an adequate modification both deformation processes. The structure of the particles controls the details of the deformation. A very important feature of rubber modification is the internal voiding of the rubber particles by which the toughness is improved. Adding of silicone oil to the rubber particles acts as nuclei for voiding and facilitates deformation processes. Cross-linking of the rubbery phase of the particles impedes voiding, which reduces toughness.

REFERENCES

1. Ramsteiner F. Zur Schlagzähigkeit von Thermoplasten. Kunststoffe 1983; 73:148–153.

2. Ramsteiner F, McKee GE, Heckmann W, Oepen S, Geprägs M. Rubber toughening of syndiotactic polystyrene and poly(styrene/diphenylethylene). Polymer 2000; 41:6635–6645.

3. Wu S. Control of intrinsic brittleness and toughness of polymers and blends by chemical structure: a review. Polymer International 1992; 29:229–247.

4. Ramsteiner F, McKee GE, Heckmann W, Fischer W, Fischer M. Rubber toughening of polystyrene-acrylonitrile copolymers. Acta Polymer 1997; 48:553–561.

5. Saam JC, Mettler CM, Falender JR, Dill TJ. Silicone-toughened poly(styrene-co-acrylonitrile). J Appl Polym Sci 1979; 24:187–199.

6. Heckmann W. Elektronenmikroskopische Methoden. In: Gause-pohl H, Gellert R, eds. Becker and Brown Kunststoff Handbuch 1996; 4:516–523.

7. Aoki Y. Dynamic viscoelastic properties of ABS polymers in molten state. 5. Effect of grafting. Macromolecules 1987; 20:2208–2213.

8. Chang MCO, Nemeth RL. Rubber particle agglomeration phenomena in acrylonitrile-butadiene-styrene (ABS) polymers I. Structure-property relationship studies on rubber particle agglomeration and molded surface appearance. J Appl Polym Sci 1996; 61:1003–1010.

9. Ahn KH. Effect of graft ratio on the dynamic moduli of acrylonitrile–butadiene-styrene copolymers. Polymer Eng and Sci 2003; 42:605–610.

10. Breuer H, Haaf F, Stabenow J. Stress whitening and yielding mechanism of rubber modified PVC. J Macromol Sci Phys 1977; B14:387–417.

11. Ramsteiner F. Einfluß von Verarbeitungs- und Produktparametern auf die Schlagzähigkeit von Syrol-Polymerisaten. Kunststoffe 1967; 67:517–522.

12. Cheng C, Hiltner A, Baer E, Soskey PR, Mylonakis SG. Cooperative cavitation in rubber-toughened polycarbonate. J Mat Sci 1995; 30:587–595.

13. Haaf F, Breuer H, Echte A, Schmitt BJ, Stabenow J. Structure and properties of rubber reinforced thermoplastics. J Sci Industrial Res 1981; 40:659–674.

14. Turley SG, Keskkula H. Effect of rubber-phase volume fraction in impact polystyrene on mechanical behaviour. Polymer 1980; 21:466–468.

15. Okamoto Y, Miyagi H, Kagugo M, Takahashi K. Impact improvement mechanisms of HIPS with bimodal distribution of rubber particle size. Macromolecules 1991; 24:5639–5644.

16. Wrotecki C, Charentenay FX. The effect of a bi-population of particle size on the impact properties of HIPS. In: Churchill Conference on Deformation, Yield, and Fracture of Polymers VII (1988):51/1–51/3.

17. Hobbs SY. The effect of particle size on the impact properties of high impact polystyrene (HIPS) blends. Polym Eng and Sci 1986; 26:74–80.

18. McKee GE, Ramsteiner F, Heckmann W, Gausepohl H. "Super polystyrene"- styrene diphenylethylene copolymers. In: Scheirs J, Priddy D, eds. Modern Styrenic Polymers. New York: John Wiley & Sons, 2003:581–603.

19. Michler G. Bruchzähigkeit kautschukmodifizierter Thermoplaste. Kunststoffe 1991; 81:548–550.

20. Cho K, Yang JH, Park CE. The effect of rubber particle size on toughening behaviour of rubber modified poly(methyl methacrylate) with different test methods. Polymer 1998; 39:3073–3081.

21. Schirrer R, Fond C, Lobrecht. Volume change and light scattering during mechanical damage in poly(methyl methacrylate) toughened with core-shell rubber particles. J Mater Sci 1996; 31:6409–6422.

22. Béguelin P, Plummer CJG, Kausch HH. Deformation mechanisms in toughened PMMA. In: Shonaike GO, Simon GP, eds. Polymer Blends and Alloys. New York: Marcel Dekker, 1999:549–573.

23. Michler GH. Micromechanics of polymers. J Macromol Sci Phys 1999; B38:787–802.

24. Ayre DS, Bucknall CB. Particle cavitation in rubber-toughened PMMA. Experimental testing of the energy balanced criterion. Polymer 1998; 39:4785–4791.

25. Cigna G, Matarrese S, Biglione GF. Effect of structure on impact strength of rubber reinforced polystyrene. J Appl Polym Sci 1976; 20:2285–2295.

26. Steenbrink AC, Litvinov VM, Gaymans RJ. Toughening of SAN with acrylic core-shell rubber particles: particles size effect or cross link density. Polymer 1998; 39:4817–4825.

27. Ramsteiner F, Heckmann W, McKee GE, Breulmann M. Influence of void formation on impact toughness in rubber modified styrenic polymers. Polymer 2002; 43:5995–6003.

28. Morbitzer L, Humme G, Ott KH, Zabrocki K. Struktur und Eigenschaften von ABS. XIV Additiv-Effekte in ABS. Angewandte Makromol Chem 1982; 108:123–140.

29. Bucknall CB, Ayre DS, Dijkstra DJ. Detection of rubber particle cavitation in toughened plastics using thermal contraction tests. Polymer 2000; 41:5937–5947.

30. Yamashita T, Nabeshima Y. A study of the microscopic plastic deformation process in poly(methyl methacrylate)/acrylic impact modifier compound by means of small angle x-ray scattering. Polymer 2000; 41:6067–6079.

31. Michler GH. Kunstsoff-Mikromechanik: Morphologie, Deformations- und Bruchmechanismen von Polymerwerkstoffen. München, Wien: Carl Hanser Verlag, 1992.

12

Deformation Mechanisms and Toughness of Rubber and Rigid Filler Modified Semicrystalline Polymers

C. HARRATS and G. GROENINCKX

Katholieke Universiteit Leuven, Department of
Chemistry, Laboratory for Macromolecular
Structural Chemistry, Heverlee, Belgium

CONTENTS

I. INTRODUCTION

The plastic deformation of polymeric materials can be grouped
into two main mechanisms: crazing and shear yielding [1–3].
These two modes of deformation are controlled by the molecular characteristics of the polymers such as chain flexibility
and chain entanglement density, and the testing conditions,
including specimen geometry, deformation speed and test
temperature. The type of specimen loading, for example tensile, flexural or compression loading, can also result in different deformation mechanisms. Among these parameters,
under standardized testing conditions, the molecular characteristics of the polymer remain the prevailing parameter that
directly or indirectly affects the deformation mechanism.

The mechanical performance of a given polymeric material depends on the type and the amount of the deformation
mode (either crazing or shear yielding) that prevails during
sample loading up to ultimate failure. Many commercial polymers, including glassy and semicrystalline thermoplastics
and thermosets, suffer from a deficiency in toughness. They
are not employed without improvement of their fracture resistance in applications where ductility is required. The strategy

of rubber toughening, which involves the blending of small amounts of a rubbery component with rigid polymers aiming at increasing their fracture resistance, has already been used since the late 1940s.

The main objective of the present chapter is to provide an in-depth understanding of the toughening of semicrystalline polymers such as thermoplastic polyesters, polyamides, and polyolefins in relation to their molecular and morphological parameters. The toughening approach that consists of dispersing a rubber phase in a brittle or a less ductile matrix will be exclusively considered. The influencing parameters that include the rubber particle size, the interparticle distance and the matrix molecular and morphological characteristics will be discussed in relation to the toughening performances.

II. TOUGHENING PRINCIPLES AND MECHANISMS

The two recognized modes of deformation, crazing and shear yielding, are considered as the structural change that occurs in a polymer or a polymer blend under deformation, which when reaching a critical extent, leads to its ultimate fracture. Depending on the chain flexibility and chain entanglement density of the polymer, crazing and shear yielding can occur separately or simultaneously at various proportions in the deforming matrix. Small changes in test parameters may produce shear yielding instead of crazing or vice-versa. Crazing and shear yielding are assumed to be independent processes, and the mechanism that requires the lowest stress will be the dominant mode of deformation that leads to material failure.

A. Multiple Crazing

A craze results from a microvoid generated under a tensile stress usually at locations of high stress concentrations such as flaws, defects, and foreign particles [1–7]. The microvoids are developed in a plane perpendicular to the direction of the largest tensile stress. A craze has the particularity of reflecting light and eventually gives rise to fracture if stressed sufficiently [8–10]. Unlike cracks, crazes are load bearing as a result of a web of microfibrils plastically stretched between the walls of what would otherwise be a crack. Crazes are important for many reasons:

- Microcracks initiate in crazes
- During the plane strain crack propagation, more crazes are formed at the crack tip constituting a plastic zone that blunts the crack
- They are a major source of ductility in rubber modified amorphous polymers

However, crazes have the great disadvantage of being frequently a precursor for brittle fracture. This is because the large plastic deformation and local energy absorption involved in crazes are often localized and confined to a very small volume of the material. A craze can thus be considered as a thin layer of polymer in which plastic deformation (~60%) and elastic deformation (~200%) in the stress direction have occurred without lateral contraction on a gross scale. As a result, the void fraction in the craze is about 40 to 75% or larger [11]. The thickness of a craze tip can be of the order of 10 nm while a mature craze body may be 10^2 to 10^4 times thicker. Far below the glass transition temperature of the polymer, a craze thickening occurs primarily by involving more polymer at the bulk interface. This will maintain the void content fairly uniformly throughout the craze. Thickening at high temperatures involves a coarsening of the structure through craze fibril breakdown [11]. Chain scission is assumed to be the primary cause of craze initiation [12–15]. In polymers having high entanglement density, the probability of breaking a chain is low because the load is distributed over different entanglements and different chains. It is therefore expected that the higher the entanglement density, the more different the craze initiation will be [16].

A craze develops and propagates by two processes: by craze tip advance that allows fibrils to be generated and by craze width growth, a normal separation of the two craze interfaces behind the craze tip. The accepted mechanism of craze tip advance is the Taylor meniscus instability [17]. This phenomenon can be observed, e.g., when separating apart two plates by forcing between them a liquid or when peeling an adhesive tape from a solid surface.

The craze width growth is much more important in generating most of the fibril structure than the tip advance. Several models describing the craze growth width exist, among them the one derived by Kramer and Berger [7]. The surface tension of the void surface (Γ) was considered by the authors in understanding the effect of the molecular chain parameters on the craze widening stress:

$$\Gamma = \gamma + \pi d V_e U \tag{1}$$

where γ is the van der Waals surface energy, V_e is the entanglement density, d is the entanglement mesh size and U is the polymer backbone bond energy. The term ($\pi d V_e U$) is an extra energy resulting from the chain scission of the entanglements crossing the interface; it is quite significant. The model predicts that increasing the entanglement density of the network leads to a substantial increase in Γ, and therefore to an increasing craze developing stress. The inter-relation between the entanglement density and the craze microstructure [18], which includes craze extension ratio [14,15,19,20], means fibril spacing can now be predicted by the model. Furthermore, the phenomenon of high temperature crazes and the transition from chain scission-dominated crazing to chain disentanglement-dominated crazing with increasing temperature are also elucidated.

Another simpler equation relates the craze initiation stress σ_z to the entanglement density V_e, (2):

$$\sigma_z \propto f_z V_e^{1/2} \tag{2}$$

where f_z is a function of the free volume reflecting the effect of the physical aging on the crazing stress. Note that σ_z is weakly dependent on temperature. A low entanglement density should favor crazing as illustrated in Figure 1 of Wu's plot showing the craze initiation stress as a function of entanglement density for a series of homopolymers and miscible

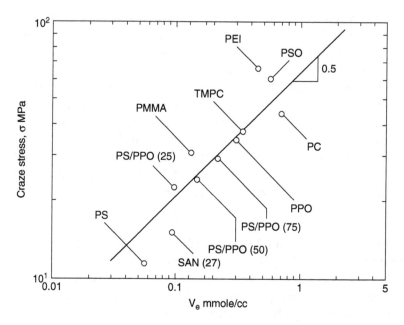

Figure 1 Craze initiation stress (σ_{craze}) versus entanglement density (v_e). (From S. Wu, Polym. Int., 29, 229, 1992. With permission.)

blends of polystyrene (PS) and polyphenylene oxide (PPO) [21–23]. For example, crazing is initiated at very low stress for polystyrene that exhibits the lowest entanglement density. Adding PPO to polystyrene at increasing concentration results in a significant resistance to crazing, i.e., a much larger stress is required to initiate crazing in blends containing a large amount of PPO (ca. 75 wt%).

Brittle polymers such as polystyrene and polymethyl methacrylate, where crazes are developed at strains as low as 0.3 to 1%, can absorb a larger amount of energy if the crazes are extended over a large volume. This can be achieved by greatly increasing the concentration of the crazes formed during deformation. This leads to the phenomenon of multiple crazing which is now recognized to be the dominant mechanism by which rubber-modified glassy polymers withstand deformation. This mechanism has been demonstrated, particularly at low testing temperatures, in high-impact polystyrene (HIPS) [24], acrylonitrile-butadiene-styrene copolymer (ABS) and in rubber-toughened polymethylmethacrylate (PMMA). The existence of multiple crazing was evidenced by both optical and transmission electron microscopy [5], as well as by using real time small-angle x-ray scattering [25,26]. Magalhaes and Borggreve have also used x-ray scattering technique to determine the fractional contribution of crazing to both the volume strain and the extension of the test bar [27]. It was concluded that crazing of the matrix in HIPS does not exceed 50% of the volumetric expansion of the specimen. Both x-ray patterns and microscopy observation revealed that rubber particle cavitation and matrix crazing are important in yielding of HIPS and ABS.

In rubber-toughened glassy polymers, the crazes were found to be initiated as a result of stress concentration at the equator of the rubber particle [28]. Multiple crazing involves a large volume of the rubber-modified polymer matrix, which results in a significant increase in toughness.

The toughening efficiency of the rubber phase depends on its particle size. A particle size window of 1–5 μm has been proposed for optimum toughness [1,27]. Rubber particles of smaller size are not able to initiate crazes and are less effective in terminating a growing craze [1]. An empirical relation (Equation 3) is found between the optimum diameter ($d_{optimum}$) of the rubber particle and the entanglement density (V_e) of the matrix.

$$Log(d_{optimum}) = 1.19 - 14.1V_e \qquad (3)$$

B. Multiple Shear Yielding

Bucknall [29] described shear yielding as the process by which most ductile materials extend to high strains in standard tests. In shear yielding, a displacement of matter (molecules and atoms slip past each other) takes place in the material under deformation. In contrast to crazing, in shear yielding no change in volume or density is associated with deformation. Also, no loss in cohesion occurs in a shear yielding deformation mechanism. Von Mises proposed a criterion for shear yielding to occur which is based upon a critical value reached by the effective stress σ_e, defined as:

$$\sigma_e \equiv \left(\frac{(\sigma_1 - \sigma_2)^2 + (\sigma_2 - \sigma_3)^2 + (\sigma_3 - \sigma_1)^2}{2} \right)^{1/2} \geq \sigma_0 \quad (4)$$

where σ_1, σ_2 and σ_3 are the principal stresses, acting normal to the 1, 2 and 3 directions of a cube of material, respectively. If σ_1 is taken as the tensile yield stress σ_{ty} and $\sigma_2 = \sigma_3 = 0$, then σ_0 is constant and equal to σ_{ty}.

Equation 4 is valid for many metals but requires modification for polymers since the critical effective stress is not constant. It varies almost linearly with the mean stress σ_{m}, defined also as negative pressure. A modified Von Mises equation for the critical strain energy density was proposed:

$$\sigma_e > \sigma_0 - \mu\sigma_m \equiv \sigma_0 - \mu \, (\sigma_1 + \sigma_2 + \sigma_3)/3 \quad (5)$$

The process of shear yielding has not been well explained on a molecular scale as is the crazing deformation mechanism, although many attempts have been made over the years. The characteristic ratio (Equation 6) which is a measure of the chain rigidity is thought to be the predominant controlling molecular parameter for shear yielding behavior.

$$C_\infty = \lim_{n \to \infty}(R_0^2 / nl^2) \quad (6)$$

where R_0^2 is the mean square end-to-end distance of an unperturbed chain, n is the number of statistical skeletal units, and l^2 is the mean-square length of a statistical unit.

The yield stress σ_y is proportional to the chain rigidity C_∞ (Equation 7):

$$\{\sigma_y\} \propto f_y C_\infty \quad (7)$$

where f_y is a function of the free volume that accounts for the effect of the physical aging. A normalized yield stress dependent strongly on temperature is defined as follows (Equation 8):

$$\left\{\sigma_y\right\} = \sigma_y / \left[\delta^2\left(T_g - T\right)\right] \tag{8}$$

where $(T_g - T)$ is the difference between the glass transition temperature of the sample and the testing temperature, and δ^2 is the cohesive energy density. It can thus be deduced that the higher the chain rigidity, the higher the initiation stress for shear yielding will be. In Figure 2 a plot of the normalized yield stress as a function of the characteristic ratio is presented for a series of polymers and miscible blends. For example, polystyrene homopolymer and styrene-acrylonitril copolymer exhibit a high chain rigidity which results in a high initiation stress for shear yielding and will prematurely deform by a crazing deformation mechanism.

Shear yielding which can be considered as a type of viscous flow is more temperature dependent than crazing.

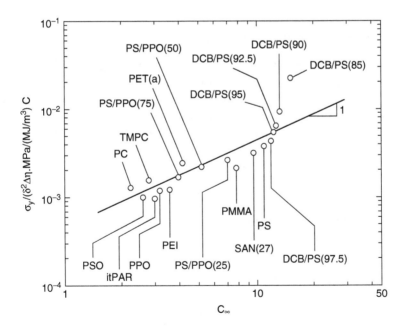

Figure 2 Normalized yield stress [σ_{yield} / δ^2 (Tg-T)] versus characteristic chain stiffness (C_∞). (From S. Wu, Polym. Int., 29, 229, 1992. With permission.)

Eyring's theory [30] describes the viscous flow in liquids. Adaptations have been made on that theory to express the yield stress as a function of temperature and strain rate as follows [31,32]:

$$\frac{\sigma_{yield}}{T} = \frac{4\sqrt{3K}}{V\eta_0} \ln\left[\frac{\sqrt{3}}{2\gamma_0 J_0} \frac{d\varepsilon}{dt} + \frac{\Delta G_\circ}{RT} \right] \qquad (9)$$

where σ_{yield} is the yield stress, T the absolute temperature, K the Boltzmann constant, V the activation volume, $(d\varepsilon/dt)$ the strain rate, ΔG_o the activation energy, R the universal gas constant, γ_0 an elementary shear strain and J_0 a rate constant.

Using this approach, the yield stress of polymers was shown to depend on the strain rate as well as on temperature [30–38]. This equation fits the experimental data well over a broad temperature interval and over several decades of strain rate.

C.   Intrinsic Brittle/Ductile Behavior

The transition between a crazing mechanism (brittle behavior) and shear yielding mechanism of deformation (ductile behavior) is a crucial phenomenon in considering the toughness of polymers. Equation 10 expresses the molecular criterion that controls the predominant deformation mode [21–23]:

$$\frac{\sigma_z}{\{\sigma_y\}} \propto \frac{V_e^{1/2}}{C_\infty} \qquad (10)$$

This equation can also be expressed as:

$$\frac{\sigma_z}{\{\sigma_y\}} \propto \left(\frac{3M_v}{\rho_a} \right)^{1/2} V_e \qquad (11)$$

where ρ_a is the amorphous mass density and M_v is the average molar mass per statistical unit. Henkee and Kramer have shown that the chain entanglement density in the polymer is a critical parameter that determines whether the polymer deforms via crazing or shear yielding [39]. A low entanglement density favors a crazing mechanism of deformation, whereas increasing the number of entanglements above a critical value results in a shear yielding behavior. Based on that, a majority of polymers can be classified as:

1. Brittle polymers when V_e < ~0.15 mmol/cm^3 and C_∞ > ~7.5. This class of polymers fails by a dominant crazing mechanism; they both have a low crack initiation energy (low unnotched toughness) and a low crack propagation energy (low notched toughness). Polystyrene, polymethylmethacrylate and styrene-acrylonitrile copolymer are typical examples of brittle polymers.

2. Pseudo-ductile polymers when the following conditions are fulfilled: V_e > ~0.15 mmol/cm^3 and C_∞ < ~7.5. The pseudo-ductile polymers tend to fail by shear yielding deformation mechanism. These polymers exhibit a high crack initiation energy (high unnotched toughness) and a low crack propagation energy (low notched toughness). Polyamides (PA6 and PA66), polyesters (polyethylene terephthalate and polybutylene terephthalate) and polycarbonate are common examples of pseudo-ductile polymers.

3. An intermediate class includes polymers that fulfill the critical criterion of V_e ~ 0.15 mmol/cm^3 and C_∞ ~ 7.5. These polymers exhibit combined crazing and shear yielding deformation mechanisms. Examples of these polymers include the less brittle PMMA of the first class and the less ductile polymers such as polyoxymethylene and polyvinylchloride of the pseudo-ductile class.

Although both V_e and C_∞ provide consistent prediction of the deformation behavior, the entanglement density V_e remains the primary parameter that controls the crazing behavior whereas C_∞ the characteristic ratio, is the predominant factor that controls the shear yielding behavior. More details on molecular criteria for craze/yield behavior in relation to chain structure parameters were summarized by Dompas and Groeninckx elsewhere [40].

D.   Deformation and Fracture of Polymer Blends

The rubber-toughening technique involves the dispersion of a rubbery component in a rigid polymer matrix. The aim of toughening is mainly to increase the impact strength beyond a given target with a tolerated loss in stiffness (modulus), yield stress and creep resistance. The fracture behavior of rubber-toughened polymers depends on a number of parameters among which are the blend composition, the phase morphology developed and the testing conditions.

Rubber toughening has been applied to amorphous polymers such as polystyrene [41–43], polycarbonate [44,45] and polyvinylchloride [46–50], and to semicrystalline thermoplastics including polyamides [51–53], polypropylene [54,55], and thermoplastic polyesters such as polyethylene terephthalate and polybutylene terephthalate [44,56,57]. Epoxies as brittle thermosets have also been rubber toughened [58,59].

It has been proven that the use of dispersed particles of rubber having a given particle size are able to involve as much as possible of the deforming matrix into the process of energy absorption. Simultaneously, there is a need to limit the growth and breakdown of voids and crazes in order to prevent premature crack initiation.

The early hypothesis attributed the enhancement of toughness to dissipation of energy in the rubbery phase [60]. This hypothesis did not hold long since it was found that the total energy associated with the rubber deformation was too small to explain the observed fracture energy of the toughened polymer. It is accepted nowadays that the deformation processes are initiated by the rubber particles from many sites of the matrix [29,61,62]. The rubber particles alter the stress field in the surrounding matrix, inducing an extensive plastic deformation.

Multiple crazing was first proposed by Bucknall [61] as the dominant toughening mechanism in high impact polystyrene HIPS, and was later generalized as the mechanism of deformation in toughened brittle polymer matrices. The rubber particle was found to be responsible for the initiation and termination of craze growth. Crazes are initiated at the equator of a rubber particle where the stresses built up to reach their maximum level. The craze propagates until it is stopped by an adjacent rubber particle (long craze formation is prevented). Multiple crazing is the process of forming a large number of short crazes rather than forming a small number of long crazes as is the case in the absence of rubber particles.

The particle size of the rubber phase is found to be of utmost importance in controlling the toughening mechanisms. Bucknall [61] proposed a rubber particle size range of 1–5 μm where HIPS exhibits a tough behavior. Out of this range, HIPS is brittle. Small rubber particles are ineffective for craze initiation and cannot efficiently control the craze growth and its termination [43]. Kramer and Donald [19] reported also that particles smaller than 1 μm were unable to initiate crazes in either ABS or HIPS.

1.   Multiple Shear Yielding in Toughened Polymers

A brittle fracture will result if the shear yielding mechanism is prohibited mainly in the middle of the sample or in front of the crack tip. This occurs mainly in samples with a large thickness or in Charpy notched samples. The surrounding less highly stressed material resists the lateral contraction needed to maintain a constant volume.

To resist the high level of stresses having a large degree of triaxiality, the rubber particles cavitate or debond, creating a void. Upon the void formation, the triaxiality is relieved ahead of the notch or the running crack [61–63]. The stress state in thin ligaments of the matrix between rubber particles is converted from a triaxial to a biaxial one. The stress state in the matrix material in the neighborhood of the cavity is thus effectively altered via the creation of an elastic stress field overlap of the cavitated particles. Shear yielding deformation is favored by a biaxial stress state, whereas crazing is enhanced under a triaxial stress state. Multiple shear yielding is favored by the presence of neighboring rubber particles in the matrix.

The deformation behavior of rubber-toughened thermo-plastics is commonly investigated using tensile dilatometry [45,64,65], light scattering [46–48,66], or x-ray scattering [67] measurements. These techniques are applied in real time during the uniaxial tensile deformation. In addition, the fracture zone (stress-whitened area) obtained after the loading operation is studied via direct observation using microscopic tools. Unfortunately all these techniques do not provide information on smaller scale and earliest stages of the cavitation process. To understand the initial stages of the rubber cavitation apart from the matrix yielding, Bucknall and coworkers [68,69] have developed a dynamic mechanical-thermal spectrometer (DMTS) which allows the measurement of the glass-transition temperature of the rubber-toughened polymer under deformation.

Many morphological parameters influence to a different extent the shear yielding deformation behavior of toughened polymers. The rubber concentration and rubber particle size are the most predominant.

2.   Effect of Rubber Concentration

The impact strength of ductile polymers was found to increase as a function of the rubber concentration [51,53,70]. Unfortunately a decrease of the modulus and yield strength is associated with the improvement in impact strength [52,71]. The brittle-ductile transition, which is a crucial parameter aimed at in toughened polymers, is shifted toward lower temperatures as the rubber content is increased in the blend [44,57]. That statement is well illustrated in Figure 3, as reported by Gaymans [72]. Model calculations [42,73] indicate that the process of rubber particle cavitation is strongly influenced by the neighboring particles. This means that a few percent of rubber is not sufficient to relieve all of the volume strain [65]. On the other hand, the multiaxial stress state will be more difficult to develop if the rubber particles are present at high concentration. Furthermore, the particle size distribution affects the local concentration and the local matrix yielding [74]. In the regions of high rubber concentration, yielding

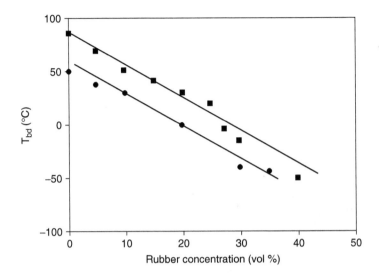

Figure 3 Brittle-ductile transition temperature as a function of rubber content, as measured by SENT, on a PP-EPDM blend at the indicated test speeds. (From R. J. Gaymans, Polymer Blends 2: Performance, eds., D. R. Paul and C. B. Bucknall, New York: John Wiley & Sons, 2000. With permission.)

takes place at a lower stress than in the regions of lower rubber concentration.

The pure polymers that deform by shear yielding do not show significant volume strain during their tensile loading. In contrast, in the blends, the onset of rubber cavitation occurs at low strains (2–4%), and before the yield point of the neat polymer is reached [65,75].

3.   Role of Particle Size of the
      Rubber Phase

It is now commonly recognized that the rubber particles play two major roles in the toughening of ductile polymers: to change the stress state in the matrix around the particles by cavitation, and to generate a local stress concentration [1–10]. The particles should not themselves be initiation sites for the fracture process; they should be sufficiently small that

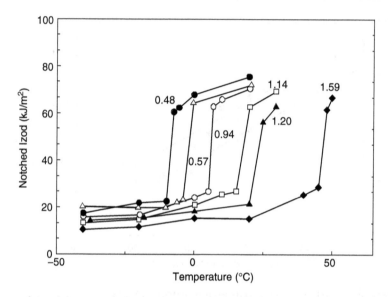

Figure 4 Notched Izod measurements as a function of temperature for PA-6 and PA-6-EPDM-g-MA blend (26 vol%) at the indicated weight-average rubber particle sizes (μm). (From R. J. M. Borggreve et al., *Polymer*, 28, 1489, 1987. With permission.)

they do not grow (during cavitation) to a critical volume that causes crack initiation to take place. A decrease in the diameter of the rubber particle at a constant rubber content results in a decrease in the brittle-ductile transition temperature (Figure 4).

Wu [76] demonstrated that the brittle-ductile transition occurs at a critical interparticle distance in pseudo-ductile matrices. The critical interparticle distance remains thus a crucial morphological parameter contributing to the toughening efficiency of rubber-modified pseudo-ductile matrices. The percolation concept remains the only plausible explanation of this phenomenon [77,78]. If the rubber particles are able to cavitate internally and if the generated voids are close enough, then the thin matrix ligaments are interconnected and the yielding process propagates over the entire sample (measure shear yielding), promoting ductile deformation behavior. This will occur when the thickness of the matrix ligament is smaller than the critical interparticle distance. For a given volume fraction, this is achieved by decreasing the particle size and by enhancing the state of dispersion. The decrease in particle size results in a shift of the brittle-ductile transition temperature (Tbd) to lower values [76,79]. A minimum has been reported to exist below which decrease in brittle-tough transition is no longer observed [46,66,70,80–83]. The reason is that particles that are too small are not able to cavitate and, as a consequence, do not relieve the hydrostatic tension in the material to promote a ductile shear yielding behavior.

The criteria for rubber cavitation have been studied by Lazerri and Bucknall [84] and also by Dompas and Groeninckx [46]. The cavitation ability of the rubber particles was found to depend on the volume strain imposed upon sample loading. The model explains the increasing resistance against cavitation with decreasing rubber particle size. The model of Groeninckx and Dompas leads to the following equation:

$$d_o = \frac{12(\gamma_r + \Gamma_{sc})}{K_r \Delta^{4/3}} \tag{11}$$

where d_o is the particle diameter, γ_r the van der Waals surface tension, Γ_{sc} the surface energy per unit area, K_r the rubber bulk modulus and Δ the relative volume strain. Equation 11 predicts that large particles will cavitate in the early stages of the deformation process, while small particles will cavitate at a later stage. As a consequence, the cavitation resistance of the rubber particles increases as the particle size decreases. The minimum particle size required for cavitation therefore depends on the maximum volume strain that can be attained in the material.

The interparticle distance (ID) which is defined as the distance between two adjacent rubber particles, referred to as the matrix ligament thickness, is a crucial morphological parameter which affects the toughening efficiency in rubber modified matrices. Wu [23] derived the equation (12) which allows for calculating ID as a function of the rubber particle size (D) and the rubber concentration (Φ_r).

$$ID = D \left[K \left(\frac{\pi}{6\Phi_r} \right)^{1/3} - 1 \right] \qquad (12)$$

K is a measure for the lattice packing arrangement with $K = 1$ for a cubic lattice and $K = \sqrt{3/2}$ for a body-centered lattice. It was proposed that a fracture is tough when ID is below a critical value, independent of whether the decrease results from an increase in rubber content or a decrease of the particle size. Figure 5 illustrates the interrelation between the impact strength and the interparticle distance ID in a rubber modified nylon 6,6. The fracture deformation behavior is found to drastically change from brittle to ductile at the same D,

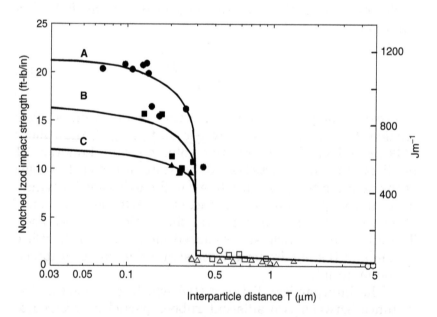

Figure 5 Notched Izod impact strength vs. interparticle distance (T) in PA 6,6/reactive rubber blends (curve A: 10 wt% rubber, Curve B: 15 wt%, Curve C: 20 w% rubber). (From S. Wu, Polymer, 26, 1855, 1985. With permission.)

regardless of the dispersed rubber concentration. The physical interpretation of the interparticle distance is not well understood, although Wu et al. were the first ones to propose that the interparticle distance affects the overlap of stress concentration fields around the particles. According to a percolation model used by Margolina and Wu [85], the brittle-tough transition takes place when the yielding process propagates through the connecting thin matrix ligaments (< IDc).

III. TOUGHENING OF SEMICRYSTALLINE POLYMERS USING RUBBER AND RIGID FILLERS

Semicrystalline polymers exhibit low-impact strength when deformed notched, under high deformation rates and usually in low-temperature applications. They rarely break in their unnotched form. This implies that crack initiation is very difficult due to the presence of crystalline lamellae, whereas the crack propagation is a fairly low-energy failure process.

Polyamides and thermoplastic polyesters are important engineering polymers used in a very broad range of applications. Unfortunately, they suffer from a lack of toughness when required for applications where this property is of crucial importance. Consequently, an intensive research activity was devoted to their toughening using rubbers of various natures. The most commonly encountered rubbers employed as toughening agents for polyamides or polyesters include ethylene-propylene-diene-monomer (EPDM), polybutadiene (PB), ethylene-propylene rubber (EPR), acrylonitrile-butadiene-styrene (ABS), styrene-ethylene-butylene-styrene (SEBS), and styrene-butadiene-styrene (SBS). The choice of a rubber phase is also crucial for the toughening performance. Borggreve et al. [86] showed that the impact behavior improves with decreasing modulus of the impact modifier. The brittle-ductile transition temperature was found to increase with increasing the modulus of EPDM rubber by higher crosslinking [87]. Those findings were ascribed to the ability of the rubber phase to cavitate (the lower the modulus, the higher the extent of cavitation) [5,7,8]. Jiang et al. recently modeled the effect of elastomer stiffness on the brittle-tough transition in the toughening of thermoplastics [88]. A good correlation between brittle-tough transition temperature (Tbd) and the stiffness of the elastomer has been established. The model calculations reveal that the modulus of the elastomer must be one-tenth or less that of the matrix in order to produce a tough blend combination at low temperature.

Because of the high interfacial tension, blending an apolar rubber with the highly polar polyamides or polyester homopolymers results in a gross phase dispersion of rubber which does not meet the requirements of particle size for an efficient toughening. In order to be able to tailor and control the particle size of the rubber phase, a "compatibilization" approach that allows for the reduction of the particle size to an optimum extent is necessary. The addition of premade compatibilizing agents or their generation *in situ* via a chemical reaction of reactive precursors during the melt-blending process of the rubber phase with the matrix are the two main techniques of compatibilization.

Polyamides have been toughened using EPDM-g-MA, EPR-g-MA or SEBS-g-MA graft copolymers that are able to react with polyamide via the readily reacting amine or amide groups of polyamide 6 with the maleic anhydride groups of the graft rubber copolymer. Unmodified rubber was also added to prevent the consumption of a large amount of expensive reactive maleic anhydride-grafted copolymer [89–93]. A mixture of polypropylene with maleic anhydride-grafted rubber was successfully employed [94–96]. An ABS copolymer and other core-shell rubbers compatibilized with styrene-maleic anhydride copolymer (SMA) were also used [97–100].

A.   Thermoplastic Polyesters

The most commonly investigated thermoplastic polyesters for the enhancement of their toughening behavior are polyethylene terephthalate (PET) and polybutylene terephthalate (PBT), which are engineering thermoplastics used for a wide range of applications. They exhibit almost similar chemical and physical behavior, but they substantially differ in their thermal behavior, i.e., melting temperature, glass-transition temperature and crystallization kinetics. PET exhibits a higher melting temperature and slower crystallization kinetics than PBT.

To our knowledge, the major part of the open literature reporting about the toughening of thermoplastic polyesters mainly deals with PBT rather than PET. Different types of rubber or rubber-based polymers have been used for the toughening of PBT or PET. These include core-shell particles [56,101–104], ABS functionalized or not [57,105–108] and functionalized rubber [109–118]. Reactive compatibilization of the rubber phase and the PBT matrix is necessary for particle size reduction of the rubber phase and for insuring a sufficient interfacial adhesion necessary for the toughening of pseudo-ductile matrices. Hobbs et al. [101,102] reported on the modification of PBT with core-shell modifiers with or without the presence of polycarbonate which ensures compatibilization (PC is miscible with PBT) of the matrix with the rubbery phase. The addition of polycarbonate was found to shift the brittle-ductile transition to lower temperatures and enhance particle dispersion. The increased shear yielding behavior observed in toughened PBT under deformation was attributed to the increased ductility of the matrix via a modification of its amorphous interlamellar region by the presence of polycarbonate. A similar approach used by Brady and co-workers [103] also led to similar results; however, no significant impact improvement was observed when the core-shell rubber particles were employed without modification. Addition of a small amount of polycarbonate resulted in improved stress-strain behavior, improved impact strength and a substantial decrease of the Tbd.

ABS has also been used as an impact modifier for PBT [57,107,108]. In the absence of compatibilizer, 30 wt% ABS were necessary to produce a tough PBT. Extrusion and molding conditions were found strongly to affect the toughness of the ABS-modified PBT. When a terpolymer of methyl methacrylate, glycidyl methacrylate or ethyl acrylate was used as compatibilizer, a more stable phase morphology with much better dispersed ABS particles were obtained. A minimum of

30 wt% of ABS containing high rubber fraction and having a low melt viscosity was necessary to obtain improved impact strength and low temperature toughness. The reactive compatibilizer was used at concentrations not exceeding 1 wt%.

The second approach of toughening PBT consists of using common elastomers such as EPR and EPDM. In most cases, a reactive compatibilization process is required to control the rubber dispersion, its particle size and its adhesion to the matrix. Reactive maleic anhydride-grafted EPR or EPDM have been used in PBT [109,110]. A critical interparticle distance IDc of 0.4 μm was determined for PBT/EPR-MA blends containing 20 wt% EPR-MA rubber phase. Similar value (0.43 μm) was reported for PBT/polyarylate miscible mixture toughened using maleic anhydride-grafted poly(ethylene-octene) copolymer [119].

A correlation between the impact strength of the blend and the interparticle distance was claimed to be independent of the matrix viscosity (molecular weight), the rubber content and the extent of interfacial adhesion between the rubber phase and the PBT matrix. Maleic anhydride-grafted styrene/ethylene-butylene/styrene triblock copolymer (SEBS-g-MA) has also been used as a toughening agent [110]. A critical interparticle distance of 0.16 μm has been reported to result in an optimum impact improvement. Sanchez-Solis [116] employed a styrene-butadiene-styrene grafted maleic anhydride (SBR-g-MA) copolymer in PET. A 2.5-fold impact strength increase was reported compared to the unmodified PET. A lack of compatibilization reaction, as indicated by the poor particle size reduction, was claimed to be at the origin of the insufficient toughness improvement of the blend.

Glycidyl methacrylate (GMA) functional groups used in reactive compatibilization of PBT with rubber were found to be more efficiently reactive towards the hydroxyl groups of PBT than maleic anhydride groups. Recently Loyens and Groeninckx published an interesting study on the toughness and ultimate mechanical properties of rubber-modified polyethylene terephthalate [120–122]. Ethylene-co-propylene rubber with and without reactive functional groups were melt-blended with PET. The reactive modifiers include maleic anhydride-grafted EPR (EPR-g-MA), glycidyl methacrylate-grafted EPR (EPR-g-GMA$_x$) and ethylene-glycidyl methacrylate copolymers (E-GMA$_x$). It was found that the most efficient route of designing a toughened PET is provided by dispersing in it a pre-blend of EPR and E-GMA$_x$. A minimum concentration of EPR-dispersed phase is needed to obtain a significant enhancement of the impact strength and to induce a brittle-ductile transition. As expected, the impact behavior of toughened PET was primarily controlled by the rubber interparticle distance. A critical interparticle distance of 0.1 μm was determined experimentally for the GMA reactively compatibilized blends. This interparticle distance was found to be independent of the GMA content, the chemical technique of attaching GMA to EPR (i.e., via copolymerization or grafting) and of the nature of the reactive compatibilizer used. An ethylene-glycidyl methacrylate copolymer containing 8 wt% of GMA functional groups used as a compatibilizer in EPR/PET blend provided the best ultimate mechanical properties; a 15-fold impact strength increase with respect to unmodified PET was observed (Figure 6). In Table 1 the weight average particle diameter of the rubber phase and ultimate mechanical properties of PET/(EPR/E-GMA8) blends as a function of the rubber concentration and composition are given.

The deformation mechanisms of rubber-toughened semicrystalline polyethylene terephthalate were investigated by Loyens and Groeninckx using fractography on impact fractured samples and tensile dilatometry [122]. The blends investigated consisted of EPR/E-GMA8 rubber dispersed in the PET matrix. The weight average particle diameter (Dw), the interparticle distance (ID) and brittle-ductile transition temperature (Tbd) of PET/(EPR/E-GMA8) blends as a function of the dispersed phase concentration and composition are presented in Table 2. A strong correlation between ID and the brittle-ductile transition temperature was found. The blends containing 30 wt% EPR/E-GMA8 exhibit the lowest brittle-ductile transition temperatures. An interparticle distance of 0.1 μm is required in order to obtain a room temperature tough PET. It has to be noted that interparticle distances below this value can only be reached in efficiently compatibi-

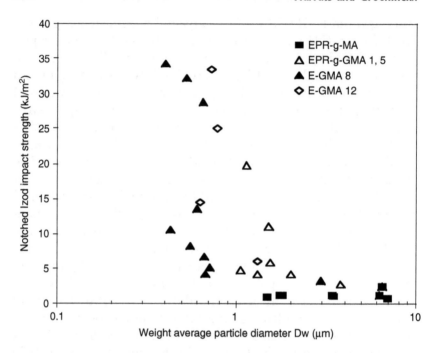

Figure 6 Notched Izod impact strength as a function of the dispersed phase particle size for the various rubber-modified PET systems. (From W. Loyens and G. Groeninckx, Polymer, 43, 5679, 2002. With permission.)

lized blends. Critical particle sizes of 0.19, 0.44 and 1.1 μm have been calculated for rubber concentrations of 10, 20 and 30 wt%, respectively, using Wu's equation and an interparticle distance of 0.1 μm.

It is also important to emphasize that the interparticle distances are not necessarily the same for different rubbers in the same matrix. Kanai et al. [110] have reported two different IDs for PBT/EPR-g-MA and PBT/SEBS-g-MA blends. The ID certainly depends indirectly on the extent of interfacial compatibilization between the dispersed phase and the matrix. Efficient compatibilization reduces the particle size of the dispersed phase and enhances the quality of phase dispersion, which directly affects the average particle size of the dispersed phase and thus the interparticle distance. The

TABLE 1 Weight Average Particle Diameters and Mechanical Properties of the PET/EPR/E-GMA8) Blends as a Function of the Dispersed Phase Concentration and Composition

Weight Fraction of Compatibilizer in Dispersed Phase Composition	D_w (μm)	Impact Strength (kJ/m^2)	Modulus (MPa)	Elongation at Break (%)
		10 wt%		
0	2.92 ± 0.18	3.2 ± 0.2	2635 ± 182	2.5 ± 1.3
0.1	0.71 ± 0.07	5.1 ± 0.2	2274 ± 23	6.9 ± 1.1
0.25	0.67 ± 0.06	4.2 ± 1.2	2235 ± 59	7.5 ± 1.4
0.4	0.66 ± 0.02	6.7 ± 0.3	2269 ± 80	4.6 ± 0.4
0.75	—	5.0 ± 1.5	2388 ± 100	8.4 ± 1.9
1	—	3.7 ± 0.5	2389 ± 125	2.5 ± 0.4
		20 wt%		
0	6.45 ± 0.61	2.4 ± 0.3	1820 ± 69	3.6 ± 0.7
0.1	0.55 ± 0.08	8.3 ± 2.8	1888 ± 57	4.5 ± 0.5
0.25	0.61 ± 0.10	13.6 ± 2.1	1853 ± 65	9.6 ± 1.9
0.4	0.43 ± 0.08	10.6 ± 1.3	1844 ± 87	9.5 ± 3.5
0.75	—	14.0 ± 1.4	1723 ± 120	13.3 ± 1.4
1	—	11.7 ± 2.2	1793 ± 60	37.9 ± 14.5
		30 wt%		
0	9.49 ± 0.91	1.3 ± 0.2	1423 ± 114	1.9 ± 0.2
0.1	0.65 ± 0.09	28.8 ± 5.3	1384 ± 59	4.9 ± 1.0
0.25	0.40 ± 0.03	34.3 ± 3.8	1295 ± 66	6.2 ± 0.9
0.4	0.53 ± 0.04	32.2 ± 9.3	1302 ± 18	18.0 ± 4.6
0.75	—	20.5 ± 1.8	1223 ± 72	18.4 ± 5.5
0.9	—	17.3 ± 4.8	1183 ± 56	28.9 ± 13.9
1	—	44.8 ± 13.2	1145 ± 129	4.7 ± 2.4

Source: W. Loyens and G. Groeninckx, Polymer, 43, 5679, 2002. With permission.

interparticle distance is not a molecular parameter; it is a geometrical factor that does not necessarily and directly depend on molecular characteristics of either the matrix or the dispersed phase. PET/EPR blends compatibilized using GMA functionalized ethylene (E-GMA) were found to exhibit the same critical interparticle distance IDc independent of the content of GMA functionality (refer to Figure 7). When

TABLE 2 Weight Average Particle Diameter (D_w), Interparticle Distance (ID) and Brittle-Ductile Transition Temperature (T_{bd}) of the PET/(EPR/E-GMA8) Blends as a Function of the Dispersed Phase Concentration and Composition

Weight Fraction of E-GMA8 in the Dispersed Phase (–)	D_w (µm)	ID (µm)	I_{bd} (°C)
	10 wt%		
0	2.92 ± 0.18	1.49	80
0.1	0.60 ± 0.07	0.31	80
0.25	0.67 ± 0.06	0.34	80
0.4	0.80 ± 0.02	0.40	80
	20 wt%		
0	6.45 ± 0.61	1.45	80
0.1	0.55 ± 0.08	0.12	40
0.25	0.61 ± 0.09	0.14	31
0.4	0.43 ± 0.08	0.10	50
	30 wt%		
0	9.5 ± 0.9	0.56	80
0.1	0.65 ± 0.09	0.058	–5
0.25	0.41 ± 0.03	0.036	–15
0.4	0.53 ± 0.04	0.047	–10

Source: W. Loyens and G. Groeninckx, Polymer, 44, 4929, 2002. With permission.

the type of copolymer is considered, no significant differences are observed between a copolymerized GMA or a grafted one. Kanai et al. [110], Borggreve and Gaymans [91] and Wu [76] also have reported that the IDc was independent of the functionality of the compatibilizer or the rubber phase content in polyamide matrices. This is of course true if all the compatibilizing agents (copolymers) generate an optimum (required) particle size reduction. The effect of the molecular weight of PET matrix on the toughening characteristics of blends of PET/EPR containing 8 wt% E-GMA was also investigated. As shown in Figure 8, the IDc is the same (0.1 µm) for all blends

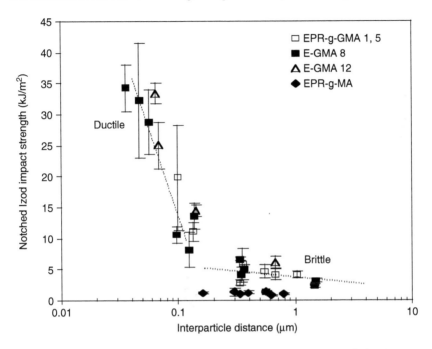

Figure 7   Notched Izod impact strength as a function of the inter-particle distance for the various rubber modified PET systems. (From W. Loyens and G. Groeninckx, Polymer, 43, 5679, 2002. With permission.)

independent of the molecular weight of the PET matrix. The blends based on the low molecular weight (L-MW) PET were unable to exhibit high-impact toughness and ductile fracture mode. In contrast, the medium molecular weight (M-MW) and the high molecular weight (H-MW) PET based blends exhibit a ductile fracture mode with a clear transition between the brittle and the ductile fracture mode at the IDc of 0.1 μm. Below IDc, the impact strength increases linearly with a decrease in the interparticle distance. It is obvious that the effect of PET matrix molar mass on the impact behavior of rubber-toughened PET originates from its direct effect on the phase morphology development, rather than from an intrinsic effect of the molecular weight itself.

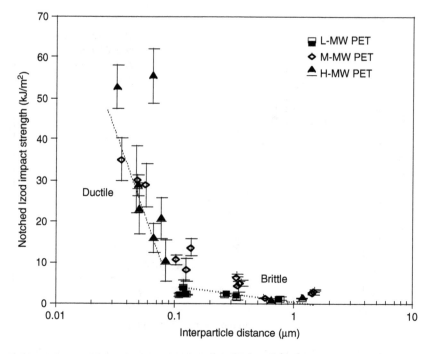

Figure 8   Notched Izod impact strength as a function of the inter-particle distance for the investigated PET (EPR/E-GMA) blends with varying PET matrix molar mass. (From W. Loyens and G. Groeninckx, Polymer, 44, 123, 2002. With permission.)

Another correlation that needs to be mentioned in this context is the effect of the interparticle distance on the brittle-ductile transition temperature for various molar masses of the PET matrix. In Figure 9 the Tbd is plotted as a function of interparticle distance for a PET/EPR compatibilized using E-GMA8 reactive copolymer having an EPR/E-GMA8 content of 30 wt%. A medium molecular weight (M-MW) PET is compared to a high molecular weight (H-MW) PET. A decrease of the interparticle distance results in a significant decrease of the Tbd.

The phenomenon responsible, among others, for the toughening of rubber-modified blends is the rubber particle cavitation. It has been evidenced in many toughened blend

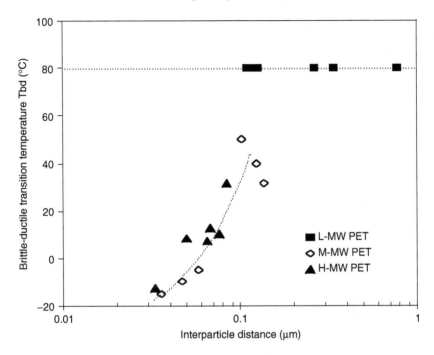

Figure 9 Brittle-ductile transition temperature as a function of the interparticle distance for the M-MW and H-MW PET blends. (From W. Loyens and G. Groeninckx, Polymer, 44, 123, 2002. With permission.)

systems using different methods, including direct observation via microscopic tools. Loyens and Groeninckx [122] investigated via microscopic examination on fractured surfaces the cavitation process in E-GMA8-modified EPR/PET blends. The cavitation process by which a ductile fracture has been induced is clearly evidenced in Figure 10 on SEM pictures of a smoothed specimen taken perpendicular to the fracture plane in a notched and impact fractured sample. The dark holes result from voiding of the rubber particles.

Furthermore, the effect of the test temperature on the deformation and fracture of rubber-toughened PET has also been investigated using SEM on fractured samples. In Figure 11 the SEM micrographs of the impact fracture surfaces

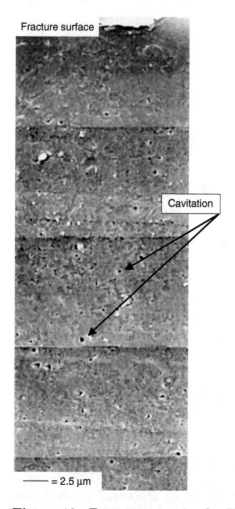

Figure 10 Fracture zone in the YZ plane of the PET/(EPR/E-GMA8) 80/(15/5) blend (smoothed and non-etched), fracture at room temperature. The fracture plane is located at the top, and the fracture propagates from right to left. (From W. Loyens and G. Groeninckx, Polymer, 44, 4929, 2002. With permission.)

(notch plane) of EPR toughened PET at different temperatures are presented. Note that at −25°C (Figure 11A-B), the blends are fractured well below their Tbd. They exhibit a limited amount of plastic deformation, directly behind the

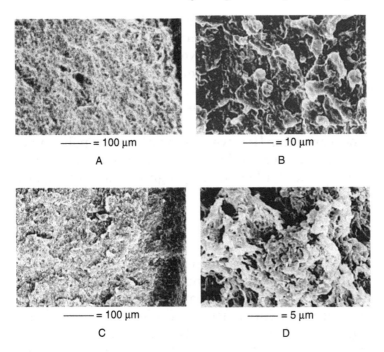

Figure 11 SEM micrographs of the fracture surfaces in the XY plane of the PET/(EPR/E-GMA8) 70/(22.5/7.5) blend fractured at various test temperatures: [A,B] –25°C; [C,D] 60°C. The left column presents the structure directly behind the notch whereas the right column presents the structure at +500 μm. (From W. Loyens and G. Groeninckx, Polymer, 44, 4929, 2002. With permission.)

notch. The crack propagates, however, in an unstable manner, resulting in an overall brittle fracture mode. At 60°C, a higher temperature than the Tbd, Figure 11C-D clearly reveals matrix shear yielding and rubber cavitation over the entire fracture surface, an evident ductile behavior.

B.   Polyamides

Polyamides can be considered as the most important class of semicrystalline engineering thermoplastics commercially available, because of their high melting characteristics, their good mechanical strength, high abrasion resistance and their excellent chemical resistance to solvents. The most commonly encountered polyamides are polyamide 6 and polyamide 66 because of their large production volume and widespread use in fibers, packaging, film production and in high added-value applications such as automotive, electrical components and power tool industries. In numerous applications below a temperature of about 70°C, they are potential substitutes for metals and ceramics. However, as neat resins, polyamides could not satisfy in applications where high toughness, high-notch resistance and a low brittle-tough transition temperature are required. There has been an urgent need to toughen polyamides by dispersing in them materials such as rubbers or elastomers [72].

Rubber toughened polyamide 66 is the first commercialized super-tough engineering blend [93,124–137]. Because of the polar nature of polyamides and the apolar nature of rubbers, toughening of polyamides requires an adequate compatibilization process, mainly reactive, in order to produce blends containing small rubber particles. Historically, Ide and Hasegawa [138] were the first to report about reactive compatibilization of a polyamide and a polyolefin (polypropylene). They employed a maleic anhydride-grafted polypropylene, the maleic anhydride groups of which are able to react with amine end groups of polyamides forming a thermally stable imide link. As presented in Table 3, many other techniques exist where various compatibilizer precursors or reactive rubbers are blended with polyamides [123]. Maleic anhydride, glycidyl methacrylate and zinc-neutralized acrylic acid are the main functionalities used for reactive compatibilization at concentrations not exceeding 8 wt%. The rubbers employed are EPR, EPDM, SEBS elastomer and ethylene-based copolymers.

Figure 12 illustrates the effect of blending various types of reactive olefinic rubbers containing various reactive groups on the notched Izod impact strength of polyamide at −40°C. Maleated EPR is by far the most efficient impact toughener for polyamide 6 above the minimum concentration of 15 wt%. This indicates the efficiency of the imidation reaction between the amine end groups of polyamides and grafted anhydride groups of EPR-g-MA. According to reports published by Paul

TABLE 3 Some Common Reactive Rubbers and Tougheners for Polyamides

Reactive Rubber/Toughener	Functionality	Reactivity	Other Features
Maleic anhydride-grafted ("maleated"), ethylene-propylene rubber (m-EPR)	Anhydride 0.3–0.9% MA	High reactivity with the amine (NH_2) end group of PA	Amorphous rubber, low T_g leads to high impact toughness down to –40°C
Maleated, styrene-ethylene/butylene-styrene block copolymer rubber (m-SEBS)	Anhydride 0.5–2% MA	High reactivity with the amine (NH_2) end group of PA	Amorphous rubber, low T_g leads to high-impact toughness down to –40°C
Ethylene-ethyl acrylate-maleic anhydride (E-EA-MA) terpolymer	Anhydride 0.3–3% MA	High reactivity with the amine (NH_2) end group of PA	Moderate T_g limits, low temperature toughness
Zinc neutralized, ethylene-methacrylic acid copolymer ionomer (E-MAA, Zn)	Zinc carboxylate, carboxylic acid	Low reactivity with amine but good polar interaction of Zn with amide and amine groups (interfacial complexation)	T_g, hardness limit low temperature toughness, good solvent resistance
Zinc-neutralized, ethylene-butyl acrylate methacrylic acid terpolymer ionomer (E-BA-MAA, Zn)	Zinc carboxylate, carboxylic acid	Same as above	Low T_g, high-impact modification efficiency

TABLE 3 Some Common Reactive Rubbers and Tougheners for Polyamides (Continued)

Reactive Rubber/Toughener	Functionality	Reactivity	Other Features
Ethylene-glycidyl methacrylate copolymer (E-GMA)	Epoxide 3–8% GMA	Moderate high reactivity with carboxyl group of PA	T_g hardness limit achievable toughness, cross-linking tendency
Ethylene-ethyl acrylate-glycidyl methacrylate terpolymer (E-EA-GMA)	Epoxide 1–8% GMA	Moderate high reactivity with carboxyl group of PA	Lower T_g better impact, high viscosity
Acrylate core-shell rubber, functionalized	Carboxyl	Low reactivity with amine	Small rubber particles (<0.5 μ) aggregation
Ethylene-acrylic acid copolymers (E-AA)	Carboxyl	Low reactivity with amine	Not rubbery enough, modest impacts
Ethylene-ethyl acrylate or butyl acrylate Copolymers (E-EA or E-BA)	Ester	No reactivity with amine	No impact improvement
			Used only as codiluent

Source: K. Akkapeddi, Reactive Polymer Blending, eds., W. Baker, C. Scott and G.-H. Hu, Munich: Hanser Publishers, 2001. With permission.

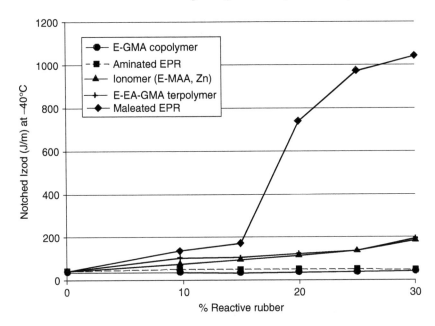

Figure 12 Effect of blending various types of reactive olefinic rubbers on the notched Izod impact strength of polyamide 6 at –40°C ("Balanced end group" PA6 of M_w = 34,000; [NH_2] = [COOH] = 48 µeq/g). (From K. Akkapeddi, Reactive Polymer Blending, eds., W. Baker, C. Scott, and G.-H. Hu, Munich: Hanser Publishers, 2001, Chap. 8. With permission.)

and coworkers, super-tough polyamide/rubber blends can be designed based on a rubber content of 20 wt%, provided that the rubber particle size is comprised within a range of 0.1–0.7 µm [139–141]. Blends of 80% nylon 6/20% (mixture of maleated and non-maleated EPR) have been investigated for their fracture toughness with respect to the particle size of the rubber phase [142]. As illustrated in Figure 13, the room temperature notched impact strength of the blends is the highest at concentrations of EPR-g-MA (in the rubber phase) larger than 50 wt%; the obtained particle size is below 0.61 µm. For intermediate EPR-g-MA concentrations (particle size up to 1.1), the blends exhibit an intermediate toughness. For larger rubber particles, the blends were brittle.

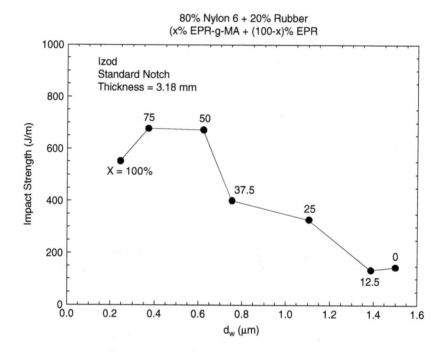

Figure 13 Izod impact strength as a function of average rubber particle diameter for blends of 80% nylon 6 and 20% maleated EPR mixture. (From O. Okada, H. Keskkula and D. R. Paul, Polymer, 41, 8061, 2000. With permission.)

The brittle-to-ductile transition temperature of the same blends is presented in Figure 14. It is lower for higher content EPR-g-MA or the smaller the particle size of the rubber phase. Blends containing less than 37.5% of the maleated EPR in the rubber phase (particle size above 0.75 μm) exhibit a relatively brittle behavior at room temperature, since the brittle-to-ductile transition temperature is close to or higher than room temperature.

Polyamide 66 has been toughened using maleated low density polyethylene (MA-g-LDPE) containing 0.4 wt% of grafted maleic anhydride [143]. A tenfold improvement in notched Izod impact strength with respect to neat PA66 was reached in a toughened PA66 containing 20–30 wt% MA-g-

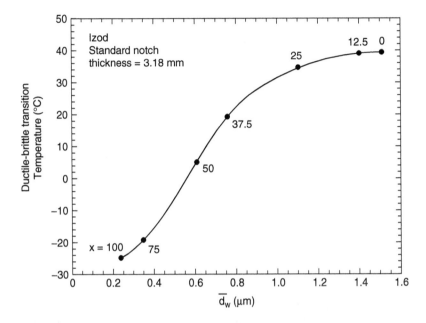

Figure 14   Ductile-brittle transition temperature as a function of average rubber particle diameter for blends of 80% nylon 6 and 20% maleated EPR mixture. (From A. J. Oshinski, H. Keskkula and D.R. Paul, Polymer, 37, 4919, 1996. With permission.)

LDPE. The particle size of the LDPE phase was about 0.4 μm. Akkappeddi et al. have shown that PA6 was efficiently toughened using maleated LDPE containing 0.5 wt% grafted maleic anhydride groups [144]. In Figure 15 the notched Izod impact strength of PA6/maleated LDPE blends is presented as a function of the percentage of maleated LDPE at a fixed maleic anhydride content of 0.5 wt%. A maximum of about 900 J/m is reached at a maleated LDPE content of 40 wt%. The authors ascribed the decrease of the impact strength above that concentration to the onset of the development of phase-inverted morphology.

Polyamide 6 has been toughened using a reactive epoxidized EPDM copolymer [145]. Epoxide groups react with amine end-groups of polyamide 6 to form *in situ* a PA6-g-EPDM copolymer that compatibilizes the PA6/EPDM blend.

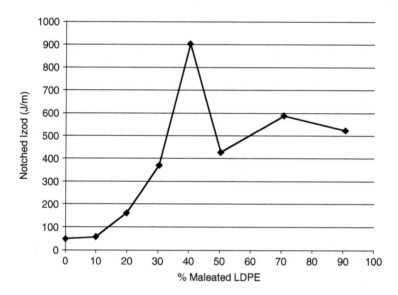

Figure 15 Effect of blending maleated LDPE (0.5% MA) on the notched Izod impact strength of PA6/ LDPE blends at room temperature. (From K. Akkapeddi, Reactive Polymer Blending, eds., W. Baker, C. Scott and G.-H. Hu, Munich: Hanser Publishers, 2001. With permission.)

The EPDM particle size can be efficiently reduced, allowing access to a phase morphology necessary for toughening of the PA6 matrix. The impact strength of polyamide 6/EPDM blends was investigated as a function of the EPDM content. An 18-fold improvement of toughness with respect to pure polyamide 6 was reported when the rubber concentration is 24 wt%; below that concentration, only fair impact strength enhancement was observed (less than 5-fold). Acrylonitrile-butadiene-styrene (ABS) containing 40 wt% polybutadiene mixed with a small amount of maleated EPR has also been reported to successfully toughen polyamides [146]. ABS has been grafted with 0.5 wt% maleic anhydride in the presence of 500 ppm of peroxide initiator using a reactive extrusion process. Addition of 5 wt% maleated EPR resulted in a further toughness improvement. Polyamide/polypropylene blends have been investigated by Wong and Mai with respect to toughening mechanisms [147]. Blends containing 60 parts of nylon 6,6, 20 parts of polypropylene and 20 parts of styrene-ethylene/butylene-styrene (SEBS; 26 wt% styrene) containing various levels of grafted maleic anhydride were considered. The reaction between the anhydride groups (grafted to SEBS) and the amine groups (end groups of nylon 6,6) was responsible for PP/PA6,6 compatibilization improvement. The rubber phase necessary for toughening is the EB phase in SEBS. The phase morphology claimed consists of core spherical domains of polypropylene surrounded by SEBS shell in a nylon matrix. Figure 16a-c, the addition of maleated SEBS containing 0.37, 0.92 and 1.84 wt% MA, respectively, results in a much finer plastic deformation in front of a sharp crack tip (machined in the specimen) compared to unmodified nylon 6,6/PP blends containing non-maleated SEBS which exhibits a large interfacial crack in front of the crack tip. SEBS containing 0.92 wt% of maleic anhydride was the most efficient in imparting high toughness in nylon 6,6/PP blends. Microscopic techniques carried out at various positions in front of a crack on a fractured specimen revealed evidence that under triaxial stress cavitation of SEBS-g-MA interphase occurred prior to multiple crazing in the blend. This was directly followed by massive shear yielding at the crack tip. Because of the larger PP particles observed in the case of SEBS-g-MA containing 0.37 wt% MA, plane-strain constraints were not able to relieve, and subsequently no shear yielding was observed. In that case the blends exhibit a reduced fracture toughness.

A study of a completely different approach of toughening focused on the use of glass-particles in the nylon 6,6 matrix [148]. The reported work deals with the influence of the debonding damage on fracture toughness and crack-tip field in these composites. Nylon 6,6 composites containing various volume fractions of glass particles as well as differing with respect to the extent of interfacial treatment between the particles and the matrix were considered. The interface-treated composites were superior in tensile strength and inferior in toughness compared to the interface-untreated composites. Finite element analysis numerical calculations were

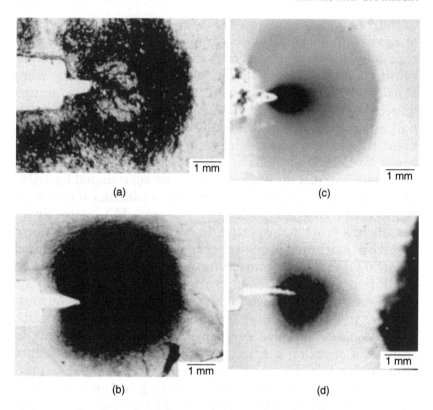

Figure 16 Crack-tip subfracture deformation zones for: (a) non-maleated; (b) 0.37-; (c) 0.92-; and (d) 1.84%-maleated blends. (From S.-C. Wong and Y.-W. Mai, Polymer, 41, 5471, 2000. With permission.)

carried out on both composites. The damage development around a crack tip depends on the interfacial strength between the particle and the matrix, and on the particle volume fraction. It was reported that the debonding damage reduces the stress level around the crack tip and is subsequently responsible for the toughening mechanism.

Laura et al. recently investigated the effect of rubber particle size and rubber type on the mechanical properties of glass-fiber reinforced, rubber (EPR/MA-g-EPR or SEBS/MA-g-SEBS) toughened nylon 6 [149]. The particle size was controlled by varying the ratio between the maleated to unmaleated rubber. Nylon 6 toughened with SEBS/SEBS-g-MA, which is tougher in the absence of glass fibers, had a lower fracture energy when reinforced with 15 wt% glass fibers. This was ascribed to high values of dissipative energy density in the absence of glass fibers. The dissipative energy vanished when glass fibers were added. TEM observations on fractured specimens indicate that the presence of glass fibers decreases the size of the damage zone of rubber-toughened nylon 6. Shear yielding was observed in blends with SEBS/SEBS-g-MA or EPR/EPR-g-MA rubbers, but the size of this shear yielded zone was larger for EPR/EPR-g-MA.

The effect of rubber particle size on the toughness of polyamides has been the subject of numerous investigations [53,65,76,127,150]. The ductile-brittle transition in rubber-modified polyamides has been found to strongly depend on the rubber particle size. Wu's plots [76] relating Izod impact strength of PA6,6/rubber blends to rubber particle size and to interparticle distance reveal that:

- For a given volume fraction of rubber, a critical particle size exists below which PA6,6 rubber blend exhibits a ductile behavior. Critical particle sizes of 0.5, 0.8 and 2 μm were found for blends containing 10, 15 and 20 wt% rubber, respectively (Figure 17).
- Independent of the volume fraction and particle size of the rubber phase, Wu found that a critical interparticle distance of 0.3 μm was needed to obtain an efficient toughening effect (Figure 5). The small particles suppress crazing and crack growth and allow for stress fields overlapping around the adjacent rubber particles which results in shear yielding behavior necessary for toughening of semicrystalline polymers.

To identify the mechanism behind the effect of the interparticle distance parameter, Wu first proposed that a strong overlap of the stress fields around the rubber particles induces shear yielding and crazing in polyamide 6,6 matrix, rendering the blend ductile. Later on Wu recognized the inadequacy of the model in explaining the particle size effect on toughening,

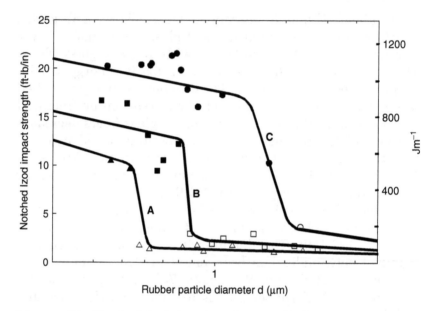

Figure 17 Notched Izod impact strength vs. rubber particle diameter (T) in PA6,6/reactive rubber blends; (curve A: 10 wt% rubber, curve B: 15 wt%, curve C: 20 wt% rubber). (From S. Wu, Polymer, 26, 1855, 1985. With permission.)

since the local stress level depends on the ratio of the center-to-center distance to the diameter of the particle [127]. As this ratio remains constant at a given volume fraction of particles regardless of their size, toughening should be unaffected by the presence of large particles at any given l/d ratio (according to the stress field overlap model). This interrelation does not agree with experimental results that clearly show that smaller particles are more effective in toughening [151]. A second proposed model is based on a transformation of the matrix material from a state of plane strain to plane stress when the volume fraction of rubber particles increases and the inter-particle distance goes through the critical size. The model fails to explain the phenomenon because it directly attributes the embrittlement to the presence of high triaxial stresses. Note that the triaxial stresses in local regions between particles can be affected only by changes in geometrical ratios. That does not change for a given volume fraction and spatial dispersion of the rubber particles regardless of their size. Muratoğlu et al. [132] have proposed an approach based on the orientation/random distribution of the crystalline lamellae of polyamide 6,6. The toughening mechanisms of maleic anhydride-grafted EPR-toughened polyamides was investigated as an alternative to the available explanations. The model considers that in a tough blend where the rubber particles are closer (compared to brittle blends where the effects of the interfaces do not overlap), the crystalline morphology is oriented. Crystalline lamellae were found to grow perpendicular to the rubber/matrix interface, inducing a real anisotropy in the interparticle zone. This clearly means that a considerable fraction of polyamide 66 has hydrogen-bonded planes parallel to the interface in the interparticle zone. The model which was supported by microscopic observations of morphological features in the stress-whitened zones clearly evidences that the oriented planes result in a local reduction of the flow stress that results in the hindrance of premature fracture.

Cavitation in polyamide/rubber blends has been quantitatively modeled by Lazzeri and Bucknall [84]. The model predicts that only particles having a size above a critical value of 0.25 μm were able to cavitate. That agrees with the concept of reaching an optimum particle size below which toughening is efficient. Formation of shear bands necessary for a ductile behavior results from the voiding induced by cavitation of the rubber particles. The compatibilizer is necessary in polyamide-based blends since the cavitation process is efficient only if it is associated with an energy dissipation process resulting from interfacial debonding.

C.   Polyolefins

Isotactic polypropylene and high-density polyethylene are the two polyolefins that are most often subjected to toughening investigations. They exhibit attractive strength and ductility at room temperature and under moderate deformation rates. Their low impact strength in low-temperature applications or under high-strain rates motivated research groups to develop approaches to alleviate the deficiency. Polyolefins also suffer from low resistance to crack propagation at low temperatures (below the glass-transition of their amorphous phases).

The incorporation of a rubbery phase in the polypropylene matrix can be achieved by copolymerization or by melt-blending using a compatibilization technique which allows the achievement of reduced particle size and enhanced interfacial adhesion.

Toughening of polyethylenes, mainly the high density grade, using rigid fillers has been reported by several authors [152–156]. Rubber-toughened HDPE was found to undergo brittle-ductile transition when the thickness of the matrix ligaments between adjacent rubber particles is below a critical value of approximately 0.6 μm. This critical thickness was claimed to be exclusively a property of the matrix alone. It does not depend on the type of rubber (amorphous as well as semicrystalline EPDMs and ethylene-octene rubbers were used) or its concentration in the blend or on the size of the rubber particles. The same fundamental reason of a transcrystalline layer of the matrix perpendicular to the rubber/matrix interface as observed in polyamide, was revealed by the authors. The transcrystalline layer is built up of polyethylene crystals oriented with their low-energy easy-to-shear (100) planes parallel to the interface. When the ligament thickness decreases below a critical value, the oriented layers around the rubber particles merge into a percolating material component of reduced plastic resistance, which results in a substantial toughening behavior.

The new approach that is being adopted since the "revolutionary" model of Muratoğlu et al. [132] consists of using rigid fillers as toughening agents instead of rubber. Bartczak et al. used calcium carbonate filler particles for toughening of HDPE [152]. $CaCO_3$ fillers of three different sizes, 3.5, 0.7 and 0.44 μm weight average diameter were employed at various volume fractions. Combination of particle size and volume fraction of the filler allowed the condition of interparticle ligament thickness below a value of 0.6 μm. As shown in Figure 18, a critical interparticle ligament thickness within a range of 0.18–0.4 μm results in the maximum jump in HDPE

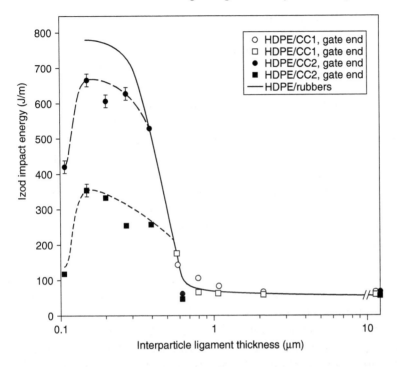

Figure 18 The dependence of the notched Izod impact energy on the interparticle matrix ligament thickness for samples of HDPE/CC1 (particle size 0.85 μm) and HDPE/CC2 (particle size 0.41 μm) blends. The solid line represents the curve determined for the blends of the same HDPE with various EPDM various EOR rubbers. (From Z. Bartczak, A.S. Argon, R.E. Cohen and M. Weinberg, Polymer, 40, 2347, 1999. With permission.)

toughness (e.g., for a volume fraction of $CaCO_3$ of 0.22, an Izod impact strength of 800 J/m was obtained compared to 50 J/m for neat HDPE). That achievement really confirms the role of the highly oriented lamellar crystallites having reduced plastic resistance in certain orientations. SEM micrographs illustrate the extent of stress-whitening, a signature of toughening mechanism via shear yielding upon addition of calcium carbonate at the particle size and volume fractions generating a critical ligament thickness below 0.6 μm. Of course, as no cavitation is expected (rigid particles), debonding at the interface as well as low yield stress characteristics of the transcrystallized layer of HDPE around the filler particle are responsible for the toughening.

Toughening of polypropylene is extensively covered in literature [72]. Although it possesses a ductile behavior at ambient temperature, it is brittle and notch sensitive at lower temperatures and also under high deformation rates [157]. Before the development of the new toughening concept of using rigid fillers such as calcium carbonate, polypropylene was most often toughened using rubbers like EPR or EPDM [158,159]. Elastomers such as SBS, SEBS, or pure rubbery homopolymers such as polybutadiene or polyisoprene also showed satisfying toughening performances. The disadvantage resulting from rubber toughening of polypropylene lies in the undesired modulus decrease.

A huge amount of work has been devoted to understanding the phase morphology developed in multicomponent polymer/elastomer composites [160–170]. Understanding which component is where in a multicomponent composite is the key to the understanding of toughening and stiffening mechanisms and performances. In polypropylene, as in most brittle polymers, the challenge faced is how to achieve an acceptable extent of toughness without adversely affecting its stiffness [171–173]. One technique consists of compensation by incorporating rigid fillers in addition to the rubber phase.

Under impact testing, unnotched neat polypropylene exhibits a brittle-to-ductile transition temperature within a range of 0 to 20°C [174]. The notched polypropylene has been reported to exhibit a Tbd temperature of 100°C [175]. Van der Wal et al. [175] investigated the effect of EDPM content on the deformation and impact behavior of polypropylene/EPDM blends. As expected, the modulus and the yield strength decrease linearly with increasing the rubber content from 0 to 40 vol%. The brittle-ductile transition temperature (T_{bd}) decreases from 85°C for pure PP to –50°C for a 40 vol% EPDM rubber blend — a considerable T_{bd} shift of 135°C.

The effect of the loading rate on the fracture resistance of isotactic PP and rubber-modified PP were studied by Gensler et al. [157]. A blend containing 15 wt% EPR was studied

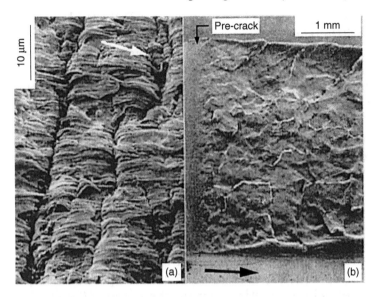

Figure 19 SEM of the fracture surfaces of iPP/EPR samples deformed at (a) 10 mm/s and (b) 5.8 m/s; the crack-propagation direction is indicated by the arrows. (From R. Gensler, C. J. G. Plummer, C. Grein and H.-H. Kausch, Polymer, 41, 3809, 2000. With permission.)

at test speeds varying from very low (0.1 mm/s) to very high rates (14 m/s). Pure polypropylene exhibits a ductile-brittle transition as the test speed was increased. This was ascribed to a transition from shear deformation to crazing. The EPR/iPP blends exhibit a stable crack propagation over the whole range of test speeds. Stress-whitening at the crack tip was observed, indicative of the formation of extensive shear lips on the fracture surface. Dramatic differences are visible between an EPR/PP blend sample deformed at a speed of 10 mm/s and a sample fractured at 5.8 m/s (Figure 19). It has been concluded that the iPP homopolymer displayed a ductile-to-brittle transition with increasing the test speed. A highly dissipative shear process is dominant at low speed, whereas multiple crazing is the mechanism at intermediate test speed (50–1000 mm/s). At speeds higher than 2 m/s, the crack tip damage was limited to a single localized deformation zone (single crack-tip craze). In the case of EPR/iPP blends, no ductile-brittle transition was observed in the range of testing speeds used. Shear processes were the dominant mechanism of deformation.

Toughening of polypropylene via the use of rigid particles such as calcium carbonate is also being reported in literature [176–179]. As in polyamides and HDPE, polypropylene is also susceptible to toughening via fillers. The condition is that the formation of a trans-crystallized layer of PP around the filler is sufficiently thick and well adhered to the particles to be able to influence the overall deformation process of the matrix within the interparticle zone. The recent investigations of Thio et al. [180] and Zuiderduin et al. [181] on toughening of polypropylene using calcium carbonate are quite interesting. In the former report, three calcium carbonate types having particle sizes of 0.07, 0.7 and 3.5 μm, respectively, were employed at volume fractions within the range of 0.05 to 0.3. Failure in toughness improvement of polypropylene observed for the 0.07 and 3.5 μm size fillers was ascribed to the presence of agglomerates in the sample which initiate brittle behavior. No information is provided regarding the interparticle ligament thickness, neither indication is given for the existence or not of a trans-crystallized PP layer around the calcium carbonate particles. Particles that are too large (3.5 μm) initiate brittle fracture, whereas particles that are too small (0.07 μm) are difficult to disperse and the presence of agglomerates is not prevented which also initiate brittle fracture. Calcium carbonate particles having an average size of 0.7 μm improved Izod impact energy up to four times that of the unfilled polypropylene. The toughening mechanism claimed was plastic deformation of interparticle ligaments, following particle-matrix debonding with additional contribution resulting from crack deflection toughening.

Zuiderduin et al. [181] investigated carefully the toughening of calcium carbonate-filled polypropylene using a combination of filler particle size (0.07–1.9 μm), volume fraction of filler within a range of 0 to 32 vol% and a matrix of polypropylene having a melt-flow index within a range of 0.3 to 24 g/10 min. The concept of dispersing calcium carbonate in

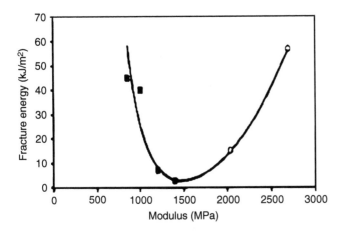

Figure 20   Fracture energy as a function of modulus, 20°C (square: PP-EPDM; open circle: PP-CaCO$_3$). (From W. C. J. Zuiderduin, C. Westzaan, J. Huétink and R. J. Gaymans, Polymer, 44, 261, 2003. With permission.)

polypropylene matrix leads to a tough and stiff composite (Figure 20). In the case of polypropylene having higher molecular weight, the brittle-ductile transition is shifted towards lower temperatures (Figure 21). The maximum improvement in toughness was achieved with stearic acid-treated filler having 0.7 μm size (Figure 22). It is therefore an established fact that rigid particles can play a toughening role provided that an optimum combination of filler particle size, matrix properties, interfacial adhesion control and composition is used.

IV.   GENERAL CONCLUSIONS AND FUTURE OUTLOOK

Toughening of brittle and less-ductile semicrystalline polymers continues to be a challenging research area for many groups. Rubber-toughening agents such as EPR, EPDM, polybutadiene rubbers and elastomers, including SEBS, SBS, SEPS, SAN, ABS, etc. are most frequently used in semicrystalline polymer matrices. The interrelation between the characteristics of the phase morphology developed in the blends

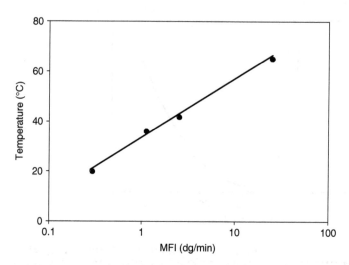

Figure 21 Brittle-to-ductile transition temperature as a function of matrix molecular weight, 30 wt% PP-CaCO$_3$ composites, type A. (From W. C. J. Zuiderduin, C. Westzaan, J. Huétink and R. J. Gaymans, Polymer, 44, 261, 2003. With permission.)

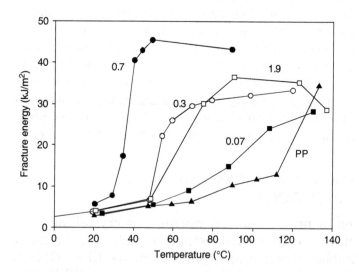

Figure 22 Fracture energy as a function of temperature, notched Izod, CaCO$_3$ particles, stearic acid treated (particle size indicated on the plot). (From W. C. J. Zuiderduin, C. Westzaan, J. Huétink and R. J. Gaymans, Polymer, 44, 261, 2003. With permission.)

and their toughening performance continues to be the route to control for a successful toughening operation. On the fundamental side and since the publication of Wu's model correlating the interparticle ligament thickness to the toughening properties (notched and unnotched impact strength, brittle-ductile transition temperature), scientific debate is still animated on the physical interpretation of the mechanism behind it. The most recent development deals with the existence of anisotropy within the interparticle zone. A layer of a significant thickness was found to crystallize normal to the particle/matrix interface. Such layer was found to exhibit a low yielding stress in the matrix and therefore contributes to the toughening of the matrix. Based on that finding, it was deduced that a rubbery dispersion can well be replaced by a rigid phase such as fillers, provided that Wu's criterion is fulfilled. It is the tendency now to use rigid particles as toughening agents in semicrystalline matrices such as polyamides, polyesters and polyolefins. The question which remains less explored is the role the extent of interfacial adhesion plays for an efficient toughening. In the rigid particles toughening concept, interfacial debonding is the equivalent mechanism to particle cavitation occurring in rubber-toughened blends.

If the success of using rigid fillers instead of rubbery components is further confirmed and its toughening efficiency generalized, then that would definitely be the approach to follow for the future because, contrary to rubber toughening, a significant gain in stiffness is associated with toughening.

From the mechanistic view, this concept is still in its "embryonic" stage because some uncertainties still exist, e.g., the crucial importance to achieve percolation (dominating effect) of the transcrystallized layers around the particles, the role of the interfacial debonding and its impact on the total toughening process (which implies control of the filler surface treatment using suitable coupling agents, exact definition and determination of the interparticle ligament thickness since fillers are not always monodisperse in size, nor do they have a well-defined shape), and viscosity built up during the melt processing of the composite (more than 30 vol% of filler is required).

ACKNOWLEDGMENTS

The authors are indebted to the Research Fund of the KULeu-ven (GOA 98/06), and to the Fund for Scientific Research-Flanders (Belgium) for the financial support given to the MSC Laboratory.

REFERENCES

1. C. B. Bucknall, Toughened Plastics, London: Applied Science Publishers, 1977.

2. A. J. Kinlock and R. J. Young, Fracture Behaviour of Polymers, London: Applied Science Publishers, 1983.

3. G. Groeninckx and D. Dompas, Advanced Polymer Science – Eupoco Course, Module 2, KULeuven: Acco, Leuven 1992, vol 3.

4. H. H. Kausch, Polymer Fracture, Berlin: Springler-Verlag, 1987.

5. R. P. Kambour, A review of crazing and fracture in thermoplastics, J. Polym. Sci. Macromol. Rev., 7, 1, 1973.

6. E. J. Kramer and L. L. Berger, Crazing in polymers, Adv. Polym. Sci.: 52/53, 1983.

7. E. J. Kramer and L. L. Berger, Crazing in polymers, vol. 2, Adv. Polym. Sci.: 91/92, 1990.

8. S. Rabinowitz and P. Beardmore, Craze formation and fracture in glassy polymers, in: Critical Reviews in Macromology Science, Vol. 1, ed., E. Baer, Cleveland, Ohio: CRC Press, March 1972.

9. E. J. Kramer, Environmental cracking of polymers, in: Developments in Polymer Fracture-1, ed., E. H. Andrews, London: Applied Science Publishers, 1979.

10. E. J. Kramer, Microscopic and molecular fundamentals of crazing, in: Crazing in Polymers, eds., H. H. Kausch, Berlin: Adv. in Polym. Sci., 52/53, Springer-Verlag, 1983.

11. R. P. Kambour, Crazing and cracking in glassy homopolymers, in: Polymer Blends and Mixtures, eds., D. J. Walsh, J. S. Higgins and A. Maconnachie, London: Nato ASI Series, 89, 1985.

12. A. M. Donald and E. J. Kramer, Deformation zones and entanglements in glassy-polymers, Polymer, 23, 1183, 1982.

13. D. Maes, G. Groeninckx, J. Ravenstijn and L. Aerts, A molecular-model for structural-changes during shear yielding and crazing in amorphous polymers, Polym. Bull., 16, 363, 1986.

14. D. Maes, PhD Thesis, Catholic University of Leuven, Belgium, 1989.

15. H. H. Kausch, Molecular mechanisms contributing to the strength and toughness of glassy-polymers, Makromol. Chem. Macromol. Symp., 41, 1, 1991.

16. S. M. Aharoni, Correlations between chain parameters and failure characteristics of polymers below their glass-transition temperature, Macromolecules, 18, 2624, 1985.

17. A. S. Argon and M. M. Salama, Mechanism of fracture in glassy materials capable of some inelastic deformation, Mater. Sci. Eng., 23, 219, 1976.

18. G. H. Michler, Makromol. Chem. Macromol. Symp. 41, 39, 1991.

19. A. M. Donald and E. J. Kramer, Effect of molecular entanglements on craze microstructure in glassy-polymers, J. Polym. Sci. Polym. Phys. Ed., 20, 899, 1982.

20. C. Creton, E. J. Kramer and G. Hadziioannou, Craze fibril extension ratio measurements in glassy block copolymers, Colloid. Polym. Sci., 270, 399, 1992.

21. S. Wu, Proceedings ACS Div. Polym. Mater. Sci., Eng., Fall Meeting, Washington, D.C., 1990.

22. S. Wu, Chain structure, phase morphology, and toughness relationships in polymers and blends, Polym. Eng. Sci., 30, 753, 1990.

23. S. Wu, Control of intrinsic brittleness and toughness of polymers and blends by chemical-structure — a review, Polym. Int., 29, 229, 1992.

24. C. B. Bucknall, Toughened Plastics, London: Applied Science Publishers, 1977.

25. R. A. Bubeck, D. J. Buckley, E. J. Kramer and H. R. Brown, Modes of deformation in rubber-modified thermoplastics during tensile impact, J. Mater. Sci., 26, 249, 1991.

26. D. J. Buckley, Toughening Mechanisms in the High Strain Rate Deformation of Rubber-Modified Polymer Glasses, PhD Thesis, Cornell University, 1993.

27. A. M. L. Magalhaes and R. J. M. Borggreve, Contribution of the crazing process to the toughness of rubber-modified polystyrene, Macromolecules, 28, 5841, 1995.

28. J. N. Goodier, Trans. M. Chem. Soc. Mech. Eng. TASMA, 55, 39, 1933.

29. C. B. Bucknall, Deformation mechanisms in rubber-toughened polymers, in: Polymer Blends, Vol. 2: Performance, eds., D. R. Paul and C. B. Bucknall, NY: John Wiley & Sons, 2000.

30. H. Eyring, J. Chem. Phys., 4, 283, 1936.

31. J. C. Bauwens, J. Polym. Sci., A2 5, 1145, 1967.

32. J. A. Roetling, Yield stress behaviour of polymethylmethacrylate, Polymer, 6, 311, 1965.

33. C. Bauwens-Crowet, J. C. Bauwens and G. Homès, J. Polym. Sci., A2 7, 735, 1969.

34. J. C. Bauwens, Relation between compression yield stress and mechanical loss peak of bisphenol-a-polycarbonate in beta-transition range, J. Mater. Sci., 7, 577, 1972.

35. C. Bauwens-Crowet, J. M. Ots and J. C. Bauwens, Strain-rate and temperature-dependence of yield of polycarbonate in tension, tensile creep and impact tests, J. Mater. Sci. Lett., 9, 1197, 1974.

36. S. Havriliak and T. J. Shortridge, Effect of k-value on the tensile yield properties of polyvinyl-chloride, J. Appl. Polym. Sci., 37, 2827, 1989.

37. S. Havriliak, S. E. Slavin and T. J. Shortridge, Tensile yield parameters for poly(vinyl chloride) blends with a methyl-methacrylate butadiene styrene impact modifier, Polym. Int., 25, 67, 1991.

38. F. Ramsteiner, Impact strength of thermoplastics, Kunstoffe, 73, 148, 1983.

39. C. S. Henkee and E. J. Kramer, Crazing and shear deformation in crosslinked polystyrene, J. Appl. Polym. Sci.: Polym. Phys. 22, 721, 1984.

40. G. Groeninckx and D. Dompas, Plastic deformation mechanisms of polymers and rubber-modified thermoplastic polymers: molecular and morphological aspects, in: Structure and Properties of Multiphase Polymeric Materials, eds., T. Araki, Q. Tran-Cong and M. Shibayama, New York: Marcel Dekker, 1998.

41. A. M. L. Magalhaes, PhD Thesis, University of Minho, Portugal, 1995.

42. C. B. Bucknall, I. Davies and I. K. Partridge, Rubber toughening of plastics. 10. Effects of rubber particle-volume fraction on the kinetics of yielding in HIPS, J. Mater. Sci., 21, 307, 1986.

43. C. B. Bucknall, I. Davies and I. K. Partridge, Rubber toughening of plastics. 11. Effects of rubber particle-size and structure on yield behavior of HIPS, J. Mater. Sci., 21, 1341, 1987.

44. D. J. Hourston and S. Lane, Rubber Toughened Polymers, ed., A. A. Collyer, Cambridge: Chapman & Hall, 1994.

45. D. S. Parker, H. J. Sue, J. Huang and A. F. Yee, Toughening mechanisms in core shell rubber modified polycarbonate, Polymer, 31, 2267, 1990.

46. D. Dompas and G. Groeninckx, Toughening behavior of rubber-modified thermoplastic polymers involving very small rubber particles. 1. A criterion for internal rubber cavitation, Polymer, 35, 4743, 1994.

47. D. Dompas, G. Groeninckx, G. Isogawa, T. Hasegawa and M. Kadokura, Toughening behavior of rubber-modified thermoplastic polymers involving very small rubber particles. 2. Rubber cavitation behavior in poly(vinyl chloride) methyl-methacrylate butadiene styrene graft copolymer blends, Polymer 35, 4750, 1994.

48. D. Dompas, G. Groeninckx, G. Isogawa, T. Hasegawa and M. Kadokura, Toughening behavior of rubber-modified thermoplastic polymers involving very small rubber particles. 3. Impact mechanical-behavior of poly(vinyl chloride) methyl-methacrylate butadiene styrene graft copolymer blends, Polymer, 35, 4760, 1994.

49. A. Tse, E. Shin, A. Hiltner, E. Baer and R. J. Laakso, The stress-whitened damage zone of PVC blends, J. Mater. Sci., 26, 2823, 1991.

50. H. Breuer, F. Haaf and J. Stabenow, Stress whitening and yielding mechanism of rubber-modified PVC, J. Macromol. Sci.: Phys. B, 14, 387, 1977.

51. R. J. Gaymans, Rubber Toughened Polymers, ed., A. A. Collyer, Cambridge: Chapman & Hall, 1994.

52. A. J. Oshinski, H. Keskkula and D. R. Paul, Rubber toughening of polyamides with functionalized block copolymers, 2. Nylon-6,6, Polymer, 33, 284, 1992.

53. R. J. M. Borggreve, R. J. Gaymans and A. R. Luttmer, Influence of structure on the impact behavior of nylon-rubber blends, Makromol. Chem., Macromol. Symp., 16, 195, 1988.

54. A. Van der Wal, J. J. Mulder and R. J. Gaymans, Polypropylene-rubber blends: 1. The effect of the matrix properties on the impact behaviour, Polymer, 39, 6781, 1998.

55. A. Van der Wal, R. Nijhof and R. J. Gaymans, Polypropylene-rubber blends: 2. The effect of the rubber content on the deformation and impact behaviour, Polymer, 40, 6031, 1999.

56. D. J. Hourston, S. Lane and H. X. Zhang, Toughened thermoplastics. 2. Impact properties and fracture mechanisms of rubber modified poly(butylene terephthalates), Polymer, 32, 2215, 1991.

57. W. R. Hale, L. A. Pessan, H. Keskkula and D. R. Paul, Effect of compatibilization and ABS type on properties of PBT/ABS blends, Polymer, 40, 4237, 1999.

58. A. F. Yee, J. Du and M. D. Thouless, Polymer Blends, Vol. 2: Performance, eds., D. R. Paul and C. B. Bucknall, New York: John Wiley & Sons, 2000.

59. S. J. Shaw, Rubber Toughened Polymers, ed., A. A. Collyer, Cambridge: Chapman & Hall, 1994.

60. E. H. Merz, G. C. Claver and M. Baer, J. Polym. Sci., 22, 325, 1956.

61. C. B. Bucknall, Toughened Polymers, Essex: Applied Science Publishers LTD, 1977.

62. A. M. Donald, Rubber Toughened Polymers, ed., A. A. Collyer, Cambridge: Chapman & Hall, 1994.

63. A. J. Kinloch and R. J. Young, eds., Fracture Behaviour of Polymers, Essex: Applied Science Publishers LTD, 1983.

64. M. Okamoto, Y. Shinoda, T. Kojima and T. Inoue, Toughening mechanism in a ternary polymer alloy: PBT/PC/rubber system, Polymer, 34, 4868, 1989.

65. R. J. M. Borggreve, R. J. Gaymans and H. M. Eichenwald, Impact behaviour of nylon-rubber blends: 6. Influence of structure on voiding processes; toughening mechanism, Polymer, 30, 78, 1989.

66. K. Dijkstra, PhD Thesis, University of Twente, The Netherlands, 1993.

67. R. A. Bubeck, D. J. Buckley and E. J. Kramer, Modes of deformation in rubber-modified thermoplastics during tensile impact, J. Mater. Sci., 26, 6249, 1991.

68. C. B. Bucknall, R. Rizzieri and D. R. Moore, Detection of incipient rubber particle cavitation in toughened PMMA using dynamic mechanical tests, Polymer, 41, 4149, 2000.

69. C. S. Lin, D. S. Ayre and C. B. Bucknall, A dynamic mechanical technique for detecting rubber particle cavitation in toughened plastics, J. Mater. Sci. Lett., 17, 669, 1998.

70. A. J. Oshinski, H. Keskkula and D. R. Paul, Rubber toughening of polyamides with functionalized block copolymers: 1. Nylon-6, Polymer, 33, 268, 1992.

71. I. Walker and A. A. Collyer, Rubber Toughened Polymers, ed., A. A. Collyer, Cambridge: Chapman & Hall, 1994.

72. R. J. Gaymans, Toughening of semicrystalline thermoplastics, in: Polymer Blends 2: Performance, eds., D. R. Paul and C. B. Bucknall, New York: John Wiley & Sons, 2000.

73. A. Lazzeri and C. B. Bucknall, Applications of a dilatational yielding model to rubber-toughened polymer, Polymer, 36, 2895, 1995.

74. R. J. M. Smit, Toughness of Heterogeneous Polymeric Systems, PhD. Thesis, Technical University of Eindhoven, 1998.

75. C. B. Bucknall, P. S. Leather and A. Lazzeri, Rubber toughening of plastics. 12. Deformation mechanisms in toughened nylon 6,6, J. Mater. Sci., 16, 2255, 1989.

76. S. Wu, Phase structure and adhesion in polymer blends: a criterion for rubber toughening, Polymer, 26, 1855, 1985.

77. S. Wu, Polym. Int., 29, 229, 1992.

78. S. D. Sjoerdsma, The tough-brittle transition in rubber-modified polymers, Polym. Commun., 30, 106, 1989.

79. R. J. M. Borggreve, R. J. Gaymans and J. Schruijer, Brittle-tough transition in nylon-rubber blends: effect of rubber concentration and particle size, Polymer, 28, 1489, 1987.

80. M. Morton, M. Cizmecioglu and R. Lhila, Model studies of rubber additives in high-impact plastics, Adv. Chem. Ser., 206, 221, 1984.

81. H. Breuer, F. Haaf and J. Stabenow, Stress whitening and yielding mechanism of rubber-modified PVC, J. Macromol. Sci. Phys. B 14, 387, 1977.

82. A. J. Oostenbrink, K. Dijkstra, S. Wiegersman, A. Van der Waal and R. J. Gaymans, PRI-International Conference on Deformation, Yield & Fracture of Polymers, Cambridge, 1990.

83. D. L. Dunkelberger and E. P. Dougherty, Experimental-design and analysis of the effects of structure, particle-size, and refractive-index on toughness-clarity of MBS impact modified PVC, Abstr. Pap. Am. Chem. Soc., 12, 212, 1990.

84. A. Lazzeri and C. B. Bucknall, Dilatational bands in rubber-toughened polymers, J. Mater. Sci., 28, 6799, 1993.

85. A. Margolina and S. Wu, Percolation model for brittle-tough transition in nylon/rubber blends, Polymer, 29, 2170, 1988.

86. R. J. M. Borggreve, R. J. Gaymans and J. Schuijer, Impact behaviour of nylon-rubber blends: 5. Influence of the mechanical properties of the elastomer, Polymer, 30, 71, 1989.

87. W. Jiang, Z. Wang, C. Liu, H. Liang, B. Jiang, X. Wang and H. Zhang, Effect of γ-irradiation on brittle-tough transition of PBT/EPDM blends, Polymer, 38, 4275, 1997.

88. W. Jiang, L. An and B. Jiang, Brittle–tough transition in elastomer toughening thermoplastics: effects of the elastomer stiffness, Polymer, 42, 4777, 2001.

89. A. J. Oshinski, H. Keskkula and D. R. Paul, The role of matrix molecular weight in rubber toughened nylon 6 blends: 1. Morphology, Polymer, 37, 4891, 1996.

90. M. Abbate, V. Di Liello, E. Martuscelli, P. Musto, G. Ragosta and G. Scarinzi, Molecular and mechanical characterization of reactive ethylene-propylene elastomers and their use in PA6-based blends, Polymer, 33, 2940, 1992.

91. R. J. M. Borggreve and R. J. Gaymans, Impact behaviour of nylon-rubber blends: 4. Effect of the coupling agent, maleic anhydride, Polymer, 30, 63, 1989.

92. P. Maréchal, G. Coppens, R. Legras and J.-M. Dekoninck, Amine anhydride reaction versus amide anhydride reaction in polyamide anhydride carriers, J. Polym. Sci., 33, 757, 1995.

93. Y. Takeda, H. Keskkula and D. R. Paul, Toughening of phase-homogenized mixtures of nylon-6 and poly(m-xylene adipamide) with a functionalized block copolymer, Polymer, 33, 3394, 1992.

94. A. Gonzalez-Montiel, H. Keskkula and D. R. Paul, Impact-modified nylon 6/polypropylene blends: 1. Morphology-property relationships, Polymer, 36, 4587, 1995.

95. A. Gonzalez-Montiel, H. Keskkula and D. R. Paul, Impact-modified nylon 6/polypropylene blends: 2. Effect of reactive functionality on morphology and mechanical properties, Polymer, 36, 4605, 1995.

96. A. Gonzalez-Montiel, H. Keskkula and D. R. Paul, Impact-modified nylon 6/polypropylene blends: 3. Deformation mechanisms, Polymer, 36, 4621, 1995.

97. D. M. Otterson, B. H. Kim and R. E. Lavengood, The effect of compatibilizer level on the mechanical-properties of a nylon 6/ABS polymer blend, J. Mater. Sci., 26, 1478, 1991.

98. V. J. Triacca, S. Ziaee, J. W. Barlow, H. Keskkula and D. R. Paul, Reactive compatibilization of blends of nylon 6 and ABS materials, Polymer, 32, 1401, 1991.

99. B. K. Kim and S. J. Park, Reactive melt blends of nylon with poly(styrene-co-maleic anhydride), J. Appl. Polym. Sci., 43, 357, 1991.

100. J. C. Angola, Y. Fujita, T. Sakai, T. Inoue, Compatibilizer-aided toughening in polymer blends consisting of brittle polymer particles dispersed in a ductile polymer matrix, J. Polym. Phys. Ed., 26, 807, 1988.

101. S. Y. Hobbs, M. E. J. Dekkers and V. H. Watkins, Toughened blends of poly(butylene terephthalate) and BPA polycarbonate. 1. Morphology, J. Mater. Sci., 23, 1219, 1988.

102. M. E. J. Dekkers, S. Y. Hobbs and V. H. Watkins, Toughened blends of poly(butylene terephthalate) and BPA polycarbonate. 2. Toughening mechanisms, J. Mater. Sci., 23, 1225, 1988.

103. A. J. Brady, H. Keskkula and D. R. Paul, Toughening of poly(butylene terephthalate) with core-shell impact modifiers dispersed with the aid of polycarbonate, Polymer, 35, 3665, 1994.

104. I. A. Abu-Isa, C. B. Jaynes and J. F. O'Gara, High-impact-strength poly(ethylene terephthalate) (PET) from virgin and recycled resins, J. Appl. Polym. Sci., 59, 1957, 1996.

105. P.-C. Lee, W.-F. Kuo and F.-C. Chang, *In-situ* compatibilization of PBT/ABS blends through reactive copolymers, Polymer, 35, 5641, 1994.

106. W. D. Cook, T. Zhang, G. Moad, G. Van Deipen, F. Cser, B. Fox and M. O'Shea, Morphology-property relationships in ABS/PET blends. 1. Compositional effects, J. Appl. Polym. Sci., 62, 1699, 1996.

107. E. Hage, W. Hale, H. Keskkula and D. R. Paul, Impact modification of poly(butylene terephthalate) by ABS materials, Polymer, 38, 3237, 1997.

108. W. Hale, H. Keskkula and D. R. Paul, Fracture behavior of PBT–ABS blends compatibilized by methyl methacrylate–glycidyl methacrylate–ethyl acrylate terpolymers, Polymer, 40, 3353, 1999.

109. A. Cecere, R. Greco, G. Ragosta, G. Scarzini and A. Tagliatela, Rubber toughened polybutylene terephthalate: influence of processing on morphology and impact properties, Polymer, 31, 1239, 1990.

110. H. Kanai, V. Sullivan and A. Auerbach, Impact modification of engineering thermoplastics, J. Appl. Polym. Sci., 53, 527, 1994.

111. M. K. Akkapedi, B. Buskirk, C. D. Mason, S. S. Chung and X. Swamikannu, Performance blends based on recycled polymers, Polym. Eng. Sci., 35, 72, 1995.

112. M. Penco, M. A. Pastorino, E. Ochiello, F. Garbassi, R. Braglia and G. Giannotta, High-impact poly(ethylene-terephthalate) blends, J. Appl. Polym. Sci., 57, 329, 1995.

113. J.-G Park, D.-H. Kim and K.-D. Suh, Blends of polyethylene-terephthalate with EPDM through reactive mixing, J. Appl. Polym. Sci., 78, 2227, 2000.

114. D. E. Mouzakis, N. Papke, J. S. Wu and J. Karger-Kocsis, Fracture toughness assessment of poly(ethylene terephthalate) blends with glycidyl methacrylate modified polyolefin elastomer using essential work of fracture method, J. Appl. Polym. Sci., 79, 842, 2001.

115. V. Tanrattanakul, A. Hiltner, E. Baer, W. G. Perkins, F. L. Massey and A. Moet, Toughening PET by blending with a functionalized SEBS block copolymer, Polymer, 38, 2191, 1997.

116. A. Sanchez-Solis, M. R. Estrada, J. Cruz and O. Manero, On the properties and processing of polyethylene terephthalate/styrene-butadiene rubber blend, Polym. Eng. Sci., 40, 1216, 2000.

117. G. Groeninckx, C. Harrats and S. Thomas, Reactive blending with immiscible functional polymers: molecular, morphological, and interfacial aspects, in: Reactive Polymer Blending, eds., W. Baker, C. Scott and G.-H. Hu, Munich: Hanser Publishers, 2001.

118. M. Hert, Tough thermoplastic polyesters by reactive extrusion with epoxy-containing copolymers, Angew. Macromol. Chemie, 196, 89, 1992.

119. A. Aróstegui and J. Nazábal, Critical inter-particle distance dependence and super-toughness in poly(butylene terephthalate)/grafted poly(ethylene-octene) copolymer blends by means of polyarylate addition, Polymer, 44, 5227, 2003.

120. W. Loyens and G. Groeninckx, Ultimate mechanical properties of rubber toughened semicrystalline PET at room temperature, Polymer, 43, 5679, 2002.

121. W. Loyens and G. Groeninckx, Rubber toughened semicrystalline PET: influence of the matrix properties and test temperature, Polymer, 44, 123, 2002.

122. W. Loyens and G. Groeninckx, Deformation mechanisms in rubber toughened semicrystalline polyethylene terephthalate, Polymer, 44, 4929, 2002.

123. K. Akkapeddi, Rubber toughening of polyamides by reactive blending, in: Reactive Polymer Blending, eds., W. Baker, C. Scott and G.-H. Hu, Munich: Hanser Publishers, 2001.

124. B. N. Epstein, U.S. Patent 4,172,895 (to E.I. duPont) 1979.

125. E. A. Flexman, Impact behavior of nylon-66 compositions — ductile-brittle transitions, Polym. Eng. Sci., 19, 564, 1979.

126. S. Wu, Impact fracture mechanisms in polymer blends — rubber-toughened nylon, J. Polym. Sci., Polym. Phys. Ed., 21, 699, 1983.

127. S. Wu, A generalized criterion for rubber toughening — the critical matrix ligament thickness, J. Appl. Polym. Sci., 35, 549, 1988.

128. D. W. Gilmore and M. J. Modic, J. Soc. Plast. Eng. ANTEC, 47, 1371, 1989.

129. C. B. Bucknall, P. S. Heather and A. Lazzeri, Rubber toughening of plastics. 12. Deformation mechanisms in toughened nylon 6,6, J. Mater. Sci., 16, 2255, 1989.

130. D. F. Lawson, W. L. Hergenrother and M. G. Matlock, Preparation and characterization of heterophase blends of polycaprolactam and hydrogenated polydienes, J. Appl. Polym. Sci., 39, 2331, 1990.

131. M. J. Modic and L. A. Pottick, Modification and compatibilization of nylon-6 with functionalized styrenic block-copolymers, Polym. Eng. Sci., 13, 819, 1993.

132. O. K. Muratoğlu, A. S. Argon, R. E. Cohen and M. Weinberg, Toughening mechanism of rubber-modified polyamides, Polymer, 36, 921, 1995.

133. O. K. Muratoğlu, A. S. Argon, R. E. Cohen and M. Weinberg, Microstructural processes of fracture of rubber-modified polyamides, Polymer, 36, 4771, 1995.

134. O. K. Muratoglu, A. S. Argon, R. E. Cohen and M. Weinberg, Microstructural fracture processes accompanying growing cracks in tough rubber-modified polyamides, Polymer, 36, 4787, 1995.

135. C. E. Scott and C. W. Macosko, Compounding and morphology of nylon ethylene-propylene rubber reactive and nonreactive blends, Int. Polym. Process, 10, 36, 1995.

136. S. V. Nair, S. C. Wong and L. A. Goettler, Fracture resistance of polyblends and polyblend matrix composites. 1. Unreinforced and fibre-reinforced nylon 6,6/ABS polyblends, J. Mater. Sci., 32, 5335, 1997.

137. S. V. Nair, A. Subramaniam and L. A. Goettler, Fracture resistance of polyblends and polyblend matrix composites. 2. Role of the rubber phase in nylon 6,6/ABS alloys, J. Mater. Sci., 32, 5347, 1997.

138. F. Ide and A. Hasegawa, Studies on polymer blend of nylon-6 and polypropylene or nylon-6 and polystyrene using reaction of polymer, J. Appl. Polym. Sci., 18, 963, 1974.

139. A. J. Oshinski, H. Keskkula and D. R. Paul, The role of matrix molecular weight in rubber toughened nylon 6 blends: 2. Room temperature Izod impact toughness, Polymer, 37, 4909, 1996.

140. A. J. Oshinski, H. Keskkula and D. R. Paul, The role of matrix molecular weight in rubber toughened nylon 6 blends: 3. Ductile-brittle transition temperature, Polymer, 37, 4919, 1996.

141. Y. Kayano, H. Keskkula and D. R. Paul, Evaluation of the fracture behaviour of nylon 6/SEBS-g-MA blends, Polymer, 38, 1885, 1997.

142. O. Okada, H. Keskkula and D. R. Paul, Fracture toughness of nylon 6 blends with maleated ethylene/propylene rubbers, Polymer, 41, 8061, 2000.

143. S. Y. Hobbs, R. C. Bopp and V. H. Watkins, Toughened nylon resins, Polym. Eng. Sci., 15, 482, 1982.

144. M. K. Akkapeddi, S. S. Chung and M. May, U.S. Patent 5,814,384.

145. X.-H. Wang, H.-X. Zhang, W. Jiang, Z.-G. Wang, C.-H. Liu, H.-J. Liang and B.-Z. Jiang, Toughening of nylon with epoxidised ethylene propylene diene rubber, Polymer, 39, 2697, 1998.

146. M. K. Akkapeddi, B. van Buskirk and J. H. Glans, in: Advances in Polymer Blends and Alloys Technology, ed., K. Finlayson, Vol. 4, Lancaster, PA: Technomic Publishing Co., 1993.

147. S.-C. Wong and Y.-W. Mai, Effect of rubber functionality on microstructures and fracture toughness of impact-modified nylon 6,6/polypropylene blends. Part II. Toughening mechanisms, Polymer, 41, 5471, 2000.

148. K. Tohgo, F. D. Fukuhara and A. Hadano, The influence of debonding damage on fracture toughness and crack-tip field in glass-particle-reinforced nylon 66 composites, Comp. Sci. Techol., 61, 1005, 2001.

149. D. M. Laura, H. Keskkula, J. W. Barlow and D. R. Paul, Effect of rubber particle size and rubber type on the mechanical properties of glass fiber reinforced, rubber-toughened nylon 6, Polymer, 44, 3347, 2003.

150. R. J. M. Borggreve, R. J. Gaymans and J. Schuijer, Impact behavior of nylon rubber blends. 5. Influence of the mechanical-properties of the elastomer, Polymer, 30, 71, 1989.

151. R. J. M. Borggreve, R. J. Gaymans, J. Schuijer and J. F. Ingen Housz, Brittle-tough transition in nylon-rubber blends: effect of rubber concentration and particle size, Polymer, 28, 1489, 1987.

152. Z. Bartczak, A. S. Argon, R. E. Cohen and M. Weinberg, Toughness mechanism in semi-crystalline polymer blends: II. High-density polyethylene toughened with calcium carbonate filler particles, Polymer, 40, 2347, 1999.

153. Y. Wang, J. Lu and G. J. Wang, Toughening and reinforcement of HDPE/CaCO3 blends by interfacial modification interfacial interaction, J. Appl. Polym. Sci., 64, 1275, 1997.

154. H. Hoffmann, W. Grellmann and H. Martin, Investigations of the effect of fillers on the toughness properties of HDPE, polymer composites, New York: Walter de Gruyter, 1986.

155. B. M. Badran, A. Galeski and M. Kryszewski, High-density polyethylene filled with modified chalk, J. Appl. Polym. Sci., 27, 3669, 1982.

156. Z. H. Liu, K. W. Kwok, R. K. Y. Li and C. L. Choy, Effects of coupling agent and morphology on the impact strength of high density polyethylene/CaCO$_3$ composites, Polymer, 43, 2501, 2002.

157. R. Gensler, C. J. G. Plummer, C. Grein and H.-H.Kausch, Influence of the loading rate on the fracture resistance of isotactic polypropylene and impact modified isotactic polypropylene, Polymer, 41, 3809, 2000.

158. L. A. Utracki, M. M. Dumoulin, in: Polypropylene: Structure, Blends and Composites, 2, ed., J. Karger-Kocsis, London: Chapman & Hall, 1995.

159. S. M. Dwyer, O. M. Boutni and C. Shu, in: Polypropylene Handbook, ed., E.P. Moore, Munich: Hanser, 1996.

160. K. C. Dao, Mechanical-properties of polypropylene crosslinked rubber blends, J. Appl. Polym. Sci., 27, 4799, 1982.

161. J. E. Stamhuis, Mechanical-properties and morphology of polypropylene composites — talc-filled, elastomer-modified polypropylene, Polym. Compos. 5, 202, 1984.

162. J. E. Stamhuis, Mechanical-properties and morphology of polypropylene composites. 2. Effect of polar components in talc-filled polypropylene, Polym. Compos. 9, 72, 1988.

163. C. Scott, F. H. J. Maurer, Composite Interfaces, eds., H. Ishida and J. L. Koening, New York: Elsevier, 1986, p. 177.

164. J. Kolarik and F. Lednicky, Polymer Composites, ed., Sedlacek, Berlin: Walter de Gruyter, 1986, p. 537.

165. B. Pukanszky, J. Kolarik and F. Lednicky, Polymer Composites, ed., Sedlacek, Berlin: Walter de Gruyter, 1986.

166. H. Kitamura, Progress in Science and Engineering of Composites, eds., T. Hayashi, K. Kawata and S. Umekawa, Tokyo: Japanese Society of Composite Materials, 1982, p. 1787.

167. H. Nakagawa and H. Sano, Improvement of impact resistance of calcium-carbonate filled polypropylene and propylene ethylene block copolymer, Abstr. Pap. Am. Chem. Soc., 26, 249, 1985.

168. M. Sumita, K. Sakata, S. Asai, K. Miyasaka and H. Nakagawa, Dispersion of fillers and the electrical-conductivity of polymer blends filled with carbon-black, Polym. Bull., 25, 265, 1991.

169. C. O. Hammer and F. H. J. Maurer, Barium sulfate-filled blends of polypropylene and polystyrene: microstructure control and dynamic mechanical properties, Polym. Compos., 19, 116, 1998.

170. J. E. Stamhuis, Mechanical-properties and morphology of polypropylene composites. 3. Short glass-fiber reinforced elastomer modified polypropylene, Polym. Compos., 9, 280, 1988.

171. B. Pukanszky, F. Tüdös, A. Kallo and G. Bodor, Effect of multiple morphology on the properties of polypropylene/ethylene-propylene-diene terpolymer blends, Polymer, 30, 1407, 1989.

172. W.-Y. Chiang, W.-D. Yang and B. Pukanszky, Polypropylene composites. 2. Structure-property relationships in 2-component and 3-component polypropylene composites, Polym. Eng. Sci., 32, 641, 1992.

173. H. K. Asar, M. B. Rhodes, R. Salovey, Multiphase Polymers, Advances in Chemistry Series, Vol. 176, eds., S. L. Cooper and G. M. Estes, Washington, D.C.: ACS, p. 489.

174. F. Ramsteiner, Structural changes during the deformation of thermoplastics in relation to impact resistance, Polymer, 20, 839, 1979.

175. A. Van der Wal, J. J. Mulder, H. A. Thijs and R. J. Gaymans, Fracture of polypropylene 1. The effect of molecular weight and temperature at low and high test speed, Polymer, 39, 5467, 1998.

176. K. Mitsuishi, S. Kodama and H. Kawasaki, Mechanical-properties of polypropylene filled with calcium-carbonate, Polym. Eng. Sci., 25, 1069, 1985.

177. G. Levita, A. Marchetti and A. Lazzeri, Fracture of ultrafine calcium-carbonate polypropylene composites, Polym. Compos. 10, 39, 1989.

178. J. Jancar, A. T. DiBenedetto and A. Dianselmo, Effect of adhesion on the fracture-toughness of calcium carbonate-filled polypropylene, Polym. Eng. Sci., 33, 559, 1993.

179. Z. Demjen, B. Pukanszky and J. Nagy, Evaluation of interfacial interaction in polypropylene surface treated $CaCO_3$ composites, Composites: Part A, 29, 323, 1998.

180. Y. S. Thio, A. S. Argon, R. E. Cohen and M. Weinberg, Toughening of isotactic polypropylene with $CaCO_3$ particles, Polymer, 43, 3661, 2002.

181. W. C. J. Zuiderduin, C. Westzaan, J. Huétink and R. J. Gaymans, Toughening of polypropylene with calcium carbonate particles, Polymer, 44, 261, 2003.

13

Structure-Property Relationships in Nanoparticle/Semicrystalline Thermoplastic Composites

J. KARGER-KOCSIS and Z. ZHANG

Institut für Verbundwerkstoffe GmbH
(Institute for Composite Materials), University
of Kaiserslautern, Germany

CONTENTS

I. INTRODUCTION

Nanoparticle-modified polymer composites (also termed polymeric nanocomposites, inorganic/organic hybrid materials) have attracted great scientific and technological interest owing to their exceptional physico-mechanical, thermal and other properties achieved at very low nanoparticle content (<5 wt% or <2~3 vol%). Nanoparticle means that the size of the related inorganic filler — at least in one dimension — is on nanometer scale. Note that the size of traditional fillers and reinforcements is in micrometer range (ca. 10 μm and more). Although the term nanocomposite sounds like a current one, nanocomposites have been produced industrially for more than half a century. In this respect attention should be drawn to the reinforcement of rubbers by nanometer scale carbon black. In addition, many natural and artificial products can be considered as nanocomposites based on their build-up.

A major specific feature of nanocomposite materials is their huge interfacial surface area. This can reach up to 1000 m^2/g filler. As a consequence, the interface/interphase properties may become the controlling parameters of the macroscopic response of polymer nanocomposites. Unlike the two-dimensional (2-D) interface, the interphase (3-D) concept considers that the molecular mobility changes from the particle surface toward the bulk in several nanometers range. A further aspect that has to be considered is that with decreasing mean particle size, the average distance between the nanoparticles also decreases when keeping the volume fraction of the filler constant. This may activate filler-filler interactions and result in a peculiar physical network structure.

Nanoparticles may be grouped upon their shape in 1-D (e.g., nanotubes), 2-D (platelets, disks) and 3-D (spheres) fillers. The 1-D and 2-D fillers of anisometric nature are usually characterized by the aspect ratio (length/thickness or length/diameter ratios). The aspect ratio of 3-D spherical fillers is per definitionem 1.

Polymer nanocomposites are very promising materials for various applications. They are expected to replace polymers, polymer blends and their traditional composites in products produced by melt processing techniques (injection, extrusion, blow and rotational molding). This prediction is justified by the improvements in properties (mechanical, thermal, barrier etc.) without sacrificing the melt rheological properties. Note that due to the low amount added, the rheology of the nanocomposites did not differ much from the neat polymers, at least in the nonviscoelastic range, which is important to the processability of these materials.

The major aim of this chapter is to survey the preparation, build-up and detection of the hybrid morphology, experimental results and theoretical predictions for the structure-property relationships of nanocomposites. Our interest was focused on semicrystalline polymer-based melt compounded or melt-produced systems containing platelet-type (2-D) and quasi-spherical (3-D) nanoparticles. Note that interested readers may find valuable further information to these topics in some recent monographs [1–4].

II. MANUFACTURING OF NANOCOMPOSITES

There are numerous ways to produce nanocomposites. As their grouping is quite a great challenge, only a distinction between *in situ*-generated nanocomposites and those made by incorporation of preformed particles will be made in this chapter.

Nanoparticles and nanocomposites can be produced *in situ* by various synthesis techniques such as sol-gel process, self assembly, coordination chemistry, bulk polymerization, polymerization inside templates, and biomimetic synthesis, which were reviewed recently [5–6]. However, we are considering among the *in situ* techniques the polymer intercalation in 2-D layered structures and 3-D frameworks. Polymer intercalation can be achieved either using polymers or monomers. Note that in the latter case, the monomers are polymerized subsequently. Why are these methods listed among the *in situ* techniques? Usually microscopic particles (>10 µm) are added in the polymer or monomer in which the particles disintegrate in their nanoscale constituents. Polymer intercalation occurs not only in the melt. For that purpose, solution, aqueous and non-aqueous dispersions may also be used after eliminating the related carrier (solvent, water, etc.)

The other basic way of nanocomposite production is the incorporation of preformed nanoparticles (such as SiO_2, TiO_2, $CaCO_3$, CdS) and clusters. This happens usually by the same routes as mentioned for polymer/monomer intercalation. Among the clusters, silsesquioxane cubes have to be mentioned due to their practical relevance. Polyhedral oligomeric silsesquioxane (POSS) has been successfully incorporated in many semicrystalline polymers [7–9]. This occurred either by physical blending or chemical grafting via suitable functionality. The functional groups are created at a corner Si atom of POSS. According to the authors' feelings, POSS represents a straightforward extension of the sol-gel route (extensively used already for thermosetting resins) for thermoplastics.

Since giving an exhaustive survey on the numerous production methods is beyond the scope of this chapter, emphasis will be put on polymer intercalation in 2-D and 3-D structures and incorporation of preformed particles in polymers. This selection is justified by the short-term market penetration of the related nanocomposites.

A.   Intercalation in 2-D and 3-D Structures

2-D and 3-D Nanoparticles

The commonly used 2-D reinforcements are 2:1 layered silicates (phyllosilicates) of natural (e.g., bentonite, montmorillonite, often termed clays) and artificial (e.g., fluorohectorite) origin. They contain two tetrahedral silicate sheets fused to an edge-shared octahedral one, resulting in an overall thickness of ca. 1 nm. The lateral dimension (and thus the aspect ratio) of the layered silicates varies in a very broad range from several ten nanometers to several micrometers. Isomorphic substitution of higher valance cations (Al^{3+} and Mg^{2+}) in the silicate framework by lower valence ones (Fe^{2+}, Mg^{2+} and Li^+, respectively) generated negative charges on the layers, which are counterbalanced usually by alkaline cations (Na^+, Ca^{2+} — generally in hydrated forms). As a consequence, such layered silicates exhibit a cation exchange capacity (CEC) and the intergallery cations can be replaced by suitable organic cationic surfactants. For that purpose primary, secondary, tertiary and quaternary ammonium compounds are used today. By this cation exchange the hydrophilic silicate is rendered organophilic and at the same time the interlayer spacing (basal or d spacing) increases. The latter is tuned by the chemical build-up of the onium intercalant (often containing a long alkyl chain) the further role of which may be to support the chemical interaction with the matrix (bearing, for example, hydroxyethyl groups) [1,2,10]. Note that the interlayer spacing should be larger than ca. 1.5 nm in organophilic clay (i.e., the interlamellar distance >0.5 nm). Needless to say, the price of the organophilic silicates is considerably higher (more than threefold) than that of the purified pristine ones.

It is worth noting that layered silicates with anion exchange capacity (their layers have a positive surface charge which is compensated by intergallery anions) are also available [11]. However, the overwhelming majority of the works done in the past were dealing with layered silicates of cation exchange capacity. Recently, other layered minerals and even graphite were also tried as nano-reinforcements in polymer composites [12–13].

Unlike 2-D layered structures which can be delaminated during compounding, the 3-D frameworks are stable and do not change their size. Thus, the related nanocomposites can be considered as host (3-D structure)–guest (polymer) hybrid materials. Among the 3-D frameworks natural zeolites, synthetic molecular sieves and mesoporous glasses have to be mentioned. Zeolites are crystalline aluminosilicates with well-defined pore size (less than 2 nm). Like zeolites, molecular sieves possess also a crystalline structure, a larger pore size and a build-up of non-aluminosilicate nature. Crystalline mesoporous silicates are very versatile, especially in respect to their pore size (ranging from 2 to 10 nm). A unique and general feature of the above-mentioned 3-D silicates is to discriminate between molecules upon their size and shape [5].

B.   Use of Preformed Nanoparticles

To incorporate micron-size preformed inorganic particles into a polymer matrix is a well-known method for improving the modulus of such composites. However, a reduction in the ductility of the material may take place. Furthermore, either by diminishing the particle size or by enhancing the particle volume fraction, the flexural strength and even the tensile strength can be enhanced. On the other hand, the fracture toughness and modulus remain fairly independent of the particle size, even when going down to the nanoscale. Recently, researchers demonstrated that inorganic nanoparticles could be of benefit for an increased tensile elongation. Many researchers also reported about an increase of the glass transition temperature (T_g) of polymers by the addition of various preformed nanoparticles, which may be due to a good bonding between the nanoparticles and the polymers, thus restricting the motion of the polymer chains. However, the potential of property improvements is still not fully explored.

As already mentioned, preformed inorganic nanoparticles, e.g., SiO_2, TiO_2, $CaCO_3$, Al_2O_3, and CdS, are frequently applied as nanofillers into polymer matrices by melt compounding techniques. These nanoparticles are commercially available, for example Degussa and Nanophase supply various sizes and surface-treated nanoparticles for different application purposes. Numerous surface treatment approaches of preformed nanoparticles either by physical interaction or by chemical reaction were reviewed recently by Zhang et al. [4].

C.   Preparation Routes

The intercalation methods of polymers in 2-D and 3-D inorganic hosts, as well as the incorporation of preformed particles in polymers can be divided in three major groups:

1. *In situ* polymerization
2. Solvent-assisted techniques
3. Melt compounding

1.   *In situ* Polymerization

In this case the nanoparticle-generating filler is dispersed or swollen (2-D layered structures) in a liquid (may be melt) monomer or oligomer in the presence or absence of additional solvent. The interlayer polymerization may be started by the usual methods (thermal polymerization, UV or electron beam irradiation, dose of initiators and catalysts). In the case of 2-D and 3-D structures this method is termed intercalative polymerization. A very promising way is to render the surface of the particle catalytic for the subsequent polymerization. The *in situ* polymerization is widely used for manufacturing nanoparticle- (any kind) reinforced thermoplastics, also on an industrial scale (e.g., polyamides). The *in situ* synthesis of clay-reinforced polyamide 6 (PA-6), credited to researchers at the

Toyota Central Research Labs, Japan (e.g., [14–15]), gave the impetus to R&D activities in this field. As a consequence numerous studies were devoted to polymers, including PA-6 (e.g., [16]), PA-12 (e.g., [17]), polycaprolactone (PCL, [18]), thermoplastic [12,19–20] and liquid crystalline polyesters [21] and polyolefins (e.g., [22–24]). Note that PAs and PCL were produced by ring opening polymerization of the related monomers which entered in the interlamellar space via swelling the organoclay (containing frequently α,ω-amino acids as organophilic intercalant). To synthesize linear polyesters via *in situ* polycondensation, the clay is usually dispersed in the glycol phase, which swells the clay [2]. Another option is to exploit the ring opening polymerization of cyclic oligomers as shown on the example of poly(butylene terephthalate) (PBT) [19]. Free radical-induced bulk polymerization is also a suitable technique as shown on the example of vinyl acetate monomer [25]. To produce polyolefins, viz. polyethylenes (PEs) and polypropylenes (PPs), via coordination polymerization the layered silicates are made to (co)catalysts. This can be achieved by various concepts [2]. In the case of heterogeneous Ziegler-Natta catalysts, the Ti-based active compound can be fixed on the clay surface. However, it is also feasible to use clay for metallocene catalysis (termed also homogeneous Ziegler-Natta catalysts) via intercalating the catalyst inside the silicate layers.

Practically the same techniques can be used to insert (spherical) preformed particles in the polymer. In this case it is also straightforward to make the particle surface organophilic, which can be done for example by the grafting of suitable monomers [26]. Irradiation grafting of another monomer than the matrix forming one on the silica nanoparticles also proved to be useful to avoid aggregation, agglomeration phenomena [27]. This prepolymerization/grafting procedure can be treated as a "masterbatch process." This becomes of paramount interest for fully compatible thermoplastic blends. Polymerizing methyl methacrylate in the presence of layered silicate and exploiting the compatibility of the resulting polymer (viz. polymethylmethacrylate [PMMA] with polyvinylidene fluoride PVDF]), the whole blend can be made "nanostructured" [28].

2.   Solvent-Assisted Techniques

Preparation of nanocomposites via solution dispersion is mostly of academic interest (to study selected structure-property relationships) for polymers soluble only in organic solvents. Such techniques have been used for poly(trimethylene terephthalate) (PTT) [29], poly(ethylene terephthalate) (PET) [30], syndiotactic polystyrene (sPS) [31–32], poly(lactic acid) (PLA) [33], polypeptides [34], PA-6 [35]. The scenario is, however, completely different for water-soluble polymers such as poly(vinyl alcohol) (PVAL) [36]. Note that both pristine and organophilic clay "swell" in water supporting the intercalation. A water-assisted method was used to produce PVAL-based visually transparent films with broadband UV filter effect. This was achieved by dispersing nanoscale TiO_2 particles of the rutile crystal modification in water-soluble polymers [37].

This is the right place to call attention to the combination of the above-listed techniques, viz. *in situ* polymerization, solution techniques and melt compounding. As layered silicates swell in water they can be incorporated in the molten polymer during compounding via inserting a clay slurry. During compounding the water carrier has to be evaporated. The beauty of this method is that cheap pristine clay is used instead of a more expensive organophilic one. Further, it has been demonstrated that rubber intercalates in pristine clays from aqueous lattices [38–39]. By adding a clay-containing rubber latex in a thermoplastic polymer during melt, compounding may result in toughened, nano-reinforced semicrystalline thermoplastics. Note that the mean size of the rubber in the latex matches very well with the requirements [40]. The reader interested in this issue is kindly referred to the related patent literature.

3.   Melt Compounding

Melt compounding (designated as melt intercalation for 2-D and 3-D silicate hosts) is the most attractive way to produce commercial nanocomposites. This is owing to: (1) fast dispersion of the nanoparticles in the melt, (2) available industrial melt compounding capacities, and (3) environmentally friendly preparation. Similar to solvent intercalation, melt intercalation is also governed by thermodynamical (compatibility) and kinetical (diffusivity) parameters. It is intuitive that polymer molecules intercalating in 2-D layers or penetrating in 3-D frameworks lose their conformational freedom in the confined space. This is associated with entropy loss. The terms "nano" and "nanoscale" already suggest that the formation of polymer nanocomposites has many similarities with (im)miscible polymer blends and thus the related rules can also be adopted. So, in order to get molecular, i.e., nanoscale, dispersion, the Gibb's free energy must be negative. As entropy loss produces an adverse effect, it has to be "overcompensated." This may occur by the entropy gain of molecules of the initial organophilic modifier due to interdiffusion with molecules of the matrix polymer and/or by energetically favored interactions between the polymer molecules and silicate particles. Energetically favored interactions involve acid/base and chemical reactions, H-bonding etc, all of them affecting the term enthalpy of mixing. This aspect has to be considered when selecting the surface modification of silicates.

It was also experimentally proven that the layer disintegration in 2-D or intercalation in 3-D inorganic structures depend on the polymer diffusivity [1–2]. A further parameter which affects the composite formation is linked to the locally acting shear and elongational flow fields. As the majority of the experimental work was done by extrusion melt compounding, the related research focused on the effects of shear stresses. They were varied by different ways, including equipment selection (type, screw design [41–43]), processing parameters such as residence time [41] and polymer characteristics [43]. Note that by changing the mean molecular mass of the melt compounded polymer — by keeping all other experimental parameters constant — the effects of shear stresses can be studied separately [43]. The outcome of these studies was that the dispersion state of the layered silicates is controlled by the shear and residence time (kinetics). Extensive shear stresses shear the layered stacks and peel apart their constituting layers [41,43]. It is worth noting that the above thermodynamical, kinetical and processing-related effects are often interrelated.

Similar rules, as listed above for 2-D layered silicates, hold also for the melt dispersion of preformed particles. Energetically favored interactions can be achieved by physical (coating the particles' surfaces by suitable surfactants) and chemical ways (use of coupling agents, surface grafting of monomers, etc.). Further, the processing conditions of the related dispersive mixing are of vital importance. According to a model explanation, the following consecutive steps can be distinguished during melt compounding: incorporation, wetting, agglomerate break-up and aggregate spatial distribution [4]. Incorporation and wetting are likely governed by thermodynamical aspects. For agglomerate break-up, the cohesive forces between the particles have to be surpassed by the melt flow-induced hydro-dynamical ones. Recall that the latter depends on type and processing conditions of the melt mixing equipment. Useful guidelines to avoid agglomeration and aggregation during melt compounding with preformed particles are listed in References [4,44] and references therein.

Nevertheless, agglomeration is a general problem, especially at increased nanofiller content. Figure 1 illuminates the correlation among particle diameter, distance and volume content. Here the spherical particles were assumed at a cubic distribution situation with perfect dispersion in a polymer matrix. It should be noted that agglomeration may easily happen for smaller particles at higher filler content due to the reduced distance between nanoparticles.

Nanocomposites have been produced using all thermoplastic polymers. Next, an attempt will be made to survey some general rules on how to produce polymer nanocomposites via melt compounding. It is noteworthy that *in situ* polymerization is not superior to melt compounding. Figure 2 demonstrates the dynamic-mechanical thermal analysis (DMTA) spectra of PA-6 nanocomposites containing 4 wt% organoclay and produced by different methods. Note that the stiffness temperature traces are practically independent on the production method.

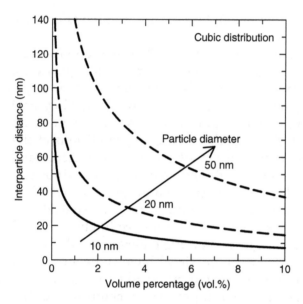

Figure 1 Correlation among particle diameter, interparticle distance and volume percentage based on assumptions of spherical particle, cubic distribution and ideal dispersion.

As most nanocomposites are of a polar nature, their incorporation in apolar polymers (e.g., polyolefins) is a great challenge. Recall that micro- and macro-fillers are often "coated" by surfactants (tensides). Their role is to improve the compatibility between the filler and polymer via their long alkyl chains. The same philosophy can be followed for the intercalation of 2-D and 3-D inorganic structures. The related polymers, oligomers are called compatibilizers. They are usually grafted copolymers due to economical reasons. Maleic anhydride and acrylic acid-grafted (polymer-g-MA and -g-AA, respectively) versions are preferred compatibilizers for polyolefin-based systems (e.g., [45–49]. It was also established that lower molecular mass compatibilizer favors the organoclay exfoliation in contrast to that of higher molecular mass [49]. Generally, in the presence of compatibilizer a higher degree of intercalation/exfoliation was found than in its absence. This was well reflected in the related mechanical properties.

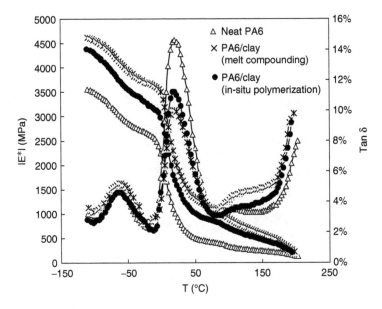

Figure 2 Comparison of the DMTA behavior of PA-6 with (4 wt%) and without organoclay. Notes: Organoclay was dispersed either by melt compounding or during *in situ* polymerization. Samples were conditioned according to ISO 1110 prior to the DMTA measurement.

Using blends, attention should be drawn to the fact that the silicate may be preferentially embedded in one of the blend components. In uncompatibilized PA-6/PP blend (70/30 parts) the organoclay was located in the PA-6 phase as shown by transmission electron microscopy (TEM) (Figure 3) [50–51]. Adding maleated PP (PP-g-MA) to the above blend, the morphology changed substantially. First, the minor phase, i.e., PP, became more finely dispersed. Second, the clay layers were covered by an interphase layer. The latter formed between PP-g-MA and PA-6 by reaction of the MA group with primary and secondary amine functionalities of the PA-6.

If the silicate is well dispersed in miscible blends at a given composition ratio, a change in the latter may induce re-aggregation (also termed confinement, de-intercalation). This was shown on the example of PA-6/ethylene vinylalcohol (EVOH) blends [52].

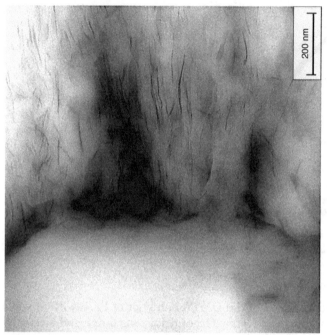

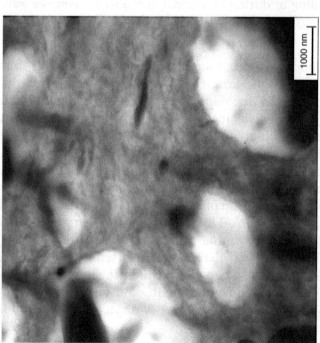

Figure 3 TEM pictures taken from a melt compounded PA-6/PP/organoclay (70/30/4 parts) system. Notes: the intercalated/exfoliated clay layers are exclusively located in the PA-6 phase (matrix). White large spots are due to the PP particles present in coarse dispersion. (From Chow WS, Mohd Ishak ZA, Ishiaku US, Karger-Kocsis J, Apostolov AA. J Appl Polym Sci 2004; 91:175–189. With permission.)

From the viewpoint of the interlayer distance (2-D) and pore size (3-D) of the silicate hosts, the general rule is the larger the size, the easier the intercalation is. This is considered by tuning the d spacing of the 2-D layered silicates via suitable organic intercalants. On the other hand, strong interaction between the filler or its intercalant and the polymer does not necessarily improve the filler dispersion. In some cases an adverse effect may appear, as will be shown later.

III. STRUCTURE DEVELOPMENT AND CHARACTERIZATION

The structure of polymer nanocomposites is very complex as it covers the following domains: dispersion state of the nano-particles, changes on molecular and supermolecular level in the matrix (bulk), interphase formation between the surface of the nanoparticles and bulk material. Again, some characteristics of the above fields are interrelated. As a consequence, it is not an easy task to find those structural parameters which control a given property.

A. Particle Dispersion

It is obvious that nanoscale-sensitive experimental techniques have to be used to detect the dispersion state of the nanoparticles. For that purpose TEM is preferred. Figure 4 displays a satisfactory dispersion of TiO_2 nanoparticles on the example of a melt compounded PA-6,6/nano-TiO_2 system. It should be borne in mind, however, that the view field at high magnifications may not represent that of the whole sample. Further, it is essential to describe the dispersion state. Albeit some trials were made to make use of image analysis codes, this issue is not yet solved properly. Another straightforward technique is the atomic force microscopy (AFM).

Apart from TEM, polymer intercalation in 2-D layers is usually evidenced by x-ray diffraction (XRD) performed at both small- and wide-angle scattering (SAXS and WAXS, respectively). In WAXS, pattern intercalation manifests in a shift toward lower scattering angles in the range $2\Theta = 1–10°$. Note

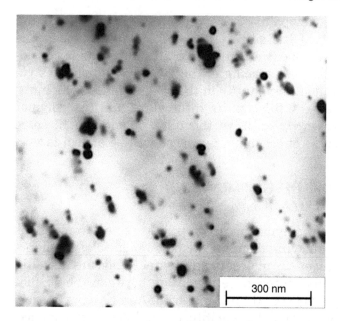

Figure 4   TEM picture taken of a PA-6,6/TiO$_2$ (21 nm, 2 vol%) nanocomposite produced by melt compounding in a twin-screw extruder. (From Zhang Z, Yang JL, Friedrich K. Polymer 2004; 45(10), 3481–3485. Copyright 2004, with permission from Elsevier.)

that $2\Theta = 1°$ corresponds to a basal spacing 8.8 nm using CuK$_\alpha$ radiation. In contrast to the frequently quoted claim that missing peak in the XRD spectra is due to exfoliation, it is not at all correct. In order to get a reliable picture on the silicate dispersion, XRD and TEM always should be combined [53–54].

Under certain conditions instead of intercalation/exfoliation, just the opposite occurs. A scanning electron microscopic (SEM) picture in Figure 5 shows the development of large montmorillonite particles which were likely formed via de-intercalation (i.e., the original organophilic intercalant was "extracted" from the interlamellar space, and in addition the clay layers stacked together at their edges). It is the right place to emphasize again that albeit the properties of nanocomposites strongly depend on the dispersion state of the particle, no general guideline is given on how to characterize it.

20 kV ×10000 ⊢———3 µm———⊣

Figure 5 SEM picture taken from clay particles which were assembled (stacking and flocculation) from platelets of an organo-clay (length ca. 200 nm) via chemically induced de-intercalation. Note: arrow indicates a clay tactoid of the initial size. (From Gatos KG, Thomann R, Karger-Kocsis J. Polym Int 2004; 53(8):1191–1197. Copyright 2004 Society of Chemical Industry. With permission from John Wiley & Sons Ltd.)

B.   Matrix Polymer (Bulk)

Changes in the matrix morphology owing to the presence of nanoparticles occur at different levels. Like some micro- and macroscopic fillers and reinforcements, nanoparticles also act as heterogeneous nucleation agents. High nucleation density on the filler surface may generate trans-crystalline growth as demonstrated on the example of organoclay-filled PA-6 [55]. Transcrystallization is caused by dense nuclei on the hetero-geneous surface due to which the spherulitic crystallization is laterally hindered. So, growth occurs in one direction, viz.

perpendicular to the filler surface. It is believed that the transcrystalline layer supports the stress transfer from the weak matrix to the "strong" nano-reinforcement [56]. It is noteworthy that transcrystallization is of epitaxial origin for which there is a good chance between PA-6 molecules and the clay surface via H-bonding [2,55].

The heterogeneous nucleating effect of nanoparticles has been demonstrated for many semicrystalline thermoplastics. The basic difference in the related reports is whether or not the linear spherulitic growth and the overall crystallization rates changed in the presence of nanoparticles. For PA-6 a dramatic increase in the crystallization rate was found at least at low organoclay content. In contrast, at higher organoclay content the crystallization was retarded [57]. In another report the nonspherulitic crystallization of PA-6 due to organoclay was established [58]. All works related to PA-6/layered silicate nanocomposites agree, however, that the silicate layers act as selective nucleants for the γ-modification (γ-nucleants) [57–61]. Preferred formation of the γ-phase was also observed in other polyamides (e.g., [62]). Note that the γ-phase is inherently more prone to ductile deformation than the usual α-modification. This is an important issue with respect to the toughness which is often deteriorated upon silicate incorporation. It is noteworthy that fast crystallization owing to heterogeneous nucleation may result in fine spherulitic morphology and lower overall crystallinity. Selective nucleation of given polymorphs due to nanoparticles has been reported also for PVDF [63–64], poly(1-butene) [65] and sPS [31–32,66–67]. Studies devoted to the crystallization (isothermal, nonisothermal) behavior of various semicrystalline thermoplastics in the presence of nanoparticles showed that the usual descriptions (Avrami, Ozawa, etc.) are valid. The related results indicated, however, for some changes in the crystallite growth geometry [68–72].

Very interesting results were achieved by investigating PP/mesoporous silicate nanocomposites. It was shown that isotactic PP confined in the mesopores does not crystallize [73]. This fact highlights the reason of the very recent research trend dealing with the crystallization behavior of polymers under spatial constraints.

C.   Interphase

To examine the interactions between nanoparticles and polymers, various techniques can be used. Fourier-transform infrared (FTIR) spectroscopy, solid-state nuclear magnetic resonance (NMR), calorimetry, thermo-gravimetric analysis (TGA), chromatographic and electrophoretic measurements, all can contribute to get a better insight in structure-property relationships and interphase properties [10]. A rather simple and informative method is the DMTA. Strong absorption of polymer molecules on nanoparticles possessing very high specific surface area yields a change in the T_g peak (shape alteration, intensity reduction, shift toward higher temperature). This — representing a close analogy to carbon black filled rubbers — was detected in several polymers, in fact (e.g., [74–77]). In some polymers even an additional mechanical damping peak, beyond that of T_g, may appear. This was shown for intercalated rubbers [78] but not yet for semicrystalline polymers. It is, however, not yet clear whether the onset of such a peak represents some confined fraction of the polymer or is related to the interphase in analogy to multiphase polymer blends [79]. A direct evidence for changes in the interphase characteristics may deliver the AFM. 3-D topography contour plots taken by AFM from physically etched PA-6/PP/organoclay (70/30/4 parts) systems with and without compatibilizer (PP-g-MA) are depicted in Figure 6. Based on Figure 6a, it is obvious that PA-6 is eroded faster than PP (see the large particle in the bottom corner in Figure 6a). As shown before by TEM (cf. Figure 3), the organoclay is located in the PA-6 phase — it is visible as sharp protrusions in this AFM scan. It was shown that the maximum length of these protrusions agrees with that of the mean length of the organoclay, indeed [80]. The morphology is changing substantially after introducing PP-g-MA compatibilizer. In Figure 6b no sharp protrusion but less eroded small domains ("humps") are visible. The height of these humps is closely matched again

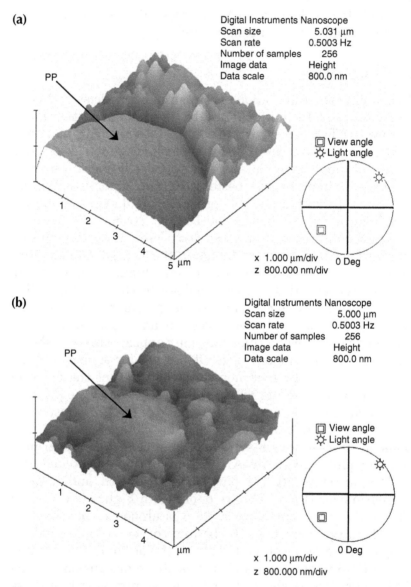

Figure 6   AFM contour plots from the polished physically etched surface of uncompatibilized (a) and with PP-g-MA compatibilized (b) PA-6/PP/organoclay systems (composition:70/30/4 parts). Note: the compatibilizer (PP-g-MA) content was 5 parts.

with that of the mean clay length. So, the intercalated clay layers are "buried" in these domains. The strong resistance to Ar^+ bombardment (used for physical etching in this case) is obviously due to the formation of a PP-rich interphase. This interphase, consisting of a PA-6 grafted PP, was developed owing to reactions between the primary and secondary amines of the PA-6 and anhydride groups of the PP-g-MA compatibilizer [80].

D. Effects of Processing

Recall that intercalation/exfoliation phenomena are governed by thermodynamics and kinetics. In addition, thermodynamic principles also play an important role for the dispersion of preformed nanoparticles. As a consequence, the thermal stability of the silicate dispersion with respect to (further) processing should be addressed. It has to be taken into account that the usual amine-type intercalants have limited thermal stability; that is the major driving force to search for more stable organophilic modifiers (tensides, surfactants, intercalants) for 2-D layered silicates. The thermal stability of the dispersion in nanocomposites can be studied by different ways. One option is to produce an exfoliated version and then study the alteration in its dispersion state upon melt processing, annealing, etc. The idea behind this method is that a "forced" exfoliation is not necessarily the thermodynamically stable one. So, an exfoliated nanocomposite produced by solution technique was thermally processed in various equipments and the accompanied changes in the dispersion were assessed via the usual techniques (TEM, XRD) [81]. Note that *in situ* polymerization of nanocomposites is not a panacea, as the clay layers may reaggregate upon melt processing. This is due to the thermodynamical incompatibility between the polymer and silicate which was set "out of limits" during polymerization but was "reactivated" upon processing [82]. Reaggregation (confinement) of 2-D layered silicates may be observed also when they are introduced in substantially higher amount, which can be intercalated/exfoliated, by the matrix polymer itself. For complete confinement (de-intercalation), however, the compatibility should change dramatically. This can be triggered for example by chemical reactions occurring during processing [82–83]. Thermal degradation of the organophilic modifier may also affect the compatibility between polymer and silicate, and the related alteration in the dispersion becomes detectable in the properties. It is intuitive that to study such effects, nanocomposites with high melting temperature matrices (such as PA-6) have to be selected. It was found that the molecular mass of PA-6 decreases, and the color of the related nanocomposite changes upon melt processing. The related changes depended on the type of the organoclay. This finding was attributed to surfactant-induced reactions (viz. Hofmann elimination reaction for quaternary ammonium compounds) and to the attack of the polymer by their by-products [83]. In addition, the water, present always in the organoclay, may attack the peptide bonds (–CO–NH–) and cause their scissions [84].

Recall that the aspect ratio of the 2-D layered silicates in their fully exfoliated stage is between 200 to 1000. The aspect ratio of the intercalated stacks is much less, usually below 100. Note that the latter value is closely matched with that of discontinuous fibers in injection-moldable composites [85]. In analogy to the alignment and layering of short fibers in injection-molded parts, one would expect that the 2-D stacks and platelets also orient during injection molding. The orientation, as for discontinuous fibers, is controlled by the shear and elongational flow fields which are superimposed during processing. In fact, the injection molding-induced orientation and layering of 2-D silicates were detected by several groups [57,86–87]. As expected, the platelets were oriented along the mold flow direction in the skin and more or less perpendicular to the flow in the core regions. It was also demonstrated that the skin was formed of the γ-phase whereas in the core both α- and γ-modifications were present [57]. The chain axis of the γ-lamellae, grown on the silicate surface, is parallel to the silicate layers [88]. A strong orientation of the clay platelets was observed under elongation flow conditions, including fiber spinning (e.g., [89]). Elongational flow conditions were suitable to study the effects of the aspect ratio of the 2-D layered silicates. This can be treated as a strong argument to adopt composite rules valid for aligned, short fiber-reinforced thermoplastics. At present, the barrier performance of nanocomposites is exploited commercially in the form of films and foils. As a consequence, considerable research efforts were made to assess the dispersion state (layering, orientation) in films produced by various methods (e.g., [90–92]).

IV. PROPERTIES AND THEIR PREDICTION

It was shown before that the structure of nanocomposites is highly complex and partly of hierarchical nature. Therefore it is a great challenge to trace those structural parameters which affect the desired property. In respect to the structure-property relationships, the basic question we have to answer is: do these issues belong to polymer physics or continuum (composite) mechanics? In the former case the bulk and inter-phase, whereas in the latter reinforcement-related character-istics should govern the properties. Unfortunately, no definite answer can be given to the above question. In certain condi-tions, grouped in low frequency mechanical tests (creep, fatigue), aspects of polymer physics may dominate. In tests of high frequency loading (dynamic, impact) the use of com-posite analogies (i.e., continuum mechanics) seems to be straightforward. With other wording, composite rules are more promising to describe the elastic (linear elastic, linear mechanic), whereas polymer physics principles are more suited to assess the relations between structure and anelastic (nonlinear elastic) properties.

A. Mechanical Response

1. Stiffness and Ultimate Properties for 2-D Intercalated Nanocomposites

The simplest way to predict stiffness (Young's or E-modulus) is to check whether or not the rule of mixtures holds. The Voigt upper bound of the E-modulus is based on a two-phase laminate model (matrix and reinforcement), according to which the reinforcing laminates are aligned along the load direction and thus all constituents experience the same strain (parallel coupling). The Reuss lower bound of stiffness reflects a serial coupling, according to which both matrix and reinforcing phases are under the same stress. On the example of injection-molded organoclay reinforced PA-6, it was shown that the tensile E-modulus follows the Voigt estimate, irrespective of whether the clay was intercalated or exfoliated [93]. It was speculated that the reason for the high E-moduli is that the γ-phase crystalline lamellae, grown in the clay surface, ensure the "parallel coupling" between the phases. Recall that the disk-shaped silicates are well oriented along the mold flow direction during injection-molding, provided that the thickness of the specimens is not very thick. The pioneering work of Fornes and Paul [94] was aimed at checking the applicability of composite theories for PA-6/organoclay systems. To predict the stiffness of nanocomposites, the Halpin-Tsai and Mori-Tanaka theories were used after assessing the dispersion state of the nanocomposite accordingly (major emphasis was put on the determination of the aspect ratio and alignment of the clay stacks and layers). The results showed that both approaches can be used, although they treat the effect of filler geometry differently. The authors demonstrated that this PA-6/organoclay nanocomposite outperforms the related short fiber-reinforced versions because of the higher aspect ratio and 2-D reinforcing effect of the clay layers. It is noteworthy that the Mori-Tanaka approach gave a slightly better fit with the experiments than the Halpin-Tsai prediction [94]. The former was the favored prediction also for organoclay-reinforced thermosets [95]. Nevertheless, the usability of the Halpin-Tsai approach was shown by other researchers, too [96]. Brune and Bicerano [96] explained why the compression modulus (and thus also the flexural one) may be lower than the tensile one (owing to buckling phenomenon), which was often found (e.g., [50–51]). The authors also emphasized that the reinforcing efficiency of the 2-D platelets strongly depends on their orientation with respect to the loading direction [96].

The scenario is far less clear for the ultimate properties. Clay exfoliation raises the stiffness, which is accompanied by reduced strain (ductility). So, the nanocomposites are becoming "harder" but more "fragile." The ultimate properties depend not only on the intercalation/exfoliation state, but also on characteristics of the interphase (wetting, adhesion) and the bulk (polymorphism, crystallinity, spherulite size). It was reported by Kim et al. [97] that the first failure is due to void formation inside of silicate stacks. This is followed by splitting, opening and sliding of the silicate stacks, depending on their relative orientation to the loading [97]. Voiding and cavitations were concluded to be first failure events also by other researchers [98]. Needles to say, these deformations (showing clear analogies with the deformation modes of crystalline lamellae in semicrystalline polymers) are accompanied with substantial changes in the interphase and bulk morphology. This was shown by *in situ* x-ray synchrotron measurements during uniaxial deformation of films of PA-6/clay hybrid [99].

2.   Stiffness and Ultimate Properties of
       Nanocomposites with Preformed
       Quasi-Spherical Particles

Here also the rule of mixture is the first approximation to predict the stiffness response. For macroscopically filled systems, the Kerner equation is widely used. The disagreement between Kerner's prediction and the experimental results obtained for nanocomposites forced the researchers to consider the interphase. It was treated as an immobilized layer which increased the effective filler volume fraction. Results suggested that the thickness of the immobilized layer may be much larger than the size of the particles [4]. This was corroborated by results derived from pressure-volume-temperature (PVT) measurements applied for PA-6/layered silicate nanocomposites [100].

The strength of traditionally filled systems decays according to a power law function. This means that the strength of the composite is always below that of the neat matrix polymer, as the filler does not bear any part of the external load. In contrast, considerable strength increase was measured for nanoparticle-reinforced thermoplastic systems. The related functions were treated by empirical models as listed in Reference [4]. On the other hand, the reinforcing effect of the nanoparticle is not yet clarified. Obviously, the reason behind the strength increase should have some analogy with nanoparticle- (e.g., carbon black) reinforced rubbers [3].

3.   Creep and Fatigue Behavior

Few papers were published on the low strain rate yield, creep and fatigue behavior of nanocomposites. This is quite surprising as the above long duration tests are very sensitive to changes in the interphase and bulk properties. Truss and Lee [101] concluded that in a poorly intercalated PE/montmorillonite composite, the yield is related to crystalline deformation mechanisms similar to that of neat PE. Studying the temperature and strain rate sensitivity of organoclay-modified PA-6 and PP nanocomposites, Mallick and Zhou [102] found that the Eyring equation works well also for these composites. The related activation volume depended on whether the test was performed below or above the T_g of the matrix polymer. Unfortunately, the authors did not comment on the possible rationale behind the difference in the activation volume and energy between the PA-6- and PP-based nanocomposites.

Creep tests were recently performed on TiO_2/PA-6,6 nanocomposites by Zhang et al. [103]. Creep is a time-dependent plastic deformation, which takes place under stresses lower than the yielding stress of materials. Relatively poor creep resistance and dimensional stability of thermoplastics are generally a deficiency. Neat PA-6,6 exhibited a high creep strain under a constant load of 80% of its ultimate tensile strength at room temperature, as shown in Figure 7. Under a similar loading condition, the incorporation of 1 vol% 21 nm TiO_2 particles significantly reduced the creep strain of PA-6,6 over all creep stages, although the final creep life was not very much different from that of the neat polymer. It is clear that, in practice, the reduction of the creep strain is

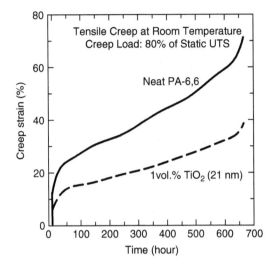

Figure 7 Tensile creep strain vs. test-duration curves under 80% of the static ultimate tensile strength (UTS) at room temperature. Note: the constant creep stress for the neat PA-6,6 and PA-6,6/TiO$_2$ (21 nm, 1 vol%) nanocomposite was 60 and 59 MPa, respectively. (From Zhang Z, Yang JL, Friedrich K. Polymer 2004; 45, 3481–3485. Copyright 2004, with permission from Elsevier.)

even more important for polymers than the extension of the creep life, since the former relates to the dimensional stability of the materials. Creep tests under higher loading situation, i.e., 90% of the ultimate tensile strength, and evaluated temperatures, i.e., 50°C, showed a similar reduction in the strain behavior. It was the authors' opinion that nanoparticles may restrict the slippage, reorientation and motion of polymer chains. In this way, the nanoparticles influence the stress transfer, which finally results in the improvement observed.

Normalized stress-cycles (S-N) curves derived from tension-tension fatigue showed some unexpected results [102]. Normalization occurred by dividing the maximum cyclic stress (S) with the respective yield one. In such representation, the fatigue resistance increased according to the following ranking:

PA-6/clay (3 wt%) < PP/clay (5 wt%) < PP/talc (40 wt%)

It was concluded that the fatigue failure is initiated by re-agglomerated particles in the materials [102]. Bellemare et al. [104] concluded that the fatigue life of organic clay/PA-6 composites depended on whether the cyclic tension-tension tests were performed at a given stress or strain amplitude. For the former case the fatigue life increased, whereas the latter decreased when compared to neat PA-6. This finding is in concert with results obtained on filled polymers with microscale fillers. Unexpectedly, the resistance to fatigue crack propagation decreased in the presence of organoclay. This is at odds with the effect of short-fiber reinforcement in thermoplastics [85] to which clays are often compared. The increase in the fatigue crack growth rate at the same stress intensity amplitude was traced to enhanced microvoid formation ahead of the crack tip.

4.   Toughness

Toughness of nanocomposites deserves a separate treatise, owing to highly contradictory findings in the open literature. Usually there is a trade-off between stiffness, strength and toughness. Accordingly, nanoparticles with reinforcing effect should result in toughness reduction. In many cases, however, the opposite tendency was found. It should be emphasized here that fracture mechanical studies on polymer composites are very scarce [4,105–108]. On the other hand, only fracture mechanical methods provide toughness values which can be collated being material parameters (e.g., [109]). The usual rule of thumb is that at very low nanoparticle content (< 2–3 wt%), the toughness does not alter compared to the matrix. At higher nanoparticle content, however, a strong decrease in the toughness can be observed. To explain the toughness improvement first the "percolation theory" of Wu was adopted [26]. It was soon recognized that this theory couldn't account for the toughness upgrade, as the matrix ligament between the particles is too large to create the necessary stress overlapping. To overcome this problem, a double percolation model was proposed [4]. According to the authors' feeling, even this model needs refinement. Specific effects of the interphase (immobilized layer, transcrystallinity, crystalline polymorph, etc.) which affect the matrix deformation (being temperature and frequency dependent) have to be considered. Unfortunately the model explanations lack in giving information as to why and how cavitation occurs in the nanocomposites with intercalated/exfoliated, finely and coarsely dispersed structures. Note that the toughness in rigid particle-filled polymers is linked to cavitation, which is followed by stretching of the interparticle matrix ligaments [110]. So, a straightforward model should account for both cavitation (initiation — controlled by the interphase) and matrix ligament stretching (propagation — controlled by the interphase and bulk properties). It is noteworthy that toughness reduction in nanocomposites is likely the rule and not the exception. This claim is in concert for example with the ultimate tensile properties and PVT data. Recall that the latter technique evidenced a considerable decay in the free volume upon exfoliation [100], which means restricted molecular motion and thus suggests toughness reduction.

The essential work of fracture (EWF) approach was applied recently on TiO_2/PA-6,6 nanocomposites by Yang and Zhang et al. [106]. Deeply double edge notched tension (DDENT) specimens were produced by injection molding as disclosed in Reference [103]. Various ligament length specimens were tested under a constant tensile speed of 1 mm/min. The specific total work of fracture, w_f, was then calculated by the work done to break the specimen, and plotted versus the ligament length, l (cf. Figure 8). The linear relationship between w_f and l can be fitted by the following equation,

$$w_f = w_e + \beta w_p l$$

in which w_e and w_p are specific essential and plastic work of fracture, respectively. β is a dimensionless shape factor. w_e can be obtained as an intercept of w_f when $l = 0$, assumed as an intrinsic property of the material for a given sheet thickness. The slope, βw_p, represents the plastic deformation ability of the material. It is interesting to note that the incorporation of only 1 vol% 21 nm TiO_2 improved w_e by about 70% compared

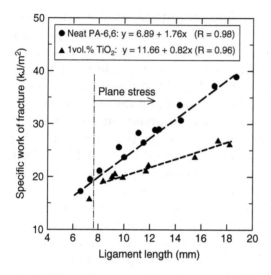

Figure 8 The specific total work of fracture versus the ligament length for neat PA-6,6 and PA-6,6/TiO$_2$ (21 nm, 1 vol%) nanocomposite on DDENT specimens.

to neat PA-6,6. However, at the same time the plastic work of fracture, βw_p, was reduced. In other words, the incorporation of inorganic nanoparticles improved the resistance of crack initiation, w_e, at a cost of the resistance of crack propagation, βw_p.

Fractographic inspection shed light on this — quite general — fracture behavior. The SEM image in Figure 9a shows a typical dimple type fracture surface at the crack tip of a pre-notched PA-6,6 specimen. This fracture morphology develops after void formation and void coalescence due to rupture of the intervening plastically deformed polymer. On the other hand, the size of dimples is much smaller and the dimple density is markedly higher in the nanocomposite (cf. Figure 9b) compared to the matrix. So, nanoparticle-induced molecular and morphological immobilization may be responsible for the enhanced essential work of fracture term. The enhanced secondary cracking (reflected by the dimple density) is due to the stress concentration effect of the nanoparticles which

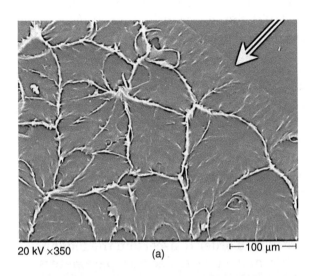

20 kV ×350   (a)   ⊢—100 μm —⊣

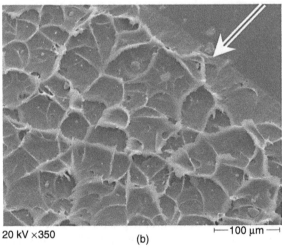

20 kV ×350   (b)   ⊢—100 μm —⊣

Figure 9 SEM pictures taken from the crack-tip region of fractured DDENT specimens of (a) neat PA-6,6 and (b) PA-6,6/TiO$_2$ (21 nm, 1 vol%) nanocomposite. Note: razor blade induced notch is indicated by arrow.

favors the crack propagation. This failure scenario is in harmony with the reduction in the plastic work of fracture term.

Bureau et al. [107] found recently that PP/clay compounds showed very good fracture toughness with increased EWF parameters. The void nucleation density was controlled by the clay particle size, which finally determined the fracture toughness. Some attractive results were also reported by Chan et al. [108] on PP/CaCO$_3$ nanocomposites. It was found that nano-CaCO$_3$ could significantly improve not only the tensile modulus but also the toughness of PP. *J*-integral tests showed a dramatic increase (ca. 500%) in the notched fracture toughness. Nanoparticles were believed to be able to promote cavitation at interphase regions, which can release the plastic constraints and trigger mass plastic deformation of the polymer. However, it should be noted that further works are needed in order to ascertain the toughness-controlling molecular and supermolecular parameters in semicrystalline polymers.

B.   Thermal Behavior

For many applications it is of great importance to know the linear thermal expansion coefficient of nanoparticle-reinforced composites. Yoon et al. [111] investigated this behavior of injection-molded organoclay-modified PA-6 composites. The expansion was measured in flow direction, as well as transverse and normal to it. Addition of organoclay (up to 7 wt%) reduced the thermal expansion in both flow and transverse directions, whereas an increase was noticed for normal direction. In addition, a PA-6 matrix of high molecular mass showed reduced thermal expansion compared to a PA-6 of lower molecular mass. The authors demonstrated that the composite model of Chow which accounts for the anisotropic feature (i.e., aspect ratio) of the filler, can well be used to predict the thermal expansion [111].

Another property of engineering relevance is the heat distortion temperature (HDT). Note that HDT was always enhanced by incorporation of 2-D layered silicates. According to the data sheets of Ube Industries, the HDT values determined by the standardized methods, A and B, increased from the initial 180/75 to 197/140°C (this representation means the HDT A/B methods, respectively) for PA-6 by adding 2 wt% organoclay [10]. Fornes and Paul [94] proved that the HDT-B value can be calculated by adopting the Halpin-Tsai composite theory for the DMTA properties when coupled with the method of Scobbo. The so predicted HDT-B values and experimentally measured data exhibited a very good agreement.

C.   Rheological Behavior

It was early recognized that the structure of nanocomposites strongly influences the rheological behavior, especially in the low frequency range (linear viscoelasticity) [112]. In this range the melt viscosity increases monotonically with increasing silicate content irrespective to its shape (intercalated 2-D layered or dispersed quasi-spherical preformed particles) and incorporation method [112–113]. Polymer nanocomposites with 2-D layered silicates exhibit pronounced shear-thinning behavior. Shear thinning starts at markedly lower frequencies than for the related matrix polymer (e.g., [112, 114]). However, at very low shear rates, characteristic for injection molding operations, practically no difference in the melt viscosity between the polymer and its nanocomposite can be found (e.g., [115]).

The relationship between the shear viscosity and shear rate (in the range of 10^{-3} to 2 s^{-1}) could be well described by the Carreau model [116]. Lim et al. [116] speculated that the crossover between the Newtonian plateau (linear viscoelasticity) and the power law function valid for high shear rate (> 2 s^{-1}) occurs at $\lambda \gamma_c = 1$, where l is the relaxation time and γ_c is the critical shear rate. λ depends on the clay volume fraction and its dispersion structure. According to Utracki and Kamal [10], the zero-shear relative viscosity as a function of clay volume fraction obeys the modified Einstein's description for PA-6 based nanocomposites. Information derived from dynamic oscillatory shear, steady shear and elongational flow measurements, eventually combined with "superimposed" techniques like transient/intermittent ones can deliver a deeper insight into the structure of the nanocomposites and its alteration owing to shear and elongational flows [2,112,117–118]. It was shown, for example, that the morphological stability of nanocomposites can be successfully studied in rheological measurement [119]. Solomon et al. [120] reported that the course of the storage modulus vs. frequency in the viscoelastic range reflects well effects of the intercalation (caused by amine surfactants of various chemical buildup). A similar conclusion was drawn by Feng et al. [121] and Chow et al. [122] in respect with the compatibilizers.

D.   Barrier Properties

Improvement in the transport properties of layered silicate-reinforced semicrystalline thermoplastics (PA-6, polyolefins) is the major driving force of commercialization of the related nanocomposites at present. Reduction in the permeability is usually attributed to the fact that the diffusing molecules have to bypass the impermeable silicate platelets ("tortuous path," labyrinth effect) which are more or less well oriented normal to the diffusion direction. It was found that the Nielsen's model works well to predict the gas barrier properties in such systems [2,117]. Recently, more advanced theories were developed addressing changes in the alignment, interphase and bulk properties, as well. Recall that the theoretical models generally consider the silicate layers as perfectly aligned and exfoliated showing a large aspect ratio [2]. The prominent effect of aspect ratio was proved experimentally, too [123]. However, gas permeabilities measured are usually markedly below the theoretical predictions (e.g., [124–126]), especially for polyolefin-based nanocomposites. Attention should be called to the fact that permeability is a product of the diffusivity (diffusion coefficient) and equilibrium sorption of the penetrant under given conditions. So, the outcome does not represent a "design parameter." Nevertheless, by studying this behavior, useful information can be deduced indirectly even for the structure of the nanocomposites. Table 1 lists the equilibrium moisture content and diffusion coefficient for organoclay-reinforced PA-6/PP blends

TABLE 1 Effects of Organoclay and Compatibilizer (PP-g-MA) Contents on the Diffusion Coefficient (D) and Equilibrium Moisture Content (M) on PA-6/PP-based Nanocomposites

Blend, Nanocomposite Composition	Ratio [parts]	M [%]	D [× 10^{-10}m^2/s]
PA-6/PP	70/30	6.82	1.70
PA-6/PP/Organoclay	70/30/2	8.53	3.39
	70/30/4	8.25	2.97
	70/30/6	8.01	2.75
	70/30/8	7.57	2.16
	70/30/10	7.07	1.94
PA-6/PP/Compatibilizer/Organoclay	70/30/5/4	5.60	1.60
	70/30/10/4	5.34	1.09

Note: for testing, specimens were immersed in water at 60°C.

Source: Chow WS, Mohd Ishak ZA, Karger-Kocsis J. The 4th Asian-Australasian Conference on Composite Materials, Sydney, Australia, July 6–9, 2004. With permission.)

with and without compatibilizer (PP-g-MA) [127]. One can recognize that changes in both parameters are not monotonic which should be linked to structural changes (alteration in the clay dispersion, development of an interphase). This was evidenced, in fact, cf. Figure 6.

E. Fire Retardant Properties

Many reports quoted that the thermal stability (usually studied by thermo-gravimetric analysis, TGA) of the polymers increased when containing dispersed nanoparticles. The temperature linked to the maximum mass loss as well as the amount of the char residue increased with increasing clay content. The related increase depended on the clay dispersion, which was controlled by the organophilic surfactant of the clays [128]. Zanetti et al. [129] concluded that the slowdown in the thermal degradation is due to hampered diffusion of the degradation products from the bulk towards the gas phase. They traced it to a labyrinth effect discussed above in respect with the barrier properties. In addition, the thermal degradation of the organophilic surfactant created some acidic sites of the clay that affected the polymer degradation (volatilization/charring) scheme. It is worth noting that all inorganic fillers exhibit some flame retardant effect in their composites. Its manifestation, however, depends on the method selected. Limiting oxygen index (LOI) data showed that there is a large difference between fillers as a function of their dispersion state (micro- or nanoscale) and aspect ratio [130]. Note that in an LOI test, the flame spreads from the top of the specimen downwards. So, in this test the char formation has a great influence. Needless to say, a vertical burn test (e.g., according to UL 94 descriptions) may deliver completely different results. Today, the ultimate method to check the fire retardance is the use of cone calorimetry. In the related tests the heat release rate (HRR) and mass loss rate are registered as a function of time. Numerous works using this method indicated that in the presence of intercalated/exfoliated layered silicates, the peak HRR is efficiently reduced, however, with some extension in the overall burning time [131–133]. This was traced to the formation of a carbonaceous-silicate char of thermal-insulating properties on the specimen surface [133]. It is still the object of discussion whether the clay layers should be exfoliated or intercalated for an optimum flame retardancy. To improve the very poor fire resistance of polyolefins, the preferred concept is to make them intumescent [134]. Intumescent formulations are halogen-free and produce a charred cellular layer upon heating. The related layer is acting as a heat shield by protecting the underlying material from the heat flux of the flame. This concept was adopted for polymer nanocomposites due to two effects: a) char yielding behavior and b) the reinforcing effect of the clay (e.g., [135–136]). Note that the reinforcing effect of the nanoclay is of vital importance as the usual intumescent formulations result in materials of poor mechanical performance. Based on the above behavior, layered clay containing nanocomposites should have outstanding resistance to ablation, too.

F.   Other Properties

Inorganic-organic hybrid materials may show further interesting properties (optical transparency, UV-screening, ionic conductivity, template for polymerization, etc.) which are far less often addressed by research activities than the above listed properties. Exploiting some of these unique properties, "smart" nanocomposites can be produced in the future.

V.   SUMMARY AND OUTLOOK

The unique combination of some key properties already paved the way for polymeric nanocomposites for industrial applications (packaging — due to barrier properties, automotive — due to stiffness, strength, density and HDT). As surveyed above, a great amount of work has been done in the past in this field. However, still much research is needed to clarify the structure-property relationships, especially in semicrystalline thermoplastic nanocomposites, and to understand the mechanisms behind the property improvements, e.g., creep resistance and toughness. Substantial contributions to open issues can be expected from molecular modelling (dispersion state, intercalation in 2-D and 3-D structures) and various experimental techniques, which can be used *in situ* for studying given properties. For example, the mechanical loading induced changes in nanocomposites can well be studied by synchrotron x-ray diffraction methods, by high-voltage electron (HVEM) and atomic force microscopy (AFM). For large-scale production, the *in situ* polymerization (intercalation) methods will be preferred instead of the melt compounding methods in the future. Considerable research interest will likely focus on the development of "smart" nanocomposites (for sensors, actuators, etc.), as well as on the biomedical applications of nanocomposites.

ACKNOWLEDGMENTS

This work was supported by grants of the German Science Foundation (DFG) and Fonds der Chemischen Industrie for

J. Karger-Kocsis. Z. Zhang is grateful to the Alexander von Humboldt Foundation for his Sofja Kovalevskaja Award, financed by the German Federal Ministry of Education and Research (BMBF) within the German government's "ZIP" program for investment in the future.

REFERENCES

1. Pinnavaia TJ, Beall GW, eds. Polymer-Clay Nanocomposites, Chichester: Wiley, 2000.

2. Sinha Ray S, Okamoto M. Polymer/layered silicate nanocomposites: a review from preparation to processing. Progr Polym Sci 2003; 28:1539–1641.

3. Chazeau L, Gauthier C, Vigier G, Cavaillé JY. Relationships between microstructural aspects and mechanical properties of polymer-based nanocomposites. In: Nalwa HS, ed., Handbook of Organic-Inorganic Hybrid Materials and Nanocomposites, Vol 2: Nanocomposites, Los Angeles: American Scientific Publ, 2003:63–111.

4. Zhang MQ, Rong MZ, Friedrich K. Processing and properties of nonlayered nanoparticle reinforced thermoplastic composites. In: Nalwa HS, ed., Handbook of Organic-Inorganic Hybrid Materials, Vol 2: Nanocomposites, Los Angeles: American Scientific Publ, 2003:113–150.

5. Kickelbick G. Concepts for the incorporation of inorganic building blocks into organic polymers on a nanoscale. Progr Polym Sci 2003; 28:83–114.

6. Liu T, Burger C, Chu B. Nanofabrication in polymer matrices. Progr Polym Sci 2003; 28:5–26.

7. Zheng L, Farris RJ, Coughlin EB. Novel polyolefin nanocomposites: synthesis and characterization of metallocene-catalyzed polyolefin polyhedral oligomeric silsesquioxane copolymers. Macromolecules 2001; 34:8034–8039.

8. Fu BX, Yang L, Somani RH, Zong SX, Hsiao BS, Phillips S, Blanski R, Ruth P. Crystallization studies of isotactic polypropylene containing nanostructured polyhedral oligomeric silsesquioxane molecules under quiescent and shear conditions. J Polym Sci Part B: Polym Phys 2001; 39:2727–2739.

9. Fu BX, Gelfer MY, Hsiao BS, Phillips S, Viers B, Blanski R, Ruth P. Physical gelation in ethylene-propylene copolymer melts induced by polyhedral oligomeric silsesquioxane (POSS) molecules. Polymer 2003; 44:1499–1506.

10. Utracki LA, Kamal MR. Clay-containing polymeric nanocomposites. Arab J Sci Eng 2002; 27:43–67.

11. Mülhaupt R, Engelhardt T, Schall N. Nanocomposites — auf dem Weg zur Anwendung. Kunststoffe-German Plastics 2001; 91:178–190.

12. Saujanya C, Imai Y, Tateyama H. Structure and thermal properties of compatibilized PET/expandable fluorine mica nanocomposites. Polym Bull 2002; 49:69–76.

13. Fukushima H, Drzal LT. A carbon nanotube alternative: graphite nanoplatelets as reinforcements for polymers. SPE-ANTEC 2003; 2230–2234/

14. Okada A, Usuki A. The chemistry of polymer-clay hybrids. Mater Sci Eng C 1995; 3:109–115.

15. Kojima Y, Usuki A, Kawasumi M, Okada A, Kurauchi T, Kamigaito O. Synthesis of nylon 6-clay hybrid by montmorillonite intercalated with β-caprolactam. J Polym Sci Part A: Chem 1993; 31:983–986.

16. Cho S-J. Study on the polymerization of -caprolactam in the interlamellar spaces of [TEACOOH]-montmorillonite intercalations complex and its characterization. J Appl Polym Sci 2003; 88:1904–1910.

17. Reichert P, Kressler J, Thomann R, Mülhaupt R, Stöppelmann G. Nanocomposites based on a synthetic layer silicate and polyamide-12. Acta Polym 1998; 49:116–123.

18. Lepoittevin B, Pantoustier N, Devalckenaere M, Alexandre M, Calberg C, Jérôme R, Henrist C, Rulmont A, Dubois P. Polymer/layered silicate nanocomposites by combined intercalative polymerization and melt intercalation: a masterbatch process. Polymer 2003; 44:2033–2040.

19. Tripathy AR, Burgaz E, Kukureka SN, MacKnight WJ. Poly(butylene terephthalate) nanocomposites prepared by *in situ* polymerization. Macromolecules 2003; 36:8593–8595.

20. Chang J-H, An YU, Ryu SC, Giannelis EP. Synthesis of poly(butylene terephthalate) nanocomposite by *in situ* interlayer polymerization and characterization of its fiber (I). Polym Bull 2003; 51:69–75.

21. Zhang G, Jiang C, Su C, Zhang H. Liquid-crystalline copolyester/clay nanocomposites. J Appl Polym Sci 2003; 89:3155–3159.

22. Kuo S-W, Huang W-J, Huang S-B, Kao H-C, Chang F-C. Syntheses and characterizations of *in situ* blended metallocene polyethylene/clay nanocomposites, Polymer 2003; 44:7709–7719.

23. Du Z, Zhang W, Zhang C, Jing Z, Li H. A novel polyethylene/palygorskite nanocomposite prepared via *in situ* coordination polymerization. Polym Bull 2002; 49:151–158.

24. Yang F, Zhang X, Zhao H, Chen B, Huang B, Feng Z. Preparation and properties of polyethylene/montmorillonite nanocomposites by *in situ* polymerization. J Appl Polym Sci 2003; 89:3680–3684.

25. Yu Y-H, Lin C-Y, Yeh J-M, Lin W-H. Preparation and properties of poly(vinyl alcohol)-clay nanocomposite materials. Polymer 2003; 44:3553–3560.

26. Ou Y, Yang F, Yu Z-Z. A new conception on the toughness of nylon 6/silica nanocomposite prepared via *in situ* polymerization. J Polym Sci Part B: Phys 1998; 36:789–795.

27. Zhang MQ, Rong MZ, Zeng HM, Schmitt S, Wetzel B, Friedrich K. Atomic force microscopy study on structure and properties of irradiation grafted silica particles in polypropylene-based nanocomposites. J Appl Polym Sci 2001; 80:2218–2227.

28. Moussaif N, Groeninckx G. Nanocomposites based on layered silicate and miscible PVDF/PMMA blends: melt preparation, nanophase morphology and rheological behaviour. Polymer 2003; 44:7899–7906.

29. Ou C-F. Nanocomposites of poly(trimethylene terephthalate) with organoclay. J Appl Polym Sci 2003; 89:3315–3322.

30. Ou CF, Ho MT, Lin JR. Synthesis and characterization of poly(ethylene terephthalate) nanocomposites with organoclay. J Appl Polym Sci 2004; 91:140–145.

31. Wu T-M, Hsu S-F, Wu J-Y. Polymorphic behavior in syndiotactic polystyrene/clay nanocomposites. J Polym Sci Part B: Phys 2002; 40:736–746.

32. Idem. Nonisothermal crystallization behavior of syndiotactic polystyrene/montmorillonite nanocomposites. J Polym Sci Part B: Phys 2003; 41:560–570.

33. Chang J-H, An YU, Sur GS. Poly(lactic acid) nanocomposites with various organoclays. I. Thermomechanical properties, morphology, and gas permeability. J Polym Sci Part B: Phys 2003; 41:94–103.

34. Krikorian V, Kurian M, Galvin ME, Nowak AP, Deming TJ, Pochan DJ. Polypeptide-based nanocomposite: structure and properties of poly(L-lysine)/Na+-montmorillonite. J Polym Sci Part B: Phys 2002; 40:2579–2586.

35. Zheng J, Siegel RW, Toney CG. Polymer crystalline structure and morphology changes in nylon-6/ZnO nanocomposites. J Polym Sci Part B: Phys 2003; 41:1033–1050.

36. Chang J-H, Jang T-G, Ihn KJ, Lee W-K, Sur GS. Poly(vinyl alcohol) nanocomposites with different clays: pristine clays and organoclays. J Appl Polym Sci 2003; 90:3208–3214.

37. Nussbaumer RJ, Caseri WR, Smith P, Tervoort T. Polymer-TiO2 nanocomposites: a route towards visually transparent broadband UV filters and high refractive index materials. Macromol Mater Eng 2003; 288:44–49.

38. Varghese S, Karger-Kocsis J. Natural rubber based nanocomposites by latex compounding with layered silicates. Polymer 2003; 44:4921–4927.

39. Karger-Kocsis J, Wu C-M. Thermoset rubber/layered silicate nanocomposites. Status and future trends. Polym Eng Sci 2004; 44:1083–1093.

40. Karger-Kocsis J. Reinforced polymer blends. In: Paul DR, Bucknall CB, eds. Polymer Blends, Vol 2: Performance. New York: Wiley, 2000:395–428.

41. Dennis HR, Hunter DL, Chang D, Kim S, White JL, Cho JW, Paul DR. Effect of melt processing conditions on the extent of exfoliation in organoclay-based nanocomposites. Polymer 2001; 42: 9513–9522.

42. Schönfeld S, Lechner F. Nanocomposites. Kunststoffe/German Plastics 2003; 7:28–33.

43. Fornes TD, Yoon PJ, Keskkula H, Paul DR. Nylon 6 nanocomposites: the effect of matrix molecular weight. Polymer 2001; 42:9929–9940.

44. Móczó J, Fekete E, László K, Pukánszky B. Aggregation of particulate fillers: factors, determination, properties. Macromol Symp 2003; 194:111–124.

45. Reichert P, Nitz H, Klinke S, Brandsch R, Thomann R, Mülhaupt R. Poly(propylene)/organoclay nanocomposite formation: influence of compatibilizer functionality and organoclay modification. Macromol Mater Eng 2000; 275:8–17.

46. Kato M, Okamoto H, Hasegawa N, Tsukigase A, Usuki A. Preparation and properties of polyethylene-clay hybrids. Polym Eng Sci 2003; 43:1312–1316.

47. García-López D, Picazo O, Merino JC, Pastor JM. Polypropylene-clay nanocomposites: effect of compatibilizing agent on clay dispersion. Eur Polym J 2003; 39:945–950.

48. Wanjale SD, Jog JP. Effect of modified layered silicates and compatibilizer on properties of PMP/clay nanocomposites. J Appl Polym Sci 2003; 90:3233–3238.

49. Koo CM, Kim MJ, Choi MH, Kim SO, Chung IJ. Mechanical and rheological properties of the maleated polypropylene-layered silicate nanocomposites with different morphology. J Appl Polym Sci 2003; 88:1526–1535.

50. Chow WS, Mohd Ishak ZA, Karger-Kocsis J, Apostolov AA, Ishiaku US. Compatibilizing effect of maleated polypropylene on the mechanical properties and morphology of injection molded polyamide6/polypropylene/organoclay nanocomposites. Polymer 2003; 44:7427–7440.

51. Chow WS, Mohd Ishak ZA, Ishiaku US, Karger-Kocsis J, Apostolov AA. The effect of organoclay on the mechanical properties and morphology on injection-molded polyamide6/polypropylene nanocomposites. J Appl Polym Sci 2004; 91:175–189.

52. Ellis TS. Reverse exfoliation in a polymer nanocomposite by blending with a miscible polymer. Polymer 2003; 44:6443–6448.

53. Vaia RA, Liu W. X-ray powder diffraction of polymer/layered silicate nanocomposites: model and practice. J Polym Sci Part B: Phys 2002; 40:1590–1600.

54. Morgan AB, Gilman JW. Characterization of polymer-layered silicate (clay) nanocomposites by transmission electron microscopy and x-ray diffraction: a comparative study. J Appl Polym Sci 2003; 87:1329–1338.

55. Maiti P, Okamoto M. Crystallization controlled by silicate surfaces in nylon6-clay nanocomposites. Macromol Mater Eng 2003; 288:440–445.

56. Karger-Kocsis J, Varga J. Interfacial morphology and its effects in polypropylene composites. In: Karger-Kocsis J, ed. Polypropylene: An A-Z Reference, Dordrecht: Kluwer, 1999:348–356.

57. Fornes TD, Paul DR. Crystallization behavior of nylon 6 nanocomposites. Polymer 2003; 44:3945–3961.

58. Devaux E, Bourbigot S, El Achari A. Crystallization behavior of PA-6 clay nanocomposites. J Appl Polym Sci 2002; 86:2416–2423.

59. Lincoln DM, Vaia RA, Wang Z-G, Hsiao BS, Krishnamoorti R. Temperature dependence of polymer crystalline morphology in nylon6/montmorillonite nanocomposites. Polymer 2001; 42:9975–9985.

60. Medellin-Rodriguez FJ, Burger C, Hsiao BS, Chu B, Vaia R, Phillips S. Time-resolved shear behavior of end-tethered nylon 6-clay nanocomposites followed by non-isothermal crystallization. Polymer 2001; 42:9015–9023.

61. Liu TX, Liu ZH, Ma KX, Shen L, Zeng KY, He CB. Morphology, thermal and mechanical behavior of polyamide 6/layered silicate nanocomposites. Compos Sci Technol 2003; 63:331–337.

62. Zhang G, Li Y, Yan D. γ-Crystalline form of nylon-10,10 in nylon-10,10-montmorillonite nanocomposites. Polym Int 2003; 52:795–798.

63. Priya L, Jog JP. Polymorphism in intercalated poly(vinylidene fluoride)/clay nanocomposites. J Appl Polym Sci 2003; 89:2036–2040.

64. Priya L, Jog JP. Intercalated poly(vinylidene fluoride)/clay nanocomposites. J Polym Sci Part B: Phys 2003; 41:31–38.

65. Wanjale SD, Jog JP. Poly(1-butene)/clay nanocomposites: preparation and properties, J Polym Sci Part B: Phys 2003; 41:1014–1021.

66. Tseng C-R, Wu S-C, Wu J-J, Chang F-C. Crystallization behavior of syndiotactic polystyrene nanocomposites for melt- and cold-crystallizations. J Appl Polym Sci 2002; 86:2492–2501.

67. Wang ZM, Chung TC, Gilman JW, Manias E. Melt-processable syndiotactic polystyrene/montmorillonite nanocomposites. J Polym Sci Part B: Phys 2003; 41:3173–3187.

68. Xu W, Liang G, Wang W, Tang S, He P, Pan W-P. Poly(propylene)-poly(propylene)-grafted maleic anhydride-organic montmorillonite (PP-PP-g-MAH-OrgMMMT) nanocomposites II. Nonisothermal crystallization kinetics. J Appl Polym Sci 2003; 88:3093–3099.

69. He J-D, Cheung MK, Yang M-S, Qi Z. Thermal stability and crystallization kinetics of isotactic polypropylene/organomontmorillonite nanocomposites. J Appl Polym Sci 2003; 89:3404–3415.

70. Qian J, He P. Non-isothermal crystallization of HDPE/nano-SiO$_2$ composite. J Mater Sci 2003; 38:2299–2304.

71. Zhu W, Zhang G, Yu J, Dai G. Crystallization behavior and mechanical properties of polypropylene copolymer by *in situ* copolymerization with a nucleating agent and/or nano-calcium carbonate. J Appl Polym Sci 2004; 91:431–438.

72. Wang Y, Shen C, Li H, Li Q, Chen J. Nonisothermal melt crystallization kinetics of poly(ethylene terephthalate)/clay nanocomposites. J Appl Polym Sci 2004; 91:308–314.

73. Nakajima H, Yamada K, Iseki Y, Hosoda S, Hanai A, Oumi Y, Teranishi T, Sano T. Preparation and characterization of polypropylene/mesoporous silica nanocomposites with confined polypropylene. J Polym Sci Part B: Phys 2003; 41:3324–3332.

74. Priya L, Jog JP. Poly(vinylidene fluoride)/clay nanocomposites prepared by melt intercalation: crystallization and dynamic mechanical behavior studies. J Polym Sci Part B: Phys 2002; 40:1682–1689.

75. Tjong SC, Meng YZ, Xu Y. Preparation and properties of polyamide 6/polypropylene-vermiculite nanocomposite/polyamide 6 alloys. J Appl Polym Sci 2002; 86:2330–2337.

76. McNally T, Murphy WR, Lew CY, Turner RJ, Brennan GP. Polyamide-12 layered silicate nanocomposites by melt blending. Polymer 2003; 44:2761–2772.

77. Hu X, Lesser AJ. Effect of silicate filler on the crystal morphology of poly(trimethylene terephthalate)/clay nanocomposites. J Polym Sci Part B: Phys 2003; 41:2275–2289.

78. Varghese S, Karger-Kocsis J. Melt-compounded natural rubber nanocomposites with pristine and organophilic layered silicates of natural and synthetic origin. J Appl Polym Sci 2004; 91:813–819.

79. Colombini D, Maurer FHJ. Interfacially induced additional damping peaks in dynamic mechanical spectra - micromechanical transitions in multipolymeric materials. Macromol Symp 2003; 198:83–90.

80. Chow WS, Karger-Kocsis J, Mohd Ishak ZA. An atomic force microscopy study on the blend morphology and clay dispersion in polyamide6/polypropylene/organoclay systems. J Polym Sci Part B: Phys 2005 (in press).

81. Manias E, Touny A, Wu L, Strawhecker K, Lu B, Chung TC. Polypropylene/montmorillonite nanocomposites. Review of the synthetic routes and materials properties. Chem Mater 2001; 13:3516–3523.

82. Vohra VR, Schmidt DF, Ober CK, Giannelis EP. Deintercalation of a chemically switchable polymer from a layered silicate nanocomposite. J Polym Sci Part B: Phys 2003; 41:3151–3159.

83. Fornes TD, Yoon PJ, Paul DR. Polymer matrix degradation and color formation in melt processed nylon 6/clay nanocomposites. Polymer 2003; 44:7545–7556.

84. Davis RD, Gilman JW, VanderHart DL. Processing degradation of polyamide 6/montmorillonite clay nanocomposites and clay organic modifier. Polym Degrad Stab 2003; 79:111–121.

85. Karger-Kocsis J. Fracture of short-fibre reinforced thermoplastics. In: Friedrich K. ed. Application of Fracture Mechanics to Composite Materials, Amsterdam: Elsevier Appl Sci, 1989:189–247.

86. Uribe-Arocha P, Mehler C, Puskas JE, Altstädt V. Effect of sample thickness on the mechanical properties of injection-molded polyamide-6 and polyamide-6 clay nanocomposites. Polymer 2003; 44:2441–2446.

87. Yu Z-Z, Yang M, Zhang Q, Zhao C, Mai Y-W. Dispersion and distribution of organically modified montmorillonite in nylon-66 matrix. J Polym Sci Part B: Phys 2003; 41:1234–1243.

88. Varlot K, Reynaud E, Kloppfer MH, Vigier G, Varlet J. Clay-reinforced polyamide: preferential orientation of the montmorillonite sheets and the polyamide crystalline lamellae. J Polym Sci Part B: Phys 2001; 39:1360–1370.

89. Pavlíková S, Thomann R, Reichert P, Mülhaupt R, Marcinčin A, Borsig E. Fiber spinning from poly(propylene)-organoclay nanocomposite. J Appl Polym Sci 2003; 89:604–611.

90. Bafna A, Beaucage G, Mirabella F, Mehta S. 3D hierarchical orientation in polymer-clay nanocomposite films. Polymer 2003; 44:1103–1115.

91. Malwitz MM, Lin-Gibson S, Hobbie EK, Butler PD, Schmidt G. Orientation of platelets in multilayered nanocomposite polymer films. J Polym Sci Part B: Phys 2003; 41:3237–3248.

92. Wang KH, Koo CM, Chung IJ. Physical properties of polyethylene/silicate nanocomposite blown films. J Appl Polym Sci 2003; 89:2131–2136.

93. Masenelli-Varlot K, Reynaud E, Vigier G, Varlet J. Mechanical properties of clay-reinforced polyamide. J Polym Sci Part B: Phys 2002; 40:272–283.

94. Fornes TD, Paul DR. Modeling properties of nylon 6/clay nanocomposites using composite theories. Polymer 2003; 44:4993–5013.

95. Luo J-J, Daniel IM. Characterization and modeling of mechanical behavior of polymer/clay nanocomposites. Compos Sci Technol 2003; 63:1607–1616.

96. Brune DA, Bicerano J. Micromechanics of nanocomposites: comparison of tensile and compressive elastic moduli, and prediction of effects of incomplete exfoliation and imperfect alignment on modulus. Polymer 2002; 43:369–387.

97. Kim G-M, Lee D-H, Hoffmann B, Kressler J, Stöppelmann G. Influence of nanofillers on the deformation process in layered silicate/polyamide-12 nanocomposites. Polymer 2001; 42:1095–1100.

98. Gloaguen JM, Lefebvre JM. Plastic deformation behaviour of thermoplastic/clay nanocomposites. Polymer 2001; 42:5841–5847.

99. Medellin-Rodríguez FJ, Hsiao BS, Chu B, Fu BX. Uniaxial deformation of nylon 6-clay nanocomposites by *in situ* synchtron measurements. J Macromol Sci Part B-Phys 2003; B42:201–214.

100. Utracki LA, Simha R, Garcia-Rejon A. Pressure-volume-temperature dependence of poly-ε-caprolactam/clay nanocomposites. Macromolecules 2003; 36:2114–2121.

101. Truss RW, Lee AC. Yield behaviour of a melt-compounded polyethylene-intercalated montmorillonite nanocomposite. Polym Int 2003; 52:1790–1794.

102. Mallick PK, Zhou Y. Yield and fatigue behavior of polypropylene and polyamide-6 nanocomposites. J Mater Sci 2003; 38:3183–3190.

103. Zhang Z, Yang JL, Friedrich K. Creep resistant polymeric nanocomposites. Polymer 2004; 45, 3481–3485.

104. Bellemare SC, Bureau MN, Denault J, Dickson JI. Fatigue crack initiation and propagation in polyamide-6 and polyamide-6 nanocomposites. Polym Compos 2004; 25:433–441.

105. Chen L, Wong S-C, Pisharath S. Fracture properties of nanoclay-filled polypropylene. J Appl Polym Sci 2003; 88:3298–3305.

106. Yang JL, Zhang Z, Zhang H. The work of fracture of thermoplastic nanocomposites. To be submitted.

107. Bureau MN, Perrin-Sarazin F, Ton-That M-T. Polyolefin nanocomposites: essential work of fracture analysis. Polym Eng Sci 2004; 44:1142–1151.

108. Chan C-M, Wu J, Li J-X, Cheung Y-K. Polypropylene/calcium carbonate nanocomposites. Polymer 2002; 43:2981–2992.

109. Karger-Kocsis J. Dependence of the fracture and fatigue performance of polyolefins and related blends and composites on microstructural and molecular characteristics. Macromol Symp 1999; 143:185–205.

110. Argon AS, Cohen RE. Toughenability of polymers. Polymer 2003; 44:6013–6032.

111. Yoon PJ, Fornes TD, Paul DR. Thermal expansion behavior of nylon 6 nanocomposites. Polymer 2002; 43:6727–6741.

112. Giannelis EP, Krishnamoorti R, Manias E. Polymer-silicate nanocomposites: model systems for confined polymers and polymer bruches. In: Advances in Polymer Science. Berlin: Springer, 1999, 138:107–147.

113. Privalko VP, Shumsky VF, Privalko EG, Karaman VM, Walter R, Friedrich K, Zhang MQ, Rong MZ. Viscoelasticity and flow behavior of irradiation grafted nano-inorganic particle filled polypropylene composites in the melt state. Sci Technol Adv Mater 2002; 3:111–116.

114. Lee CH, Lim ST, Hyun YH, Choi HJ, Jhon MS. Fabrication and viscoelastic properties of biodegradable polymer/organophilic clay nanocomposites. J Mater Sci Letters 2003; 22:53–55.

115. Akkapeddi K, Facinelli J, Worley D. Effects of organoclay structure on the rheology and crystallization behavior of *in situ* polymerized PA6 nanocomposites. SPE-ANTEC 2003: 2240–2244.

116. Lim ST, Lee CH, Choi HJ, Jhon MS. Solidlike transition of melt-intercalated biodegradable polymer/clay nanocomposites. J Polym Sci Part B: Phys 2003; 41:2052–2061.

117. Sinha Ray S, Okamoto M. Biodegradable polylactide and its nanocomposites: opening a new dimension for plastics and composites. Macromol Rapid Commun 2003; 24:815–840.

118. Lee KM, Han CD. Rheology of organoclay nanocomposites: effects of polymer matrix/organoclay compatibility and the gallery distance of organoclay. Macromolecules 2003; 36:7165–7178.

119. Reichert P, Hoffmann B, Bock T, Thomann R, Mülhaupt R, Friedrich C. Morphological stablity of poly(propylene) nanocomposites. Macromol Rapid Commun 2001; 22:519–523.

120. Solomon MJ, Almusallam AS, Seefeldt KF, Somwangthanaroy A, Varadan P. Rheology of polypropylene/clay hybrid materials. Macromolecules 2001; 34:1864–1872.

121. Feng W, Ait-Kadi A, Riedl B. Morphology and linear viscoelasticity of poly[ethylene-co-(1-octene)]/layered silicate nanocomposites. Macromol Rapid Commun 2002; 23:703–708.

122. Chow WS, Abu Bakar A, Mohd Ishak ZA, Karger-Kocsis J, Ishiaku US. Effect of maleic anhydride grafted ethylene/propylene rubber on the mechanical, rheological and morphological properties of organoclay reinforced polyamide 6/polypropylene nanocomposites. Eur Polym J 2005; 41(4):687–696.

123. LeBaron PC, Wang Z, Pinnavaia TJ. Polymer-layered silicate nanocomposites: an overview. Appl Clay Sci 1999; 15:11–29.

124. Ellis TS, D'Angelo JS. Thermal and mechanical properties of a polypropylene nanocomposite. J Appl Polym Sci 2003; 90:1639–1647.

125. Gorrasi G, Tortora M, Vittoria V, Kaempfer D, Mülhaupt R. Transport properties of organic vapors in nanocomposites of organophilic layered silicate and syndiotactic polypropylene. Polymer 2003; 44:3679–3685.

126. Tomova D, Reinemann S. Nanopartikel erweitern Anwendungspotenzial. Kunststoffe/German Plastics 2003; No.7:18–20.

127. Chow WS, Mohd Ishak ZA, Karger-Kocsis J. Moisture absorption and hygrothermal aging of organomontmorillonite reinforced polyamide 6/polypropylene nanocomposites. In: Ye L, Mai Y-W, Su Z, eds. Composites Technologies for 2020. Cambridge: Woodhead Publ. Ltd. 2004: 790–795.

128. Tang Y, Hu Y, Song L, Zong R, Gui Z, Chen Z, Fan W. Preparation and thermal stability of polypropylene/montmorillonite nanocomposites. Polym Degr Stab 2003; 82:127–131.

129. Zanetti M, Camino G, Reichert P, Mülhaupt R. Thermal behavior of poly(propylene) layered silicate nanocomposites. Macromol Rapid Commun 2001; 22:176–180.

130. Wagenknecht U, Kretzschmar B, Reinhardt G. Investigations of fire retardant properties of polypropylene-clay-nanocomposites. Macromol Symp 2003; 194:207–212.

131. Wang D, Wilkie CA. *In situ* reactive blending to prepare polystyrene-clay and polypropylene-clay nanocomposites. Polym Degr Stab 2003; 80:171–182.

132. Wang S, Hu Y, Li Z, Wang Z, Zhuang Y, Chen Z, Fan W. Flammability and phase-transition studies of nylon 6/montmorillonite nanocomposites. Coll Polym Sci 2003; 281:951–956.

133. Gilman JW, Jackson CL, Morgan AB, Harris R, Manias E, Giannelis EP, Wuthenow M, Hilton D, Phillips SH. Flammability properties of polymer-layered-silicate nanocomposites. Polypropylene and polystyrene nanocomposites. Chem Mater 2000; 12:1866–1873.

134. Le Bras M, Bourbigot S. Intumescent fire retardant polypropylene formulations. In: Karger-Kocsis J., ed., Polypropylene: An A-Z Reference. Dordrecht: Kluwer, 1999:357–365.

135. Bourbigot S, Le Bras M, Dabrowski F, Gilman JW, Kashiwagi T. PA-6 clay nanocomposite hybrid as char forming agent in intumescent formulations. Fire Mater 2000; 24:201–208.

136. Tang Y, Hu Y, Wang S, Gui Z, Chen Z, Fan W. Intumescent flame retardant-montmorillonite synergism in polypropylene-layered silicate nanocomposites. Polym Inter 2003; 52:1396–1400.

14

Carbon Nanotube and Carbon Nanofiber-Reinforced Polymer Composites

K. SCHULTE and M.C.M. NOLTE

Polymer Composites, Technical University
Hamburg-Harburg (TUHH), Hamburg, Germany

CONTENTS

I.   INTRODUCTION

One topic of major interest in the field of scientific research today is the combination of polymer matrices and fillers with dimensions on the nanometer scale to so-called "polymer nanocomposites." Though the expression "nanocomposite" appeared just in the past two decades, these nanocomposites have already existed in the form of rubber carbon black composites for about one century. Additionally, nature formed many examples of materials with nanodimensional structures such as bones, implying a certain superiority concealed in this design.

However, fundamental understanding of the mechanisms which make the properties of this class of materials so remarkable have arisen slowly by intensive scientific research in recent years. Polymer nanocomposites, which are defined to consist of a polymeric matrix containing particles with at least one of their dimensions being around or below 100 nm, can be tailored to become stiffer and stronger at extremely low filler loading fractions while at the same time maintaining the toughness and ductility of the bulk host material. This reinforcement occurs for particles of spherical shape as well as for plate-like and fibrous nanofillers, although the latter with their increased aspect ratio exhibit a stronger effect. As a result, carbon nanotubes and carbon nanofibers with their unique combination of excellent mechanical properties and a very high aspect ratio theoretically are the ideal candidates for reinforcing a polymeric matrix.

In practice, a number of challenges arises due to the small size of the second phase. As a matter of fact, the specific surface area increases with decreasing dimensions of the reinforcing particles and, hence, the forces per unit mass resulting from interactions between this surface and surrounding media become more pronounced. As a result, carbon nanofibers and carbon nanotubes cannot easily be dispersed in substances of different surface energy such as polar liquids (e.g., water) or most polymers. In addition, they cannot be positioned by simple means, leading to difficulties in orienting carbon nanotubes and nanofibers parallel to the load direction, which would give large benefits to the reinforcing effect. Finally, a strong interface between the reinforcing phase and the host matrix is always desirable in order to achieve a good load transfer.

Strong efforts have been made in scientific research in the last two decades and many advances toward industrial applications have been made so far, but still many key issues are not fully solved and a better understanding of the mechanisms underlying the reinforcing effect in polymer nanocomposites has to be found. The state of the art will be displayed in the following.

II. CARBON NANOTUBES AND NANOFIBERS

A completely new field of research in carbon materials has developed since the discovery of the C_{60} structure in 1985 [1], which is a hollow sphere made out of carbon atoms arranged in pentagons and hexagons. The hexagonal structure is also known from graphite sheets and carbon fibers, which provides them with a high strength and stiffness at a low density. The ideal carbon nanotube (CNT) is a small particle that can be regarded as a graphite plane rolled up to a tube with a hemispherical cap at each end. Its dimensions reach from 1 to several 100 nm in diameter and up to more than 100 μm in length so that an exceptionally high aspect ratio of more than 10,000 can be attained.

Two different types of CNTs exist: single wall and multiwall carbon nanotubes (SWCNT and MWCNT, respectively). They differ, as the names already imply, in the number of concentric graphene layers. While SWCNTs only possess one single layer rolled up to form a tube, MWCNTs consist of 2 to ~50 layers of different diameters set up in concentric cylinders. These are held together by secondary van der Waals bondings. SWCNTs mostly occur in forms of bundles or ropes, parallel arrangements of individual nanotubes also held

Figure 1   High resolution TEM micrograph of a multi-wall carbon nanotube.

together by van der Waals forces. In Figure 1 and Figure 2 TEM micrographs of a MWCNT and bundles of SWCNTs are displayed, respectively. These images reveal another large difference between the two general types of nanotubes. Since the diameter of SWCNT is again much smaller than the one of MWCNT and due to their extremely large specific surface

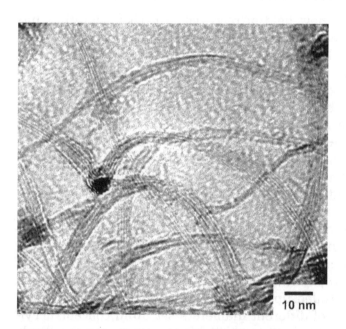

Figure 2   High resolution TEM micrograph of single-wall carbon nanotube bundles.

area, they exhibit large interaction forces which makes it difficult to separate them.

Carbon nanofibers (CNF) do not exhibit such a perfect crystallographic structure as compared to carbon nanotubes. They are built up by separate graphite layers which can be oriented at various angles from perpendicular to parallel to the fiber axis [2]. Figure 3 shows a TEM micrograph of two CNFs, exhibiting different structures with the one on the right being straight and the left one of the so-called "herring-bone structure." The diameters of CNFs range from 4 to 200 nm, and their lengths vary between 5 and 100 µm [3]. Next to straight CNFs helical, twisted, branched, and bidirectional forms exist, and it is understandable that the mechanical performance of carbon nanofibers should vary as a function of the structural differences. Similar differences in the mechanical properties also occur for the various structures of carbon nanotubes. Since for both types of particles the synthesis process determines the resulting morphology, the most common production routes are described in the following.

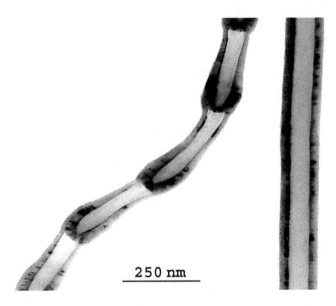

250 nm

Figure 3 TEM micrograph of two carbon nanofibers.

A.   Production

The main processes for the synthesis of carbon nanotubes are the electric arc discharge, laser ablation, electrolytic synthesis, gas phase catalytic growth from carbon monoxide and chemical vapor deposition from hydrocarbons. In the electric arc-discharge process, two graphite rods of high purity are used as electrodes in an inert helium or argon atmosphere. They are brought together until a direct current arc is formed creating a plasma of a temperature of ~4000°C. The carbon is removed from the anode and deposited at the cathode, building an outer shell of heat-fused material and an inner core consisting of nearly 70% of MWCNTs and 30% other graphitic nanoparticles [4]. In order to produce SWCNTs, the anode is doped with metal particles which act as the catalyst. The quality of the obtained material strongly depends on the experimental conditions such as voltage, temperature and composition of the graphite rods [4].

In the laser ablation, CNTs are synthesized by vaporizing graphite due to laser irradiation in an inert atmosphere in a furnace at about 1200°C. The vaporized carbon particles are swept away by the gas flow and deposited on a cooled copper collector. A CNT containing soot forms which can be collected at the graphite target, the copper collector and the furnace walls. MWCNTs as well as SWCNTs can be produced by this method, the latter, again, by doping the graphite target with metal particles [5].

The electrolytic production of CNT is based on passing an electric current of 1 to 30 A through a molten salt (e.g., lithium chloride) between two graphite electrodes. The cathode is being consumed, and a huge variety of nanoparticles are formed in the melt from where they can be washed out and purified. Generally, MWCNTs are synthesized by this method and not more than 45% of the obtained particles are CNTs [6], but also SWCNTs were recently produced with this method [7].

Finally, there is the possibility to grow CNTs by catalytic vapor deposition (CVD). Mainly, this method is based on decomposing a hydrocarbon gas (e.g., acetylene or natural gas) in a furnace flowing at about 700°C over a mono- or bi-metallic catalyst, for example Fe, Ni, Co, or Cr. The catalyst can be placed in a ceramic container, and the decomposing gas is forming carbon particles at the surface of the catalyst particles. The control over length, diameter and structure is given by adjusting the conditions of the process such as temperature, pressure, type of carbon source and catalyst, gas flow and many others. Mainly, MWCNTs are produced by this method, but since 1996 also SWCNTs have been synthesized. Very similar to this process is the gas-phase catalytic growth with carbon monoxide as the carbon source. Since the decomposition of this gas takes place at more elevated temperatures than that of hydrocarbons, the furnace has to be heated up to 1200°C. Furthermore, the effectiveness of the method rises with increasing pressure [8,9].

A special variant of the CVD process is the growth of aligned CNTs on a flat substrate. Here, a double stage furnace is used. In the first stage a precursor is sublimed and forms together with the carbon source gas nanoparticles suitable for the production of CNTs. In the second stage, the MWCNTs grow on the substrate, e.g., silicate where the catalyst is deposited. This method creates the possibility of producing large quantities of nanotubes which are rather straight, aligned and, therefore, not entangled, with a uniform and controllable length and diameter [10,11]. In Figure 4 a SEM micrograph of such material is shown. The alignment of the outermost layers of the MWCNTs are disturbed due to the sample preparation, but it can be seen that the majority of the CNTs forms a carpet-like structure.

One large step towards the industrial mass production of carbon nanotubes is the so-called "floating catalyst" method. Here, the catalyst is not supported anymore, but the nanoparticles form directly from metallic catalyst particles in the gas phase [12]. This route has been well established for decades for the production of vapor-grown carbon fibers (VGCF) and is also the current approach towards the production for carbon nanofibers. However, carbon nanotubes seem to require more control of the diameter of the catalyst particles as well as of the growth time in the furnace than VGCFs and CNFs. One solution to overcome these difficulties might

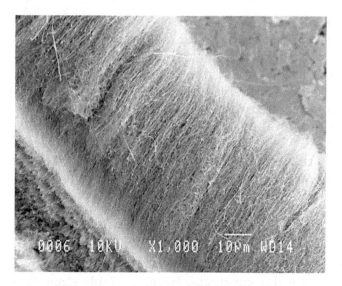

Figure 4 SEM micrograph of aligned CVD-grown multi-wall carbon nanotubes.

be to fluidize and at the same time stabilize the catalyst particles as it was already successfully applied for the synthesis of MWCNTs [13] as well as for SWCNTs [14].

In general, CVD processes have the advantage of being continuous and, therefore, allow large-scale industrial production. Furthermore, they require lower temperatures for the formation of nanostructures, making them less cost intensive. On the other hand, lower growth temperatures lead to decreased mobility of the carbon atoms and, hence, to increased misalignment of the graphitic planes and a higher defect density. Therefore, CNTs synthesized by arc-discharge or laser ablation are considered to be crystallographically more perfect. However, structural defects can be annealed during subsequent but expensive graphitization treatments.

B. Properties

It is now established that the structure of carbon nanotubes as well as of carbon nanofibers depends strongly on the route of synthesis. It is not surprising that carbon nanotubes and nanofibers do not possess a given value for most of their properties but instead cover a range between particular limits. It is important to understand this relationship between structure and properties for these nanostructures in order to choose the appropriate material for a given application.

The small dimensions of nanotubes and nanofibers make it additionally difficult to determine their intrinsic mechanical properties. In consequence, many studies were aimed at the indirect experimental determination or the theoretical prediction of Young's modulus of nanotubes. Measuring the amplitude of thermal vibrations of CNTs in a TEM yielded an average value of 1.8 TPa [15] and 1.25 TPa [16] for MWCNT and SWCNTs, respectively, but the values varied by more than one order of magnitude and the error was in a span of ±60%. A different approach was chosen by Poncharal et al. [17] investigating the resonant frequencies of MWCNTs in a TEM. A diameter dependent Young's modulus of approximately 1 TPa for CNTs smaller than 12 nm was found. Theoretical calculations tend to yield even higher stiffnesses exceeding 1 TPa [18,19]. In an atomic force microscope (AFM), the mechanical properties of MWCNTs [20] and bundles of SWCNTs [21] were directly determined. Here, the nanotubes were attached to an AFM tip and exposed to a tensile load. The resulting forces and displacements were measured and resulted in modulus values ranging from 270 to 950 GPa for MWCNTs and between 320 and 1470 GPa for SWCNTs. The large variations in mechanical properties observed most likely reflect the deviations in the defect structure as already mentioned, especially when compared to theoretical results predicted for atomically perfect nanotubes.

Besides the elastic modulus the strength and the elongation to break could also be obtained from the aforementioned direct tensile tests. Similar to Young's modulus, the evaluated strength values varied considerably but on a rather high level between 11 to 63 MPa for MWCNTs [20] and 13 to 52 GPa for SWCNTs [21].

Besides the high strength and stiffness, CNTs show a great flexibility. In computer simulations as well as transmission electron microscopy observations, Iijima et al. [22] have shown that the bending of CNT is completely reversible up to angles of 110°, despite the formation of kink shapes. Furthermore, the strain to failure seems to be extremely high, and calculated values in simulations are in the range of 30 to 40%. Due to the movement of dislocations, a highly symmetric structure is kept [23]. Experimentally 12% and 5.3% elongation at break could be observed for MWCNTs and SWCNTs, respectively [20,21].

Carbon nanofibers, in contrast to nanotubes, have not been the target of extensive research considering their comparatively high degree of structural imperfection [2]. Their macroscopic counterparts, the vapor-grown carbon fibers (VGCF), were taken as a model system. The tensile modulus of VGCFs is reported to lie in the range of 100 to 400 GPa [24,25,26]. Furthermore, a diameter dependence revealed a linear increase of Young's modulus from 120 to 300 GPa with a decreasing diameter from 32 to 6 µm [24,27]. It was assumed that a higher degree of disorder in the graphitic structure as a result of a quicker decomposition of pyrolytic carbon during the thickening period could explain the observed variations in mechanical properties [27]. Besides the modulus, the tensile strength of VGCFs shows a similar inverse linear relationship to the fiber diameter with values varying between 2.5 to 3.5 GPa for fiber diameters below 10 µm [24,26,27]. This is a general observation when comparing bulk and fibrous forms of matter and can be explained by a lower flaw population in the latter. However, it is believed that the mechanical properties of carbon nanofibers are similar or better than those of VGCFs [28] even if an extrapolation of the properties to the smaller diameters is assumed not to be valid.

III. POLYMER NANOCOMPOSITES CONTAINING CARBON NANOTUBES/NANOFIBERS

In summary, carbon nanotubes as well as carbon nanofibers are materials with strong variations in their properties on a very high level. In combination with their extremely small size and enormous aspect ratio of up to several 10,000, they seem to be ideal candidates for the reinforcement of polymer components where continuous fiber reinforcement cannot be realized. Consequently, extensive scientific research dealing with polymer nanocomposites containing carbon nanotubes or carbon nanofibers has been carried out within the last fifteen years, resulting in the recognition of some key problems arising mainly from the small size of the nanofiller. As the properties of the nanoparticles themselves already vary strongly as a result of structural variations, the mechanical performance of polymer nanocomposites hinges on a large variety of influencing factors such as filler volume fraction, degree of dispersion of the particles and their distribution, orientation of the fiber axis, the properties of the filler matrix interface as well as the matrix morphology.

Carbon nanotubes and nanofibers are characterized by an extremely high specific surface area of more than 1,000 m^2/g which can interact with the matrix material and form a potentially strong interface for a good load transfer. In turn, such a high specific surface area means that interaction forces become more important, resulting generally in increased difficulties in dispersing these nanoparticles. As a result, various methods of producing polymer nanocomposites have been established, which can be classified in three general routes: mixing in the liquid state, solution-mediated processes and *in situ* polymerization. Melt-mixing should be the most favorable for industrial applications though some industrial fiber-spinning processes are based on dissolved polymers.

The idea behind solution-mediated nanocomposite production methods is to homogeneously disperse nanotubes together with the polymer in a solvent which is subsequently evaporated off, leaving a residue of a polymer-nanoparticle mixture. This production technique seems to depend on the formation of a stable dispersion since Haggenmueller et al. reported large agglomerates in SWCNT-polymethylmethacrylate (PMMA) nanocomposites [29], while Shaffer and Windle observed good dispersions up to high loading fractions of 40 wt% producing polyvinylalcohol (PVOH)-MWCNT nanocomposites [30]. The composite stiffness in the latter study was found to increase linearly with increasing loading fraction and much more pronounced at temperatures above T_g than below.

When producing nanocomposites by the *in situ* polymerization route, the nanoparticles are mixed with the monomer. This process could be used in industrial production and can facilitate the dispersion process. One example of *in situ* polymerization is given by the study of Roslaniec et al. examining MWCNT and SWCNT nanocomposites based on a polyether ester elastomer (PEE) matrix [31]. SEM micrographs indicated a good degree of dispersion of 0.5 wt% MWCNTs in the polymer matrix. On the other hand, unforeseen effects can occur during polymerization in the presence of carbonaceous particles. Jia et al. reported a decreasing polymer chain length and an increasing consumption of the polymerization starting agent with an increasing amount of filler [32]. The latter effect was explained with the existence of π-bonds in the CNT which are broken up instead of the C=C bonds in the monomers. A further disadvantage of this method is that a small residue of monomer will always remain in the polymer.

Standard melt processing techniques such as extrusion and injection molding are preferable for economic reasons but are often inapplicable due to the lack of a sufficient amount of CNT. Solely thermoplastic-MWCNT masterbatches by Degussa-Hüls AG can be obtained in large capacities. These are based on highly entangled nanotubes by Hyperion Catalysis Inc., which are difficult to be dispersed by shear forces [33]. In contrast, CNFs are commercially available in adequate quantities without being too cost intensive. Tibbets and McHugh [34], for example, produced polypropylene/CNF nanocomposites with the help of a small injection molding machine. They reported that the occurring shear forces in processing were too low to break up all agglomerates of the CNFs. An initial ball milling step of the as-received nanofibers reduced the size of the aggregates but also significantly decreased the length of the CNFs. Other melt processing techniques involving high shear forces such as twin-screw extrusion [35] or melt mixing [36] led to improved degrees of dispersion of the nanofibers in the matrix polymer and enabled the incorporation of loading fractions up to 60 wt%.

But it has to be kept in mind that large shear forces also can shorten the length of the nanofibers [37].

As soon as aggregates are broken up, the high viscosity of thermoplastic polymers seems to be sufficient to prevent re-agglomeration. Hence, a masterbatch of a polyetheretherketone/CNF nanocomposite produced by twin-screw extrusion which was subsequently processed by injection-molding was observed to retain the high degree of dispersion [38]. Interestingly, low loading fractions of CNFs up to 10 wt% did not lead to an increase in melt viscosity of polypropylene [39], even a decrease in viscosity was reported up to the same amount of CNFs in polycarbonate [35]. In contrast, the viscosity of duromers increases strongly with the addition of nanoparticles, which is also a reason why only small loading fractions have been incorporated into such matrices. Due to the overall lower viscosity of uncured duroplastics compared to thermoplastic melts, the nanoparticles tend to re-agglomerate again, making constant stirring necessary [40].

In addition to shear forces to separate CNTs or CNFs, often ultrasonication is applied which can improve the dispersion process considerably. It has to be kept in mind that this procedure is only applicable for small batches since the power of the ultrasound decreases strongly with increasing distance from its source. Even more important seems to be the observation that ultrasonic treatment damages the CNTs and can reduce their length, yielding weaker structures with higher defect densities and lower aspect ratios [41].

Another route to improve the dispersion of nanotubes in the matrix material is to enhance the compatibility of the surfaces. A surfactant, for example, acting as a compatibilizer similar to soap for oil and water was applied to stabilize the dispersion of MWCNTs in an epoxy matrix [42]. As little as 1 wt% of dispersed MWCNTs exhibited improved thermomechanical behavior compared to a similar amount of untreated nanofiller, although the surfactant negatively influenced the stiffness of the epoxy itself by decreasing the storage modulus significantly. This effect strongly indicates the high importance of the degree of dispersion for the improvement of the mechanical properties of polymeric matrices. Furthermore, an increase in T_g indicated strong interactions between the nano-filler and the polymer chains of the epoxy. Surfactants were also used to assist the dispersion of individual SWCNTs in polyvinylpyrrolidine, which was subsequently blended with polyvinylalcohol [43]. The obtained composite films revealed stiffening as well as strengthening effects but also a reduction in ductility.

A further step for the dispersion enhancement is the chemical functionalization of the nanofiller involving the cova-lent or noncovalent attachment of functional groups to nano-tubes [44]. The end caps of both single and multi-wall carbon nanotubes can be opened under oxidizing conditions and ter-minated with carboxyl, carbonyl and hydroxyl groups. Such oxidized nanotubes have a better solubility and can form elec-trostatically stabilized colloidal dispersions in water as well as alcohols [45], but the oxidization treatment was observed to reduce the nanotube length. The oxidizing process can also assist the dispersion process in thermoplastic matrices. For example, in an amorphous polystyrene matrix, the function-alization was found to improve the degree of dispersion of SWCNT bundles [46]. Shaffer and Windle [30] introduced up to 60 wt% of catalytic grown MWCNTs by solution casting into polyvinylalcohol with the help of stabilizing the dispersion of nanotubes in a solvent with an acid treatment. Furthermore, they report an improved polymer adsorption onto the nano-particles, which was assumed to be related to the oxidization treatment as well. Figure 5 shows a SEM micrograph of the fracture surface of a polypropylene/MWCNT nanocomposite. It was assumed from this image that a good adhesion was existent between the filler and the matrix because of the poly-meric material stuck to the nanotubes.

As could be observed in the aforementioned example, the chemical functionalization approach brings up another vital issue concerning the improvement of the mechanical proper-ties of nanocomposites. Similar to composites containing mac-roscopic fiber reinforcement, the fiber-matrix interface plays an essential role in transferring an applied load. Epoxy resins generally seem to have a good adhesion to carbon fillers as it has already been established for carbon fibers. A study by

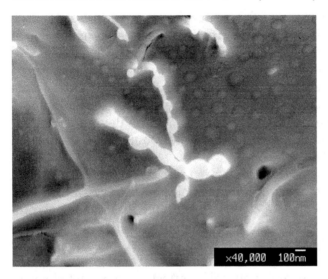

Figure 5 SEM micrograph of a fracture surface of a nanocomposite consisting of a polypropylene matrix containing CVD-grown MWCNTs.

Wagner et al. [47], for example, reported a stress transfer ability for MWCNTs in a UV-curable polymer, which is one order of magnitude higher than the one of other advanced composites. However, Ajayan et al. [48] observed alignment of arc-grown MWCNTs dispersed in epoxy after microtoming the nanocomposite. From this effect and the lack of nanotube fractures due to the cutting process, they concluded that the nanotubes were very strong and the nanotube-matrix interface rather weak. Gojny et al. [49] also reported poor adhesion between matrix and filler which could be improved noticeably by an amino-functionalization [50]. It is assumed that the amino groups attached to the nanotubes form covalent bonds to the polymer chains of the epoxy matrix and, therefore, strengthen the filler-matrix interface significantly.

When evaluating the influence of nanoparticles on the mechanical properties of polymeric matrices, possible changes in the matrix morphology have to be taken into account. The presence of nanoparticles can strongly affect the curing reaction of duromers and elastomers. An acceleration of the epoxy curing reaction was, for example, shown for untreated SWCNT bundles, with the effect being most pronounced at low curing temperatures [51]. This increase in reaction rate arises from the high thermal conductivity of the nanotubes and also depends on the specific surface area and surface chemistry, as shown by a comparative study of carbon nanofibers and carbon black [52]. The higher the degree of graphitization of the filler surface, the less pronounced the effect on the curing rate.

In the case of thermoplastic matrices extensive research to establish the effect of nanoparticles on the crystalline morphology has been conducted. One example is isotactic polypropylene which is a widely applied and well studied semi-crystalline polymer. It was already well established that the presence of macroscopic fibers such as carbon, glass, polyethyleneterephtalate and polyimide affects the morphology of polypropylene [53–56]. A solid surface for heterogeneous nucleation is provided within the matrix and effects such as transcrystallization occur. Similar effects have been reported for polypropylene-containing nanoparticles such as carbon black [57] and organoclay [58] and, indeed, a pronounced nucleating effect of CNFs in the same polymer was observed by Lozano et al. [36]. They reported additionally an effect on the resulting average crystallite size and size distribution as well as an increase in the overall degree of crystallinity with increasing filler content. The introduction of functionalized SWCNTs to polypropylene even altered the crystal structure from the α to the β structure [59]. Sandler et al. [60], incorporating CNFs as well as MWCNT in polypropylene in a comparative study, showed strong effects of the nanoparticles on the crystalline structure of the matrix as displayed in Figure 6. Here, a TEM micrograph shows the end of a carbon nanofiber in a highly oriented polypropylene film drawn from solution. The darkest material on the image reflects the metal catalyst with which most of this section of the CNF is filled while the graphitic structure of the nanofiber extends to the left. The vertical bands in the matrix correspond to the crystal lamellae of the polypropylene which have a higher density to each side of the nanofiber compared to the edges of the micrograph. When

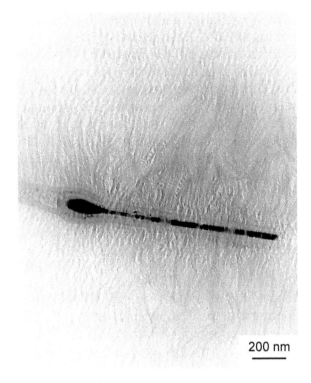

200 nm

Figure 6   TEM micrograph of carbon nanofiber in a polypropylene matrix exhibiting changes in matrix morphology.

taking into account these aforementioned effects, the large differences in the reported mechanical behavior of theromplastic-nanotube/nanofiber composites are not surprising since the mechanical performance of such materials depends strongly on the degree of crystallinity of its morphology.

A final factor strongly affecting the mechanical reinforcement of polymeric materials with nanofillers possessing a high aspect ratio is the orientation of the nanoparticle's axis into the loading direction. In fluids of low viscosity such as uncured resins, orientation of carbon nanotubes can be achieved by applying EM-fields as shown for example by Martin et al. [61]. Such methods do not work in thermoplastic polymer melts exhibiting higher viscosities. Here, the orientation of the nanotubes and nanofibers into one direction can be obtained by the introduction of shear forces for example by mechanical stretching [62,63], fiber-spinning [29,64], spin casting [65] and injection molding [38]. Especially the fiber-spinning process led to highly oriented nanotube and nanofiber composites within thermoplastic matrices. One route for the fiber production is spinning from a solution. In one successful example, SWCNTs dispersed in a surfactant were injected into a polyvinylalcohol-coagulation bath [66]. Gel fibers were obtained which were subsequently drawn yielding in fibers with a tensile modulus of 80 GPa. Compared to high-performance carbon fibers this stiffness is still rather low, but they exhibit a remarkable tensile strength of 1.8 GPa and a high toughness.

IV. CONCLUSIONS

Despite other influences, for example processing conditions, the main issues affecting the mechanical properties of polymer nanocomposites containing carbon nanotubes or nanofibers were displayed. Carbon nanotubes and carbon nanofibers are both nanoparticles possessing a unique and highly interesting combination of mechanical properties, electrical and thermal conductivity and diameters at the nanometer scale with an aspect ratio exceeding 10,000.

Many different routes of production together with various post-synthesis processes result in a huge variety of different materials possessing a wide range of properties. In a very recent study, for example, the performance of three different types of nanotubes and CNFs was compared when spun to fibers in a nylon-12 matrix [67]. Strong variations in the mechanical performance were observed for the different types of nanofiller even though the degree of alignment and of overall crystallinity were reported to be similar. Interestingly, fibers containing both types of catalytically grown CNTs as well as the CNFs exhibited a higher tensile modulus than those filled with arc-grown nanotubes which should generally possess the highest degree of crystalline perfection and the least defect density. This effect was assumed to be related to low purity and degree of dispersion of the as-received material and distinctly shows that the choice of the kind of nanofiller strongly depends on the application and is rather difficult. The purity, diameter, aspect ratio, dispersability, crystalline structure, chemical and thermal post-synthesis treatment as well as the initial degree of alignment play a vital role next to the cost and the choice of an appropriate matrix material. At the moment, carbon nanofibers and catalytically grown, multi-wall carbon nanotubes seem to be the best choices for industrial applications exploiting mainly their potential as a reinforcing phase, justified simply by their comparably uncomplicated dispersability and low cost.

Another interesting result derived from [67] is the fact that an additional cold drawing of the fibers after processing did not lead to a further increase in orientation of the nano-filler particles. It was assumed that the nanoparticles were already fully aligned before the cold drawing and that the intrinsic waviness of the nanotubes and nanofibers was the limiting element to further orientation. This finding corre-lated well to other studies reporting a similar degree of align-ment [68,69] and is supported by the study of Fisher et al. [70] investigating the effect of nanotube waviness on the mod-ulus of polymer nanocomposites. The nanotube waviness was also addressed to be the reason for the calculated effective modulus being much lower with between 2.4 and 13.2 GPa than theoretically possible [69].

However, today already several commercial applications for nanotube-polymer composites exist, but these are almost exclusively exploiting the electrical properties of carbon nan-otubes. The utilization of the promising mechanical properties of nanotubes and nanofibers in polymer composites is hin-dered by challenges that yet need to be overcome. Since great efforts have been made so far and major difficulties still remain, it is doubtful that short-fiber reinforced bulk compos-ites will be ever replaced by polymer composites containing carbon nanotubes or carbon nanofibers, but there exist vari-ous fields of application where they could be extremely advan-tageous. As examples, consider small structures where reinforcement with macroscopic fibers is impossible such as fibers or films or applications where an exclusive combination of properties is required [71]. Another issue of rising interest is the recycling of composites without losing their reinforcement which is difficult to be achieved with conventional short fiber reinforcement.

ACKNOWLEDGMENTS

The authors which to thank Ian A. Kinloch from the Department of Materials Science and Metallurgy at the University of Cambridge, U.K., and Jan K. W. Sandler from Polymer Engineering Department at the University of Bayreuth, Germany, for the fruitful discussions and kind help. They further acknowledge the financial support from the EC Thematic Network "CNT-Net" [G5RT-CT-2001-05026] and the Deutsche Forschungsgemeinschaft DFG for funding within the SFB 371 project C9.

REFERENCES

1. Kroto HW, Heath JR, O'Brian SC, Curl RF, Smalley RE. C_{60}: Buckminsterfullerene. Nature 1985; 318:162–163.

2. Rodriguez NM. A review of catalytically grown carbon nanofibers. J Mat Res 1993; 8:3233–3250.

3. Oberlin A, Endo M, Koyama T. High resolution electron microscope observations of graphitized carbon fibres. J Cryst Growth 1976; 14:133–135.

4. Ebbesen TW and Ajayan PM. Large-scale synthesis of carbon nanotubes. Nature 1992; 358:220–222.

5. Guo T, Nikolaev P, Thess A, Colbert, DT, Smalley RE. Catalytic growth of single-walled nanotubes by laser vaporization. Chem Phys Lett 1995; 243:49–54.

6. Hsu WK, Hare JP, Terrones M, Kroto HW, Walton DRM, Harris PJF. Condensed-phase nanotubes. Nature 1995; 2377:687.

7. Bai JB, Harmon AL, Marraud A, Jouffry B, Zymla V. Synthesis of SWCNTs and MWCNTs by a molte salt (NaCl) method. Chem Phys Lett 2002; 365:184–188.

8. Kukovecz A, Konya Z, Nagaraju N, Willems I, Tamasi A, Fonseca A, Nagy JB, Kiricsi I. Catalytic synthesis of carbon nanotubes over Co, Fe and Ni containing conventional and sol-gel silicate-aluminas. Phys Chem Chem Phys 2000; 2:3071–3076.

9. Joseyacaman M, Mikiyoshida M, Rendon L, Santiesteban JG. Catalytic growth of carbon microtubules with fullerene structure. Appl Phys Lett 1993; 62:657–659.

10. Li WZ, Xie SS, Qian LX, Chang BH, Zou BS, Zhou WY, Zhao RA, Wang G. Large-scale synthesis of aligned carbon nanotubes. Science 1996; 274:1701–1703.

11. Satishkumar BC, Govindaraj A, Rao CNR. Bundles of aligned carbon nanotubes obtained by the pyrolysis of ferrocene-hydrocarbon mixtures: role of the metal nanoparticles produced *in situ*. Chem Phys Lett 1999; 307:158–162.

12. Rao CNR. Synthesis of multi-walled and single-walled nanotubes, aligned-nanotube bundles and nanorods by employing organometallic precursors. Mat Res Innovat 1998; 2:128–141.

13. Venegoni D, Serp P, Feurer R, Kihn Y, Vahlas C, Kalck P. Parametric study for the growth of carbon nanotubes by catalytic chemical vapor deposition in a fluidized bed reactor. Carbon 2002; 40:1799–1807.

14. Li YL, Kinloch IA, Shaffer MSP, Geng J, Johnson B, Windle AH. Synthesis of single-walled carbon nanotubes by a fluidized-bed method. Chem Phys Lett 2004; 384:98–102.

15. Treacy MMJ, Ebbesen TW, Gibson TM. Exceptionally high Young's modulus observed for individual carbon nanotubes. Nature 1996; 381:678–680.

16. Krishnan A, Dujardin E, Ebbesen TW, Yianilos PN, Treacy MMJ. Young's modulus of single-walled nanotubes. Phys Rev B 1998; 58:14013–14019.

17. Poncharal P, Wang ZL, Ugarte D, de Heer WA. Electrostatic deflections and electromechanical resonances of carbon nanotubes. Science 1999; 283:1513–1516.

18. Overney G, Zhong W, Tománek Z. Structural rigidity and low frequency vibrational modes of long carbon tubules. Z Phys D 1993; 27:93–96.

19. Yakobson BI, Brabec CJ, Bernholc J. Nanomechanics of carbon tubes: instabilities beyond linear response. Phys Rev Lett 1996; 76:2511–2514.

20. Yu MF, Lourie O, Dyer M, Moloni K, Kelly T. Strength and breaking mechanism of multi-walled carbon nanotubes under tensile load. Science 2000; 287:637–640.

21. Yu MF, Files BF, Arepalli S, Ruoff RS. Tensile loading of ropes of single wall carbon nanotubes and their mechanical properties. Phys Rev Lett 2000; 84:5552–5555.

22. Iijima S, Brabec C, Maiti A, Bernholc J. Structural flexibility of carbon nanotubes. J Chem Phys 1996; 104:2089–2092.

23. Yakobson BI, Campbell MP, Brabec CJ, Bernholc J. High strain rate fracture and C-chain unravelling in carbon nanotubes. Comp Mat Sci 1997; 8:341–348.

24. Tibbetts GG, Beetz Jr CP. Mechnical properties of vapour-grown carbon fibres. J Phys D Appl Phys 1987; 20:292–297.

25. Tibbetts GG, Gorkiewicz DW, Alig RL. A new reactor for growing carbon-fibers from liquid-phase and vapor-phase hydrocarbons. Carbon 1993; 31:809–814.

26. Mandronero A. Strength of short carbon fibres germinated and grown under hydrogen. Mater Sci Eng A 1994; 185:L1–L4.

27. Tibbetts GG. Vapor-grown carbon fibres. In: Figueiredo JL, Bernardo CA, Baker RTK, Hüttinger KJ, eds., Carbon Fibres Filaments and Composites. NATO ASI Ser. C, Math. Phys. Sci. Netherlands: Dordrecht, 1990:73–94.

28. Patton RD, Pittman CU Jr., Wang L, Hill JR, Day A. Ablation, mechanical and thermal conductivity properties of vapor grown carbon fiber/phenolic matrix composites. Compos A 2002; 22:243–251.

29. Haggenmueller R, Gommans HH, Rinzler AG, Fischer JE, Winey KI. Aligned single-wall carbon nanotubes in composites by melt processing methods. Chem Phys Lett 2000; 330:219–225.

30. Shaffer MSP and Windle AH. Fabrication and characterization of carbon nanotube/poly(vinyl alcohol) composites. Adv Mater 1999; 11:937–941.

31. Roslaniec Z, Broza G, Schulte K. Nanocomposites based on multiblock polyester elastomers (PEE) and carbon nanotubes (CNT). Compos Interf 2003; 10:95–102.

32. Jia Z, Wang Z, Xu C, Liang J, Wei B, Wu D, Zhu S. Study on poly(methyl methacrylate)/carbon nanotube composites. Mat Sci Eng A 1999; 271:395–400.

33. Pötschke P, Fornes TD, Paul DR. Rheological behaviour of multiwalled carbon nanotube/polycarbonate composites. Polymer 2002; 11:3247–3255.

34. Tibbetts GG, McHugh JJ. Mechanical properties of vapor-grown carbon fibre composites with thermoplastic matrices. J Mater Res 1999; 14:2871–2880.

35. Carneiro AS, Covas JA, Bernardo CA, Caldeira G, Van Hattum FWJ, Ting J-M, Alig RL, Lake ML. Production and assessment of polycarbonate composites reinforced with vapour-grown carbon fibres. Comp Sci Tech 1998; 58:401–407.

36. Lozano K and Barrera EV. Nanofibre-reinforced thermoplastic composites I: Thermoanalytical and mechanical analysis. J Appl Polym Sci 2001; 79:125–133.

37. Kuriger RJ, Alam MK, Anderson DP, Jacobsen RL. Processing and characterization of aligned vapor grown carbon fiber reinforced polypropylene. Compos A 2002; 33:53–62.

38. Sandler J, Werner P, Shaffer MSP, Demchuk V, Altstädt V, Windle AH. Carbon-nanofibre-reinforced poly(ether ether ketone) composites. Compos A 2002; 33:1033–1039.

39. Lozano K, Bonilla-Rios J, Barrera EV. A study on nanofibre-reinforced thermoplastic composites II: Investigation of the mixing rheology and conduction properties. J Appl Polym Sci 2001; 80:1162–1172.

40. Lau K-T, Shi S-Q, Cheng H-M. Micro-mechanical properties and morphological observation on fracture surfaces of carbon nanotube composites pre-treated at different temperatures. Comp Sci Tech 2003; 63:1161–1164.

41. Lu KL, Lago M, Chen YK, Green MLH, Harris PJF, Tsang SC. Mechanical damage of carbon nanotubes by ultrasound. Carbon 1996; 34:814–816.

42. Gong X, Liu J, Baskaran S, Voise RD, Young JS. Surfactant-assisted processing of carbon nanotube/polymer composites. Chem Mater 2000; 12:1049–1052.

43. Zhang X, Liu T, Sreekumar TV, Kumar S, Moore VC, Hauge RH, Smalley RE. Poly(vinyl alcohol)/SWCNT composite film. Nano Lett 2003; 3:1285–1288.

44. Zhu J, Kim JD, Peng H, Margrave JL, Khabashesku VN, Barrera EV. Improving the dispersion and integration of single-wall carbon nanotubes in epoxy composites through functionalization. Nano Lett 2003; 3:1107–1113.

45. Shaffer MSP, Fan X, Windle AH. Dispersion and packing of carbon nanotubes. Carbon 1998; 36:1603–1612.

46. Mitchell CA, Bahr JL, Arepalli S, Tour JM, Krishnamoorti R. Dispersion of functionalized carbon nanotubes in polystyrene. Macrom 2002; 35:8825–8830.

47. Wagner HD, Lourie O, Feldman Y, Tenne R. Stress-induced fragmentation of multiwall carbon nanotubes in a polymer matrix. Appl Phys Lett 1998; 72:188–190.

48. Ajayan PM, Stephan O, Colliex C, Trauth D. Aligned carbon nanotube arrays formed by cutting a polymer resin-nanotube composite. Science 1994; 265:1212–1214.

49. Gojny FH, Nastalczyk J, Schulte K, Roslaniec Z. Surface modified nanotubes in CNT/epoxy-nanocomposites. Chem Phys Lett 2003; 370:820–824.

50. Gojny FH and Schulte K. Functionalisation effect on thermo mechanical behaviour of multi-wall carbon nanotub/epoxy-composites. Comp Sci Tech 2004; in press.

51. Puglia D, Valentini L, Kenny JM. Analysis of the cure reaction of carbon nanotubes/epoxy resin composites through thermal analysis and raman spectroscopy. J Appl Polym Sci 2003; 88:452–458.

52. Yin M, Koutsky JA, Barr TL, Rodriguez NM, Baker RTK, Klebanov L. Characterization of carbon microfibers as a reinforcement for epoxy resins. Chem Mater 1993; 5:1024–1031.

53. Tan JK, Kitano T, Hatakeyama T. Crystallisation of carbon fibre reinforced polypropylene. J Mat Sci 1990; 25:3380–3384.

54. Avella M, Martuscelli E, Sellitti C, Garagnani E. Crystallisation behaviour and mechanical properties of polypropylene-based composites. J Mat Sci 1987 22:3185–3193.

55. Janevski A, Bogoeva-Gaceva G, Mäder E. DSC analysis of crystallisation and melting behaviour of polypropylene in model composites with glass and poly(ethyleneterephthalate) fibres. J Appl Polym Sci 1999; 74:239–246.

56. Sukhanova TE, Lednick, F, Urban J, Baklagina YG, Mikhailov GM, Kudryavtsev VV. Morphology of melt crystallised polypropylene in the presence of polyimide fibres. J Mat Sci 1995; 30:2201–2214.

57. Mucha M, Marszalek J, Fidrych A. Crystallization of isotactic polypropylene containing carbon black as a filler. Polymer 2000; 41:4137–4142.

58. Xu W, Ge M, He P. Nonisothermal crystallization kinetics of polypropylene/montmorillonite nanocomposites. J Polym Sci B Polym Phys 2002; 40:408–414.

59. Grady BP, Pompeo F, Shambaugh RL, Resasco DE. Nucleation of polypropylene crystallisation by single-walled carbon nanotubes. J Phys Chem B 2002; 106:5852–5858.

60. Sandler J, Broza G, Nolte M, Schulte K, Lam YM, Schaffer MSP. Crystallisation of carbon nanotube and nanofibre polypropylene composites. J Macromol Sci B 2003; 42:479–488.

61. Martin CA, Sandler JKW, Windle AH, Shaffer MSP, Schulte K, Schwarz MK, Bauhofer W. Electric field-induced aligned multiwall carbon nanotube networks in epoxy composites. Polymer; submitted 2004.

62. Jin L, Bower C, Zhou O. Alignment of carbon nanotubes in a polymer matrix by mechanical stretching. Appl Phys Lett 1998; 73:1197–1199.

63. Bower C, Rosen R, Jin L, Han J, Zhou O. Deformation of carbon nanotubes in nanotube-polymer composites. Appl Phys Lett 1999; 74:3317–3319.

64. Sennett M, Welsh E, Wright JB, Li WZ, Wen JG, Ren ZF. Dispersion and alignment of carbon nanotubes in polycarbonate. Mat Res Soc Symp Proc 2002; 706:Z3.31.1.

65. Safadi B, Andrews R, Grulke EA. Multiwalled carbon nanotube polymer composites: synthesis and characterization of thin films. J Appl Polym Sci 2002; 84:2660–2669.

66. Dalton AB, Collins S, Munoz E, Razal JM, Ebron VH, Ferraris JP, Coleman JN, Kim BG, Baughman RH. Super-tough carbon-nanotube fibres. Nature 2003; 423:703.

67. Sandler JKW, Pegel S, Cadek M, Gojny F, van Es M, Lohmar J, Blau WJ, Schulte K, Windle AH, Shaffer MSP. A comparative study of melt spun polyamide-12 fibres reinforced with carbon nanotubes and nanofibres. Polymer 2004; 45:2001–2015.

68. Ma H, Zeng J, Realff ML, Kumar S, Schiraldi DA. Processing, structure and properties of fibers from polyester/carbon nanofiber composites. Comp Sci Tech 2003; 63:1617–1628.

69. Sandler J, Windle AH, Werner P, Altstädt V, van Es M, Shaffer MSP. Carbon-nanofibre-reinforced poly(ether ether ketone) fibres. J Mater Sci 2003; 38:2135–2141.

70. Fisher FT, Bradshaw RD, Brincon LC. Effects of nanotube waviness on the modulus of nanotube-reinforced polymers. Appl Phys Lett 2002; 80:4647–4649.

71. Park C, Ounaies Z, Watson KA, Pawlowski K, Lowther SE, Connell JW, Siochi EJ, Harrison JS, St. Clair TL. Polymer-single wall carbon nanotube composites for potential spacecraft applications. Mat Res Soc Symp Proc 2002; 706:Z3.30.

15

Nano- and Microlayered Polymers: Structure and Properties

T.E. BERNAL-LARA, A. RANADE, A. HILTNER and E. BAER

Department of Macromolecular Science and Engineering,
and Center for Applied Polymer Research, Case Western
Reserve University, Cleveland, OH

CONTENTS

I.   INTRODUCTION

As the field of polymer science evolves, the polymeric systems developed are becoming increasingly more complex. However, the level of structural complexity is still less than that of naturally occurring systems. In understanding complex systems, the importance of treating the system on the molecular, nano, micro and macro scales has been proposed [1]. The structure-property relationships are best understood when the hierarchical nature of the system is taken into account. In the study of biocomposites, such as soft connective tissues, "three rules of complex assemblies" have been suggested [2–4]. First, the structure is organized in discrete levels or scales. Nearly all biocomposite systems are found to have at least one distinct structural level at each of the molecular, nanoscopic, microscopic, and macroscopic scales. Second, the levels of structural organization are held together by specific interactions between components. Whatever the nature of the bonding between levels, adequate adhesion is required for system structural integrity. Third, these highly interacting levels are organized into a hierarchical composite system that is designed to meet a complex spectrum of functional requirements. It is interesting to note that as synthetic materials become more complex and multifunctional, they start to display similar hierarchical features [4].

Understanding the relationship between the hierarchical structure and the properties of a given polymer architecture is of extreme importance. New materials are rapidly being developed as nanotechnology pushes the limits of fabrication to nano and even molecular scales. The physical properties in polymeric layered systems such as the glass transition temperatures, crystallizability, and toughness are significantly altered by reducing the layer thickness down to a few tens of nanometers [5–10]. In this chapter, the concepts of hierarchy and interaction will be applied to elucidate the properties of micro- and nanolayered polymers.

II. MICROLAYER COEXTRUSION TECHNOLOGY

The synergistic combination of two or more materials in a layered structure can enhance the overall properties of the material. In the case of metals, the layered structure in antique Japanese swords made from laminated steels provided two property improvements: the toughness was enhanced to absorb energy originating from sword blows and the hardness was increased to facilitate the retention of sharp cutting edges [11,12]. On the other hand, natural biological systems as diverse as wood, seashells, tendon and even the lens of the human eye possess layered structures [2–4] designed to meet a diverse spectrum of functional requirements. Nacre, for example, consists of a layered structure of aragonite inorganic "bricks" and a tough organic matrix that glues them together. This structure exhibits superior fracture toughness compared to monolithic aragonites [2,13].

Although naturally occurring complex hierarchies found in biology are not yet obtainable synthetically, the first step in processing layered architectures was achieved in the 1960s when the Dow Chemical Company developed process technology for combining two polymers as assemblies of hundreds of continuous layers [14]. The developed technology couples coextrusion, a series of multiplying die elements and a cast film line to produce films composed of hundreds or even thousands of layers with individual layer thicknesses in the micro or nano scale. In contrast to the well-known concept of self-assembly [15], layer-multiplying coextrusion takes advantage of the viscoelastic nature of polymeric melts and uses "forced-assembly" to create micro- or nanolayered polymer films.

At Case Western Reserve University, the coextrusion system used to prepare microlayered and nanolayered materials on an experimental scale consists of two three-quarter inch single screw extruders with melt pumps, a coextrusion block, a series of layer multiplier elements and an exit die as shown in Figure 1. From the feedblock, the two layers flow through a series of multiplying elements, and each of these doubles the number of layers. In each element, the melt is first sliced vertically, then spread horizontally and finally recombined by stacking. An assembly of n multiplier elements produces an extrudate with the layer sequence $(AB)_x$ where x is equal to 2^n. Nanolayered structures with over 8000 layers have been successfully processed in our laboratory by this versatile coextrusion system. By varying the melt feed ratio, the final sheet or film thickness, and the number of layers,

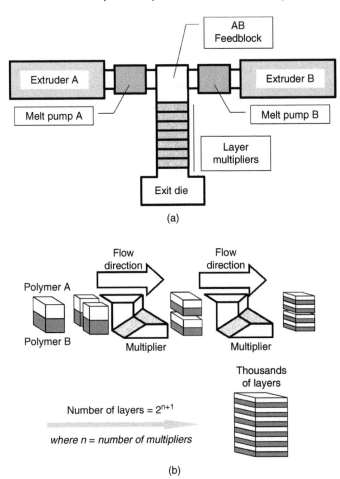

Figure 1 Two-component microlayer coextrusion system: (a) schematic of the extruder and the multiplier element setup; (b) schematic of the multiplication process.

the individual layer thickness can be precisely controlled down to the nanoscale.

The versatility of this multilayer technology has been expanded with a second coextrusion system that adds a third polymer between layers of polymer A and B. This component may be added for certain desirable properties such as barrier, strength or adhesion. Insertion of the tie layer (T) at each interface is accomplished by extruding three polymers into a feedblock that combines the melts into five layers with the sequence ATBTA as shown in Figure 2. Typically, the thickness of the tie layers is less than one tenth that of the A and B layers. The five-layer melt is then fed into the multiplying elements which repeatedly cut, spread and recombine the melt with a layer sequence $(ATBT)_xA$ where x is equal to 2^n for an assembly with n multiplier elements.

As with any coextrusion, the quality of coextruded microlayers depends on the viscosity ratio of the components. The extruder and the melt pump temperatures are controlled to optimize layer uniformity and to ensure consistent material properties.

III. MICROLAYERED POLYMERS

In nature, a mechanical system like a seashell or tendon have highly specific functions that have been optimized during bio-evolution over very long periods of time into complex hierarchical structures [2]. In most synthetic blends and composites, the hierarchical structure is most often created accidentally during the synthesis or processing of the polymer. Microlayer coextrusion offers a way of producing hierarchical structures with thousands of layers by controlling layer thickness down to the nanoscale. As with the other hierarchical systems, the mechanical properties of microlayers must be understood by examining the scale, interaction and architecture.

Enhanced mechanical properties are obtained in PC/SAN and PC/PMMA microlayer composites with decreasing layer thickness. This is attributed to the change in deformation mechanism of the brittle SAN and PMMA layers, from craze opening to shear yielding as layer thickness decreases. Interaction between the crazes in the brittle layers and micro-shear bands in the tough layers is the key to this change in mechanism [16].

Furthermore, the high surface-to-volume ratio attainable with microlayers makes them ideal for studying interfacial phenomena related to polymer blends. Several important phenomena such as the effect of compatibilizer, effect of chain

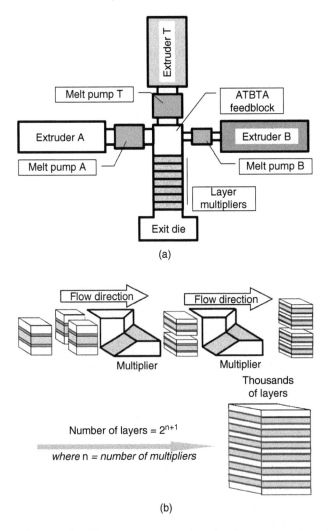

(a)

(b)

Figure 2 Three-component microlayer coextrusion system: (a) schematic of the extruder, and the multiplier element setup; (b) schematic of the multiplication process.

microstructure, and the effect of layer thickness on polymer adhesion have been elucidated [17–22].

Coextruded foam-film microlayers have been shown to replicate the microcellular structure of natural cork. Control over cell size has been demonstrated for polypropylene foam systems which are inherently difficult to control by means of conventional foam extrusion techniques. The compressive and tensile moduli of the foam-film microlayers increased with decreasing cell size. Mechanical toughness improved significantly for systems with an elastomeric film layer [23].

A.   Mechanical Properties and Irreversible Deformation Mechanisms

1.   PC/SAN Microlayers

The effect of layer thickness on the mechanical properties and irreversible deformation mechanisms has been demonstrated on microlayered composites made of poly(carbonate) (PC) and poly(styrene-acrylonitrile) (SAN). The layer thickness is varied by changing the melt feed ratios and/or the number of layers, while maintaining the same sheet thickness. The effect of layer thickness on the uniaxial tensile stress-strain behavior on 65/35 PC/SAN composite is shown in Figure 3 [16]. As the layer thickness decreased from 30 to 2 µms, the fracture strain increased by an order of magnitude. Even though SAN is a brittle polymer, all the microlayered systems with hundreds of layers showed ductile behavior, and the amount of ductility increased as the layer thickness was decreased.

This scaling effect was also evident in the impact strength measurements. Impact strength measurements normalized to 100% PC (214 J/cm) are shown in Figure 4 [24]. For a given number of layers, the impact strength increases with increasing PC content. Such a trend would be expected since PC is the tough component while SAN is brittle. The fact that the impact strength for a given PC content increases significantly with increasing numbers of layers is noteworthy. For example, for a 50/50 PC/SAN composite, the impact strength increases approximately by a factor of five in a sheet of the same thickness. Figure 4 also shows that the brittle to ductile transition, where the impact strength sharply rises, shifts steadily to a lower PC content with an increasing number of layers.

These scaling effects can be best understood by examining the interaction between the PC and SAN layers during

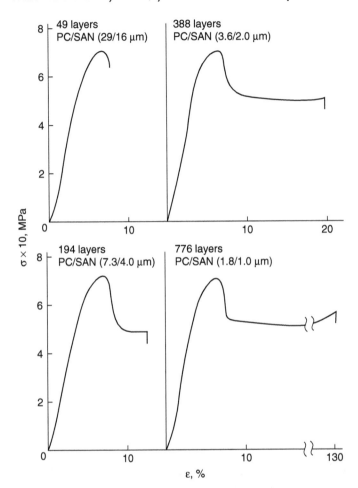

Figure 3 Effect of layer thickness on tensile stress-strain behavior of PC/SAN (65/35) microlayered composite. The layer thicknesses are given in the parentheses.

irreversible deformation. Irreversible deformation was investigated by carrying out tensile tests on notched rectangular specimens (100 mm × 20 mm) on an Instron testing machine [25]. The damage zone at the notch was photographed during deformation with a traveling optical microscope in the transmission mode. For PC/SAN (65/35), 49 layered composite (layer thickness, 29/16 μm, respectively), it is seen that crazes

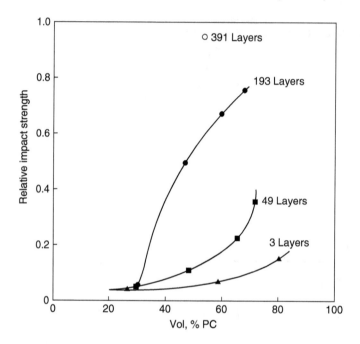

Figure 4 Relative impact strength at room temperature of PC/SAN sheet as a function of PC content and number of layers (relative impact strength of PC = 1.0 and SAN = 0.0).

initiate in SAN layers before necking (Figure 5a). At yield point, the SAN crazes opened to accommodate localized yielding of the PC layers. During the craze opening, the craze fibrils fractured with the result that the load-bearing craze was replaced by a void. Opening of the voids leads to tearing fracture of the PC layers. As a result, conventional tensile stress-strain specimens of this composite did not exhibit stable neck propagation, and during the initiation of necking, fracture occurred.

On the other hand, in the 776 layered PC/SAN (65/35) sample, the micro-shear bands grew through the PC layers and extended into several adjacent SAN and PC layers before necking. During neck propagation, the crazes did not open up into voids (Figure 5b). Instead, the SAN layers extended uniformly between the crazes and thinning of both PC and

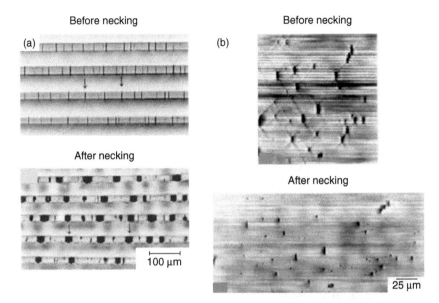

Figure 5 Optical micrographs of irreversible deformation before and after necking for (a) 49 layers PC/SAN (29/16 µms); (b) 776 layers PC/SAN (1.8/1.0 µms). Arrows indicate local delamination.

SAN layers was observed. This change in deformation mechanism in the SAN layers at the yield instability, from craze opening to shear yielding, was a consequence of impingement of the micro-shear bands on the PC/SAN interface and subsequent propagation of the micro-shear bands into and through the SAN layer. The change in deformation mechanism resulted in stable neck propagation. Conventional tensile stress-strain specimens with thinner layers were distinctly more ductile (Figure 3).

The interaction between layers is evident, and shear banding and crazing interact cooperatively as the layer thicknesses are decreased. Cooperative cavitation of SAN crazing is a function of the PC layer thickness. As the PC layer thickness is lowered, the plastic zone and subsequent elastic stress concentration at the craze tip into the PC layer can extend to the neighboring SAN layer [26]. A three-dimensional model of interactive surface crazing and micro-

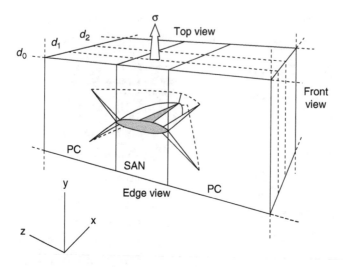

Figure 6 Three-dimensional model of a surface craze with micro-shear bands in PC/SAN.

shear banding is proposed (Figure 6) [27]. The model proposes that micro-shear bands initiate first in the PC layers and propagate rapidly along the edges of the craze. When they overtake the craze tip, the micro-shear bands penetrate through the PC/SAN interface and continue around the craze tip to entirely engulf the craze. This terminates craze growth and further strain in the SAN layer is accommodated by shear deformation.

2. Comparison between PC/SAN and PC/PMMA Systems

Effect of layer thickness on irreversible deformation and yielding was investigated in PC/PMMA microlayers and compared to the results obtained for a PC/SAN system. Microlayers with 32, 256, 1024, 2048 and 4096 layers were studied [28]. Comparison of the stress-strain curves of PC/PMMA with different numbers of layers revealed the effect of layer thickness on ductility. As in the case of PC/SAN, an increase in toughness with decreasing layer thickness was observed.

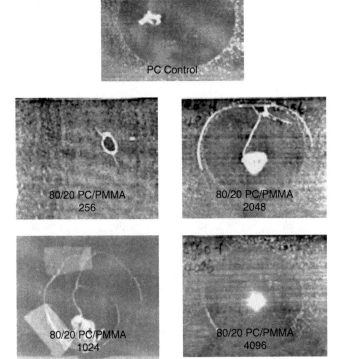

Figure 7 Effect of layer thickness on the ballistic properties of PC/PMMA microlayers.

The effect of layer thickness on ductility was also examined in ballistic tests. Figure 7 shows specimens of PC/PMMA (80/20) and the PC control after ballistic impact. The PC control had good ballistic response. The projectile did not penetrate the specimen; dissipation of the impact energy away from the impact site left a circular impression where the specimen was clamped. The PC/PMMA (80/20) specimen with 256 layers fractured upon impact, an indication of poor ballistic response. Absence of a circular impression indicated that the material did not absorb much energy. Increasing the number of layers improved the ballistic performance. The projec-

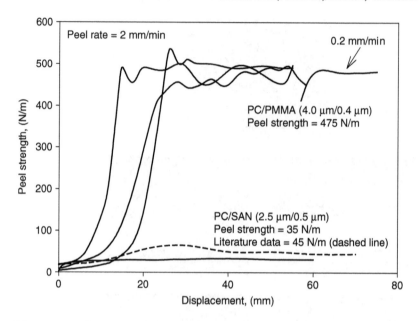

Figure 8 Comparison of delamination toughness in PC/SAN (2.5/0.5 µm) and PC/PMMA (4.0/0.4 µm) microlayers.

tile penetrated the 1024 and 2048 layer specimens, but the emergence of a circular impression showed that the material absorbed increasing amounts of impact energy. The 4096 layer specimen achieved the ballistic performance of the PC control. The projectile did not penetrate the specimen and the circular impression where the specimen was clamped was evident. The absence of delamination in the PC/PMMA specimens led to the conclusion that adhesion between PC and PMMA is stronger than the adhesion between PC and SAN.

To investigate this further, the delamination toughness of PC/SAN and PC/PMMA microlayers was measured using the T-peel test. Figure 8 compares the peel curves for PC/PMMA (with 4/0.4 µm layers thickness respectively) and PC/SAN (with 2.5/0.5 µm layers thickness respectively) tested at an extension rate of 2 mm/min. A much higher peel force was required to propagate the crack in PC/PMMA than in PC/SAN. PC/PMMA microlayer had a high delamination toughness of 950 J/m^2.

A comparison of PC/PMMA with a PC/SAN microlayer of the same composition and layer thickness revealed the role of interfacial adhesion in the microdeformation behavior. Before the yield point, microdeformation of both the systems was essentially identical with crazing in the SAN or PMMA layers, followed by micro-shear band formation in the PC layers. As the shear bands coalesced and a stable neck formed at the yield point, SAN crazes opened up into microcracks. However, the SAN cracks did not cut into the PC layers as the PMMA cracks did in PC/PMMA. Local delamination at the PC/SAN interface relieved the constraint on the SAN layers and permitted the PC layers to draw out. The interface was weak enough that the interfacial stress concentration caused local delamination as the specimen deformed as indicated by the arrows in Figure 5a. This contrasted with PC/PMMA as shown in Figure 9a, where the adhesion was strong enough to prevent even local delamination at the crack tip.

The difference in microdeformation behavior was manifested in the macroscopic stress-strain behavior. At lower strain rates, constraint imposed by PMMA on yielding of PC resulted in a diffuse neck. In contrast, local delamination of SAN relieved the constraint on yielding of PC and a sharp neck resulted (Figure 10). At low strain rates, the yield stress of PC/SAN and PC/PMMA microlayers was about the same because surface cracks in PC/SAN opened up to about the same depth as the surface cracks in PC/PMMA. At higher strain rates, the yield stress of PC/PMMA increased abruptly when deformation of the PMMA layers changed from microcracking to yielding. The yield stress of PC/SAN microlayers did not similarly increase because the deformation mechanism of thick SAN layers did not change with strain rate.

Upon deformation, PC/PMMA thin microlayers revealed craze arrays with aligned crazes (Figure 9b) in several neighboring PMMA layers and cooperative shear bands that grew across several PC layers and the intervening PMMA layers. During yielding, the PMMA crazes remained closed as both PC and PMMA layers drew out. The same mechanism was seen in yielding of PC/SAN with thin layers [25].

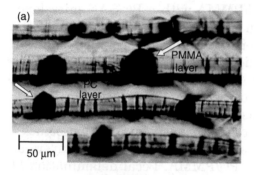

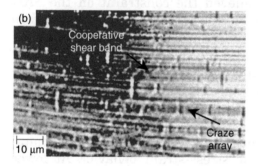

Figure 9 Optical micrographs of irreversible deformation after necking for: (a) 32-layer PC/PMMA(70/30); and (b) 256-layer PC/PMMA (70/30). Arrows indicate positions of stress concentration tested at 0.1%/min.

B.   Microlayers as Model Systems to Study Adhesion

Any situation that brings two polymers into intimate contact, whether as blends or multilayer films, achieving synergistic property combinations, depends on adhesion of the constituents. Interfacial properties are not easily examined in the dispersed domain morphology of conventional melt blends. The high surface-to-volume ratio attainable with microlayers makes them ideal for studying interfacial phenomena related to polymer blends. Adhesion of two polymers in microlayers is conveniently studied with the T-peel test. Although peel tests generally do not supply absolute adhesive energies, the delamination toughness obtained from the test provides use-

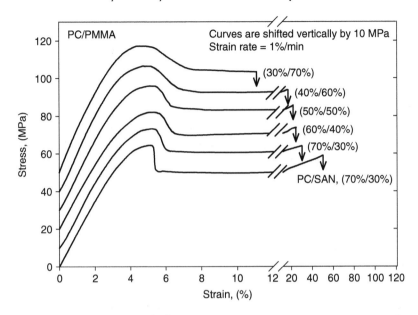

Figure 10 Stress-strain curves at 1%/min for 32-layer PC/PMMA specimens of various compositions; curves are shifted vertically by 10 MPa. A PC/SAN (70/30) specimen is included for comparison.

ful comparisons if the testing parameters and specimen dimensions are kept constant.

1. Delamination Toughness and Mechanism in PC/SAN Microlayers

The potential of microlayers to determine interfacial strength was first demonstrated with measurements of PC and SAN adhesion [17–19]. Peel curves of different PC/SAN layer thicknesses are shown in Figure 11. The delamination toughness changes drastically as the layer thicknesses of the PC and SAN are varied. As with tensile experiments, the micromechanics of the microlayer composite are crucial in understanding the delamination toughness. The method of delamination can involve crazing of the SAN layer ahead of the crack tip and the crack can also run along a single interface or several PC/SAN interfaces. Four different modes controlled by the

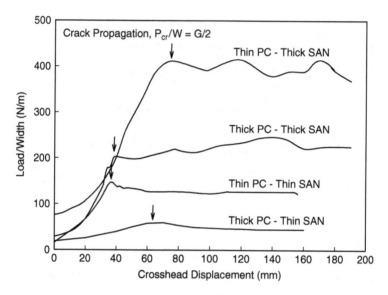

Figure 11 Normalized peel curves of four systems: thick PC-thin SAN (2.5 /0.5 μm), thin PC-thin SAN (0.5 /1.0 μm), thick PC-thick SAN (34 /18 μm), and thin PC-thick SAN (3.1/8.7 μm).

layer thickness are summarized schematically in Figure 12 as single interface, multiple interface, single-layer craze and multiple-layer craze delamination. The PC layer thickness controls single versus multiple layer delamination, as the PC layer has to be thin enough for the stress at adjacent interfaces to initiate secondary cracks and also thin enough to tear easily. Crazing in the SAN layers is controlled by the SAN layer thickness. As the SAN layer thickness is decreased, a critical layer thickness of 1.5 μm is reached below which no crazing occurs. If the SAN layers are thick and the PC layers are thin, both the mechanisms occurred with multiple-craze delamination. The effect of layer thickness on both the mechanism and delamination toughness is shown in Figure 12. Using thick PC layers and thin SAN layers, the actual interfacial delamination can be measured without additional energy-absorbing mechanisms such as crazing or layer tearing to achieve a real measure of interfacial adhesion. Controlling the layer thickness allows the measurement of actual

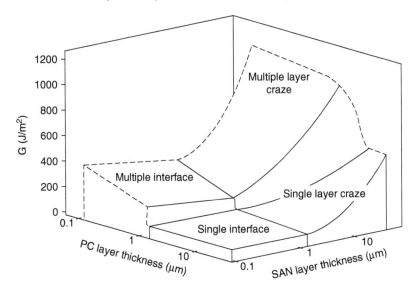

Figure 12 Map of peel level energies as a function of layer thickness and delamination mechanism in PC/SAN microlayers.

interfacial delamination using thick PC and thin SAN layers. In all the other cases, crazing or layer tearing provides additional energy-absorbing mechanisms and therefore the peel energies are much higher.

2.   Effect of Compatibilizer on Adhesion of Polypropylene (PP) and Polyamide (PA)

The benefits of compatibilization to mechanical properties of polymer blends are attributed to finer phase dispersion and improved interfacial adhesion. Measurements of particle size in the solid state reveal how well a compatibilizer improves dispersion. However, the extent to which a compatibilizer enhances interfacial adhesion cannot be measured directly in melt blends. With the versatility of the microlayer technology, a thin tie layer can be coextruded between each layer as seen in Figure 2. The effect of a thin tie layer of maleated polypropylene (mPP) on adhesion of polypropylene (PP) and polyamide-66 (PA) was studied [20]. Increasing the maleic

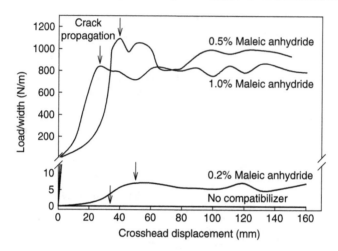

Figure 13 Effect of MA content in PP-g-MA on delamination toughness of PP/PP-g-MA/nylon microlayers.

anhydride content of the tie layer increased the interfacial toughness until the adhesion was strong enough to cause a transition from interfacial delamination to cohesive failure of one of the components as seen from Figure 13. The tests revealed a dramatic increase in toughness from 10 to 1000 J/m^2 when adhesion was strong enough for cohesive damage to initiate in one of the components.

3.   Evaluating Ethylene-Styrene
       Copolymers as Compatibilizers for
       Polyethylene and Polystyrene Blends

Combining polyethylene and polystyrene as blends has attractive applications, however, these polymers are considered incompatible. Copolymers of ethylene and styrene are potential compatibilizers. Adhesion of ethylene-styrene copolymers to low-density polyethylene was examined by measuring delamination toughness of microlayers [21]. Adhesion strongly depended on styrene content of the copolymer, the copolymer layer thickness, and the temperature. As expected, delamination toughness decreased with increasing styrene

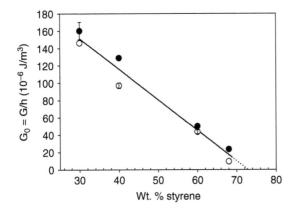

Figure 14 Effect of styrene content on delamination toughness normalized to ES layer thickness at an ambient temperature and peel rate of 10 mm/min. Closed circles represent 90/10, open circles represent 95/5. Dotted line is from extrapolation.

content and a linear correlation predicted that no adhesion would be observed with 72 wt% styrene in the copolymers (Figure 14). This approximation is close to the maximum amount of styrene incorporation in ES copolymers of 80 wt%. *In situ* observations of the damage zone showed that delamination occurred with stretching of the copolymer layer and with localized stretching and crazing of the polyethylene layer in cases of very high adhesion.

4. Effect of Chain Microstructure on Adhesion of Polyethylene to Polypropylene

The effect of chain microstructure on adhesion of ethylene copolymers to polypropylene was investigated using coextruded microlayers [22]. Good adhesion was achieved with homogeneous metallocene-catalyzed copolymers (m-PE) (Figure 15) which was attributed to entanglement bridges (Figure 16). In contrast, a heterogeneous Ziegler-Natta catalyzed copolymer (ZNPE) exhibited poor adhesion to PP due to an amorphous interfacial layer of low molecular weight and

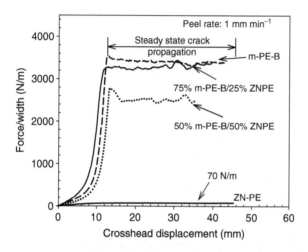

Figure 15 Effect of metallocene PE content on adhesion of polyethylene blends to polypropylene.

highly branched fractions that prevented effective interaction of ZNPE bulk chains with PP. Blending m-PE with ZNPE eliminated the interfacial layer and resulted in epitaxial crystallization of ZNPE bulk chains with some increase in delamination toughness. Phase separation of m-PE and ZNPE during crystallization produced an interfacial region with epitaxially crystallized ZNPE bulk chain and other regions of entangled m-PE chains. Entanglement bridges imparted much better adhesion than did epitaxially crystallized lamellae.

C.   Foam/Film Microlayers — A Novel Way of Controlling Foam Cell Structure

Polymeric-foamed materials are desirable because of their high strength to weight ratio and excellent acoustic and thermal insulation. Control of foam cell size is very critical in obtaining these key performance properties. Thermal insulation and compressive stress-strain behavior are particularly sensitive to cell size. Control of cell size can be very difficult to achieve in the conventional foam extrusion process. Controlling cell size in polypropylene foams is particularly diffi-

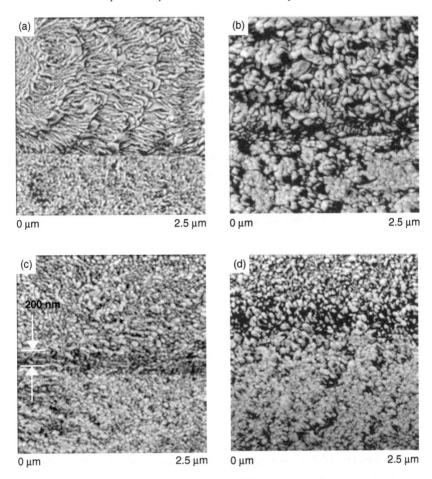

Figure 16 AFM phase images of PE/PP interface of m-PE-B blends of: (a)100% ZN-PE, (b) 25% m-PE-B, (c) 50% m-PE-B, and (d) 100% m-PE-B.

cult due to low solubility of gases in PP and inhomogeneous nucleation. Some of the previous approaches to control the cell size in PP foams include cross-linking, blending and use of supercritical fluids [29–30].

Cork, which is a natural foam, shows unique properties like outstanding insulation for heat and sound, high compressive modulus and recoverable compressive strain. These prop-

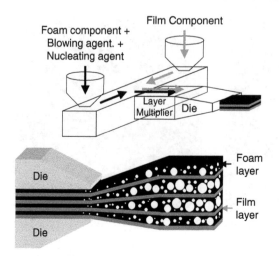

Figure 17   Schematic of microlayer foam/film coextrusion process.

erties are attributed to the small cell size and the closed cell foam/film layered structure in cork [30]. The foam/film structure of cork can be replicated synthetically using the microlayer coextrusion technology since control of cell size by this technique makes the creation of microcellular foam possible.

Recently it has been demonstrated that microcellular PP foam/film structures can be successfully produced using the microlayer coextrusion technology [23]. Different types of polymers could be used as the film layer to alter the flexibility of the foam/film composite.

A two-component microlayer coextrusion setup with layer multipliers was used to coextrude PP foam/film layered structures (Figure 17). One extruder contained foam layer polymer, chemical blowing agent and the nucleating agent, while the other extruder contained the film polymer. After merging in the two component feedblock, the foam and film layers were formed into a number of layers using the layer multipliers. PF 814 polypropylene from Basell Inc. was used as a foam layer polymer. Two types of films — polypropylene (PF 814) and ethylene-styrene copolymer were used to produce two different foam/film systems. Azodicarbamide was

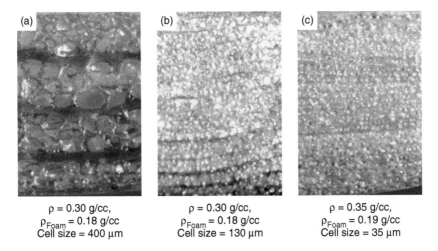

p = 0.30 g/cc,
p_{Foam} = 0.18 g/cc
Cell size = 400 µm

p = 0.30 g/cc,
p_{Foam} = 0.18 g/cc
Cell size = 130 µm

p = 0.35 g/cc,
p_{Foam} = 0.19 g/cc
Cell size = 35 µm

Figure 18 Effect of number of layers on cell morphology for PP foam (1.75% Blowing Agent)/PP film (50/50) system: (a) 8 layers, (b) 32 layers, and (c) 64 layers.

used as a chemical blowing agent which gives off N_2, CO_2 and water on decomposition.

By increasing the number of layers, we can reduce the cell size considerably. In PP foam (1.75% Azo BA)/PP film 50/50 system, the cell size gets reduced from 400 µm to 35 µm when the number of layers is increased from 8 to 64 (Figure 18). The overall density was approximately 0.30 g/cc for these systems. Overall density can be further reduced to 0.26 g/cc by increasing the amount of foam layer in the composite. The cell sizes obtained in the microlayered foam/film system are comparable to that of cork. The reduced cell size can be attributed to enhanced nucleation and/or constrained bubble growth. It has been observed previously that small cell sizes are obtained when foaming takes place under geometrical confinement between two impermeable plates [32] and in the presence of sub-micron sized rubber particles which enhance nucleation [33].

Compressive stress-strain response of PP foam/ES 69 films at different compositions (Figure 19) shows that the compressive modulus increases with addition of film, the yield

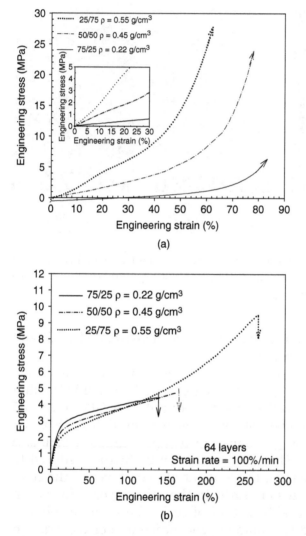

Figure 19   Effect of foam/film composition on: (a) compressive, and
(b) tensile stress-strain for PP foam (1.75% Blowing Agent)/ES 69
film system, 64 layers: 75/25 (——), 50/50 (— · — ·), and 25/75 (····).

stress increases and the densification strain decreases. Similar effects are seen in conventional extruded foam systems
with increase in density [29]. Tensile stress-strain response
of PP foam/ES 69 films at different compositions shows sig-

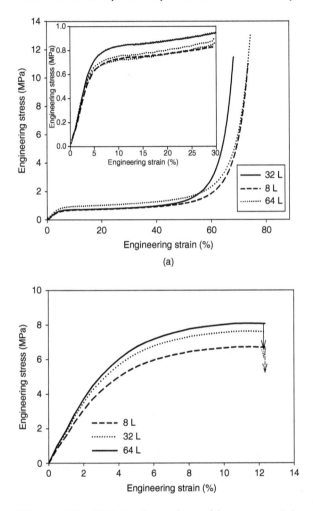

Figure 20 Effect of number of layers on: (a) compressive and (b) tensile stress-strain for PP foam (1.75% blowing agent)/PP film (50/50) system: 8 layers (— · — ·), 32 layers (····) and 64 layers (——).

nificant increase in tensile modulus, yield stress and the fracture strain with the addition of more film. The compressive and tensile moduli increase with increasing number of layers (Figure 20). This is as expected, since reduced cell size means increased total number of cells and hence increased total number of cell walls [30]. Compressive modulus, yield stress

TABLE 1   Comparison of Compressive Stress-Strain
Response: Cork and Foam/Film Microlayered
Systems

	Cork	PP Foam (1.75% BA)/ PP Film (50/50), 64 Layers
Compressive Modulus (MPa) E_1, E_2	12	12 ± 5
Collapse Stress (MPa) $(\sigma_{el})_1$ $(\sigma_{el})_2$	0.7	0.6 ± 0.2
Densification Strain ε_D	0.8	0.8 ± 0.05

and the densification strain of the PP foam/PP film system is
similar to that of cork (Table 1).

IV. NANOLAYERED POLYMERS

A.   Morphology, Deformation Behavior and Microhardness of PC/PET Nanolayered Polymers

1.   Morphology and Deformation Behavior

The morphology and micromechanical deformation behavior
of coextruded micro- and nanolayers composed of alternating
layers of polycarbonate (PC) and polyethylene terephthalate
(PET) were studied using transmission electron microscopy
(TEM) and scanning force microscopy (SFM) [34]. TEM micrographs of PC/PET micro- and nanolayered composites
revealed macroscopically aligned continuous layers having
nearly uniform thickness, Figure 21. The thickness of most of
the layers is 110–140 nm, very close to the calculated value
from the processing conditions of 120 nm. Since both polymers
are polyesters, they can react by transesterification in the
molten state to form block or random copolymers. Thus,
depending on the thermal treatment, the mutual interaction
between the PC and the PET layers can be affected to alter
the morphology and the mechanical behavior of the system.
On annealing the samples, the PET layers were found to

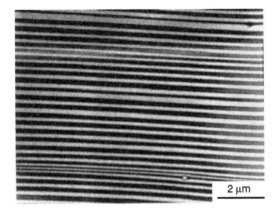

Figure 21 TEM micrograph of the cross-section of a PC/PET nanolayered film. The thickness of most of the layers is 110–140 nm which is very close to the calculated value of 120 nm. The PC layers appear dark due to staining with RuO_4.

undergo cold crystallization and to form lamellar crystals independent of the thickness of individual layers. Typical lamellar crystals are observed in a 70 nm PET layer in Figure 22. Due to the physical confinement imposed by the adjacent PC layers, PET layers are unable to develop well-defined spherulitic structures as in bulk melt-crystallized or solution-crystallized samples. Using x-ray scattering techniques, Puente et al. [35] recently found that the three-dimensional crystallization of PET is significantly hindered when one of the dimensions of the layered system is confined below 10 μm.

Preliminary studies of the micromechanical deformation behavior of PC/PET micro- and nanolayered composites under tensile loading were performed. The mechanical properties of the composites were found to be intermediate between that of the constituent homopolymers. Figure 23 shows the layered structure before and after deformation. Interestingly, it was observed that the components of the micro- and nanolayered composites deformed in a homogeneous manner at high strains without forming localized deformation zones such as shear bands or crazes (For additional details of micromechanical processes, see Chapter 10, Section IV.E).

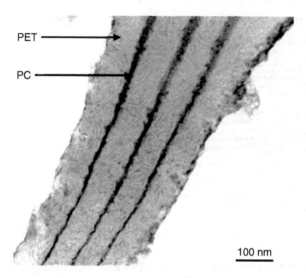

PET

PC

100 nm

Figure 22 TEM micrograph of PET/PC 70 nm layers showing the lamellar structure of the PET layers formed due to confined crystallization between adjacent PC layers. The PC layers and the amorphous regions of PET appear dark due to staining with RuO_4. Sample annealed at 160°C for 12 hrs.

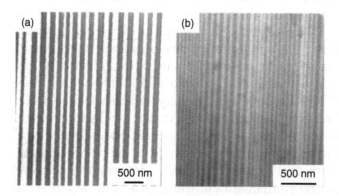

(a)

(b)

500 nm

500 nm

Figure 23 TEM images of PET/PC 120 nm layers showing the effect of tensile deformation on the layered structure: (a) before deformation, and (b) after deformation. The deformation direction is vertical. The PC layers appear dark.

2.   Microhardness

In addition to the micromechanical deformation behavior, the microhardness of microlayered PET/PC films has also been studied. The micromechanical properties have been investigated as a function of layer thickness of the single polymer components, total number of layers, film thickness and the influence of heat treatment [36].

The microhardness value was obtained from the measurement of the residual impression made by the indenter upon application of the load. Microhardness was derived from:

$$H = kP/d^2 \qquad (1)$$

where P is the applied load in N, d is the diagonal of the impression in meters and k is a geometrical factor equal to 1.854×10^{-6}. The H values were calculated from an average of at least ten indentations. Typical indentation depths for these materials are in the range of 3–15 μm using loads of 0.05 to 1 N. The traditional approach (Figure 24a) only considers the first layer penetrated by the indenter. With micro- and nanolayered systems one may distinguish between two more possibilities: the case where only the first layer is penetrated by the indenter but more than one layer is subjected to plastic deformation (Figure 24b) and the case when a few layers are penetrated and subjected to plastic deformation (Figure 24c). In all cases, the drawn shaded elliptical area under the indenter schematically represents the plastic deformation region [37]. Therefore, the most important parameter in determining the final hardness of the multi-layered films is the ratio of the penetration depth to the thickness of the layer.

By plotting the H value calculated from each indentation as a function of the ratio of indentation depth to layer thickness, Figure 25a, a straightforward picture of the microhardness changes in relation to the layer penetration by the indenter is obtained in microlayered samples. The $\delta/l = 1$ value means that the indenter tip is reaching the interphase between the first and the second layer, while a $\delta/l = 0.2$ value denotes that the plastic deformation is reaching the above mentioned interphase. Figure 25a shows the H variation cor-

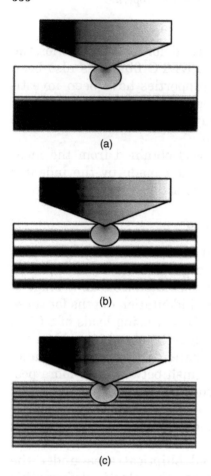

Figure 24 Schematic representation of different indentation types in microlayered systems: (a) only the first layer is penetrated by the indenter; plastic deformation (elliptical area under the indenter) is restricted to the first layer; (b) only the first layer is penetrated by the indenter, but plastic deformation exceeds the first interphase; and (c) more than one layer is penetrated and several layers are subject to plastic deformation.

responding to the indentations performed on the as processed 8 layered amorphous samples. With increasing relative penetrations, from 10 to 100% of the first layer, the H data exhibit a weak linear decrease, in which those data corresponding to

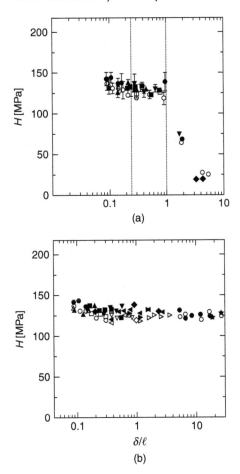

Figure 25 Plot of microhardness H, derived from single indentations, as a function of indentation depth to layer thickness ratio, δ/l, for: (a) 8-layered PET/PC system, l = 59 (○), 42, 22 (□), 15 (∇) and 4 μm (◇); and (b) micro- and submicro-layered samples, l = 10.5 (◁), 3.5 (▷), 1.02 (⊠), and 0.21 m (☆). Open symbols correspond to the PET layer, solid symbols correspond to the PC layer.

the PC surface show slightly larger *H* values. When the indenter reaches the interphase (δ/l = 1), which in this particular case also coincides with the plastic deformation zone extending beyond the whole sample thickness, the microhardness rapidly decreases.

One may suggest two different explanations for the observed H decrease. The first one concerns the nature of the interphase which could introduce a soft component to the overall deformation. On the other hand, it must be taken into account that when the penetration depth reaches the first interphase in an 8-layered film, the plastic deformation is also reaching the last layer surface of the film. To clarify the above question, additional samples with 32, 256 and 1024 layers were tested. The corresponding variation of H vs. δ/l is shown in Figure 25b. Results demonstrate that for the microlayered samples, the influence of the interphases decreases slightly in microhardness with increasing numbers of interphases penetrated by the indenter. For the as processed samples (Figure 25b), the average H value at the point where the first layer is reached by the indenter is about 132 MPa. However, after penetrating about 30 interphases, the H value drops to 125 MPa. It is clear that after the initial separate H values for PET and PC, when only the first layer is plastically deformed, a tendency towards an average common value for the rest of the indentations is observed. The strong H decrease found in Figure 24 can be related to the deformation of the soft glue layer which is used to fix the film and is located beneath the sample of thickness t. Thus for sufficiently large indentations, the low H value of the glue layer may also influence the overall hardness value. In summary, results reveal that the influence of the interphase on the microhardness values for the samples with a large number of layers is rather small.

B. Tunable Optical Properties of an Elastomer/Elastomer Nanolayered System

By using nanolayer coextrusion technology, multilayer polymer films with individual layer thicknesses of 10–100 nm can be fabricated. When the individual layer thickness in the film is at the quarter wavelength, $\lambda/4n$ and the refractive index of the alternating layers is sufficiently different, constructive reflection will be observed. By using polymers of different refractive indices and designing component layer thickness ratio, material of different refractive indices can be made.

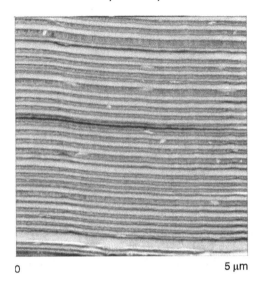

0   5 µm

Figure 26 AFM phase image of PU/EO 150 nm layers. The bright phase is polyurethane (PU), the dark phase is an ethylene-octene (EO) copolymer.

More than 100 layers are required for a highly reflective nanolayered film. Layer thickness should be around 150 nm for nanolayered films that reflect 900 nm wavelength light.

Two elastomers, polyurethane and an ethylene-octene copolymer, with refractive indices of 1.54 and 1.49, respectively, were coextruded into films with 256 layers and a calculated individual layer thickness of 150 nm. A continuous layered structure was observed using the atomic force microscope (AFM), Figure 26. The measured layer thickness of 160 nm corresponded very well with the calculated value. A selective distinct reflection peak, in this case at 965 nm, was observed in these films when the layer thicknesses were on the order of visible light wavelength. This selective reflection is tunable and reversible through loading and unloading of an applied strain, Figure 27. In addition, the reflection shift is theoretically predictable, allowing this film to be used as a tunable optical filter. The reflection spectrum is also very sensitive to low strain, allowing this film to be

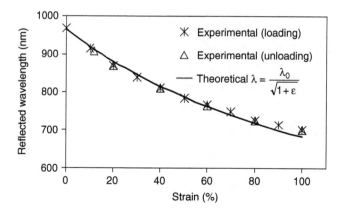

Figure 27 Reversible reflected wavelength shift due to applied strain during loading and unloading of PU/EO nanolayers. These films exhibit potential as tunable optical filters in the VIS-NIR spectrum.

used as an optical strain sensor. Figure 28 shows the sensitivity of the reflection wavelength at low strains as well as the theoretical prediction.

C.   Physical Properties of Interphase Materials

It has been recognized for some time that when two polymers are brought into intimate contact, the interface between them is not perfectly sharp. Highly localized mixing of polymer chains creates an "interphase" region. A vast literature attests to the importance of the interphase for polymer adhesion and for compatibility of polymer blends and alloys [38]. Indeed, it is reasonable to consider the interphase of immiscible polymer blends as a third phase with its own characteristic properties [39]. The interphase takes on new significance as nanotechnology and microelectronics drive the fabrication of increasingly thin polymer layers. Under these conditions, as bulk polymers become thinner and more interphase-like, departures in physical properties are expected. Understanding interphase properties is crucial if polymers are to be integrated effectively into modern technology.

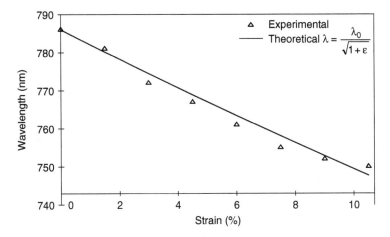

Figure 28 Sensitivity of reflection wavelength to low strains exhibited by PU/EO nanolayers. These films exhibit potential as tunable optical strain sensors in the VIS-NIR spectrum.

The origin of interfacial mixing is the entropic advantage for chains to diffuse across the boundary. Halving in the conformational entropy of chains at a surface is ameliorated if the chains cross the interface. The entropic advantage for crossing the interface is offset by the negative effect of the interaction energy between incompatible chain segments. The generalized theory developed by Helfand and Sapse is the basis of a quantitative relationship between interphase thickness and the thermodynamic interaction parameter [40]. However, experimental determination of interphase properties has challenged the field. Even measurement of the interphase dimension is difficult and requires extreme care. Nevertheless, existing theoretical and experimental results agree that the dimension of the interphase for immiscible polymers is in the range of 5–10 nm [41].

Nanolayer processing facilitates the creation of new hierarchical systems. Although the amount of material in a single interphase between two polymer layers is very small, the layer thickness can be made comparable to the interphase dimension and the properties of the interphase are multiplied

100 nm layers   25 nm layers   10 nm layers

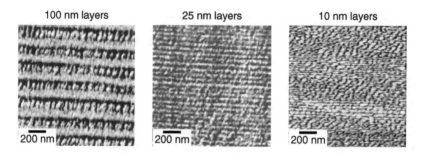

Figure 29 AFM phase images of microtomed cross-sections of PC/PMMA nanolayer films. Films with average layer thicknesses of 100 nm and 25 nm show continuous, uniform layers of PC (smooth and bright) and PMMA (hackled). The film with a 10 nm average layer thickness shows an irregular periodicity about twice the average layer thickness.

a thousand-fold by the number of identical interphases in the assembly. In this way, conventional methods of polymer analysis can be used to probe size-scale-dependent properties as nanolayers become thinner and more interphase-like.

As the layer thickness becomes comparable to the interphase dimension, the layers lose their identity and a new composition is created that is totally interphase. To test this possibility, polycarbonate (PC) and poly(methyl methacrylate) (PMMA) were chosen for forced assembly into nanolayers and studied by thermal analysis, gas transport and positron annihilation lifetime spectroscopy to probe the effect of layer thickness [9].

Continuous uniform layers are well resolved in AFM images of films with average layer thicknesses of 100 nm and 25 nm, Figure 29. The PC layers are smooth and bright; the PMMA layers are hackled due to brittle fracture during microtoming. Layer thickness measured from the images corresponds well with the calculated average thickness from processing conditions. A different texture is seen in the image of the film with an average layer thickness of 10 nm, Figure 29. Instead of distinct layers with straight boundaries, the image shows an irregular periodicity of about 20 nm or about twice the expected layer thickness.

DSC thermograms of films with layers thicker than 100 nm contain two inflections in heat capacity at 112°C corresponding to the T_g of PMMA and at 144°C corresponding to the T_g of PC, Figure 30, top. However, the glass transition temperatures gradually shift closer together as layer thickness decreases below 100 nm. When the layer thickness is 10 nm or less, the two inflections merge into a single inflection at a temperature that is intermediate between the glass transition temperatures of PC and PMMA.

The convergence to a single glass transition is not symmetrical, Figure 30, bottom. The T_g of PMMA increases only about 10°C whereas the T_g of PC decreases by 22°C. The glass transition of a miscible blend usually is described by either the Fox law ($1/T_g = w_1/T_{g1} + w_2/T_{g2}$) or the additive law ($T_g = T_{g1}w_1 + T_{g2}w_2$) where w_1 and w_2 are the weight fractions of the blend constituents. The calculated T_g of a miscible PC/PMMA (1:1 vol:vol) blend is 126°C from the Fox law and 128°C from the additive law. In nanolayers, the glass transition temperatures of PC and PMMA converge asymmetrically to 122°C, which is 4–6°C lower than the calculated values.

Gas transport, like the glass transition, is a molecular scale probe of the glassy polymer structure. Oxygen permeability (P) and diffusivity (D) of nanolayer films were extracted from the non-steady state oxygen flux following established procedures [42], and the oxygen solubility (S) was calculated as $S = P/D$. If the layers are thicker than 100 nm, P, D and S at 23°C and 1 atm pressure are independent of layer thickness, Figure 31. A substantial increase in P is observed for films with layers less than 100 nm thick. Permeability more than doubles from 0.72 to 1.62 cc(STP) cm m^{-2} day^{-1} atm^{-1} as the layer thickness decreases to 10 nm, at which point the permeability seems to level off. The change in P with layer thickness comes mainly from D whereas S remains almost constant. Apparently, the change in hole-free volume affects the dynamic component of permeability. Oxygen solubility in glassy polymers at low pressure is determined primarily by excess hole-free volume [42]. Lower free volume of nanolayers as demonstrated by higher density and smaller free volume hole size should decrease oxygen solubil-

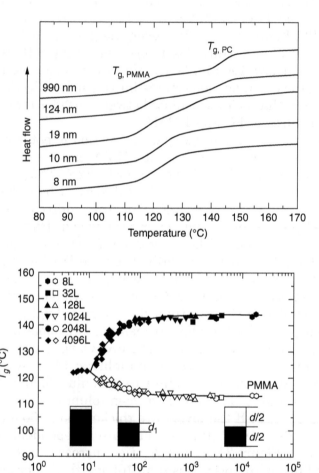

Figure 30 Glass transition behavior of PC/PMMA nanolayer films. In the upper plot, heating thermograms show convergence of the glass transition temperature of the constituent polymers into a single inflection as the layers become thinner. In the lower plot, the dependence of Tg on layer thickness is described by the three-layer interphase model as illustrated schematically in the figure.

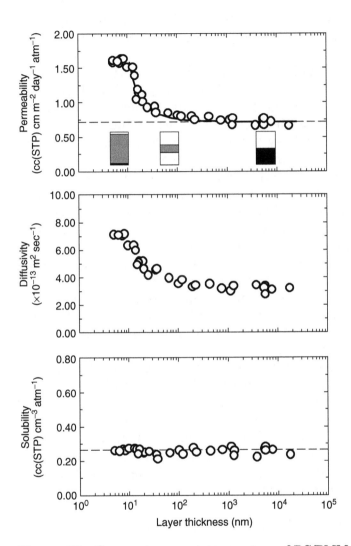

Figure 31 Oxygen transport parameters of PC/PMMA nanolayer films. The effect of layer thickness on oxygen permeability is due to a change in diffusivity rather than a change in solubility. The dependence of permeability on layer thickness is described by the three-layer interphase model as illustrated schematically in the figure.

ity. However, a density increase of 0.14 wt% produces a decrease in S of approximately 5% [9], which is comparable to experimental uncertainty (2%). Thus, S is not sensitive enough to detect the change in free volume and S is adequately described by the additive law, $S = w_1 S_1 + w_2 S_2$. Similar results are obtained with CO_2 transport. Nanolayer films become more permeable as the layer thickness decreases below 100 nm with P for CO_2 increasing from 4.5 to 9.6 cc(STP) cm m^{-2} day^{-1} atm^{-1}. Again, the increase in P is due to increasing D, whereas S remains constant.

It is proposed that when the layer thickness in nanolayer films becomes comparable to the interphase dimension, the layers lose their integrity as constituent layers and the film becomes essentially totally interphase [9]. The composition profile is not homogeneous but rather undulating with a period that is twice the layer thickness. Film properties are no longer additive combinations of constituent properties, but instead, film properties are dominated by interphase properties. For the PC/PMMA pair, nanolayers with average layer thickness of 10 nm or less present properties of the interphase.

The experimental result for the interphase thickness, d_I = 10 nm, can be compared with the theoretical prediction of Helfand and coworkers. The calculated interphase thickness of 7 nm is comparable to d_I determined experimentally from nanolayers, 10 ± 1 nm. The small discrepancy may originate from the infinite molecular weight assumption of the theory as compared to polydisperse "real" polymers which contain low molecular weight fractions.

D.   Novel Structures Produced by Confined Crystallization in Nanolayers

As the layer thickness decreases from the microscale to the nanoscale, the layer thickness is on the same size scale as the dimensions of the polymer molecule. In this instance, unusual crystalline structures are possible, giving rise to unique material properties. For example, it has been observed that nanolayered polyethylene produces unique row-nucleated morphologies [7] with highly anisotropic mechanical properties. The solid-state structural hierarchy of high density polyethylene (HDPE) and polypropylene (PP) was studied as the layers were made thinner and confinement at the nanoscale was approached [8]. The crystallizable polymer was coextruded as microlayers and nanolayers between thick layers of amorphous polystyrene. The crystalline layer thickness varied from 1 μm to 10 nm. A continuous and uniform layered structure was observed even at the nanoscale with our atomic force microscope (AFM), Figure 32. The crystalline layers separated easily from polystyrene, and the morphology was studied by AFM.

In microlayers, polypropylene and polyethylene crystallized in discoidal morphologies, i.e., as spherulites whose growth was restricted in the third dimension by the layer thickness, Figure 33. The discoid diameter remained almost constant, resulting in an increase in diameter-to-thickness ratio from 10 to 140 as the layer thickness decreased from 1 μm to 100 nm. Individual PP discoids had a peak in the center and sharp edges where they impinged on neighboring discoids. These effects were the result of volume contraction during crystallization. They indicated that PS layers were above the T_g when PP crystallized and therefore, accommodated the shape changes of crystallizing PP, rather than imposing constraint on the crystallizing layer. The corresponding section profiles of discoids revealed a trend toward flatter discoids as the layer thickness decreased. The morphology changed when the layer thickness was less than 50 nm. No discoids were observed. In polypropylene nanolayers, Figure 34, long stacks of very short lamellae were arranged in a fan-like array. The sharpness of the azimuthal (110) reflection in the WAXD indicated that under extreme confinement, these crystals originated from the rarely observed (110) plane growth rather than the more usual (010) growth. This growth habit was associated with a significant distortion of the crystallographic unit cell, including the loss of register along the chain axis. Crystallographic unit cell distortion correlated with a decrease in crystallinity and a decrease in crystal melting temperature in the DSC.

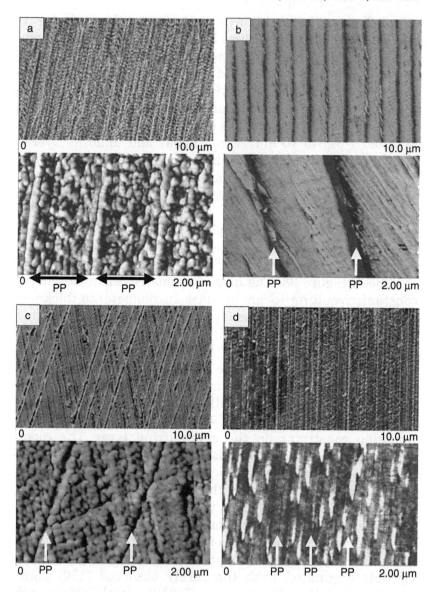

Figure 32 AFM phase images of cross-sections from layered PP/PS films: (a) PP/PS 90/10 250 μm thick film with 460 nm PP layers; (b) PP/PS 20/80 250 μm thick film with 108 nm PP layers; (c) PP/PS 10/90 250 μm thick film with 65 nm PP layers; and (d) PP/PS 10/90 40 μm thick film with 10 nm PP layers.

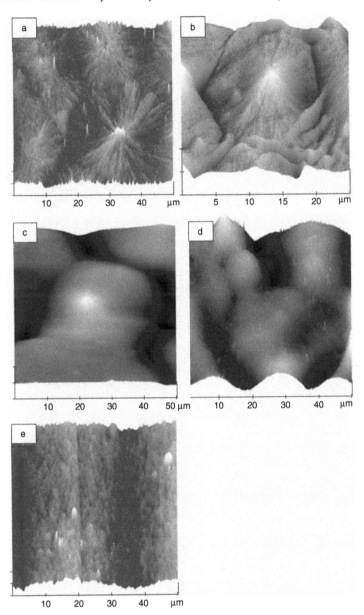

Figure 33 AFM height images: (a) a typical PP spherulite at a free film surface; (b) a discoid in a 460 nm PP layer; (c) a discoid in a 108 nm PP layer; (d) a discoid in a 65 nm PP layer; and (e) a smooth 10 nm PP layer.

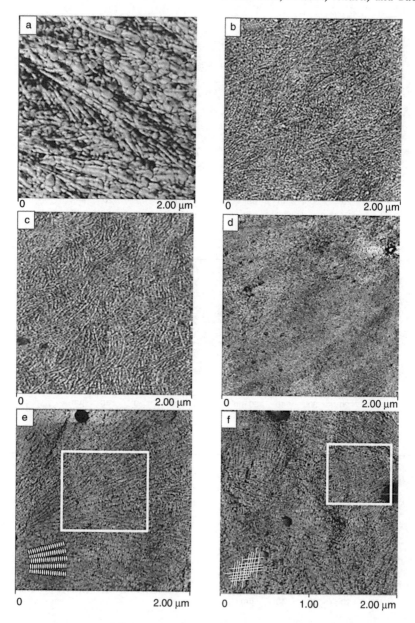

Figure 34 AFM phase images showing lamellar morphology: (a) PP control film; (b) a 460 nm PP layer; (c) a 108 nm PP layer; (d) a 65 nm PP layer; (e) a 10 nm PP layer; and (f) another area of a 10 nm PP layer.

In nanolayers of HDPE, extremely long lamellae, about 7 μm in length and 30 nm in thickness, organized in randomly oriented stacks, Figure 35. Wide angle x-ray diffraction (WAXD) revealed that constraint imposed by the nanolayer structure resulted in orientation of the a-axis in the thickness direction. If the nanolayer HDPE was oriented during coextrusion by increasing the take-off rate, the stacks of long lamellae were replaced by "shish-kebab"-like structures oriented 0–45° with respect to the extrusion direction, thus confirming previous observations of Pan et al. [7]. It was thought that constraints imposed by the layer thickness retarded relaxation of extended chain, which resulted in formation of shish-kebab structures [7].

The layer thickness in microlayers and nanolayers of HDPE and PP films confined between PS can be reduced to macromolecular dimensions by uniaxial or biaxial drawing. For example, the thickness of 50 nm PP nanolayers can be reduced to 9 nm by uniaxial orientation, Figure 36. Interestingly, the layers remained intact after orientation. This challenges the limit on how thin a nanolayer can be before it fibrillates. Preliminary results indicate that oriented fibers less than 20 nm in diameter can be produced by uniaxially orienting confined nanolayers, Figure 37.

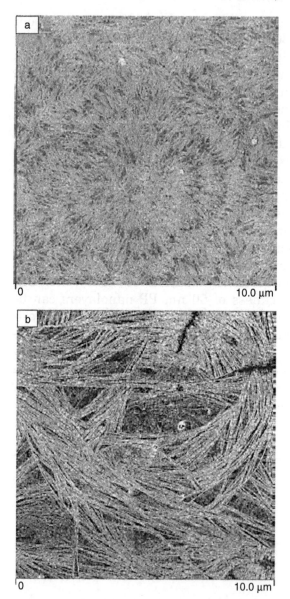

Figure 35 AFM phase images showing lamellar morphology: (a) 1000 nm HDPE layer and (b) a 40 nm HDPE layer.

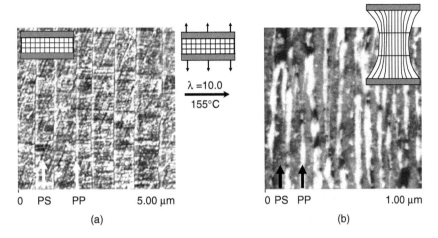

(a)   (b)

Figure 36 AFM phase images of cross-sections of the layered structure in a PP/PS nanolayered film: (a) 65 nm PP layers observed before orientation; (b) 10 nm PP layers observed after orientation of film in (a) to a draw ratio ~10 at 155°C. Note the change of scale between (a) and (b).

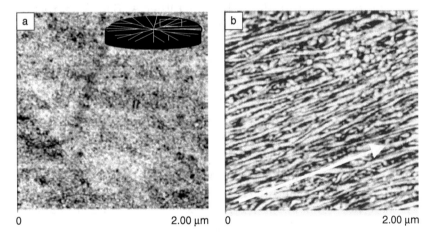

Figure 37 AFM phase images showing the lamellar morphology in a PP/PS nanolayered film: (a) with 65 nm PP layers before orientation; (b) with 10 nm PP layers after orientation of film in (a). The film in (a) was oriented to a draw ratio ~10 at 155°C.

REFERENCES

1. Bement AL, American Society for Metals and the Metallurgical Society Joint Distinguished Lecture in Materials and Society, Materials Week, October 1986.

2. Aksay A, Baer E, Sarikaya M, Tirrel DA, eds. Hierarchical Structures in Biology as a Guide for New Materials Technology. NMAB Report 464. Washington, D.C.: National Academy Press, 1994.

3. Baer E, Hiltner A, Morgan R. Biological and synthetic hierarchical composites. Phys Today 1992; October: 60–67.

4. Baer E, Hiltner A, Keith HD. Hierarchical structure in polymeric materials. Science 1987; 235: 1015–1022.

5. Balsamo V and Stadler R, Influence of the crystallization temperature on the microphase morphology of a semicrystalline ABC triblock copolymer. Macromolecules 1999; 32: 3994–3999.

6. Loo Y-L, Register RA, Ryan AJ, Dee GT. Polymer crystallization confined in one, two, or three dimensions. Macromolecules 2001; 34: 8968–8977.

7. Pan SJ, Im J, Hill MJ, Keller A, Hiltner A, Baer E. Structure of ultrathin polyethylene layers in multilayer films. J Polym Sci B 1990; 28: 1105–1119.

8. Jin Y, Rogunova M, Nowacki AR, Piorkowska E, Galeski A, Hiltner A, Baer E. The structure of polypropylene in confined nanolayers. Submitted to J Polym Sci B.

9. Liu RYF, Jin Y, Hiltner A, Baer E. Probing nanoscale polymer interactions by forced-assembly. Macromol Rapid Commun 2003; 24: 943–948.

10. Michler GH, Adhikari R, Lebek W, Goerlitz S, Weidisch R, Knoll K. Morphology and micromechanical deformation behavior of styrene/butadiene-block copolymers. I. Toughening mechanisms in asymmetric star block copolymers. J Appl Polym Sci 2002; 85: 683–700.

11. Sherby OD and Wadsworth J. Damascus steels. Sci Am 1985; 252: 112–120.

12. Wadsworth J, Kum DW, Sherby OD. Welded Damascus steels and a new breed of laminated composites. Met Prog 1986; 61–67.

13. Evans AG, Suo Z, Wang RZ, Aksay IA, He MY, Hutchinson JW. Model for the robust mechanical behavior of nacre. J Mater Res 2001; 9: 2475–2484.

14. Boyer RF, Mark HF, eds. Selected Papers of T. Alfrey. New York: Marcel Dekker, 1986.

15. Whitesides GM and Grzybowski B. Self assembly at all scales. Science 2002; 295: 2418–2421.

16. Ma M, Vijayan K, Im J, Hiltner A, Baer E. Thickness effects in microlayer composites of polycarbonate and poly(styrene-acrylonitrile). J Mater Sci 1990; 25: 2039–2046.

17. Ebeling T, Hiltner A, Baer E. Effect of peel rate and temperature on delamination toughness of PC-SAN microlayers. Polymer 1999; 40: 1525–1536.

18. Ebeling T, Hiltner A, Baer E. Effect of acrylonitrile content on the delamination toughness of PC-SAN microlayers. Polymer. 1999; 40: 1985–1992.

19. Ebeling T, Hiltner A, Baer E. Delamination failure mechanisms in microlayers of polycarbonate and poly(styrene-co-acrylonitrile). J Appl Polym Sci 1998; 68: 793–805.

20. Ebeling T, Norek S, Hasan A, Hiltner A, Baer E. Effect of a tie layer on the delamination toughness of polypropylene and polyamide-66 microlayers. J Appl Polym Sci 1999; 71: 1461–1467.

21. Ronesi V, Cheung Y, Hiltner A, Baer E. Adhesion of ethylene-styrene copolymers to polyethylene in microlayers. J Appl Polym Sci 2003; 89: 153–162.

22. Poon B, Chum S, Hiltner A, Baer E. Adhesion of polyethylene blends to polypropylene. Polymer 2004; 45: 893–903.

23. Ranade A, Bland D, Hiltner A, Baer E. Structure-property relationships in coextruded foam-film microlayers. Society of Plastics Engineers: ANTEC 2004 Chicago proceedings. In Press.

24. Im J, Hiltner A, Baer E. Microlayer composites. In: Baer E, Moet A., eds. High Performance Polymers. New York: Hanser Pubs, 1991: 175–198.

25. Sung K, Hiltner A, Baer E. Mechanisms of interactive crazing in PC/SAN microlayer composites. J Appl Polym Sci 1994; 52: 147–162.

26. Haderski D, Hiltner A, Baer E. Crazing phenomena in PC/SAN microlayer composites. J Appl Polym Sci 1994; 52: 121–133.

27. Sung K, Hiltner A, Baer E. Three dimensional interaction of crazes and micro-shearbands in PC/SAN microlayer composites. J Mater Sci 1994; 29: 5559–5568.

28. Kerns J, Hsieh A, Hiltner A, Baer E. Comparison of irreversible deformation and yielding in microlayers of polycarbonate with poly(methyl methacrylate) and poly(styrene-co-acrylonitrile). J Appl Polym Sci 2000; 77: 1545–1557.

29. Sha H and Harrisson IR. CO2 permeability and amorphous fractional free volume in uniaxially drawn HDPE. J Polym Sci. B: Polym Phys 1992; 30: 915–922.

30. Shimbo M, Baldwin DF, Suh NP. The viscoelastic behavior of microcellular plastics with varying cell size. Polym Eng Sci 1995; 35, 17: 1387–1393.

31. Gibson L and Ashby J. Cellular Solids: Structure and Properties. Cambridge: Cambridge University Press, 1997.

32. Siripurapu S, Desimone JM, Spontak RJ, Khan SA. Generation of nano-porous thin polymer films via foaming with carbon-dioxide. Preprints of symposia — American Chemical Society Division of Fuel Chemistry 2003; 48: 262–263.

33. Ramesh NS, Rasmussen DH, Campbell GA. The heterogeneous nucleation of microcellular foams assisted by the survival of microvoids in polymers containing low glass transition particles: experimental results and discussion. Polym Eng Sci 1994; 34, 22: 1698–1706.

34. Adhikari R, Lebek W, Godehardt R, Henning S, Michler GH, Baer E, Hiltner A. Investigating morphology and deformation behavior of multilayered PC/PET composites. Polym Adv Tech 2004, in press.

35. Puente I, Ania F, Baltá Calleja FJ, Funari SS, Baer E, Hiltner A, Bernal T. Confined Crystallization in PET/PC Micro and Nanolayers: Influence of Layer Thickness. Research Report of Desy (Hasy) Lab, 2002.

36. Puente Orench I, Ania F, Baer E, Hiltner A, Bernal T and Baltá Calleja FJ. Basic aspects of microindentation in multi-layered PET/PC films. Phil Mag, in press.

37. Rikards R, Flores A, Ania F, Kushnevski V, Baltá Calleja FJ. Numerical-experimental method for the identification of plastic properties of polymers from microhardness tests. Comp Mater Sci 1998; 11: 233–244.

38. Paul DR and Bucknall CB. Polymer Blends, Vol. 1 and 2. New York: Wiley, 2000.

39. Utracki LA. Polymer Alloys and Blends: Thermodynamics and Rheology. Munich: Hanser Publishers, 1990: 118–124.

40. Helfand E and Sapse AM. Theory of unsymmetric polymer-polymer interfaces. J Chem Phys 1975; 62: 1327–1331.

41. Merfeld GD and Paul DR. Polymer-polymer interactions based on mean field approximations. In: Paul DR, Bucknall CB, eds. Polymer Blends, Vol. 1. New York: Wiley, 2000: 55–91.

42. Sekelik DJ, Stepanov EV, Nazarenko S, Schiraldi D, Hiltner A, Baer E. Oxygen barrier properties of crystallized and talc-filled poly(ethylene terephthalate). J Polym Sci Part B: Polym Phys 1999; 37: 847–857.

43. Polyakova A, Liu RYF, Schiraldi DA, Hiltner A, Baer E. Oxygen-barrier properties of copolymers based on ethylene terephthalate. J Polym Sci Part B: Polym Phys 2001; 39: 1889–1899.

16

High Stiffness and High Impact Strength Polymer Composites by Hot Compaction of Oriented Fibers and Tapes

P.J. HINE and I.M. WARD

IRC in Polymer Science and Technology, School of
Physics and Astronomy, University of Leeds, UK

CONTENTS

I.   INTRODUCTION

The hot compaction of oriented fibers to produce large section products brings together two research themes that have been pursued at Leeds University for many years. First, there is the development of manufacturing procedures for high stiffness and high strength oriented polymers [1]. The routes chosen by Leeds have always involved deformation in the solid state either by tensile drawing [2] or die drawing [3] to very high draw ratios or by hydrostatic extrusion [4]. These developments have been matched by similar developments elsewhere, notably the gel spinning process invented by Smith and Lemstra [5] for high strength polyethylene fibers and the extensive studies of Porter and Weeks [6] using ram extrusion to produce very high stiffness and strength oriented polymers. Second, at Leeds there has been a program of research devoted to studies of oriented fiber composites, where the fibers have been either glass, carbon or high modulus polyethylene fibers [7].

      The hot compaction development seeks to overcome limitations identified in each of these research themes. Hydrostatic extrusion is a batch process operating at comparatively slow speeds, so that it is only commercially viable for specialist applications such as bone replacement where oriented products with similar stiffness to bone have been obtained [8,9]. Die drawing has been developed as a continuous process, operating at commercially acceptable speeds but usually limited to products of lower stiffness and strength than can be achieved in fibers by tensile drawing processes. It has been successfully developed for wires, ropes, pipes and products with comparatively simple cross-sections [10,11].

Fiber composites have the advantages of being able to utilize fibers with very high strength and high stiffness, but there is the major issue of recyclability because the matrix is conventionally of a different chemical composition to the fibers. Also, it is desirable to be able to make complex shapes by rapid post forming of flat sheets, rather than producing parts *in situ* during a resin-curing process. These requirements suggested the advantages of manufacturing thermoplastic composites with only one chemical composition.

In this chapter we describe the science and technology of hot compaction. The science followed from the discovery that it was possible in a simple process to make single polymer composites where the fibers and the matrix are of identical chemical composition [12]. It was found that an array of oriented fibers could be heated under suitable conditions of temperature and pressure so that only a thin layer on the surface of each fiber melts. On cooling, this material forms the matrix of the composite. This process of hot compaction has been shown to be very generally applicable to crystalline thermoplastic fibers, to liquid crystalline fibers and even to amorphous fibers. It is possible to find a suitable temperature window for commercial production of the composites where sufficient material is melted to form a cohesive structure with a strong bond between the oriented tapes or fiber, while retaining a substantial fraction of the original oriented phase, so that the composites can be of sufficiently high stiffness and strength.

The scientific studies have focused primarily on the melting and recrystallization during the hot compaction process and on the morphology and mechanical properties of the products. The technological developments were initiated by setting up a small incubation company under the auspices of University of Leeds Innovations. Vantage Polymers was a small limited company whose membership included the initial inventors I.M. Ward (Managing Director), P. J. Hine and K.E. Norris, together with D.E. Riley (Sales Director), M. Bonner, B. Brew and others. The commercial developments were initially supported financially by Hoechst-Celanese and then by Ford Motor Company who joined a consortium consisting of Ford, BP Amoco, BI Composites and Vantage Polymers. It was determined that polypropylene was a cost-effective polymer for the hot compaction technology and that automotive parts presented a potentially viable market. Other applications under development include sports goods and loudspeaker cones and (in polyethylene), radomes and protective helmets. Since 2001, large scale production facilities have been in operation at Gronau, Germany, under the auspices of AMOCO FABRICS Gmbh, who have taken a license for the technology from the British Technology Group, and the products are being marketed under the trade name CURV™.

The scientific studies detailed in the remainder of this chapter summarize our research work on a range of oriented semicrystalline polymers, including polyethylene, polypropylene and polyesters.

II. SEMICRYSTALLINE HOT COMPACTION RESEARCH STUDIES

A. Polyethylene

The discovery of ultra-high modulus (UHMPE) polyethylene fibers stimulated much research on the application of these fibers to produce fiber composites. Early research in this area focused on the melt spinning/high draw route using fibers produced commercially by the Celanese Fibers Company. Ladizesky and Ward [13] found that it was necessary to subject the fibers to a plasma-etching treatment in order to obtain adequate adhesion between the fibers and the epoxy or polyester resin matrix. It was shown that the combination of high strength and high extension to break of the UHMPE fibers leads to very high energy absorption of the UHMPE composites [14–17]. With the invention of the high-strength/high-modulus gel spun polyethylene (GSPE) fibers, GSPE fiber composites are now widely used in ballistic applications [18].

It was recognized by Capiati and Porter [19] that there could be merit in replacing the epoxy or polyester resin in polyethylene fiber composites by polyethylene to eliminate the problems due to lack of compatibility between the chemically inert polyethylene and conventional resins. There are several methods which have been reported to make such polyethylene fiber/polyethylene composites. Capiati and Porter utilized the difference between the melting points of highly oriented PE fibers and isotropic PE to embed a PE filament in a block of high-density polyethylene. For this model structure, high interfacial shear strengths were observed which could be attributed to the development of an epitaxially crystallized transcrystalline layer on the fiber surface.

Marais and Feillard [20], and later Marom and coworkers [21], followed this principle of different melting points for fiber and PE polymer to make unidirectional Dyneema fibers/PE composites and Spectrafibre/PE composites, respectively, by using HDPE films as the matrix and the gel spun fibers as the fabric reinforcement. In later developments, Hinrichsen et al. [22] and Lacroix et al. [23] produced similar composites by introducing the matrix by a powder impregnation process or by solvent impregnation of the fibers, respectively. Choy and coworkers [24] have also made PE fiber/PE composites by using HDPE fibers with a low melting, low density PE as the matrix.

The hot compaction process developed at Leeds University is unique in converting highly oriented fibers or tapes into a single polymer composite using only one starting component of single chemical composition. The initial discovery showed that it was possible to take an array of melt-spun high-modulus polyethylene fibers and, under suitable conditions of temperature and pressure, melt a thin layer on the surface of individual fibers which, on cooling, recrystallizes to form the matrix phase of a polyethylene fiber/polyethylene composite [12]. Because there is molecular continuity between the oriented-fiber phase (the core of the fibers which does not melt) and the recrystallized polyethylene-matrix phase, there is an excellent bond between the fiber and matrix phases so that the fiber matrix interface does not produce a line of weakness.

The first hot compaction studies were undertaken using the melt-spun fibers produced commercially by SNIA-Fibre under the trade name Tenfor [25] (these fibers are identical to those produced subsequently by Hoechst-Celanese under the trade name Certran). For the initial experiments the fibers were arranged unidirectionally in a matched metal mold and heated at a low pressure of 0.7 MPa (100 psi) to the appropriate temperature which can be called the hot compaction temperature. A higher pressure (21 MPa) was applied for 10 s before removing the mold from the press and allowing it to cool to ambient temperature. With the hot compaction temperature as the only variable, the products were monitored in several ways, using both mechanical measurements and structural measurements by DSC, wide angle x-ray diffraction and electron microscopy.

It was very quickly established that providing the comparatively low containment pressure was adequate and sufficient time was permitted to heat to the compaction temperature — the time under pressure could be standardized to 10 s and the compaction temperature was then the key variable. Figures 1 and 2 show the density and longitudinal flex modulus versus the compaction temperature, and Figure 3 shows DSC melting endotherms at 138°C, 140°C and 142°C. The DSC endotherms show very clearly that a fraction of the original fiber melts and recrystallizes. A subsidiary experiment showed that the lower melting peak is exactly in the position for the melted and recrystallized fiber, i.e., the peaks do not arise from two different chemical species in terms of molecular weight.

The results for density and longitudinal flex modulus taken together show that there is an optimum compaction temperature ca 138°C where there is excellent consolidation and a high retention of the fiber properties. The transverse strength, as might be expected, rises continuously with

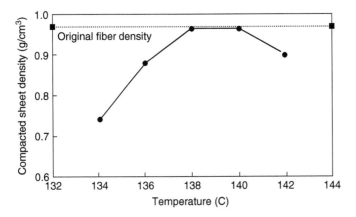

Figure 1 Compacted sheet density vs. compaction temperature for melt-spun polyethylene fibers.

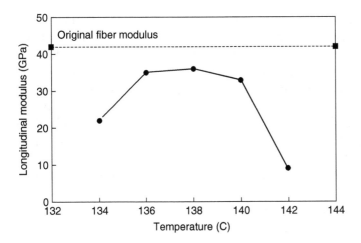

Figure 2 Longitudinal modulus vs. compaction temperature for unidirectionally arranged melt-spun polyethylene fibers. (From P.J. Hine, I.M. Ward, R.H. Olley and D.C. Bassett, J. Mater. Sci., 28, 316–324, 1993. With permission.)

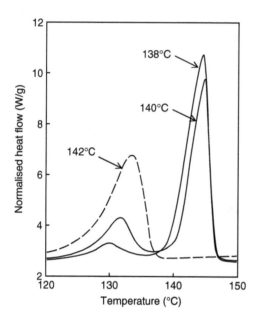

Figure 3 Typical DSC endotherms for compacted melt-spun samples made at 138, 140, and 142°C.

increasing temperature as more material is melted and recrystallized. The DSC results show that at 138°C only about 8% of the virgin fiber has been melted. This is very close to the theoretical fraction of material required to fill the gaps in a close-packed hexagonal arrangement of cylinders which is 9.3%. This encouraging result showed that only a comparatively small fraction of the original fiber had to be melted to give adequate compaction. It was subsequently found that when fabrics were subjected to hot compaction, a much greater fraction has to be melted to give a sufficiently coherent structure.

There was particular interest in studying the morphology of the hot-compacted materials and this was undertaken in collaboration with Professor David Bassett, Richard Olley and colleagues at Reading University [26,27]. Compacted sheets of 2.5 mm thickness were cut with a diamond knife at temperatures slightly below 0°C to expose internal surfaces, usu-

Figure 4 A scanning electron micrograph of a transverse section through an optimum compacted sheet of unidirectionally arranged melt-spun Certran PE fibers.

ally planes parallel or transverse to the fiber direction in the unidirectional fiber composites. These exposed surfaces were etched with a permanganic reagent and a two-stage replication process used, first making an impression of the etched surface in softened cellulose acetate and then shadowing this replica with tantalum-tungsten, followed by deposition of a carbon film and extraction of the shadowed carbon replica. Figure 4 is a scanning electron micrograph of a specimen cut transverse to the fiber direction. The original fibers can be seen to be mostly retained and the recrystallized material is neatly packed into the interstices of a closely packed fiber arrangement. A transmission electron micrograph of a section cut parallel to the fibers is even more revealing (Figure 5). There is recrystallized material both between and within the fibers and, in some instances as shown in Figure 6, the lamellae grow in three different directions, with epitaxial recrystallization from the surfaces of the fibers, two sets from the two longitudinal fibers seen in the micrograph and a third set growing normal to the plane of the photograph. It appears that the lamellae form from molten polymer in strain-free

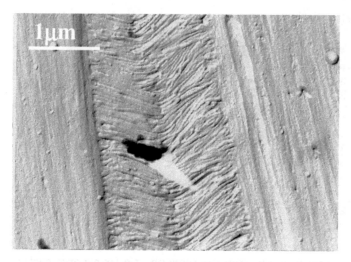

Figure 5 A transmission electron micrograph of a section cut parallel to the fibers in a compacted melt-spun PE sheet.

conditions but sharing the c-axis orientation of the fibers in which they nucleate giving good lateral properties to the compacted composite.

Further studies showed the differences between these fibers compacted at the optimum temperature and those at lower or higher temperatures [27]. At lower temperatures the original circular fibers were compressed towards regular hexagons with little interstitial recrystallized material. At higher temperatures, too much material was melted and crystallized so that although the lamellae start growing in a plane normal to the fiber axis and sharing its chain direction, banding develops as in the case of a conventional spherulitic structure.

Although the unidirectional fiber composites produced by hot compaction have a very high level of axial stiffness for most applications, it is better to produce the composites from a woven cloth so as to produce more balanced in-plane properties. Woven cloths with a plain weave construction produced from similar HMPE fibers (Certran™ supplied by Hoechst-Celanese) were subjected to the hot compaction process [28].

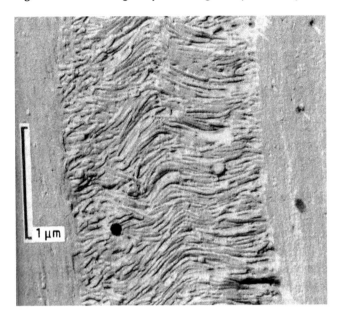

Figure 6 Longitudinal section of interstice between oriented fibers: the lamellar details identify the lamellae as growing in three different directions from three distinct fibers, under strain-free conditions. (From R.H. Olley, D.C. Bassett, P.J. Hine and I.M.Ward, J. Mater. Sci., 28, 1107–1112, 1993. With permission.)

As in the studies of unidirectional fiber arrays, the pressures and procedure times were held constant and the compaction temperature varied. In addition to measuring modulus, the peel strengths were determined by T-peel tests (following ASTM D1876). It was found that an adequate peel strength (~6N/10 mm) required about 20% of the melt and recrystallized phase, much greater than the 10% required to give a consolidated unidirectional fiber composite.

The tensile modulus data showed an optimum compaction temperature for the woven cloths about 2°C higher than that for the unidirectional composite. There is therefore a correspondingly narrower processing window because the temperature is closer to the melting point. Furthermore, because more of the initial fiber is melted, the maximum

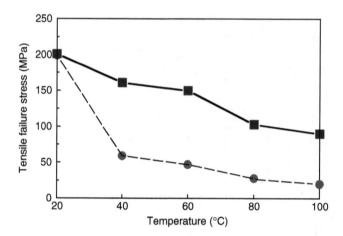

Figure 7   The relationship between tensile failure stress and testing temperature for compacted sheets of woven PE fibers: a) as
received fibers (--●--); b) crosslinked fibers (-■-).

modulus for the compacted woven cloth was only 9.9 GPa
compared with 36 GPa for the unidirectional composite.

It is well known that one of the limitations of UHMPE
fibers is their high creep and loss of stiffness and strength at
temperatures appreciably above ambient. Even for non-load
bearing applications such as an aircraft radome, it is necessary to withstand stresses in the range of 10 MPa at 70°C.
Previous research at Leeds University has shown that a good
method for reducing creep in UHMPE fibers is to subject them
to gamma irradiation in an atmosphere of acetylene which
promotes a chain reaction for cross-linking and hence reduces
the amount of chain scission [29,30]. Bonner et al. showed
that the cross-linking can be carried out on compacted sheets
or on the fabric prior to compaction and in both cases there
is a marked improvement in the elevated temperature performance [31] (see, for example, Figure 7). There is also an
additional advantage in cross-linking the UHMPE, as illustrated by Figure 8, which shows the percentage of the melting
peak remaining as a function of compaction temperature.
Whereas there is a very sharp onset of total melting in the
untreated fiber, the cross-linked material retains its fiber

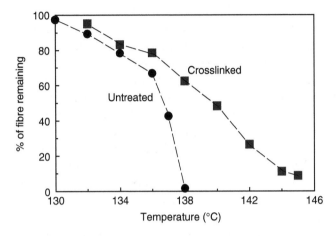

Figure 8   Percentage of fiber remaining versus compaction temperature for sheets of woven PE fibers: a) as received fibers (--●--); b) cross-linked fibers (-■-).

structure to much higher temperature, which has the considerable advantage of giving a much wider processing window. In fact, radomes have been very satisfactorily produced by hot compaction of cross-linked fabric in an autoclave at suitable temperature and pressure.

Following these studies of hot compaction of the melt-spun UHMPE fibers, it was of considerable interest to undertake similar research on the high-strength gel-spun UHMPE fibers Spectra® and Dyneema and on a more recent development, the melt-spun high-strength Tensylon® fibers. Leeds research on these materials is presented in several publications [32–34] and it is appropriate to describe the results in historical sequence, contrasting the different behavior of each of these fibers also with that of the melt-spun Certran (or Tenfor) fibers.

The hot compaction of gel-spun Spectra fibers was studied by DSC, scanning electron microscopy (SEM) and broad line nuclear magnetic resonance (NMR) over the temperature range 142–155°C [32]. The flexural properties of unidirectional fiber composites showed no change with increasing temperature up to 154°C, and the DSC and SEM studies revealed that no evident surface melting and recrystallization occurred within this range, although the rigid crystalline fraction measured by NMR was significantly lower than that for the original fiber. Significant transverse strength was also developed at these lower compaction temperatures, but this only increased markedly in the temperature range 154–155°C. In the temperature range 142–153°C the fibers do not melt but achieve reasonable transverse strength by a combination of mechanical interlocking and intermittent fusion at the fiber surfaces. At the high temperatures of 154–155°C, the fibers soften and the welding points become more evenly distributed. This localized welding is distinctly different from the general melting and recrystallization observed at the surfaces of the melt-spun Certran fibers.

The hot compaction of the Spectra and Certran is reviewed in a later paper [33] which also shows results for the Dyneema and Tensylon fibers.

In the case of the Dyneema fibers compaction at the optimum temperature requires appreciable melting of material in the interior of the fibers, and SEM-etched micrographs of these fibers reveal a higher melting skin. Similar to the Spectra fibers, successful hot compaction involves mechanical interlocking and "spot welding," but in this case there is considerable loss of mechanical properties due to internal melting.

Hot compaction of the Tensylon material, which is in the form of highly oriented tapes, is much more successful than in the case of the gel-spun fibers and is more akin to that in the melt-spun Certran fibers. SEM photographs (Figure 9) show that the tapes are well fused together with melted and recrystallized material between, with the recrystallized lamellae crystallizing epitaxially on to the oriented tapes as in the case of the melt-spun fibers. Although there is evidence of significant melting and recrystallization in the interior of the tapes, there is still good retention of the initial properties. Table 1 summarizes the properties of the optimum compacted woven samples for all four high modulus polyethylenes. As can be anticipated from the above discussion, Tensylon produces the best results and Dyneema the

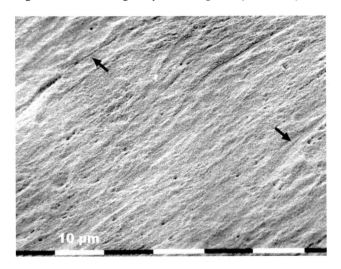

Figure 9 An SEM micrograph of a transverse section through an optimum compacted sheet of unidirectionally arranged melt-spun TENSYLON tapes (the black arrows indicate the tape/tape junctions. (From P.J. Hine, I.M.Ward, N.D. Jordan, R.A. Olley and D.C. Bassett, J. Macromol. Sci., Phys. B40, 959–989, 2001. With permission.)

TABLE 1 Properties of the Optimum Compacted Woven PE Samples

Reinforcement Name		Certran	Dyneema	Spectra	Tensylon
Optimum compaction temperature	°C	138.5	150.5	151	153
Original fiber/tape modulus	GPa	42	70	70	88
Compacted sheet tensile modulus	GPa	10.0	7	21	30
Compacted sheet ultimate tensile strength	MPa	160	250	460	400
Peel strength	N/10 mm	8	5.2	7.4	9

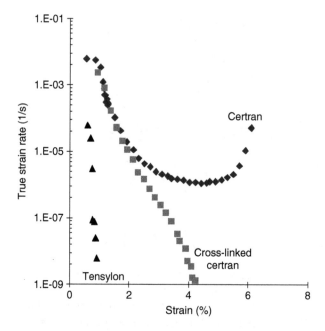

Figure 10 Sherby-Dorn creep curves for Certran (◆), cross-linked Certran (■) and Tensylon (▲). (From P.J. Hine, I.M. Ward, N.D. Jordan, R.A. Olley and D.C. Bassett, J. Macromol. Sci. Phys. B40, 959–989, 2001. With permission.)

worst, with the other two fibers in between but for different reasons. This comparison ignores the important issue of creep performance. Figure 10 shows the Sherby-Dorn creep curves for Certran, cross-linked Certran and Tensylon. Although both cross-linked Certran and Tensylon show satisfactory creep behavior, Tensylon has the major advantage of much higher stiffness, hence lower deformation. Tensylon also showed an optimum compaction temperature 2°C below its crystalline melting point, compared with 1°C for the other three UHMPE fibers. This gives a wider processing window for this material which can only be broadened by crosslinking in the case of the other fibers.

In a further study of the hot compaction of Tensylon tapes and woven cloths [34], it was shown by scanning electron microscopy on permanganic-etched samples that the Tensylon tapes are made up of ribbon-like units which show a two-component morphology both before and after hot compaction with both internal and surface melting. These results confirm that the Tensylon melting behavior is more akin to that of Dyneema and Spectra than the melt-spun Certran.

B.   Polypropylene

Although the hot compaction of UHMPE fibers was the starting point for all the compaction studies, it became clear that from a commercial viewpoint, the process would be much more valuable if it could be applied to lower-cost fibers or tapes. Polypropylene is an ideal candidate for producing cost-effective hot-compacted products, so there has been much attention given to both the scientific and technological aspects of hot compaction of oriented polypropylene fibers and tapes. As in the case of the PE fibers, the research has involved a combination of hot compaction in a heated press with the compaction temperature as the key variable but with somewhat higher pressures (at least 400 psi) to ensure that the shrinkage forces developed did not permit shrinkage of the fibers or tapes so that the significant loss of stiffness and strength occurred due to relaxation of the molecular orientation. Mechanical measurements and DSC measurements were combined with electron microscopy of etched samples, the latter studies being undertaken at Reading University by Professor David Bassett, Robert Olley and their colleagues.

The first exploratory studies of polypropylene hot compaction were somewhat disappointing [35]. High tenacity polypropylene (PP) fibers were compacted in a unidirectional array in an open-ended matched metal mold. It was found that at the comparatively low temperature of 174°C the fibers became too soft and extruded from the mold so that surface melting and recrystallization could not be achieved. However, a later study [36], using a woven PP-fibrillated tape commercialized by Milliken Industries USA, for geotextile applications, was very successful and has formed the basis for the future technological developments, as well as initiating further very interesting science. Whereas the uniaxial array of fibers in an open-ended mold could only be heated to 174°C, the woven fibrillated tape could be heated in a closed mold up to temperatures of 190°C. Mechanical measurements of in-plane flexural stiffness, combined with the morphological studies showed that at an optimum compaction temperature of 182°C, good tape-to-tape bonding and good interlayer bonding between the tape layers could be achieved, with good retention of the stiffness and strength of the initial tapes. At this temperature, a homogeneous and well-bonded sheet was produced by a process of surface melting and recrystallization while retaining a significant fraction of the original structure. The morphological studies confirmed that the good mechanical properties are due, as in the case of the melt-spun Certran PE fibers, to the development of a transcrystalline layer between the tapes and tape bundles in the woven cloth (Figure 11).

A further study [37] explored the morphology of the hot-compacted woven tapes over a range of hot-compaction temperatures. With increasing temperature, the interior structure of the tapes was seen to undergo progressively greater melting and recrystallization in the form of shish-kebab structures. The regions of recrystallized extra-fibrillar melt showed effects of flow-induced crystallization in the form of row structures (Figure 12). These effects are, of course, only seen at very much higher temperatures than the optimum. Optimum compaction is achieved when sufficient polymer melts just to fill all the cavities in the structure.

After these initial studies on the hot compaction behavior of PP, the research focused on compacted sheets of woven oriented PP tapes and fibers [38] because it was clear that for practical applications, the woven cloths were much easier to handle and the compacted sheets showed a satisfactory balance of properties because these are essentially orthotropic. An interesting aspect of the compaction of PP is the beneficial effects of heating to the compaction temperature with respect to annealing the initial structure. Figure 13 shows a comparison between the melting behavior of the original PP tape and a hot-compacted sheet. There is an overall increase in crystallinity, and it is also important that in the sheet there is significant superheating due to the melting

Figure 11 Etched SEM micrograph of compacted polypropylene showing details of epitaxial crystallization (the arrow here indicates the boundary between an original oriented tape and the recrystallized material). (From J. Teckoe, R.H. Olley, D.C. Bassett, P.J. Hine and I.M.Ward, J. Mater. Sci., 34, 2065–2073, 1999. With permission.)

under constraint which gives a significant increase in the melting point.

A more substantial study of the hot compaction of woven-oriented PP fabric and tapes is described in two complementary publications [39,40] where the behavior of five different woven cloths is compared. The key variables include fabric weight, weave style, tape count (per 10 cm) and degree of crimp, in addition to properties of the tapes such as modulus, strength and molecular weight. The key properties of the compacted sheets include initial tensile modulus, ultimate tensile strength and the interlayer adhesion (usually called

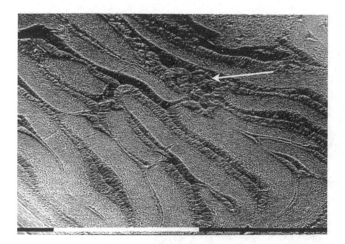

Figure 12 Etched SEM micrograph of compacted polypropylene fibrillated tapes showing crystallization nucleating between the tapes (white arrow) as well as at the tape surfaces. (From J. Teckoe, R.H. Olley, D.C. Bassett, P.J. Hine and I.M.Ward, J. Mater. Sci., 34, 2065–2073, 1999. With permission.)

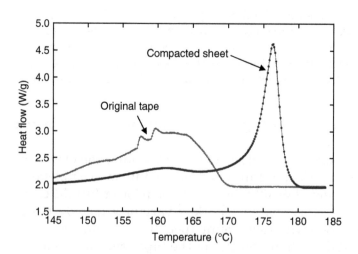

Figure 13 A comparison of the melting endotherms of an oriented PP tape and a hot-compacted sheet.

the peel strength) which is a measure of the practical deformability of the sheets.

One of the most important conclusions drawn from this study was that the hot-compacted sheets can be considered to be a fiber composite, with the unmelted fraction of the tapes or fibers as the fiber phase bound together by a matrix phase, formed by melting and crystallization of the surfaces of the original tapes or fibers. From a structural viewpoint, the principal effect of the compaction process is the melting and recrystallization of the matrix phase and the structure of this phase has important consequences for the behavior of the compacted sheet. To study this, films were produced by melting out samples of the original cloths and adopting three different cooling procedures; first, quenching into cold water; second, cooling at ~20°C/min; and third, slow cooling at ~2°C/min. It was found that with regard to the subsequent mechanical properties of the sheets, molecular weight was important, as well as the cooling procedure. Most importantly, fast cooling produces lower crystallinity and more tie molecules, leading to higher strengths as established previously for PE [41]. Also, as can be anticipated, high molecular weight is required to ensure a high degree of ductility.

The ductility of the matrix phase was found to be reflected in the peel strengths of the hot compacted sheets. For example, compacted samples based on low molecular weight showed lower average peel strengths than those based on ductile higher molecular weight polymers (at the same cooling rate used in the production of the composite). The hot compacted PP composites are very different from conventional fiber or tape composites because the failure strain of the oriented fibers or tapes is higher than that of the matrix phase. This means that it is vital to have a high strain to failure in the matrix phase so that the composite retains its integrity up to the failure point of the reinforcement. This is achieved by using a high molecular weight polymer and adopting a fast cooling rate.

The mechanical properties of the sheets can be considered in terms of a simple rule of mixtures, adding the properties of the fiber and matrix phase in their relative proportions. This implies that for polypropylene where the properties of the fiber phase are comparatively modest (compared with, e.g., PE), both phases make a significant contribution to the stiffness.

The morphological studies were especially informative with regard to differences due to polymer molecular weight. Lower molecular weight PP grades tend to exude much more material than high molecular weight during the hot compaction process. This material recrystallizes to give regions of lower mechanical strength than the remaining fiber, especially where there are opposing transcrystalline layers with a junction which offers a path of low resistance to peeling. Higher molecular weight materials tend to weld tape to tape. Peel cracks then disrupt the tapes themselves, giving extensive fibrillation and superior mechanical properties.

Weave style and fiber geometry are important, especially the fact that tapes present a greater surface for compaction than fibers which have to be distorted by compression to make optimum contact.

C.   Polyesters

1.   Polyethylene Terephthalate (PET)

Although the studies on the compaction of PET are at an earlier stage compared to both polypropylene and polyethylene, PET is of interest because it is a commercially available oriented polymer with good mechanical properties: it was therefore very appropriate to examine its hot-compaction behavior.

The initial studies on PET were carried out on unidirectional arrangements of fibers [42]. Tyre cord yarn supplied by ICI plc (modulus 15 GPa) was compacted using a two-stage pressure process in an open-ended matched metal mold. In this procedure the compaction assembly was placed into the hot press set at the compaction temperature and a pressure of 1.85 MPa was applied: once the assembly reached the compaction temperature, it was left to dwell for 15 min at which time the pressure was increased to 32.4 MPa and then cooled. As with the first polyethylene studies, the initial target

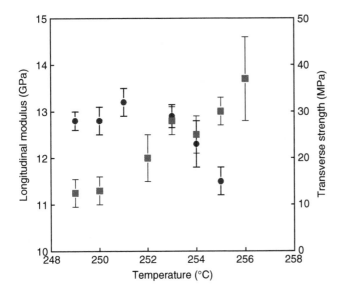

Figure 14 Longitudinal modulus (●) and transverse strength (■) vs. compaction temperature for unidirectional compacted samples of PET fibers.

here was to establish the optimum temperature for compaction by contrasting the changes in the longitudinal modulus with the increase in transverse strength, as the amount of melted and recrystallized matrix material increases with increasing compaction temperature. The results, shown in Figure 14, indicate that as the compaction temperature is increased the longitudinal modulus passes through a maximum (as also seen with compacted unidirectionally arranged polyethylene fibers – Figure 2) whereas the transverse strength increases monotonically. The trend of these results is very similar to that measured for the melt-spun polyethylene fibers described earlier in section II.A. In fact it was possible to show [43] that by normalizing the temperature axis to the melting range of each fiber, the transverse strength results for PE and PET were very similar (Figure 15).

The results in Figure 14 indicate that the optimum temperature for the compaction of these particular PET fibers is 254 ± 1°C, where a good level of bonding is achieved without

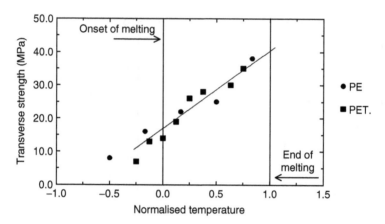

Figure 15 Transverse strength of unidirectional compacted samples of PE and PET fibers normalized with respect to the melting range of the fibers. (From J. Rasburn, P.J. Hine, I.M. Ward, R.H. Olley, D.C. Bassett and M.A. Kabeel, J. Mater. Sci., 30, 615–622, 1995. With permission.)

losing too much of the properties of the original oriented fibers. Morphological investigation of the PET-compacted samples, using the same techniques as for the polyolefins, initially proved unsuccessful, as it was found that the PET melted matrix material was completely dissolved by the permanganic-etching reagents: for this reason, cut but unetched transverse sections were then prepared. Figure 16 shows a picture of such a section from a sample compacted at 255°C, imaged using Nomarski interference contrast optics. The picture clearly illustrates that the morphology is similar to that of the compacted polyolefins described earlier, with the regions between the fibers filled with melted and recrystallized material.

A more recent series of experiments investigated the hot-compaction behavior of woven PET multifilaments supplied by Hoechst-Celanese [44]. This study highlighted the major challenge with the hot compaction of PET, in contrast to the polyolefins, which is the well-known mechanism of hydrolytic degradation, researched previously both in this department and by other workers [45–48]. Importantly, at the tempera-

Figure 16 Cut but unetched transverse section through a PET compacted sample (255°C). (From J. Rasburn, P.J. Hine, I.M. Ward, R.H. Olley, D.C. Bassett and M.A. Kabeel, J. Mater. Sci., 30, 615–622, 1995. With permission.)

tures required for successful hot compaction of PET, significant chain degradation can occur, often very rapidly: for example, at a temperature of 255°C, the molecular weight can drop by 50% in 20 min compared to 18 h for the same percentage fall at 140°C.

Compacted samples were made using a temperature of 255°C with a range of dwell times from 1 to 15 min. Subsequently, intrinsic viscosity measurements were carried out on the compacted samples, and the results (Figure 17) showed a significant fall in the number average molecular weight, M_n, with dwell time. Mechanical tests on the same series of samples showed that this decrease in molecular weight, due to hydrolytic degradation, was allied to a significant embrittlement of the samples: a 2-min dwell time showed a failure strain of 12%; for a 15-min dwell time the failure strain was only 2%. The key issue with the compaction of PET is therefore to keep the oriented elements at the compaction temperature for the shortest time to minimize hydrolytic degradation while still attaining even melting throughout the assembly of oriented elements: 2 min was measured to be the optimum dwell time.

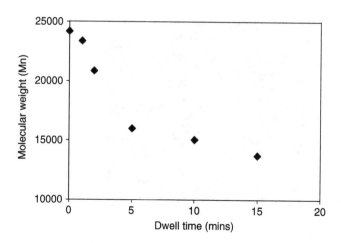

Figure 17 M_n vs. dwell time for compacted woven PET sheets. (From P.J. Hine and I.M. Ward, J. Appl. Polym. Sci., 91, 2223–2233, 2004. With permission.)

Having established the optimum compaction conditions, more detailed studies were carried out using DSC and DMTA techniques. Measurements of the glass transition temperature of the isotropic, oriented and compacted PET polymer provided confirmation of the ideas developed during the polypropylene studies, of considering the hot compacted sheets as a polymer/polymer composite whose properties depend on a mix of the properties of the constituents. It is well known that the glass transition temperature of PET is raised during the drawing process used to make the oriented filaments, due to changes in chain conformation and crystallinity. During the hot compaction process, a fraction of the skin of these oriented elements is melted, and this material then reverts to the isotropic configuration. Figure 18 shows the glass transition (T_g) of three PET samples, an oriented fiber, a piece of completely melted fiber (taken as equivalent to the "matrix" material of the hot-compacted sheet) and a piece of the hot-compacted sheet. It is seen that the original oriented fibers show a higher T_g compared to the isotropic (or melted) material (120 and 67°C, respectively) and that the compacted sheet shows two peaks, one for the matrix material

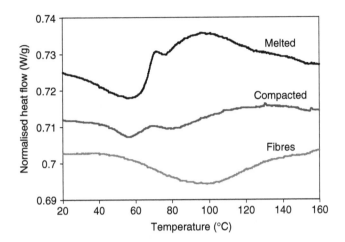

Figure 18 Glass transitions of the various PET materials. (From P.J. Hine and I.M.Ward, J. Appl. Polym. Sci., 91, 2223–2233, 2004. With permission.)

component (62°C) and one for the remaining oriented fibers (99°C). The ratio of the specific heats of the two peaks seen in the compacted material, compared to the 100% values, can give a value for the fraction of each phase, which in this case was determined to be 24% matrix material and 74% remaining oriented fibers, which is a typical ratio for optimum hot-compacted woven samples. Interestingly the glass transition temperature of the oriented fiber peak in the compacted sheet is lower than that measured for the original fibers suggesting some shrinkage and morphological changes in the oriented fibers during the compaction procedure.

    Another key finding of the PET studies was confirmation of the importance of ductility of the matrix phase. As with the polypropylene work, ductility of the melted and recrystallized matrix phase was found to be promoted by fast cooling after compaction, reducing crystallinity and hence promoting ductility.

    After carrying out a full set of mechanical tests, and assessing the results in comparison to previous PP studies, the final conclusion of the PET studies was that the optimum compacted PET samples (255°C, 2-min dwell time, fast cooled) have roughly comparable mechanical properties to compacted PP at room temperature (+20°C) and at +120°C, but much better properties at temperatures between these two limits, offering the potential of improved creep performance in this range. The mechanical tests also showed that compacted PET sheets had a more linear stress-strain curve compared to PP such that at 2% strain, PET had almost twice the stiffness of PP. In addition, compacted PET also had a lower thermal expansion than compacted PP.

2.   Polyethylene Naphthalate

Polyethylene naphthalate (PEN) has the potential for an even greater improvement in performance over PP because oriented fibers can be produced with a significantly higher stiffness compared to PET [49] and PEN also has a much higher glass-transition temperature. In addition, previous work in this department [45] has shown that PEN has improved hydrolytic stability compared to PET due to a combination of lower water absorption and lower ester linkage concentration. For this combination of reasons it was logical to assess the potential of hot-compacted PEN. Sheets were manufactured by compacting bidirectional arrangements of fibers (supplied by KOSA Gmbh) over a range of temperatures around the melting range (bidirectional fiber arrangements are made by winding around a metal plate in a 0/90 configuration to give balanced in-plane properties when compacted). Tensile tests were used to determine the mechanical properties of the samples made at various temperatures in order to determine the optimum processing conditions, and DSC was used to determine the percentage of the original oriented fibers lost during the process, the fraction of matrix material produced and any changes to the morphology as a result of the process.

    The results obtained in this study, described in detail in a recent publication [50], showed that the compaction behavior was very similar to the other semicrystalline polymers studied, particularly PET, as would be expected. The optimum compaction temperature, where good interfiber bonding was achieved without significant loss of the oriented phase, was

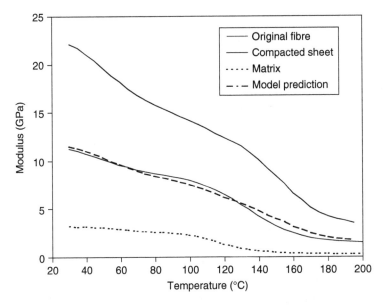

Figure 19   A comparison of the measured and predicted temperature performance of the compacted PEN sheet modulus. (From P.J. Hine, A. Astruc and I.M. Ward, J. Appl. Poly. Sci., 2004 in press. With permission.)

271°C. DMTA measurements of the glass transition temperature, taken as the peak in tanδ, gave very similar results to PET, with the compacted sheet showing a glass transition temperature between that of the original oriented fibers and the melted "matrix" material. The analogy with traditional composite ideas, where the properties of the composite material are predicted based on the properties and fractions of the component phases using simple mixing rules, was further strengthened by these studies on the compacted PEN fibers. Figure 19 shows DMTA measurements on the original oriented PEN fiber, the melted PEN matrix material and the compacted PEN composite. As expected the properties of the PEN composite lie between the properties of the two components. Moreover, it is possible to make a quantitative comparison. If we assume strain continuity between the phases, then a simple mixing rule can be developed as shown below by

$$E_{compacted} = \left(\frac{E_L}{2}\right) \cdot \left(V_O\right) + E_M \frac{\left(1 + V_M\right)}{2} \tag{1}$$

where E_L is the longitudinal modulus of the oriented fibers, E_M the modulus of the melted and recrystallized matrix and V_O and V_M the fractions of the oriented and melted phases. Taking the fraction of the two phases from DSC tests, the model predictions are shown as the dotted line on Figure 19, which is seen to be in excellent agreement with the measured compacted sheet values.

A key aspect of this study was to compare the properties of compacted PEN with PET. Table 2 shows a comparison of the important mechanical properties of these two compacted polyester sheets. Compacted PEN is seen to have significantly improved modulus, strength and temperature performance compared to PET. The failure strain is lower than PET, suggesting that there could be potential drawbacks in terms of impact performance and thermoformability which are yet to be addressed in any detail. However the results indicate that PEN shows great potential as a compacted sheet, with a significant performance improvement over both PP and PET.

D.   Nylon 66

As with the polyesters, oriented nylon fibers are widely commercially available, so it was a logical step to assess the hot-compaction behavior of this polymer [52]. Woven cloth, made from nylon 66 multifilaments, was obtained from a commercial source. Compaction experiments were carried out by placing layers of the woven nylon cloth between metal plates and compacting in a hot press at a pressure of 400 psi using a dwell time of 2 min to restrict any possible degradation as with the polyester studies described earlier. The optimum compaction temperature for the nylon 66 cloth was found to be 261°C. For nylon 66 the peak melting positions of the original fiber and the melted and recrystallized material were found to be quite close together, at 261 and 265°C, respectively. For this reason, DSC melting endotherms of the compacted materials did not show the clear demarcation

TABLE 2   A Comparison of the Properties of the Optimum
Compacted Samples of PEN and PET

	Compaction Temperature (°C)	Tensile Modulus (GPa)	Tensile Strength (MPa)	Tensile Failure Strain (%)	Modulus at +120°C (DMTA) (GPa)	Tg (°C)
PEN	271	9.6	207	5	6.53	149
PET	257	5.8	130	11	1.58	110

between the two components seen with other materials such
as polyethylene and polypropylene (Figures 3 and 13, respectively). It is well known that nylon absorbs significant
amounts of water due to hydrogen bonding. For this reason
an optimum sample of compacted nylon 66 was tested immediately and then after equilibriating at room temperature
and humidity (50% RH) for 2 weeks, which resulted in ~2%
water uptake. Figure 20 shows a comparison of the tensile
properties of the dry and wet compacted nylon sheets. As
expected, the water uptake affects those properties which

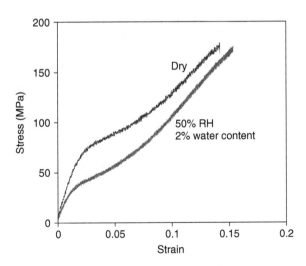

Figure 20 A comparison of the tensile stress-strain behavior of
dry and wet compacted nylon 66 sheet.

depend on local chain interactions (e.g., modulus and yield strength) but has less effect on those properties that depend on large scale properties of the molecular network (e.g., strength). The modulus and strength of the compacted wet nylon sheet (2.8 GPa and 150 GPa) are comparable to compacted polypropylene sheets and the peel strength, a measure of the bonding between the woven cloth layers, was measured as among the highest ever measured at a value of 23N/10mm, probably a consequence of the high cohesive strength of isotropic nylon. The drawback of the compacted wet nylon sheet, is the elevated temperature performance, with the modulus dropping to almost zero at a temperature of 80°C. However, if elevated temperature is not an issue, compacted nylon sheet shows good mechanical properties.

E. Other Semicrystalline Polymers: Polyetheretherketone (PEEK), Polyphenylene Sulphide (PPS) and Polyoxymethylene (POM)

Throughout the hot-compaction studies, the target has always been to produce a compacted sheet with the ultimate combination of mechanical properties in terms of stiffness, strength and elevated temperature performance. To this end we have evaluated the hot-compaction behavior of three higher performance oriented engineering thermoplastics, namely polyphenylene sulphide (PPS), polyetheretherketone (PEEK) and polyoxymethylene (POM).

1. Polyphenylene Sulphide (PPS)

PPS was chosen for evaluation because it is perceived as having good elevated temperature performance, excellent chemical resistance, very low shrinkage and does not show hydrolytic degradation. The grade chosen for the study was FORTRAN 0320, manufactured by Ticona. Commercial oriented material was not available in this case, so drawn filaments were produced in-house following conditions established in a previous study [52]. Drawn tapes, with a modulus of 5.7 GPa, were compacted using the bidirectional (0/90) arrangement of fibers as used in the PEN studies described earlier, by winding them around a metal plate in a 0/90 configuration: the optimum compaction temperature was found to be 288°C. Samples of isotropic material were also tested to establish the properties of the matrix phase of the compacted PPS sheet. Isotropic sheets showed a very high modulus, of 4.2 GPa, suggesting that the matrix phase will make a significant contribution to the properties of the compacted composite. Measurements on the optimum compacted sheets gave a tensile modulus of 5.2 GPa, which is a good value, but a tensile strength of only 80 MPa and brittle behavior at a failure strain of 8% [53]. It is conceivable that the grade chosen for these preliminary studies is not optimum, but these results do give an indication of the potential of compacted PPS. DMTA tests showed the expected temperature performance, with a tensile modulus similar to compacted polypropylene at +20°C and a tensile modulus similar to PET at 100°C. However, like PET, PPS has a glass transition temperature around 100°C so temperature performance above 120°C is limited.

2.   Polyetheretherketone (PEEK)

Another polymer which offers the prospect of elevated temperature performance is PEEK. Compaction experiments were carried out on woven cloth supplied by Victrex, UK [54]. As in previous compaction studies using woven cloth, layers of the cloth were placed into a matched metal mould and compacted at a range of temperatures around the melting temperature of PEEK, which DSC indicated as ~345°C. Compaction experiments were carried out at a pressure of 700 psi and a dwell time of 10 min. The optimum compaction temperature was found to be 347°C, which gave a composite comprising 17% melted and recrystallized matrix phase and 83% of the remaining original oriented woven fibers (measured by DSC).

Tensile tests on the optimum samples showed an initial modulus of 3.65 GPa, with a long linear region up to a strain of a few percent followed by a yield point and ultimate failure at a stress of ~100 MPa and a failure strain of >20%. DMTA tests showed that the tensile modulus fell only slightly until the glass transition region of ~163°C. In fact PEEK has the highest glass transition of all the polymers evaluated in this research program, which is reflected in the properties of the compacted sheet.

3.   Polyoxymethylene (POM)

POM is a suitable candidate for a hot-compacted sheet because highly oriented tapes and filaments can be produced which have properties which are only bettered by polyethylene. These compaction studies [55] drew on the considerable experience in this department in this research area [56,57] both in the choice of the grade of polymer (Delrin 500 NC010) and the drawing procedures. The best oriented tapes, with a modulus of ~22 GPa, were produced by a two-stage drawing process: the first stage at 145°C and a second stage at 155°C with a total draw ratio of ~13. As with the studies on PEN and PPS, where woven material was not available, compacted sheets were produced by winding the drawn tapes around a metal plate in a 0/90 configuration. Tests over a range of temperatures showed that the compaction processing range for POM, which is highly crystalline, is quite narrow and that the optimum compaction temperature was 182°C.

Tensile tests on the 0/90 compacted POM sheets showed excellent properties, with a tensile modulus of 10 GPa and a strength of 280 MPa, reflecting the excellent properties of the original oriented POM tapes. With such a high room temperature modulus, it was no surprise that the elevated temperature performance of the compacted POM sheets was excellent, with a value of the modulus at +120°C of 6.7 GPa, higher than most of the other compacted polymer sheets studied at 20°C. However, above this temperature the properties dropped significantly due to the lower melting temperature of this polymer.

III. CONCLUSIONS

The main conclusion from the various studies detailed above is that the properties of the final compacted sheet are determined by the properties of the two component phases (the original oriented phase and the created melted and recrystallized matrix phase) and the proportions of each. The goal is always to discover the best combination of processing conditions, most notably the compaction temperature, in order to produce enough melted matrix material to bind the oriented elements together and form a homogeneous composite material, while retaining as high a percentage as possible of the oriented component. The optimum amount of melted material is seen to lie between 20 and 30% (depending on the geometry of the oriented component), which in the worst case gives a composite with a 70% reinforcement fraction, higher than can be achieved with any other processing technique, maximizing the contribution from the oriented elements.

There is also a clear link between the properties of the two phases and the properties of the final compacted sheet. If the oriented elements are highly drawn, and have a high stiffness and strength, then the resulting compacted sheet will also be stiff and strong. Conversely if the isotropic polymer has a high cohesive strength, then it will be a good "matrix" material or glue to bind the structure together. While it is an advantage to produce the matrix material from the surfaces of the oriented elements, forming a composite made from chemically the same polymer and with molecular continuity between the phases, perhaps the weakness in this approach is that it does not allow the optimum properties to be chosen for the oriented and fiber phases, because the optimum chemical composition for drawing to the highest draw ratio, may not have the highest cohesive strength.

Table 3 summarizes the properties of the different compacted materials described in the earlier sections, and Figures 21 and 22 show details of the stress-strain behavior and dynamic temperature behavior, respectively. The results in both the table and the two figures strengthen the idea of the sheet properties being dependent on the properties of the two constituent phases. As described in the section on PEN, we have used the simple rule of mixtures (assuming continuity of strain between the components) to predict the compacted sheet properties which are given by Equation 1, where E_L is

TABLE 3 A Comparison of the Properties of the Optimum Compacted Sheets for Each Polymer Type

	PE (Tensylon)	PP (Curv™)	PET	PEN	Nylon 66 (Wet)	PPS	PEEK	POM
Oriented phase type	Tapes	Tapes	Fibers	Fibers	Fibers	Tapes	Fibers	Tapes
Oriented phase arrangement	Woven	Woven	Woven	0/90	Woven	0/90	Woven	0/90
Compaction temperature (°C)	153	191	258	271	261	288	347	182
Oriented phase modulus (GPa) E_L	88	11	14	22	5.8	5.7	7	22
Matrix phase modulus (GPa) E_M	0.5	1.2	2.8	3.3	1.9	4.2	2.5	3.2
Initial compacted sheet modulus (GPa)	30	5	5.8	9.6	2.8	5.2	3.5	10
Predicted composite sheet modulus (GPa) from rule of mixtures (strain continuity) assuming an oriented fraction of 0.75	33.1	4.9	7	10.3	3.4	4.8	4.1	10.2
Compacted sheet failure strength (MPa)	400	182	130	207	150	80	100	280
Compacted sheet failure strain	2	15	10	6	15			6
Peel strength N/10mm	9	8	18		23			
Density (kg/m³)	980	910	1400	1410	1140	1350	1310	1420

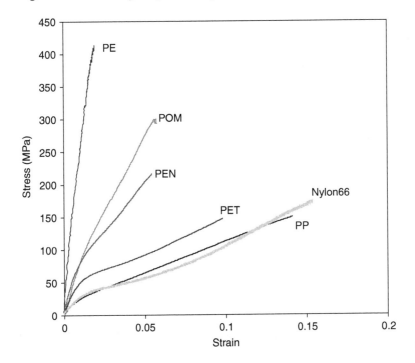

Figure 21 Stress-strain curves of optimum compacted sheets.

the longitudinal modulus of the oriented fibers, E_M the modulus of the melted and recrystallized matrix and V_O and V_M the fractions of the oriented and melted phases. It is assumed that the transverse modulus of the oriented component is the same as the matrix modulus E_M, which we have found to be a reasonable assumption for most polymers. The model also assumes that there is no crimp when woven fibers or tapes are used: as such the model gives an upper limit to the predicted compacted sheet modulus.

The fourth and fifth lines of Table 3 show the values of E_L and E_M for the various polymers. In all the systems studied, the optimum percentage of matrix materials has been found to be between 20 and 30%: the model predictions for the compacted sheets have therefore been calculated for an oriented volume fraction of 0.75. The sixth line of the table then shows the measured compacted sheet moduli while the sev-

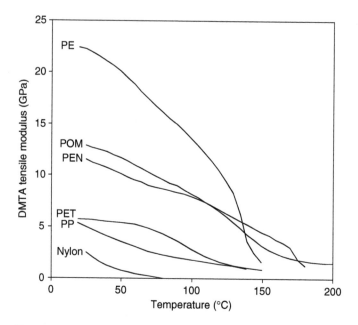

Figure 22 DMTA temperature scans on optimum hot-compacted sheets.

enth line shows the model predictions based on the phase properties. It is seen that the model predictions are in general an upper limit prediction, and that the agreement between measured modulus and predicted modulus is excellent. PET shows the greatest difference and this is thought to be due to two effects: the woven PET cloth was composed of thick multifilament bundles and so the crimp in the cloth was high; second, the results detailed in Section II.C.1 showed that the orientation and properties of the PET fibers were affected by exposure to the compaction temperature, so the input values for the fiber, based on the original properties, could be an overestimate. An identical model was used to predict the compacted sheet strength (here the oriented component dominates) and a similar good agreement was found.

Also shown in the table are the values of the sheet strength and failure strain, and Figure 21 shows the complete stress-strain curves for each polymer. Once again, the stress-strain curves are related to the properties of the component phases, with the most highly oriented polymer (PE – Tensylon), showing the highest values of stiffness and strength, accompanied by the lowest failure strain. Two other important parameters are shown in the table, the peel strength and the compacted sheet density.

Finally, Figure 22 shows the temperature performance of the various compacted sheets. The temperature performance will depend on a number of factors, including the glass transition temperature of the polymer, T_g, the melting temperature, the level of crystallinity and the degree of orientation (high crystallinity and orientation will suppress the effect of T_g). The results show that only PEN and POM have a significant modulus above 150°C.

The overall conclusion to be drawn from this research is that in terms of exploitation, then, the best choice of polymer for a hot-compacted sheet would have a low density, a high strength and stiffness and a high failure strain (giving both high impact performance and thermoformability), a high glass transition temperature, medium to high crystallinity, low melting point (for ease of processing) and a high cohesive strength (giving good bonding and a high peel strength). Because no single polymer will have all these attributes, the potential choice will depend on the performance portfolio required for a particular application. Hopefully, we have shown that the hot-compaction process is applicable to a wide range of oriented semicrystalline polymers, both commercially available and produced in-house, giving a range of new materials with a wide range of performance.

ACKNOWLEDGMENTS

The hot compaction project has been a team effort, and we wish to acknowledge the research and development input from our colleagues who have worked with us in this area: in particular K.E. Norris, D.E. Riley and M.J. Bonner for the work on polyethylene and polypropylene, J. Rasburn for the original work on PET, Celine Kermarrec for the work on PPS, Alexandre Astruc for the work on PEN and Mathieu Goode for the work on PEEK. We also wish to acknowledge financial support from EPSRC, BTG, Hoechst-Celanese, Ford Motor Company and BP Amoco Fabrics GmbH.

REFERENCES

1. I.M. Ward, Solid Phase Processing of Polymers, eds. I.M.Ward, P.D. Coates and M.M. Dumoulin, Hanser, Munich, 2000, Chapter 5.

2. G. Capaccio and I.M. Ward, Properties of ultra-high modulus linear polyethylenes, Nature Phys. Sci., 243, 130–143 (1973).

3. P.D. Coates and I.M. Ward, The plastic deformation behaviour of linear polyethylene and polyoxymethylene, J. Mater. Sci., 13, 1957–1970 (1978).

4. A.G. Gibson, I.M. Ward, B.N. Cole and B. Parsons, Hydrostatic extrusion of linear polyethylene, J. Mater. Sci., 9, 1193–1196 (1974).

5. P. Smith and P.J. Lemstra, Ultra-high strength polyethylene filaments by solution spinning/drawing, J. Mater. Sci., 15, 505–514 (1980).

6. N.E. Weeks and R.S. Porter, Mechanical properties of ultra oriented polyethylene, J. Polym. Sci., Polym. Phys. Ed., 12, 635–643 (1974).

7. P.J. Hine, N. Davidson, R.A. Duckett and I.M. Ward, Measuring the fiber orientation and modeling the elastic properties of injection-molded long-glass-fiber-reinforced nylon, Composites Sci. Technol., 53, 125 (1995).

8. N.H. Ladizesky, I.M. Ward and W. Bonfield, Hydrostatic extrusion of polyethylene filled with hydroxyapatite, Polym. Adv. Tech., 8, 496–504 (1997).

9. N.H. Ladizesky, I.M. Ward and W. Bonfield, Hydroxyapatite/high-performance polyethylene fiber composites for high-load-bearing bone replacement materials, J. Appl. Polym. Sci., 65, 1865–1882 (1997).

10. A. Richardson, B. Parsons, I.M. Ward, Production and properties of high stiffness polymer rod, sheet and thick monofilament oriented by large-scale die drawing, Plastics Rubb. Proc. Applns., 6, 347–361 (1986).

11. A. Selwood, I.M. Ward and B. Parsons, The production of oriented polymer tube by the die-drawing process, Plastics Rubb. Proc. Applns., 8, 49–58 (1987).

12. I.M. Ward, P.J. Hine and K. E. Norris, Polymeric Materials, British Patent Office GB2253420, March 1992.

13. N.H. Ladizesky and I.M. Ward, A study of the adhesion of drawn polyethylene fibre/polymeric resin systems, J. Mater. Sci., 1983, 18, 533–544.

14. N.H. Ladizesky and I.M. Ward, Ultra high modulus polyethylene fibre composites: I. The preparation and properties of conventional epoxy resin composites, Composites Sci. and Technol., 26, 129–164 (1986).

15. N.H. Ladizesky, M. Sitepu and I.M. Ward, Ultra high modulus polyethylene fibre composites: II. Effect of resin composition on properties, Composites Sci. and Technol., 26, 169–183 (1986).

16. N.H. Ladizesky and I.M. Ward, Ultra high modulus polyethylene fibre composites: III. An exploratory study of hybrid composites, Composites Sci. and Technol., 26, 199–224 (1986).

17. N.H. Ladizesky and I.M. Ward, Conference on Fibre Reinforced Composites: The preparation and properties of ultra high modulus polyethylene composites, Liverpool University, 8–10 April 1986, I. Mech. E. Proceedings 1986, p.7–12.

18. B.L. Lee, J.W. Song and J.E. Ward, Failure of Spectra® polyethylene fiber reinforced composites under ballistic impact loading, J. Composite Mat., 28, 1202–1226 (1994).

19. N.J. Capiati and R.S. Porter, The concept of one polymer composites modelled with high density polyethylene, J. Mater. Sci., 10, 1671–1677 (1975).

20. Marais and P. Feillard, Manufacturing and mechanical characterization of unidirectional polyethylene-fiber polyethylene-matrix composites, Composite Sci. Technol., 45, 247–255 (1992).

21. A. Teishev, S. Incardona, C. Migliaresi and G. Marom, Polyethylene fibers-polyethylene matrix composites: preparation and physical properties, J. Appl. Polym. Sci., 50, 503–512 (1993).

22. G. Hinrichsen, S. Kreuzberger, Q. Pan, M. Rath, Production and characterization of UHMWPE fibers LDPE composites, Proceedings of the Ninth International Conference on the Mechanics of Composite Materials, Riga, 32, 719–728, (1996).

23. F.V. Lacroix, M. Werver and K. Schutte, Wet powder impregnation for polyethylene composites: preparation and mechanical properties, Composites Part A, 29A, 369–373 (1998).

24. C.L. Choy, Y. Fei and T.G. Xi, Thermal-conductivity of gel spun polyethylene fibers, J. Poly. Sci., B, Polym. Phys., 31, 365–370 (1993).

25. P.J. Hine, I.M. Ward, R.H. Olley and D.C. Bassett, The hot compaction of high modulus melt-spun polyethylene fibers, J. Mater. Sci., 28, 316–324 (1993).

26. R.H. Olley, D.C. Bassett, P.J. Hine and I.M.Ward, Morphology of compacted polyethylene fibers, J. Mater. Sci., 28, 1107–1112 (1993).

27. M.A. Kabeel, D.C. Bassett, R.H. Olley, Compaction of High-modulus melt-spun polyethylene fibers at temperatures above and below the optimum, P.J. Hine and I.M.Ward, J. Mater. Sci., 29, 4694–4699 (1994).

28. P.J. Hine, I.M. Ward, M.I. Abo El Maaty, R.H. Olley and D.C. Bassett, The hot compaction of 2-dimensional woven melt spun high modulus polyethylene fibres, J. Mater. Sci., 35, 5091–5099 (2000).

29. D.W. Woods, W.K. Busfield and I.M. Ward, Br. Patn. No. GB2 253420B.

30. D.W. Woods, W.K. Busfield and I.M. Ward, Improved mechanical behaviour in ultra-high modulus polyethylenes by controlled crosslinking, Plast. Rubber Process. Appl., 5, 157–164 (1985).

31. M.J. Bonner, P.J. Hine and I.M. Ward, Hot compaction of crosslinked high modulus polyethylene fibres and fabrics, Plast. Rubber Process. Appl., 27, 58–64 (1998).

32. R.J. Yan, P.J. Hine, I.M. Ward, R.A. Olley and D.C. Bassett, The hot compaction of SPECTRA gel-spun polyethylene fibre, J. Mater. Sci., 32, 4821–4832 (1997).

33. P.J. Hine, I.M. Ward, N.D. Jordan, R.A. Olley and D.C. Bassett, A comparison of the hot-compaction behavior of oriented, high-modulus, polyethylene fibers and tapes, J. Macromol. Sci. Phys. B40, 959–989 (2001).

34. N.D. Jordan, R.H. Olley, D.C. Bassett, P.J. Hine, I.M. Ward, The development of morphology during hot compaction of Tensylon high-modulus polyethylene tapes and woven cloths, Polymer, 43, 3397–3404 (2002).

35. M.I. Abo El-Maaty, D.C. Bassett, R.H. Olley, P.J. Hine and I.M. Ward, The hot compaction of polypropylene fibres, J. Mater. Sci., 31, 1157–1163 (1996).

36. P.J. Hine, I.M. Ward and J. Teckoe, The hot compaction of woven polypropylene tapes, J. Mater. Sci., 33, 2725–2733 (1998).

37. J. Teckoe, R.H. Olley, D.C. Bassett, P.J. Hine and I.M. Ward, The morphology of woven polypropylene tapes compacted at temperatures above and below optimum, J. Mater. Sci., 34, 2065–2073 (1999).

38. P.J. Hine, M. Bonner, B. Brew and I.M. Ward, Hot compacted polypropylene sheet, Plast. Rubb. Comp. Proc. Applns., 27, 167–171 (1998).

39. P.J. Hine, I.M. Ward, N.D. Jordan, R.H. Olley and D.C. Bassett, The hot compaction behaviour of woven oriented polypropylene fibres and tapes. I. Mechanical properties, Polymer, 44, 1117–1131 (2003).

40. N.D. Jordan, D.C. Bassett, R.H. Olley, P.J. Hine and I.M. Ward, The hot compaction of woven oriented polypropylene fibres and tapes. II. Morphology of cloths before and after compaction, Polymer, 44, 1133–1143 (2003).

41. N. Brown and I.M. Ward, The influence of morphology and molecular weight on ductile-brittle transitions in linear poly-ethylene, J. Mater. Sci., 18, 1405–1420 (1983).

42. J. Rasburn, P.J. Hine, I.M. Ward, R.H. Olley, D.C. Bassett and M.A. Kabeel, The hot compaction of polyethylene terephthalate, J. Mater. Sci., 30, 615–622 (1995).

43. I.M. Ward and P.J. Hine, Novel composites by hot compaction of fibers, Polym. Eng. Sci., 37, 1809–1814 (1997).

44. P.J. Hine and I.M. Ward, The hot compaction of woven polyethylene terephthalate multifilaments, J. Appl. Polym. Sci., 91 2223–2233 (2004).

45. H. Zhang and I.M. Ward, Kinetics of Hydrolytic degradation of poly(ethylene naphthalene-2,6-dicarboxylate), Macromolecules 28, 7622–7629 (1995).

46. F.J. Baltá-Calleja, M.E. Cagiao, H.G. Zachmann and C.J.Vanderdonckt, Hydrolysis etching of crystalline and amorphous polyethylene terephthalate: influence of molecular weight and microstructure, J. Macromol. Sci.-Polym. Phys. Ed. 33, 333–346 (1994).

47. H. Zimmerman and N.T. Kim, Investigations on thermal and hydrolytic degradation of poly (ethylene terephthalate), Polym. Eng. and Sci. 20, 680–683 (1980).

48. D.A.S. Ravens and I.M. Ward, Chemical reactivity of polyethylene terephthalate: hydrolysis and esterification reactions in the solid phase, Trans. Faraday Society, 57, 150–159 (1961).

49. P.L. Carr, H. Zhang and I.M. Ward, The production and properties of poly(ethylene naphthalate-2,6-dicarboxylate) monofilaments, Polym. for Adv. Technol., 7, 39–46 (1996).

50. P.J. Hine, A. Astruc and I.M. Ward, The hot compaction of polyethylene naphthalate (PEN), J. Appl. Poly. Sci., 93, 796–802 (2004).

51. P.J. Hine and I.M. Ward, The hot compaction of woven Nylon 66 multifilaments, J. Appl. Poly. Sci. in press (2005).

52. P.L. Carr and I.M. Ward, Drawing behaviour, mechanical properties and structure of poly(p-phenylene sulphide) fibres, Polymer, 28 2070–2076 (1987).

53. C. Kerrmarrec, The hot compaction of polyphenylene sulphide tapes, internal IRC Report (2002).

54. M. Goude, The hot compaction of poly (ether ether ketone), internal IRC report (2003).

55. P.J. Hine and I.M. Ward, The hot compaction of polyoxymethylene, internal IRC report (2002).

56. B. Brew and I.M. Ward, Study of the production of ultra-high modulus polyoxymethylene by tensile drawing at high temperatures, Polymer, 19, 1338–1344 (1978).

57. P.S. Hope, A. Richardson and I.M. Ward, Manufacture of ultra high-modulus poly(oxymethylene) by die drawing, J. Appl. Poly. Sci., 26 2879–2896 (1981).

Index

tension, 196
ductile behavior, 273
effect of nanoparticles, 618
enhanced toughness of, 275
flow-induced crystallization, 43,
    49
half-times of crystallization for,
    45
lamellar nanostructure, 275
lamellar separation, 263, 265
melt(s)
    deformation studies of
        supercooled, 35
    development of crystallinity
        in, 40
    shear-induced crystallization
        in, 34
    step-shear, 35
melting point of, 35
mesomorphic form of, 55
microhardness linear variation,
    290
micromechanical mechanisms,
    256, 259
morphology, 253
natural draw ratio, 221
relative fragility of, 226
rolling with side constraints of,
    196
sample characteristics, 250, 257
SAXS
    images collected during
        deformation of, 44
    patterns, 58, 59
SEM micrographs, 254, 255
structural changes during
    deformation of, 54
tensile testing, 250
thermo-oxidative degradation,
    275
true stress-strain curves for, 197
uniaxial tensile testing, 258
unresolved issues, 54
WAXD
    fingerprints, 57
    identity profiles, 48

patterns, 47, 55
weight average molar masses,
    221, 222
Isothermal crystallization, 284
Isothermal lamellar thickening, 8
Isotropic polymer, cohesive
    strength, 717
Izod impact energy, 321
Izod specimen, actual toughness
    and, 361
Izod test, 198, 199

J

J-integral tests, 584

K

Kausch model, 140, 141
Kebab, 188
Kerner equation, 577

L

LABS 321, 461, 463
Lamellar habits, crystal structure
    and, 12
Lamellar morphology, AFM phase
    images, 674
Lamellar separation, 267, 268
Lamellar slip, control of, 266
Lamellar styrene-butadiene block
    copolymers, 247
Layer
    -multiplying coextrusion, 632
    tearing, 646
    thickness, effect of on polymer
        adhesion, 634–635
LDPE, *see* Low-density
    polyethylene
Leeds die-drawing process, 193
Ligament
    extension, 141

Milton Keynes UK
Ingram Content Group UK Ltd.
UKHW020002071024
449327UK00031B/2616

9 780367 392727